Geometric, Algebraic and Topological Methods for Quantum Field Theory

Proceedings of the 2013 Villa de Leyva Summer School

Geometric, Algebraic and Topological Methods for Quantum Field Theory

Proceedings of the 2013 Villa de Leyva Summer School

Villa de Leyva, Colombia 15 – 27 July 2013

Editors

Leonardo Cano
Universidad Sergio Arboleda, Colombia

Alexander Cardona
Universidad de los Andes, Colombia

Hernán Ocampo
Universidad del Valle, Colombia

Andrés F Reyes Lega
Universidad de los Andes, Colombia

World Scientific

NEW JERSEY • LONDON • SINGAPORE • BEIJING • SHANGHAI • HONG KONG • TAIPEI • CHENNAI • TOKYO

Published by

World Scientific Publishing Co. Pte. Ltd.
5 Toh Tuck Link, Singapore 596224
USA office: 27 Warren Street, Suite 401-402, Hackensack, NJ 07601
UK office: 57 Shelton Street, Covent Garden, London WC2H 9HE

Library of Congress Cataloging-in-Publication Data
Names: Villa de Leyva Summer School (8th : 2013 : Leiva, Boyacá, Colombia) | Cano, Leonardo (Leonardo A.), editor. | Cardona, Alexander, editor. | Ocampo, Hernán, editor. | Reyes-Lega, Andrés F., 1973– editor.
Title: Geometric, algebraic, and topological methods for quantum field theory : proceedings of the 2013 Villa de Leyva Summer School, Villa de Leyva, Colombia, 15–27 July 2013 / edited by Leonardo Cano (Universidad Sergio Arboleda, Colombia), Alexander Cardona (Universidad de los Andes, Colombia), Hernán Ocampo (Universidad del Valle, Colombia), Andrés F. Reyes Lega (Universidad de los Andes, Colombia).
Description: [Hackensack] New Jersey : World Scientific, 2016. | Includes bibliographical references and index.
Identifiers: LCCN 2016033480 | ISBN 9789814730877 (hardcover : alk. paper)
Subjects: LCSH: Geometric quantization--Congresses. | Quantum field theory--Mathematics--Congresses. | Topology--Congresses.
Classification: LCC QC174.17.G46 G455 2013 | DDC 530.14/30151--dc23
LC record available at https://lccn.loc.gov/2016033480

British Library Cataloguing-in-Publication Data
A catalogue record for this book is available from the British Library.

Printed in Singapore

INTRODUCTION

The eighth summer school on *"Geometric, Algebraic and Topological Methods for Quantum Field Theory"* was held in Villa de Leyva, Colombia, during the last two weeks of July 2013. This volume brings together the lectures delivered during the summer school, which we hope will give an updated overview of their respective subjects, as well as a short communication presented during the school.

There where eight invited lectures, aimed at graduate students in physics or mathematics, each one of which is self-contained and can be read independently of the rest, starting with introductory material before presenting more advanced results. Some of the lectures included here share a common field or a common language, and complement each other; the authors of these lectures have successfully strived at putting into play the many interactions between mathematics and theoretical physics.

This volume offers an introduction to some recent developments in several active topics at the interface between geometry, topology, algebra, analysis and theoretical physics discussed along the two weeks of the summer school: Operator algebras, conformal field theory, black holes, relativistic fluids, Lie groupoids and Lie algebroids, renormalization methods, spectral geometry and index theory for pseudo-differential operators.

The volume begins with the lectures on operator algebras and quantum physics by Andrés Reyes in which, from a very fundamental discussion about the distinction between the probability theory used in classical physics and the one inherent to quantum mechanics, the author takes us to some aspects of operator algebras which play a prominent role in quantum physics, explaining why certain mathematical structures are better adapted for the study of quantum theory. Several examples, which include a discussion of bipartite entanglement, quantum phase transitions in spin chains, quasi-free states in fermionic systems and other applications to quantum field theory, illustrate this point of view as well as the various mathematical concepts involved. Rich in examples, these lecture notes contain a very generous list of problems for the reader to reinforce both the acquaintance with the mathematical structures used and the physical intuition.

In her contribution to paving the way from geometry to conformal field theory, Katrin Wendland presents the background for some basic conformal field theory constructions based on Calabi–Yau geometry. After some basic mathematical material leading to the definition of Calabi–Yau manifolds and some of their fundamental properties, the author introduces a number of topological invariants used to classify these spaces. Then K3 surfaces and Kummer surfaces are considered, as well as the lifting of the Kummer construction to the level of conformal field theory and the $\mathbb{Z}_2$-orbifold constructions of toroidal superconformal field theories. These orbifold conformal field theories are interpreted as non-linear sigma models on Kummer surfaces, and the author introduces the conformal field theoretic elliptic genus as a counterpart of the geometric elliptic genus, showing that the geometric elliptic genus of K3 surfaces agrees with the CFT elliptic genus of the $\mathbb{Z}_2$-orbifold theory obtained from a toroidal superconformal field theory on a complex two-torus. The final section of this contribution motivates and discusses the definition of K3 theories, which is formulated purely within representation theory.

The lectures on geometric aspects of black holes, by Bruno Carneiro da Cunha, survey the geometric aspects of these singular objects, with an emphasis on their event horizon and global structure. From a discussion on gravitational collapse, the author takes us to the notion of a Killing horizon, discusses black hole thermodynamics and introduces the rigidity theorems in the classical (electro-)vacuum solutions of 4-dimensional General Relativity. He then considers the case of space-times of dimension different from our usual 4 space-time dimension, giving a description of higher-dimensional black holes–both in supergravity and string theory– and finishing with a description of 3-dimensional space-time gravity and the Strominger-Vafa microcanonical formulation for the Bañados-Teitelboim-Zanelli black hole.

The theory of real relativistic fluids is the topic of the lectures by Esteban Calzetta. With the final aim of describing relativistic heavy ion collisions, the author starts by explaining why the natural generalization of the theory of real fluids to the relativistic realm is not physically meaningful. A derivation of the nonrelativistic theory of ideal fluids from thermodynamical principles is presented. Its generalization to real fluid leads to the Navier–Stokes equations, but the natural covariant formulation turns out to be problematic. Therefore, a more fundamental point of view is taken, based on relativistic kinetic theory, this leading to a successful approach that is applied to the analysis of heavy ion collisions.

The theory of Lie algebroids, which has grown and been enriched in many aspects thanks to its relevance in Poisson geometry and mathematical physics, was the common root of the lectures series delivered by Kirill Mackenzie and Paul Bressler during the summer school. On the one hand, in his lecture notes on symplectic groupoids and the double structures associated with them, Kirill Mackenzie goes from the very basic structures which —in the context of Poisson manifolds— led to the notion of symplectic groupoids, to point out the relevance of this notion through the specific example of the cotangent groupoid of a symplectic groupoid. The lectures include a very detailed discussion on the Poisson geometry underlying such structures and the symplectic realizability problem they generalize. The author then moves ahead to the general case of Poisson groupoids, where he explains his diagrammatic approach used to understand the relevant relationships that led to the generalization of the results and constructions done in the symplectic case, and discusses the double structures and some integrability results known in this context. On the other hand, the theory of connections and characteristic classes for Lie algebroids is the subject of the lecture notes by Paul Bressler, in which the author illustrates how the classical constructions of algebraic topology naturally apply to Lie algebroids giving rise to such kind of invariants in this context. The theory of connections, Lie algebroid cohomology and the Chern-Weil map used to define characteristic forms on Lie algebroids are systematically presented.

The goal of Elmar Schrohe's lectures is to bring the reader, from some basic knowledge in analysis, to the K-theoretic approach to index theory for boundary value problems. After a presentation of the necessary background on analysis and pseudo-differential calculus on compact manifolds, the first part of these lecture notes is devoted to explaining the index theorem for elliptic pseudo-differential operators, the dependence of such invariants on the stable homotopy class of the principal symbol of the corresponding operator and the local expression of the index found by Atiyah and Singer using K-theoretic methods. These constructions are extended to manifolds with boundary using the Boutet de Monvel calculus associated to pseudo-differential boundary value problems. In the final part of the notes, after some fundamentals on K-theory for C^*-algebras and their use in index theory for closed manifolds, the author finalizes his lectures with a detailed account of the index problem for operators in the Boutet de Monvel calculus.

Li Guo, Sylvie Paycha and Bin Zhang present in their paper, based on Sylvie Paycha's lecture series in Villa de Leyva, a mathematical introduction to some renormalization methods commonly used in physics and used in mathematics to "count" an infinite number of points in different contexts. Starting from the issue of how to count integer points on the non-negative real axis and the one-dimensional Euler–Maclaurin formula, the authors go to the problem of evaluating meromorphic functions at poles by means of a "minimal subtraction" and the heat-kernel regularization method. Counting lattice points on product cones offers a first toy model to understand the necessity of an additional algebraic structure to organize the divergences. Algebraic tools such as the algebraic Birkhoff factorization are introduced in a simple way in this simple product case, which can be generalized to all rational polyhedral convex (lattice) cones. The full (co)algebraic framework needed to derive an algebraic Birkhoff factorization from a complement map is then displayed and applied to the renormalization of conical zeta values at non-integer arguments.

The invited lectures are followed by a short communication by Andrés Angel and Diego Duarte in which the authors compute the Batalin–Vilkovisky algebra structure of the Hochschild cohomology of the group rings of cyclic groups of prime order. These, as much as their Gesterhaber structure, are of prime interest in string topology.

We hope that these contributions will give —as much as the school itself seems to have given— young students the desire to pursue what might be their first acquaintance with some of the problems on the edge of mathematics and physics presented here. On the other hand, we hope that the more advanced reader will find some pleasure in reading about different outlooks on related topics and seeing how the well-known mathematical tools prove to be very useful in some areas of theoretical physics.

We are indebted to various organizations for their financial support for this school. Let us first of all thank *Universidad de los Andes* in Colombia, which has been from the very beginning our main source of financial support. We also warmly thank the Deutscher Akademischer Austauschdienst - DAAD and the Abdus Salam Internacional Centre for Theoretical Physics - ICTP for their contribution to the success of this school.

Special thanks to the administrative staff at the Universidad de los Andes, particularly Carl Langebaek, Vice Rector for Research, Silvia Restrepo, Dean of the School of Sciences, José Ricardo Arteaga, Director of the Mathematics Department, and Carlos Ávila, Director of the Physics Department, for their constant encouragement and support. We also would like to thank Lina Duarte and Carlos Mayorga, who did a great job for the practical organization of the school, the quality of which was very much appreciated by participants and lecturers. We are also very indebted to Luz Malely Gutiérrez, Mauricio Morales and Nohora Galán for their help in various essential tasks needed for the successful development of the school. We also want to express our gratitude to the editorial staff of World Scientific for their excellent job, and in particular to E.H. Chionh for her guidance and patience.

Without the people named here, all of whom helped in the organization in some way or another, before, during and after the school, this scientific event would not have left such vivid memories in the lecturers' and participants' minds. Last but not least, thanks to all the participants who gave us all, lecturers and editors, the impulse to prepare this volume through the enthusiasm they showed during the school, and thank you to all the contributors and referees for their participation in the realization of these proceedings.

The editors,
Leonardo Cano,
Alexander Cardona,
Hernán Ocampo and
Andrés Reyes-Lega

CONTENTS

SOME ASPECTS OF OPERATOR ALGEBRAS IN QUANTUM PHYSICS

Andrés F. Reyes-Lega
Departamento de Física, Universidad de los Andes,
Bogotá, Colombia
E-mail: anreyes@uniandes.edu.co

Motivated by the sharp contrast between classical and quantum physics as probability theories, in these lecture notes I introduce the basic notions of operator algebras that are relevant for the algebraic approach to quantum physics. Aspects of the representation theory of C*-algebras will be motivated and illustrated in physical terms. Particular emphasis will be given to explicit examples from the theory of quantum phase transitions, where concepts coming from strands as diverse as quantum information theory, algebraic quantum physics and statistical mechanics agreeably converge, providing a more complete picture of the physical phenomena involved.

Keywords: Operator Algebras; Entanglement; Quantum Phase Transitions; Identical Particles.

1. Introduction

This notes represent the written version of lectures I gave in mini-courses at Universidade de Brasília (April 3-6, 2013), Universidad Central de Venezuela (May 23-27, 2016) and at the Villa de Leyva Summer School "Geometric, Topological and Algebraic Methods for Quantum Field Theory" (July 15-27, 2013). They were mainly intended as an introduction to some aspects of operator algebras, emphasizing the prominent role they play in quantum physics. As the audience consisted of students of both physics and mathematics at different stages of their studies, my choice was to focus on the most basic structures and examples, in the hope that a good grasp of these would motivate them to go deeper into the subject. Now, something that to a physicist may appear as completely familiar (as, say, an experimental set-up with polarizers, or the distinction between a classical and a quantum field) to a mathematician may not. The same could be said of the proof that the spectrum of any element in a C^*-algebra is never empty: It is a standard result in analysis, but may look quite awkward to many physics students. Therefore, the emphasis of these lecture notes will be on explaining *why* certain mathematical structures may be useful for the study of quantum theory. This will be illustrated by means of several examples that include a discussion of bipartite entanglement, algebraic and geometric aspects of quantum phase transitions in spin chains, quasi-free states in fermionic systems and applications to quantum field theory.

Our starting point will be based on the sharp distinction between *classical probability* and the probability theory inherent to quantum mechanics. This will provide a physical motivation to the various mathematical concepts we will be dealing with. After introducing the basic concepts about C^*-algebras, we will show how an algebraic approach to entanglement can lead to a resolution of certain discrepancies appearing when we deal with systems of identical particles. Then we will focus on the study of certain specific models (quantum Ising and XY models) that turn out to be ideal in order to illustrate how entanglement, geometry and the theory of CAR algebras[a] are interrelated.

1.1. *Quantum correlations: Bell-type inequalities*

Our first objective will be to understand what are the main structural differences between *classical* and *quantum* physics, when regarded as probability theories. This will, by the way, provide a physical motivation for some of the mathematical notions we will consider in the next sections. Let us recall that classical physics is usually modeled on a configuration (or phase) space, with a dynamics governed by, *e.g.*, Hamilton's principle. On the other hand, quantum mechanics is modeled on a Hilbert space. So the first issue we want to explore is: Why do we have to use Quantum Mechanics to describe the microscopic world? Indeed, simple experiments with light polarizers make it clear that there is no way we can describe certain phenomena using classical physics or, better said, classical *probability*. Let us then explore some elementary polarization phenomena, following the presentation in [1].

First let us recall that light is just made of electromagnetic waves, their behavior being governed by Maxwell's equations:

$$\begin{aligned} \nabla \cdot E = \rho/\epsilon_0, \;\; \nabla \cdot B = 0, \\ \nabla \times E = -\frac{\partial B}{\partial t}, \;\; \nabla \times B = \epsilon_0 \mu_0 \frac{\partial E}{\partial t} + \mu_0 j, \end{aligned} \tag{1}$$

[a] A special type of C^*-algebras used to model fermionic systems

where E (resp. B) stands for the electric (resp. magnetic) field, ρ for the charge density and j for the current density.

Problem 1.1. *Show that Maxwell's equations (1) in vacuum ($\rho = 0$, $j = 0$) lead to plane-wave solutions for the electromagnetic field propagating at a speed $c = 1/\sqrt{\epsilon_0\mu_0}$ and such that E and B are always perpendicular to each other and to the propagation direction. What is the relation between the* intensity *of the wave (defined as $I = \|E\|^2$) and the energy content of the fields?*

In simple terms, a *polarizer* is a filter that only allows the transmission of light waves which have a specific polarization angle. If we let unpolarized light go through an ideal polarizer, the intensity I_1 of the transmitted wave will be found to be half the intensity I_0 of the incident wave: $I_1 = 1/2\, I_0$ (Malus' law). After passing through the polarizer, the light is said to be linearly polarized. Let us now suppose that we have a beam of linearly polarized light and we let it go through a second polarizer, such that its polarization axis has been rotated by an angle φ with respect to the axis of the first polarizer. Then, experiment tells us that the intensity of the transmitted light will be $I_2 = \cos^2\varphi\, I_1$. This is all fine if we are working with classical electromagnetic waves, which are described by Maxwell's equations. In this case we just need to consider the projection of the field onto the direction singled out by the polarization axis. But we know that light is actually made of photons and, if their number is small, we are led to regard the $\cos^2\varphi$ term as a kind of "expectation value".

In a very influential paper [2], Einstein, Podolski and Rosen presented a criticism of quantum theory in what is now known as the *EPR paradox*. This led to the development of alternative, so-called "hidden variable" theories that aimed at explaining physical phenomena using classical probability models. It was only until Bell proved his famous inequalities, and Aspect's experiments proved the former were violated, that the controversy could be resolved, showing that quantum theory provided the correct description of the phenomena. An experimental set-up, of the type studied by Aspect, consists of a source (Ca atom) located in the middle that emits simultaneously a pair of photons. One of them goes to the right, the other to the left. There are two detectors, one at each extreme. There is also a polarizer in-between each detector and the source, as depicted in Fig. 1.

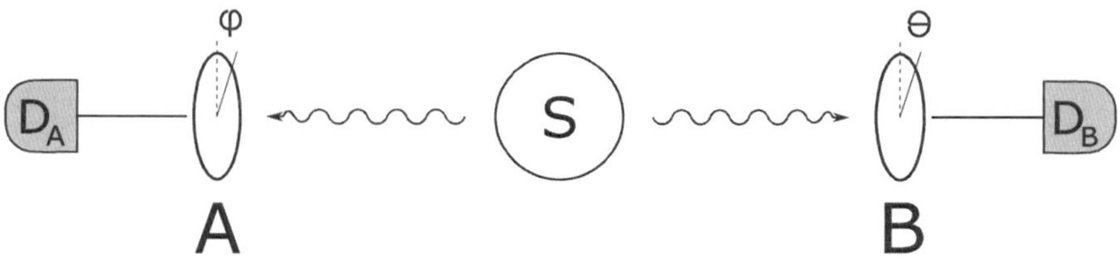

Fig. 1: A source S (Calcium atom) emits a pair of photons, each of which may be detected by detectors located in the extremes (D_A, D_B). There is a polarizer between the Ca atom and each of the detectors.

Let us now consider the following two propositions:

· A = "Left photon passes through (is detected by D_A) when polarizer's angle is φ".

· B = "Right photon passes through (is detected by D_B) when polarizer's angle is θ".

Then, what we learn from experiment is that the *joint probability* for both photons passing through the polarizers, thus being detected, is

$$p(AB) = \frac{1}{2}\sin^2(\varphi - \theta).$$

Before exploring why classical probability is in conflict with this result, let us recall how the principles of quantum mechanics allow us to predict it.

Let $\mathcal{H}$ denote a Hilbert space which, for simplicity, will be considered to be finite dimensional. In quantum mechanics, probability distributions are obtained from state vectors $|\Psi\rangle \in \mathcal{H}$ or, more generally, by density matrices ρ, that is, self-adjoint positive operators of trace one. On the other hand, observables are described by self-adjoint operators. Let A be such an observable. Let $\sigma(A) = \{a_1, \ldots, a_N\}$ denote its spectrum. We can then consider its spectral decomposition

$$A = \sum_j a_j E_j, \quad \sum_j E_j = \mathbb{1}, \quad E_i E_j = \delta_{ij} E_j. \tag{2}$$

If the state of the system is given by a density matrix ρ, the *expectation value* of A is defined as

$$\langle A\rangle_\rho := \mathrm{Tr}(\rho A). \tag{3}$$

In the case of a *pure* state, the density matrix is a rank-one projector, of the form $\rho = |\psi\rangle\langle\psi|$, $|\psi\rangle \in \mathcal{H}$, and so (3) reduces to $\langle A\rangle_\psi = \langle\psi|A|\psi\rangle$. In this way we obtain a probability distribution over $\sigma(A)$, with

$$p(a_i) = \langle E_i \rangle_\rho.$$

In fact, it follows from (2) and (3) that $0 \leq p(a_i) \leq 1$ and $\sum_i p(a_i) = 1$.

Let us use this to give a mathematical description of the experiment described above. The polarization state of a photon can be described using a 2-dimensional Hilbert space. Let $\{|x\rangle, |y\rangle\}$ denote an orthonormal basis, that can be used to describe, say, horizontal and vertical polarization states. Since we are considering a system consisting of two photons, the Hilbert space of the system can be taken to be $\mathcal{H} = \mathbb{C}^2 \otimes \mathbb{C}^2$ (as the photons are supposed to be far from each other at the time of detection, the symmetrization postulate can be ignored). The 2-photon state, as produced by the emission from an excited state of the Ca atom, can be described by the following state vector:

$$|\psi\rangle = \frac{1}{\sqrt{2}} \left(|x\rangle \otimes |y\rangle - |y\rangle \otimes |x\rangle \right). \tag{4}$$

For a polarizer with polarization axis pointing at an angle α we can use $P(\alpha)$, defined as the *projector* onto $\cos\alpha|x\rangle + \sin\alpha|y\rangle$. Since we have two polarizers, we introduce the following projectors (*cf.* Fig. 1):

$$P_A(\varphi) = P(\varphi) \otimes \mathbb{1}, \quad P_B(\theta) = \mathbb{1} \otimes P(\theta).$$

It is important to remark that these two operators correspond to compatible observables, in the sense that the measurement of one of them does not affect the result of the other, *i.e.*, they are commuting observables:

$$[P_A(\varphi), P_B(\theta)] = 0.$$

Problem 1.2. *Show that*

$$\langle\psi|P_A(\varphi)|\psi\rangle = \langle\psi|P_B(\theta)|\psi\rangle = \frac{1}{2}.$$

Also show that, for the joint *measurement of the two polarization states, one obtains*

$$\begin{aligned} \langle\psi|P_A(\varphi)P_B(\theta)|\psi\rangle &= \langle\psi|(\mathbb{1} - P_A(\varphi))(\mathbb{1} - P_B(\theta))|\psi\rangle \\ &= \frac{1}{2}\sin^2(\varphi - \theta). \end{aligned} \tag{5}$$

The previous problem shows that the predictions of quantum mechanics for this type of experiment are in accordance with what is actually measured in the laboratory. In spite of its extreme simplicity, there is an intriguing feature of this result: By appropriately arranging the polarization angles

θ and φ, we can obtain a *total anticorrelation* for the joint measurements. As we will see, this is due to the fact that the state (4) is an *entangled state*. That these kind of correlations cannot be obtained from a (local, realistic) classical theory is at the core of the original EPR controversy. For an interesting discussion of these issues in the context of an actual experimental situation, we recommend [3].

Our immediate aim will therefore be to understand where exactly classical probability fails at describing the results of such experiments. For this purpose we will assume an approach to (classical) probability based on propositions and degrees of plausibility, following Jaynes [4].

In this setting, the objects to be considered are the following:

- A set of propositions: $\{A, B, C, \ldots\}$, each one of which can take on (only) two values: true or false.
- Logical operations, that can be performed on the set of propositions:
  - Conjunction, or logical product (AND): AB. It is true if and only if both A and B true.
  - Disjunction, or logical sum (OR): $A + B$. It is true if at least one of them true.
  - Negation (NOT): $\bar{A}$, with opposite truth value as A.

These logical operations are assumed to satisfy the defining rules of a *Boolean algebra*:

- Idempotency: $AA = A \;\;,\;\; A + A = A$
- Commutativity: $AB = BA \;\;,\;\; A + B = B + A$
- Associativity:

  (A1) $A(BC) = (AB)C \equiv ABC$
  (A2) $A + (B + C) = (A + B) + C \equiv A + B + C$

- Distributivity:

  (D1) $A(B + C) = AB + AC$
  (D2) $(A + B)(A + C) = A + BC$

- Duality (De Morgan's laws): $\overline{AB} = \bar{A} + \bar{B}, \quad \overline{A + B} = \bar{A}\,\bar{B}$

Problem 1.3. *Show that the proposition "$A \Rightarrow B$ " is equivalent to "$A = A\,B$".*

Problem 1.4. *Show that (D2) follows from the other rules.*

Now, given a proposition A, a probability (or "degree of plausibility", *cf.* [4]) $p(A)$ can be assigned to it, under the following basic assumptions:

I. Let $p(A) \in \mathbb{R}$ denote the probability, or degree of plausibility, of a given proposition A. Then we assume that:
   - $p(f(A_1, A_2, \ldots, A_n)) \in \mathbb{R}$, for any logical function f of the propositions $A_1, A_2, \ldots, A_n$
   - Using the notation $p(A|B)$ for the conditional probability that A is true, given B is true, we assume that $p(A|C) > p(B|C)$ whenever $A|C$ more plausible than $B|C$.

II. Given a proposition C, let us suppose that we improve our state of knowledge, obtaining a new proposition C'. If as a result A becomes *more* plausible, *i.e.* $p(A|C') > p(A|C)$, but $p(B|AC') = p(B|AC)$, then $p(AB|C') \geq p(AB|C)$ should hold (this is dubbed the *common sense* assumption by Jaynes [4]).

From I and II above we may obtain, under very general assumptions (like consistency), the following two basic rules:

- Product Rule:

$$\begin{aligned} P(AB|C) &= P(A|BC)P(B|C) \\ &= P(B|AC)P(A|C) \end{aligned} \tag{6}$$

- Sum Rule:

$$P(A|C) + P(\bar{A}|C) = 1. \tag{7}$$

The reader is invited to consult Jaynes for a comprehensive exposition of this point of view, including many illustrative derivations and examples.

Now, in order to return to our experiment, let us define the following "coincidence" function f, for two given propositions A and B:

$$f(A, B) := A\,B + \bar{A}\,\bar{B}. \tag{8}$$

Problem 1.5. *Given 4 propositions A_1, A_2, B_1, B_2, use the product and sum rules to show that the following proposition is true:*

$$f(A_1, B_1) \Rightarrow f(A_1, B_2) + f(A_2, B_2) + f(A_2, B_1). \tag{9}$$

Since $p(A|X) \leq p(B|X)$ whenever $A \Rightarrow B$ holds (product rule) we obtain, from the previous problem,

"Bell's Inequality":

$$p(f(A_1, B_1)) \leq p(f(A_1, B_2)) + p(f(A_2, B_2)) + p(f(A_2, B_1)). \qquad (10)$$

Referring back to figure 1, let us consider the following propositions ($i, j = 1, 2$):

A_i = "Left photon passes through (is detected by D_A) when polarizer's angle is φ_i".

B_j = "Right photon passes through (is detected by D_B) when polarizer's angle is θ_j".

The inequality (10) should hold true for all choices of φ_i, θ_j. But this then implies:

$$\sin^2(\varphi_1 - \theta_1) \leq \sin^2(\varphi_1 - \theta_2) + \sin^2(\varphi_2 - \theta_2) + \sin^2(\varphi_2 - \theta_1).$$

For the choice $\varphi_1 = 0$, $\varphi_2 = \frac{\pi}{3}$, $\theta_1 = \frac{\pi}{2}$, $\theta_2 = \frac{\pi}{6}$, this means:

$$\sin^2\left(\frac{\pi}{2}\right) \leq \sin^2\left(\frac{\pi}{6}\right) + \sin^2\left(\frac{\pi}{3} - \frac{\pi}{6}\right) + \sin^2\left(\frac{\pi}{3} - \frac{\pi}{2}\right),$$

$$\sin^2\left(\frac{\pi}{2}\right) \leq \sin^2\left(\frac{\pi}{6}\right) + \sin^2\left(\frac{\pi}{6}\right) + \sin^2\left(-\frac{\pi}{6}\right),$$

$$1 \leq \frac{3}{4}.$$

This contradiction means that the assumptions we have considered above (which are at the basis of classical probability) do not apply in the quantum realm. So, definitely, quantum theory leads to a very different type of probability theory. In the next section we will explore some of the more notorious differences between classical and quantum probabilities.

1.2. *Classical versus quantum probability*

Until now we have avoided any mention of propositions in terms of set theory, which is the basis of the Kolmogorov axiomatics. In the finite dimensional case, at least, they turn out to give equivalent structures. Let us then consider the main properties of a (classical) probability theory, formulated in terms of set theory and, for simplicity, in the finite dimensional context. Let us consider a finite set $\Omega = \{x_1, x_2, \ldots, x_N\}$, regarded here as the *sample space*. The *event space* $\mathcal{E}$ is a certain collection of subsets of Ω, that must contain the empty set and be closed under complements and unions.

The set of events forms a Boolean algebra, under the standard set theoretic operations:

- AND: $A \cap B$,
- OR: $A \cup B$,
- NOT: $A^c = \Omega \setminus A$.

A probability distribution is then defined as a map $p : \mathcal{E} \to [0,1] \subset \mathbb{R}$, such that

(i) $p(\emptyset) = 0, \quad p(\Omega) = 1$,
(ii) $p(\bigcup_{k=1}^n A_k) = \sum_{k=1}^n p(A_k)$, for $A_1, A_2, \ldots, A_n \in \mathcal{E}$ pairwise disjoint.

Remark 1.1. Notice that the space of probability distributions is a convex set, for if p_1 and p_2 are two probability distributions, then $\lambda p_1 + (1-\lambda)p_2$ is again a probability distribution, provided $0 \leq \lambda \leq 1$. But not only is the space of probability distributions a convex set, it is a special kind of convex set: a *simplex*. To see this, all we have to do is to consider the following "extremal" distributions, $p^{(1)}, p^{(2)}, \ldots, p^{(N)}$, defined by

$$p^{(i)}(x_j) := \delta_{ij}.$$

It is clear that any p can be written as a *unique* convex combination of these extremal distributions, as we have

$$p = \sum_i \lambda_i p^{(i)},$$

with $\lambda_j = p(x_j)$.

A random variable (or "observable") is a function $f : \mathcal{E} \to \mathbb{R}$. We call the space of all such functions $\mathrm{Obs}(\Omega)$. Notice that $\mathrm{Obs}(\Omega)$ forms a commutative algebra.

Summarizing, in the finite-dimensional case we have the following structure:

- The event space ($\mathcal{E} \subseteq \mathcal{P}(\Omega)$) forms a Boolean algebra.
- The space of probability distributions is *convex* and furthermore has the structure of a simplex: Every probability distribution p can be uniquely written as convex combination of "extremal" distributions.
- The space of observables (random variables) has the structure of a commutative algebra.

Remark 1.2. In cases where the sample space is not a finite set (like in classical mechanics) we need a suitable generalization of the above definitions. This is afforded by measure theory, in the following way: Sample and event spaces are now replaced by a measurable space $(\Omega, \mathcal{E})$, where $\mathcal{E}$ is a suitable σ-algebra of Ω. Probability distributions are then defined as *normalized, positive measures.* In particular, we can now handle countable additivity:

$$p\left(\bigcup_{k=1}^{\infty} A_k\right) = \sum_{k=1}^{\infty} p(A_k), \quad \text{for } A_j \text{ pairwise disjoint.}$$

Finally, the space of observables is still a commutative algebra, for the product and sum of two measurable functions is a measurable function. A similar remark applies to the convex structure of the space of probability distributions.

How do these structures show up in classical physics? In classical mechanics, for example, the dynamics of a system can be described in terms of canonical variables "position" (q_i) and "momentum" (p_j), that give rise to the *phase space* of the system. It follows from Hamilton's variational principle that if $H(q,p)$ is the Hamiltonian of the system, the canonical variables will evolve along solutions to Hamilton's equations:

$$\begin{aligned} \frac{dp_i}{dt} &= -\frac{\partial H}{\partial q_i}, \\ \frac{dq_i}{dt} &= \frac{\partial H}{\partial p_i}, \quad i = 1, \ldots, n. \end{aligned} \tag{11}$$

Usually the phase space is the cotangent bundle T^*Q of some configuration space Q. The canonical variables (q,p) are then local coordinates on T^*Q and the *observables* are smooth functions on phase space: $C^\infty(T^*Q)$. Since time evolution is given by Hamilton's equations or, at the level of observables by $\frac{df}{dt} = \{H, f\}$, with $\{\cdot,\cdot\}$ denoting the Poisson bracket, one would think that there is no much space for a probabilistic model here since, given a set of initial conditions, by solving the equations of motion we are able, in principle, to predict the position and momenta of all the particles. But, as we learn from classical statistical physics, for $n >> 1$ $(n \sim 10^{23})$ we certainly need a statistical approach! Therefore we are forced to introduce probability distributions, expressed in terms of probability densities $\rho(q,p)$, so that the average value of an observable $f \in C^\infty(T^*Q)$ is given by

$$\langle f \rangle = \int f(q,p)\rho(q,p)d\mu,$$

where $d\mu = d^n q d^n p$ is the Liouville measure on phase space. For instance, for the *canonical ensemble* (used to describe a subsystem embedded in a thermal bath at temperature T), we have $\rho(q,p) \propto e^{-H(q,p)/k_B T}$. Now we can return to the case of a few point particles and interpret the state of the system at a given time, usually defined as just a point (q_0, p_0) in phase space, as an extremal probability distribution (a "pure state"), for which the probability density is just a Dirac delta distribution: $\rho(q,p) = \delta(q-q_0, p-p_0)$. In any case, we see that classical physics can be regarded as a special type of probability theory, where the three properties of classical probability highlighted above still hold, *i.e.*, every classical system of point particles can be understood as a probability theory, in the Kolmogorov sense.

What about Quantum Mechanics? As discussed above, probability distributions are obtained from state vectors $|\Psi\rangle$ in a Hilbert space $\mathcal{H}$, or more generally from density matrices, ρ, whereas observables are described by self-adjoint operators. Consider the spectral decomposition of a self-adjoint operator A, as in (2). Assuming a non-degenerate, discrete spectrum, we notice that the projectors E_i, having as spectrum the set $\{0,1\}$, can be regarded as "indicators of events". Now, there is a correspondence between projections $E : \mathcal{H} \to \mathcal{H}$ and subspaces $V \subset \mathcal{H}$. The partial order that we naturally obtain by inclusion then gives rise to the structure of an "orthocomplemented lattice of proposition", where the corresponding operations are defined as

$$\begin{aligned} \text{OR:} \quad & V_1 \vee V_2 \longleftrightarrow \operatorname{span}(V_1, V_2) \\ \text{AND:} \quad & V_1 \wedge V_2 \longleftrightarrow V_1 \cap V_2 \\ \text{NOT:} \quad & \quad V' \quad \longleftrightarrow V^{\perp}. \end{aligned}$$

An important feature of this system is that it *does not* give rise to a Boolean algebra structure. The reason for this is that the orthogonal complement is not the only possibility for a complement in this lattice. This in turn implies the breakdown of the distributive law, which is part of the definition of a Boolean algebra. As in the classical case, the state space (here the space of density matrices) is a convex space. But, in contrast to the classical case, this space is not a simplex. A consequence of this is that the representation of a density matrix as a convex sum of pure states is (highly) non-unique. Finally, in quantum theory the space of observables forms a non-commutative algebra, in contrast to the classical case, where the algebra is commutative.

It is now clear that, in principle, both approaches (classical and quantum) can be considered in order to describe physical phenomena in probabilistic terms. In fact, EPR-like arguments are in favor of a "local-realistic" point of view according to which even quantum phenomena should be explained in terms of classical probability. That this is not the case is proved by Bell's inequalities and their violation, experimentally verified by Aspect in the 80's. So quantum theory can be regarded as a kind of "non-commutative" probability theory. One of the points of these lectures is that, when formulated in the language of operator algebras, both quantum and classical physics can be described in a unified way. We will also take advantage of this formulation to discuss a similar phenomenon, occurring in topology and geometry: In non-commutative geometry [5] the generalization of topological/geometric notions to the non-commutative setting has been mainly achieved by first expressing them in algebraic terms and then realizing that dropping the commutativity assumption allows for vast generalizations.

2. Aspects of operator algebras in quantum physics

2.1. *Observable algebras and states*

Let $\mathcal{H}$ be a separable Hilbert space. Recall that, given a linear operator $T : \mathcal{H} \to \mathcal{H}$, T is said to be bounded if there is some $C > 0$ such that

$$\|T(x)\| \leq C\|x\|$$

for all x in $\mathcal{H}$, where the norm is the one induced by the inner product: $\|x\|^2 = (x, x)$. If T is a bounded operator, we define its norm as follows:

$$\|T\| := \sup_{x \neq 0} \frac{\|T(x)\|}{\|x\|}.$$

One then checks that for T and S bounded the inequalities

$$\begin{aligned} \|T + S\| &\leq \|T\| + \|S\|, \\ \|T\, S\| &\leq \|T\|\, \|S\|, \end{aligned}$$

are satisfied. From the completeness of $\mathcal{H}$ it follows that the space

$$\mathcal{B}(\mathcal{H}) = \{T : \mathcal{H} \to \mathcal{H} \,|\, T \text{ is linear and bounded}\}$$

is a complete normed space. It is also an algebra and has an involution "$*$" given by the adjoint: $T^* := T^\dagger$. We thus may call $\mathcal{B}(\mathcal{H})$ the "$*$-algebra of bounded operators on $\mathcal{H}$".

Problem 2.1. *Show that for $T \in \mathcal{B}(\mathcal{H})$ we have:*

$$\|T^*T\| = \|T\|^2.$$

We now abstract these notions and make them independent of any underlying Hilbert space. As we will see below, a lot will be gained from this, since we will then be able to study *representations* of the (abstract) operator algebras, and the equivalence/inequivalence of these representations will have a deep physical meaning.

Definition 2.1. A *Banach space* is a normed vector space $(V, \|\cdot\|)$ which is complete[b] with respect to the (metric induced by) $\|\cdot\|$.

Definition 2.2. A *Banach algebra* is a Banach space $(\mathcal{A}, \|\cdot\|)$ which is also an algebra, with the property that $\|ab\| \leq \|a\|\|b\|, \ \forall a, b \in \mathcal{A}$ (which in turn implies that multiplication is a continuous operation).

Definition 2.3. An *involution* on a (complex) algebra $\mathcal{A}$ is a map $* : \mathcal{A} \to \mathcal{A}$ such that, for any $a, b \in \mathcal{A}$ and $\mu, \nu \in \mathbb{C}$:

i. $(\lambda a + \mu b)^* = \bar{\lambda} a^* + \bar{\mu} b^*$
ii. $(ab)^* = b^* a^*$
iii. $(a^*)^* = a$

Notice that this is, basically, an abstraction of the adjoint operation on $\mathcal{B}(\mathcal{H})$.

Definition 2.4. A *C^*-algebra* is a Banach $*$-algebra $(\mathcal{A}, \|\cdot\|, *)$ with the fundamental property

$$\|a^*a\| = \|a\|^2. \tag{12}$$

Example 2.1. Let M be a compact, Hausdorff topological space and set $\mathcal{A} = C(M)$, the space of continuous complex functions on M and, for $f \in C(M)$, set $f^*(x) = \overline{f(x)}$, and $\|f\| = \sup_{x \in M} |f(x)|$. Then $(C(M), \|\cdot\|, *)$ is a C^*-algebra.

Example 2.2. Let $\mathcal{H}$ be a Hilbert space. Then, as we expect, $(\mathcal{B}(\mathcal{H}), \|\cdot\|, *)$ is a C^*-algebra, where the norm is the operator norm and the involution is given by the adjoint operation.

---

[b]"all Cauchy sequences converge"

Later we will see that the list of examples of C^*-algebras is basically exhausted by the previous two examples, a remarkable fact. A motivation to work with algebras of bounded operators comes from physics since, as can be shown, the *canonical commutation relations* (CCR)

$$[\hat{q}, \hat{p}] = i\hbar \mathbb{1}$$

cannot be implemented by means of bounded operators $\hat{q}$ and $\hat{p}$. Then, although these commutation relations have a very clear meaning from the physical point of view, mathematically they correspond to *unbounded* operators and hence issues like self-adjointness, domains, etc. come into play, which make them more difficult to deal with. One can, nevertheless, replace the CCR by their exponentiated (or Weyl) form, as follows. First we define operators $U(a)$ and $V(b)$, for $a, b \in \mathbb{R}$, acting on wave functions as follows:

$$\begin{aligned} (U(a)\psi)(x) &:= \psi(x - \hbar a), \\ (V(b)\psi)(x) &:= e^{-ibx}\psi(x). \end{aligned} \tag{13}$$

By Stone's theorem, it follows that $U(a) = e^{-ia\hat{p}}$ and $V(b) = e^{-ib\hat{q}}$.

Problem 2.2. *Show that the operators $U(a)$ and $V(b)$ satisfy the following commutation relations:*

$$\begin{aligned} U(a_1)U(a_2) &= U(a_1 + a_2), \\ V(b_1)V(b_2) &= V(b_1 + b_2), \\ U(a)V(b) &= e^{i\hbar ab}V(b)U(a). \end{aligned} \tag{14}$$

As we will see later on, these operators give rise to a C^*-algebra. In fact, we will see that to any symplectic vector space we can canonically associate a C^*-algebra (its Weyl C^*-algebra). The unitary representations of these algebras, in the case of infinite dimensional symplectic vector spaces, play a prominent role in the study of quantum field theory on curved spacetimes.

Example 2.3. Consider a (complex, involutive, with unit $\mathbb{1}$) algebra $\mathcal{A}$ generated by elements $a_1, a_2, \ldots, a_n, \mathbb{1}$, subject to the following *canonical anticommutation relations (CAR)*:

$$a_i a_j^* + a_j^* a_i = \delta_{ij}\mathbb{1}, \quad a_i a_j + a_j a_i = 0. \tag{15}$$

If we want to define a C^* norm on this algebra, it has to be such that the C^*-property (12) is satisfied. But from (15) we obtain $(a_i^* a_i)^2 = a_i^* a_i$, which along with (12) implies $\|a_i^* a_i\|^2 = \|a_i^* a_i\|$. The only option we have, then, is to define $\|a_i\| = 1$ for every $i \in \{1, \ldots, n\}$.

Problem 2.3. *Generalize the previous example to the case of a finite dimensional Hilbert space $\mathcal{H} \cong \mathbb{C}^n$, in the following way. Let $\langle \cdot | \cdot \rangle$ denote the inner product. For any pair of vectors, $u, v \in \mathcal{H}$, consider generators $a(u), a(v)$ satisfying the following relations (CAR):*

$$\{a(u), a(v)^*\} = \langle u|v\rangle \mathbb{1}, \quad \{a(u), a(v)\} = 0.$$

where $\{A, B\} \equiv AB + BA$. How must $\|a(u)\|$ be defined in this case?

The (mathematical) notion of *state* for a C^*-algebra is also very close to the notion of quantum state. To appreciate this, consider a physical system modeled by the algebra $\mathcal{B}(\mathcal{H})$ of bounded operators on a Hilbert space $\mathcal{H}$. As mentioned in the previous section, in physics we distinguish between two kinds of states:

- Pure states: These are described by normalized vectors $|\psi\rangle \in \mathcal{H}$ (actually, by rays on Hilbert space). If A is an observable, then its expectation value is defined as

$$\langle A\rangle_\psi := (\psi, A\psi) \equiv \langle \psi|A|\psi\rangle.$$

- Mixed states: They are described by operators $\rho : \mathcal{H} \to \mathcal{H}$ such that $\rho^\dagger = \rho > 0$, $\text{Tr}\rho = 1$. The expectation value is given, in this case, by

$$\langle A\rangle_\rho := \text{Tr}_{\mathcal{H}}(\rho A).$$

Problem 2.4. *Show that the condition for a mixed state (described by a density matrix ρ) to be pure is $\rho^2 = \rho$. Thus, all states can be described by density matrices. The space of all density matrices is naturally a convex space. The extremal elements then turn out to be the pure ones. Check this assertion in the simple case $\mathcal{H} = \mathbb{C}^2$.*

We therefore see that a state can be regarded as a mapping $A \mapsto \langle A\rangle$ from the observable algebra to the complex numbers. It must be possible, then, to express the properties $\rho > 0$ and $\text{Tr}\rho = 1$ only in terms of this mapping. This leads us to the general definition of state for a C^*-algebra.

Definition 2.5. Let $\mathcal{A}$ denote a (unital) C^*-algebra, with unit $\mathbb{1}$. A positive linear functional

$$\omega : \mathcal{A} \to \mathbb{C}$$

such that $\omega(\mathbb{1}) = 1$ is called a *state*. We will denote the set of all states on $\mathcal{A}$ with $\mathcal{S}_\mathcal{A}$.

Remark 2.1. In this context, positivity of ω means that $\omega(a^*a) \geq 0$ for all $a \in \mathcal{A}$.

Remark 2.2. If $\mathcal{A}$ is not unital, we can replace the normalization condition by the condition $\|\omega\| = 1$. For unital algebras, both definitions coincide.

Having in mind the example of $\mathcal{B}(\mathcal{H})$, it then makes sense to have a notion of spectrum for an abstract C^*-algebra. This leads, in turn, to the spectral theorem for C^*-algebras [6].

Definition 2.6. Let a be an element of a unital C^*-algebra $\mathcal{A}$. Then its *spectrum* is defined as the set

$$\sigma(a) := \{\lambda \in \mathbb{C} \,|\, (\lambda \mathbb{1} - a) \text{ is not invertible}\}.$$

Definition 2.7. Let a be an element of a unital C^*-algebra $\mathcal{A}$. Then its *spectral radius* is defined as

$$\rho(a) := \sup\{|\lambda| \,|\, \lambda \in \sigma(a)\}.$$

Problem 2.5. *Show that the space $M_n(\mathbb{C})$ of $n \times n$ matrices with complex entries is a C^*-algebra with respect to the norm defined as*

$$\|A\|_{max} := |\lambda_{max}|^{1/2},$$

where λ_{max} is the eigenvalue of $A^\dagger A$ with the largest absolute value. Show also that this norm coincides with the operator norm.

Remark 2.3. The previous problem illustrates the remarkable fact that, for a C^*-algebra $(\mathcal{A}, \|\cdot\|, *)$, the norm of any element normal $a \in \mathcal{A}$ coincides with its spectral radius:

$$\|a\| = \sqrt{\rho(a^*a)}. \tag{16}$$

This is an important fact, as it links topological properties of the algebra with algebraic ones. In particular, it implies that the norm in a C^*-algebra is *unique*.

The following two properties will be very useful:

Problem 2.6. *Let $\phi : \mathcal{A} \to \mathbb{C}$ be a positive linear functional on a C^*-algebra. Prove that*

(i) $\phi(a^*b) = \overline{\phi(b^*a)}$.
(ii) $|\phi(a^*b)|^2 \leq \phi(a^*a)\phi(b^*b)$ *("Cauchy-Schwarz inequality")*

Hint: Consider $\phi((\lambda a + b)^*(\lambda a + b))$ as a quadratic form on λ. As an alternative, you may try deriving these relations just by thinking of $\phi(a^*b)$ as "$\langle a|b\rangle$".

Another consequence of positivity is continuity:

Proposition 2.1. *Let ϕ be a positive linear functional on a unital C^*-algebra $\mathcal{A}$. Then ϕ is a continuous linear functional, and $\|\phi\| = \phi(\mathbb{1})$.*

Proof (*cf.* [11]). If a is positive ($a = a^*$ and $\sigma(a) \subseteq \mathbb{R}_+$) then we have, from the spectral radius formula, that $\sigma(a) \subseteq [0, \|a\|]$. This means that $\|a\| - \lambda \geq 0$ for all $\lambda \in \sigma(a)$. This can be restated as follows. Define

$$\begin{aligned} f : \sigma(a) &\longrightarrow \mathbb{C} \\ \lambda &\longmapsto f(\lambda) := \|a\| - \lambda. \end{aligned}$$

Then $f \in C(\sigma(a))$ and $f \geq 0$. But then it follows (from the continuous functional calculus) that $\sigma(f(a)) = f(\sigma(a))$, so that $f(a) \geq 0$ or, in other words, that $\|a\|\mathbb{1} - a$ is a positive operator. Since ϕ is positive and linear, this implies $\phi(a) \leq \phi(\mathbb{1})\|a\|$. Now, for an arbitrary $\alpha \in \mathcal{A}$, apply the Cauchy-Schwarz inequality from problem 2.6 with $a = \mathbb{1}$ and $b = \alpha$, and use the C^* property of the norm to obtain $\phi(\alpha) \leq \phi(\mathbb{1})\|\alpha\|$. It follows that ϕ is continuous, with $\|\phi\| \leq \phi(\mathbb{1})$. But we also have

$$\phi(\mathbb{1}) = |\phi(\mathbb{1})| \leq \|\phi\|\|\mathbb{1}\| = \|\phi\|,$$

so that $\|\phi\| = \phi(\mathbb{1})$. □

Thus, for unital C^*-algebras we conclude that the set of states over $\mathcal{A}$ is a convex subset of the continuous dual $\mathcal{A}^*$ of $\mathcal{A}$.

We gather from all this that it makes sense to formulate quantum physics in terms of C^*-algebras. The general philosophy will be to describe a given physical system in terms of its "algebra of observables", which will be here taken to be a C^*-algebra $(\mathcal{A}, \|\cdot\|, *)$. Then we will consider the following two "dual" notions:

- A *quantum state* will be defined as a state (in the sense of definition 2.5) on the algebra $\mathcal{A}$.
- An *observable* will be an element of $\mathcal{A}$.

Furthermore, the "pairing" between an observable a and a state ω will be given the physical interpretation of an *expectation value*, and we will write

$$\langle a\rangle_\omega := \omega(a).$$

Regarding the distinction between pure and mixed states, we can now call a state *pure* if it cannot be written as a convex combination of other states; otherwise we will call it a *mixed* state.

2.2. *The Gelfand-Naimark theorem and the GNS construction*

Before discussing physical applications of the notions introduced in the previous section, we will take the opportunity to briefly review the characterization of C^*-algebras due to Gelfand, Naimark and Segal. The characterization of commutative C^*-algebras (the Gelfand-Naimark theorem) is of fundamental importance, in particular because it leads to the notion of a *"noncommutative topological space"* [5]. Also, the so-called GNS-construction discussed below is relevant not only because of its role in the characterization problem for noncommutative C^*-algebras, but also because of its striking consequences for the study of quantum systems (equilibrium states, symmetry breaking, inequivalent vacua, among others) [7, 8].

Definition 2.8. A *character* of a C^*-algebra $\mathcal{A}$ is a $*$-homomorphism $\mu : \mathcal{A} \to \mathbb{C}$. Let us denote with $\mathcal{M}_{\mathcal{A}}$ the set of all characters of $\mathcal{A}$.

Now, it is easy to see that if μ is a character, then $\mu(a) \in \sigma(a)$. This, together with the spectral radius formula (16), implies that $\|\mu\| \leq 1$, so that $\mathcal{M}_{\mathcal{A}} \subset \mathcal{A}_1^*$, where

$$\mathcal{A}_1^* := \{\phi \in \mathcal{A}^* \mid \|\phi\| \leq 1\}$$

is the unit ball in the dual space. Since (by the Banach-Alaoglu theorem [9]) this ball is compact in the weak-$*$ topology (*i.e.* the one induced by the family of seminorms on $\mathcal{A}^*$: $p_\alpha(\phi) := |\phi(\alpha)|$, $\alpha \in \mathcal{A}$) then $\mathcal{M}_{\mathcal{A}}$ becomes a *compact topological space* in the subspace topology inherited from $\mathcal{A}^*$. This is quite interesting in view of example 2.1 as this is providing the converse statement. In fact, for a commutative, unital C^*-algebra $\mathcal{A}$, we define the *Gelfand transform* $\mathcal{G}$ as the map

$$\begin{aligned} \mathcal{G} : \mathcal{A} &\longrightarrow C(\mathcal{M}_{\mathcal{A}}) \\ a &\longmapsto \hat{a}, \end{aligned}$$

where $\hat{a}(\mu) := \mu(a)$.

Theorem 2.1 (Gelfand-Naimark [5, 10, 11]). *The Gelfand transform is an isometric $*$-isomorphism $\mathcal{A} \cong C(\mathcal{M}_{\mathcal{A}})$.*

Rather than reviewing the proof of this theorem, we will consider some basic examples that should provide a good intuition about the correspondence between commutative C^*-algebras and (locally compact, Hausdorff) topological spaces.

Example 2.4. Consider the unital, commutative C^*-algebra $\mathcal{A}$ generated by a unitary element u (*i.e.* $u^*u = uu^* = \mathbb{1}$), with norm fixed by the condition [12] that

$$\|\mathbb{1} + e^{-i\alpha}u\| = 2,$$

for all $\alpha \in [0, 2\pi)$. Theorem 2.1 states that this algebra must be the function algebra of a topological space which, as a set, is precisely the set $\mathcal{M}_\mathcal{A}$ of all characters of $\mathcal{A}$. Since the algebra is generated by u, a character μ is fixed by its value on u. But $|\mu(u)|^2 = \mu(u^*u) = 1$, and so every character is of the form $\mu(u) = e^{i\theta}$, *i.e.* $\mathcal{M}_\mathcal{A} \subseteq S^1$. In order to see that actually $\mathcal{M}_\mathcal{A} = S^1$, we proceed as in [12] by noticing that from example 2.1 and from the uniqueness of the norm, it follows that $\|a\| = \sup_{\mu \in \mathcal{M}_\mathcal{A}} |\mu(a)|$, so that we must have

$$\sup_{\mu \in \mathcal{M}_\mathcal{A}} |1 + e^{-i\alpha}\mu(u)| = 2.$$

But this can only happen if for every $\alpha \in [0, 2\pi)$ there is a character μ with $\mu(u) = e^{i\alpha}$.

Example 2.5. The 2-sphere S^2 can also be easily characterized in terms of its algebra of continuous functions. In this case, the algebraic structure is well-known, as $C(S^2)$ is generated by the spherical harmonics Y_{lm}, their product being given as in the usual Clebsch-Gordan decomposition. More details on how S^2 can be obtained as the character space of such an algebra (this involves an appropriate definition of the norm, as in the previous example) can be found in [12].

Example 2.6. Take $\mathcal{A} = C(S^2)$ as above (with the sup norm) and consider the subalgebra $\mathcal{A}_+$ consisting of all even functions on $\mathcal{A}$, that is, functions f such that $f(-x) = f(x)$. Then $\mathcal{A}_+ \cong C(\mathbb{R}P^2)$. This fact has been exploited in studies of the spin-statistics connection in quantum mechanics [12–15].

Remark 2.4. The isomorphism provided by theorem 2.1 is actually an equivalence between the categories of commutative C^*-algebras and locally compact, Hausdorff topological spaces. This means, among other things, that all information regarding the topology of M is encoded in the algebra

$C(M)$: An open set in M can be equivalently described in terms of an ideal in $C(M)$, closed sets are described by quotient algebras, metrizability of M amounts to separability of $C(M)$, and so on [16]. This is the reason why the Gelfand-Naimark theorem is considered one of the main sources for the development of noncommutative geometry [5, 11].

Now we turn to the GNS (Gelfand-Naimark-Segal) construction. Let $\mathcal{A}$ be a C^*-algebra, and ω a state thereon. The GNS construction furnishes a representation of $\mathcal{A}$ on some (ω-dependent) Hilbert space. The basic idea is to use multiplication in $\mathcal{A}$ in order to obtain a linear action of $\mathcal{A}$ on a vector space. So we begin by regarding $\mathcal{A}$ as a vector space $\hat{\mathcal{A}}$ (*i.e.* we just "forget" multiplication). Even though the underlying spaces are equal, it will be convenient to distinguish the *algebra*, with elements $a \in \mathcal{A}$, from the underlying *vector space*, with elements $|a\rangle \in \hat{\mathcal{A}}$. Recalling the Cauchy-Schwarz inequality (problem 2.6), we realize that it makes sense to introduce the following sesquilinear form on $\hat{\mathcal{A}}$:

$$\langle a|b\rangle_\omega := \omega(a^*b). \tag{17}$$

This is "almost" an inner product, as it may happen that $\langle a|a\rangle_\omega = 0$ for some $a \neq 0$. But ignoring that fact for a moment, we notice that the product on $\mathcal{A}$ can be used to make $\mathcal{A}$ act as a linear operator on $\hat{\mathcal{A}}$:

$$\begin{aligned} \mathcal{A} \times \hat{\mathcal{A}} &\longrightarrow \hat{\mathcal{A}} \\ a, |b\rangle &\longmapsto a \cdot |b\rangle := |ab\rangle. \end{aligned}$$

Now, in order to have a Hilbert space representation, we have to "fix" the problem with (17). This is done as follows.

Problem 2.7. *Define* $\mathcal{N}_\omega := \{a \in \mathcal{A} \,|\, \omega(a^*a) = 0\}$, *and show that it is a closed left-ideal of* $\mathcal{A}$.

We therefore obtain a Hilbert space $\mathcal{H}_\omega$ given by the completion of $\hat{\mathcal{A}}/\mathcal{N}_\omega$ with respect to the inner product $\langle [a]|[b]\rangle_\omega := \omega(a^*b)$, where we use $|[a]\rangle$ to denote the equivalence class of a in (the completion of) $\hat{\mathcal{A}}/\mathcal{N}_\omega$. With this we have obtained a $*$-representation of $\mathcal{A}$ by bounded operators acting on $\mathcal{H}_\omega$:

$$\begin{aligned} \pi_\omega : \mathcal{A} &\longrightarrow \mathcal{B}(\mathcal{H}_\omega) \\ a &\longmapsto \pi_\omega(a), \end{aligned}$$

where $\pi_\omega(a)|[b]\rangle := |[ab]\rangle$.

Problem 2.8. *Obtain the following inequality:*

$$\sqrt{\omega(a^*a)} \leq \|\pi_\omega(a)\| \leq \|a\|.$$

Although the representation π_ω is a $*$-homomorphism, it is not isometric. But it can be shown that for every $a \in \mathcal{A}$ there is a state ω such that $\omega(a^*a) = \|a\|^2$. Then, if we consider the direct sum

$$\mathcal{H} = \bigoplus_{\omega \in \mathcal{S}_\mathcal{A}} \mathcal{H}_\omega := \big\{(\xi_\omega)_{\omega \in \mathcal{S}_\mathcal{A}} \,|\, \xi_\omega \in \mathcal{H}_\omega, \sum_{\omega \in \mathcal{S}_\mathcal{A}} \|\xi_\omega\|_\omega^2 < \infty\big\},$$

where in each sequence $(\xi_\omega)_{\omega \in \mathcal{S}_\mathcal{A}}$ only countable many elements are different from zero. The representations π_ω then give rise to a representation π on $\mathcal{H}$ for which $\|\pi(a)\| = \|a\|$. The details of this construction can be found *e.g.* in [6]. The importance of this result is that *any* C^*-algebra is isometrically isomorphic to the C^*-algebra of bounded operators in some Hilbert space.

2.3. *Composite systems, entanglement*

In this section, our aim will be to study certain quantum correlations that arise as a result of *entanglement*. This is a vast subject, and here we will only consider elementary examples, corresponding to bipartite systems. Although very simple from the mathematical point of view, these examples already contain the essence that will allow us to distinguish between classical and genuinely quantum correlations. We will also discuss a recent application of the algebraic formalism to the study of entanglement for systems of identical particles [17, 18], a topic of current interest for quantum information, condensed matter, atomic physics, and quantum optics.

We thus start by considering a composite system that is the result of coupling two subsystems 1 and 2. We will first consider quantum systems described directly in terms of Hilbert spaces. But, as will become clear, the algebraic approach based only on observable algebras and states will allow us to reformulate these ideas in more generality. This will be important for the applications to systems of identical particles. Let $\mathcal{H}_1$ and $\mathcal{H}_2$ denote the corresponding (finite dimensional, for simplicity) Hilbert spaces, and assume that the subsystems are described in terms of observable algebras $\mathcal{A}_i \subseteq \mathcal{B}(\mathcal{H}_i)$, $i = 1, 2$. Subsystems 1 and 2 are then coupled to form a composite system that will be described by $\mathcal{A} = \mathcal{A}_1 \otimes \mathcal{A}_2$.

Let now $\omega : \mathcal{A} \to \mathbb{C}$ be a state of the composite system. Consider "partial", or "local" measurements performed on each subsystem. This leads us to consider the restriction of ω to $\mathcal{A}_i$, $i = 1, 2$:

$$\omega_1 := \omega\big|_{\mathcal{A}_1}, \quad \omega_2 := \omega\big|_{\mathcal{A}_2}.$$

Definition 2.9. In the present context of a bipartite composite system, if ω is a pure state on $\mathcal{A}$, we say it is a *separable* state if

$$\omega(a \otimes b) = \omega_1(a)\omega_2(b),$$

for all $a \in \mathcal{A}_1$, $b \in \mathcal{A}_2$. A (pure) state that is not separable in the above sense is called an *entangled* state.

Example 2.7. Let $\mathcal{H}_1 = \mathbb{C}^n$, with basis $\{|e_1\rangle, \ldots, |e_n\rangle\}$ and $\mathcal{H}_2 = \mathbb{C}^m$, with basis $\{|u_1\rangle, \ldots, |u_m\rangle\}$. Let $\mathcal{A}_1 = M_n(\mathbb{C})$ and $\mathcal{A}_2 = M_m(\mathbb{C})$. The composite system is then described by the matrix algebra $\mathcal{A} = \mathcal{A}_1 \otimes \mathcal{A}_2$. Pick now a vector state $|\psi\rangle \in \mathcal{H}_1 \otimes \mathcal{H}_2$ and set

$$\begin{aligned} \omega_\psi : \mathcal{A} &\longrightarrow \mathbb{C} \\ \alpha &\longmapsto \langle\psi|\alpha|\psi\rangle. \end{aligned}$$

Then, the following three conditions are equivalent:

(*i*) The vector state $|\psi\rangle$ is of the form $|\psi\rangle = |\varphi\rangle \otimes |\xi\rangle$.
(*ii*) $S(\rho_{\psi,i}) = 0$, $(i = 1, 2)$ where $S(\rho_{\psi,i})$ stands for the *von Neumann entropy* of the reduced density matrix $\rho_{\psi,i}$ (see definitions below).
(*iii*) The state ω_ψ is separable.

To see where these equivalences come from, let us expand $|\psi\rangle$ as

$$|\psi\rangle = \sum_{i=1}^{n}\sum_{j=1}^{m} A_{ij}|e_i\rangle \otimes |u_j\rangle.$$

Then $A \in M_{n,m}(\mathbb{C})$, so we may perform a singular value decomposition and write it in the form $A = UDV^\dagger$, with D a diagonal matrix in $M_{n,m}(\mathbb{C})$, $U \in M_n(\mathbb{C})$ and $V \in M_m(\mathbb{C})$. The elements of D are $D_{kl} = \delta_{kl}\sqrt{\lambda_k}$, where the λ_k, called the "Schmidt coefficients", are the eigenvalues of $A^\dagger A$. If we define vectors

$$|k\rangle_1 := \sum_{i=1}^{n} U_{ik}|e_i\rangle, \quad |k\rangle_2 := \sum_{j=1}^{m} \bar{V}_{jk}|u_j\rangle,$$

then we obtain

$$|\psi\rangle = \sum_k \sqrt{\lambda_k}|k\rangle_1 \otimes |k\rangle_2.$$

From this expression, it is clear that (*i*) holds precisely when there is only one non-vanishing Schmidt coefficient. Furthermore, from $\langle\psi|\psi\rangle = 1$ we obtain $\sum_k \lambda_k = 1$, with $0 \leq \lambda_k \leq 1$. We can, therefore, regard $\{\lambda_k\}_k$ as a probability distribution, and compute its (Shannon) entropy, defined as

$$H(\{\lambda_k\}_k) := -\sum_k \lambda_k \log \lambda_k. \tag{18}$$

Notice that this function vanishes precisely when $|\psi\rangle$ is a separable state. We can relate it to the von Neumann entropy of the restricted states $\omega_{\psi,i}$ as follows.

Problem 2.9. *Consider the restriction $\omega_{\psi,i}$ of ω_ψ to $\mathcal{A}_i \subseteq \mathcal{A}$ ($i = 1, 2$). Find density matrices $\rho_{\psi,i}$ ("reduced density matrices") acting on $\mathcal{H}_i$ and such that $\omega_\psi(a \otimes \mathbb{1}_m) = Tr_{\mathcal{H}_1}(\rho_{\psi,1} a)$ and $\omega_\psi(\mathbb{1}_n \otimes b) = Tr_{\mathcal{H}_2}(\rho_{\psi,2} b)$.*

The von Neumann entropy *of a density matrix ρ is defined as*

$$S(\rho) = -Tr(\rho \log \rho).$$

Show that $S(\rho_{\psi,1})$ and $S(\rho_{\psi,2})$ coincide and are exactly equal to $H(\{\lambda_k\}_k)$ from (18).

We therefore consider the quantity $S(\rho_{\psi,i})$ ($i = 1, 2$) as a *measure* of the "amount of entanglement" of the state ω_ψ. In order to really appreciate the meaning of this assertion, we have to understand the (very surprising, and interesting) features of quantum correlations. This can be considered as one of the starting points of quantum information theory. In quantum information theory, there are orderly ways to pose and study these type of problems where, for instance, in the (pure) bipartite case the von Neumann entropy of the reduced density matrix can be obtained as (basically) the unique entanglement measure of the state $|\psi\rangle$, when entanglement is regarded as a physical resource and defined in operational terms [19, 20]. The general problem, that of multipartite entanglement, is much more complicated and involves many mathematically as well as physically interesting problems. However, our interest here will be restricted to the (pure) bipartite case. This will be enough to illustrate one of our points, which is that the algebraic framework leads to a unified description of physical systems, no matter whether they are classical or quantum. Also, the "detachment" from an *a priori* given Hilbert space in the quantum case allows for a more clear understanding of phenomena. As it turns out, the noncommutativity of the observable algebra can be seen as the main source of (truly quantum) correlations. This fits nicely with the relation between commutative and noncommutative spaces, as hinted above.

The next example illustrates the previous remarks.

Example 2.8. Let us consider the composition of two classical systems. By this we mean that the "coupling" of two systems described by (commutative) observable algebras of the form $\mathcal{A}_1 = C(X)$ and $\mathcal{A}_2 = C(Y)$ is effected by the tensor product $\mathcal{A}_1 \otimes \mathcal{A}_2$. Tensor products of C^*-algebras have to be treated carefully, but assuming we have defined the appropriate tensor product, let us assume that $C(X) \otimes C(Y) \simeq C(X \times Y)$ (a detailed discussion about tensor products can be found in [16]). Hence, the composite system will be assumed to be described by the observable algebra $\mathcal{A} = C(X \times Y)$. Consider now a state ω on $\mathcal{A}$. It is not difficult to see that it must be given by a probability distribution $p(x, y)$ in such a way that for $h \in \mathcal{A}$,

$$\omega(h) = \int_{X \times Y} h(x, y) p(x, y) dx dy.$$

Restriction to $\mathcal{A}_1$ or $\mathcal{A}_2$ in this case leads to marginal distributions $p_1(x) = \int_Y p(x, y) dy$ and $p_2(y) = \int_X p(x, y) dx$, which in turn define states ω_i on $\mathcal{A}_i$. From the above definition we have that ω is separable if, and only if,

$$\omega(f \otimes g) = \omega_1(f)\omega_2(g), \quad \text{for all} \quad f \in \mathcal{A}_1, g \in \mathcal{A}_2.$$

But from the properties of the Gelfand transform it follows that any pure state on $C(X \times Y)$ is of the form $\omega(h) = h(x_0, y_0)$, *i.e.*, they are given by point measures: $p(x, y) = \delta_{(x_0, y_0)}(x, y)$. It follows immediately that $\omega(f \otimes g) = \omega_1(f)\omega_2(g)$. This illustrates the fact that entanglement entails purely quantum correlations, as all pure states of a classical composite (bipartite) system are separable.

Problem 2.10. *The previous example was formulated in rather loosely terms. Provide the necessary details to turn it into a rigorous proof that works for locally compact spaces.*

Example 2.9. On $\mathcal{H} = \mathbb{C}^2 \otimes \mathbb{C}^2$ consider the following family of states:

$$|\psi_\lambda\rangle := \sqrt{\lambda}\,| + -\rangle - \sqrt{1 - \lambda}\,| - +\rangle,$$

and the following types of observables, acting on $\mathbb{C}^2$:

$$P(a) := \frac{1}{2}\left(\mathbb{1}_2 + \vec{a} \cdot \vec{\sigma}\right), \quad E(a) := \vec{a} \cdot \vec{\sigma},$$

with $\vec{a}$ a *unit* vector in $\mathbb{R}^3$ and $\vec{\sigma} = (\sigma_1, \sigma_2, \sigma_3)$ (Pauli matrices). Straightforward computations give

$$\langle \psi_\lambda | E(a) \otimes E(b) | \psi_\lambda \rangle = -a_3 b_3 - 2\sqrt{\lambda(1 - \lambda)}(a_1 b_1 + a_2 b_2)$$

and

$$
\begin{aligned}
&\langle\psi_\lambda|P(a)\otimes P(b)|\psi_\lambda\rangle \\
&\quad= \frac{1}{4}\left(1+(2\lambda-1)(a_3-b_3)-a_3b_3-2\sqrt{\lambda(1-\lambda)}(a_1b_1+a_2b_2)\right).
\end{aligned}
$$

Whereas for $\lambda = 0, 1$ we obtain separable states, for $0 < \lambda < 1$ we have entangled states. In particular, for $\lambda = 1/2$ we obtain a maximally entangled state (Bell state), for which

$$\langle\psi_{1/2}|P(a)\otimes P(b)|\psi_{1/2}\rangle = \frac{1}{4}\left(1-\vec{a}\cdot\vec{b}\right).$$

In this case, the marginal distributions give

$$
\begin{aligned}
P_A(a,+) &= \tfrac{1}{2} = P_A(a,-), \\
P_B(b,+) &= \tfrac{1}{2} = P_B(b,-),
\end{aligned}
$$

whereas for the joint probabilities we obtain $(r, r' = \pm 1)$:

$$P_{rr'} = \frac{1}{4}\left(1-rr'\vec{a}\cdot\vec{b}\right).$$

In particular, for the choice $\vec{a}\cdot\vec{b} = 1$, we obtain *total anticorrelation.*

What is so special about the previous example? Consider the Clauser-Horne-Shimony-Holt (CHSH) inequality [19, 20]:

$$\langle A(B+B') + A'(B-B')\rangle \leq 2,$$

where $\langle X\rangle$ denotes the expectation value of an observable X, and where for the involved observables it is assumed that $-1 \leq A, A', B, B' \leq 1$. The inequality can be obtained if we assume some underlying (classical) probability space $(\Omega, d\mu)$, for which the correlations above take the form

$$\langle AB\rangle_\mu = \int_\Omega A(x)B(x)d\mu(x),$$

or, stated in terms of algebras of observables:

Problem 2.11. *Let $\mathcal{A}$ be a commutative C^*-algebra, ω a pure state thereon, and a, a', b, b' elements of $\mathcal{A}$, all of them with norm less or equal to one. Show that*

$$\omega(a(b+b') + a'(b-b')) \leq 2.$$

But, in contrast to this, in example 2.9 (with $\lambda = 1/2$) we obtain

$$\langle\psi_{1/2}|E(a)\otimes(E(b)+E(b'))+E(a')(E(b)-E(b'))|\psi_{1/2}\rangle$$
$$=\vec{a}\cdot\vec{b}+\vec{a}\cdot\vec{b}'+\vec{a}'\cdot\vec{b}-\vec{a}'\cdot\vec{b}'.$$

Now, it is easy to find an arrangement for the vectors a, a', b, b' such that the CHSH inequality is violated. The importance of these type of inequalities, of which Bell's inequality was the first one, cannot be overemphasized, as they have allowed for definitive experimental tests of quantum mechanics. The CHSH inequality is also relevant in the context of hidden-variable models, and related no-go theorems [21].

In the next section we will consider the problem of entanglement for systems of identical particles, for which the tools developed in the last sections will prove very useful.

2.4. *Identical particles and entanglement*

When combined, two of the most intriguing features of quantum theory -the intrinsic indistinguishability of identical particles and quantum entanglement- lead to formidable conceptual issues that have been addressed for years, and for which no generally accepted framework exists. The main source of problems comes from the fact that entanglement and related concepts have been studied mainly in cases where a subsystem decomposition coincides with a tensor product decomposition of the Hilbert space. But precisely in the case of identical particles, the Hilbert space is the antisymetric/symmetric subspace obtained from the action of the permutation group on a tensor product space. So precisely in this case decomposition into subsystems does not correspond to tensor product decomposition. This makes the use of partial trace a doubtful operation, leading to a clash with standard notions of entanglement. In [17, 18, 22], we have shown how, using the representation theory of operator algebras, it is possible to obtain a generalized notion of entanglement. The universality of the approach, which can be applied to particles obeying any kind of statistics including bosons, fermions, parafermions and particles obeying braid statistics, provides a unifying view of entanglement, with many potential applications.

The key idea is that, since subsystems can always be described in terms of subalgebras, the use of partial trace will be superseded by the more general notion of restriction of a quantum state ω to a subalgebra. Then one can make use of the GNS construction in order to find a Hilbert space representation π_ω, as above.

Problem 2.12. *Prove that the GNS representation* π_ω *is irreducible precisely when the state* ω *is pure.*

From this problem, it follows that the condition for irreducibility of the representation π_ω is precisely that the von Neumman entropy of the state ω vanishes and so, given a subsystem described in terms of a subalgebra $\mathcal{A}_0 \subset \mathcal{A}$, one may consider the restriction of the state ω to $\mathcal{A}_0$. This is a generalization of the notion of partial trace. Hence, a generalized notion of entanglement emerges, based on the von Neumann entropy of the restricted state. The properties of this entanglement measure are very closely tied to the GNS-representation of the subalgebra and the restricted state. In other words, one finds that entanglement depends *both* on the state and the subsystem of the full system. This formalism, then, lends itself to the study of problems where partial trace loses its meaning, as is the case with systems of identical particles. Let us consider a few illustrative examples.

Example 2.10. Consider the algebra $\mathcal{A} = M_2(\mathbb{C})$ of 2×2 matrices with complex entries, and let

$$\omega_\lambda(a) = \lambda a_{11} + (1-\lambda)a_{22}$$

be a state on $\mathcal{A}$. Of course, this can only happen if $0 \leq \lambda \leq 1$. Let us then see how the GNS representation is constructed, and how we can associate an entropy to the state ω_λ. For this purpose, let us consider the matrix units $e_{ij} = |i\rangle\langle j|$ $(i,j=1,2)$. They generate the algebra, and provide a basis for the underlying vector space $\hat{\mathcal{A}}$. They also fulfill the relations

$$e_{ij}e_{kl} = \delta_{jk}e_{il}. \tag{19}$$

Let us now check whether there will be null vectors. Writing a general element $a \in \mathcal{A}$ as

$$a = \sum_{ij} a_{ij}e_{ij},$$

we compute

$$\begin{aligned} \omega_\lambda(a^*a) &= \sum_{i,j,k} \omega_\lambda(\bar{a}_{ki}a_{kj}e_{ij}) \\ &= \lambda(|a_{11}|^2 + |a_{21}|^2) + (1-\lambda)(|a_{12}|^2 + |a_{22}|^2). \end{aligned} \tag{20}$$

From this expression we recognize that, if $\lambda \in (0,1)$, then $\dim(\mathcal{H}_{\omega_\lambda}) = 4$, whereas for $\lambda = 0$ or $\lambda = 1$ we obtain $\dim(\mathcal{H}_{\omega_\lambda}) = 2$. In fact, if λ is different from zero or one, we must have both $(|a_{11}|^2 + |a_{21}|^2) = 0$ and $(|a_{12}|^2 + |a_{22}|^2) = 0$. It follows that $\mathcal{N}_{\omega_\lambda} = \{0\}$. From

$$\pi_{\omega_\lambda}(a)|[b]\rangle = |[ab]\rangle$$

and (19) we see that the representation is reducible, with 2 invariant subspaces generated, respectively, by $\{|[e_{11}]\rangle, |[e_{21}]\rangle\}$ and $\{|[e_{12}]\rangle, |[e_{22}]\rangle\}$. From problem 2.12 we conclude that, in this case, the state has to be a mixed state. The cases $\lambda = 0$ and $\lambda = 1$ lead to a non-trivial null space, and from (20) we conclude that the dimension of these irreducibles is, in both cases, equal to 2. It is left to the reader to obtain a pure state on the algebra $M_2(\mathbb{C}) \otimes M_2(\mathbb{C})$ such that ω_λ is the restriction to $M_2(\mathbb{C})$.

The next question we need to address is that of how to compute the entropy of the reduced state. Since our algebras will in general be unital, there is a simple way to acomplish this, namely, given a (unital) C^*-algebra $\mathcal{A}$ and a state ω on it, it is always possible to find a *density matrix* ρ_ω acting on the GNS space $\mathcal{H}_\omega$, and such that

$$\omega(a) = \mathrm{Tr}_{\mathcal{H}_\omega}(\rho_\omega a)$$

for all a in $\mathcal{A}$. From the decomposition of the GNS space into irreducibles,

$$\mathcal{H}_\omega = \bigoplus_j \mathcal{H}_\omega^{(j)},$$

we obtain *projectors* $P^{(j)}$, with $\sum_j P^{(j)} = \mathbb{1}_{\mathcal{H}_\omega}$. From the definition of the inner product in $\mathcal{H}_\omega$ we have $\omega(a) = \langle[\mathbb{1}_\mathcal{A}]|\pi_\omega(a)|[\mathbb{1}_\mathcal{A}]\rangle$. But then, using an orthonormal basis $\{|n\rangle\}_n$ on $\mathcal{H}_\omega$, we can write:

$$\begin{aligned}
\omega(a) &= \langle[\mathbb{1}_\mathcal{A}]|\pi_\omega(a)|[\mathbb{1}_\mathcal{A}]\rangle \\
&= \langle[\mathbb{1}_\mathcal{A}]|\sum_k P^{(k)}\pi_\omega(a)|[\mathbb{1}_\mathcal{A}]\rangle \\
&= \langle[\mathbb{1}_\mathcal{A}]|\sum_k P^{(k)}\pi_\omega(a)P^{(k)}|[\mathbb{1}_\mathcal{A}]\rangle \\
&= \langle[\mathbb{1}_\mathcal{A}]|\sum_k P^{(k)}\pi_\omega(a)\sum_n |n\rangle\langle n|P^{(k)}|[\mathbb{1}_\mathcal{A}]\rangle \\
&= \mathrm{Tr}_{\mathcal{H}_\omega}(\rho_\omega \pi_\omega(a)),
\end{aligned}$$

with ρ_ω given by

$$\rho_\omega = \sum_k P^{(k)}|[\mathbb{1}_\mathcal{A}]\rangle\langle[\mathbb{1}_\mathcal{A}]|P^{(k)}.$$

Consider now a composite, bipartite system for which the Hilbert space is of the form $\mathcal{H} = \mathcal{H}_A \otimes \mathcal{H}_B$. If ω_ψ is a state on the algebra of bounded operators on $\mathcal{H}_A \otimes \mathcal{H}_B$ that is obtained from a vector state $|\psi\rangle \in \mathcal{H}$, then its restriction to the subalgebra $\mathcal{A}_1$ generated by elements of the form $K \otimes \mathbb{1}_B$ gives a state

$$\omega_1(K) \equiv \omega\,|_{\mathcal{A}_1}\,(K \otimes \mathbb{1}_B) = \mathrm{Tr}_{\mathcal{H}_A}(\rho_A K),$$

where ρ_A is the *reduced* density matrix, defined through partial trace (*cf.* problem 2.9),

$$\rho_A = \mathrm{Tr}_{\mathcal{H}_B}|\psi\rangle\langle\psi|.$$

We therefore see that, for these kind of systems, partial trace equals restriction (to some subalgebra).

Problem 2.13. *Compute de von Neumann entropy of the density matrix ρ_{ω_λ} corresponding to the GNS space of example 2.10 and provide a physical interpretation.*

The real usefulness of the algebraic approach becomes apparent only when we consider situations where the Hilbert space does not have a simple tensor product structure. This is what happens, *e.g.*, with systems of identical particles. Because of the symmetrization postulate, the Hilbert space contains only the symmetric (for bosons) or the antisymmetric (for fermions) subspaces of the many-particle Hilbert space. As a consequence, one finds that states that from a physical point of view should not *a priori* be considered to be entangled, will have reduced density matrices with non-vanishing entanglement entropy [23]. In approaches like those of [24] or [17], the description of subsystems is given by specifying suitable subalgebras. Then, the restriction of a given state to the subalgebra provides a physically sensible generalization of the notion of partial trace. Applying the GNS construction to the restricted state, it is possible to study the entropy emerging from restriction and use it as a generalized measure of entanglement.

Let us briefly discuss how this entanglement measure can be computed in concrete cases. For this purpose, consider a Hilbert space $\mathcal{H}^{(1)} = \mathbb{C}^d$, assumed to correspond to the space of 1-particle states of a fermionic system. The full Hilbert space is then the antisymmetric Fock space $\mathcal{F}$ obtained from $\mathcal{H}^{(1)}$. It decomposes as a direct sum of spaces of fixed number of particles. The k-particle Hilbert space $\mathcal{H}^{(k)}$ is just the antisymmetrized k-fold tensor product of $\mathcal{H}^{(1)}$. This is a kind of "toy model" for a fermionic quantum field theory, but many important features of a quantum field theory can be seen to appear already at this level. One of these features is the connection to the representation theory of Clifford algebras, as explained in full detail in [11]. As is well-known, the Clifford algebra of $\mathcal{H}^{(1)}$ acts naturally on the exterior algebra $\bigwedge^\bullet(\mathcal{H}^{(1)})$ which, in turn, is related to the Fock space construction in the following way.

Let $\{e_n\}_n$ denote an orthonormal basis for $\mathcal{H}^{(1)}$, and denote with $a_n^{(\dagger)}$ the corresponding annihilation (creation) operators (*cf.* problem 2.3). Then, there is a vector space isomorphism between $\mathcal{F}$ and $\Lambda^\bullet(\mathcal{H}^{(1)})$, furnished by the correspondence

$$e_{i_1} \wedge e_{i_2} \wedge \cdots \wedge e_{i_k} \longleftrightarrow a_{i_1}^\dagger a_{i_2}^\dagger \cdots a_{i_k}^\dagger |0\rangle. \tag{21}$$

This correspondence is behind the famous quote by E. Nelson: *"first quantization is a mystery, but second quantization is a functor"*. In fact, given a self-adjoint operator A on $\mathcal{H}^{(1)}$ (that is, a 1-particle observable), we obtain (by functoriality) an operator $d\Gamma(A)$ acting on Fock space, whose restriction to the k-particle sector $\mathcal{H}^{(k)}$ is given by

$$\begin{aligned} d\Gamma^{(k)}(A) &:= (A \otimes \mathbb{1}_d \otimes \cdots \otimes \mathbb{1}_d) \\ &+ (\mathbb{1}_d \otimes A \otimes \cdots \otimes \mathbb{1}_d) + \cdots + (\mathbb{1}_d \otimes \cdots \otimes \mathbb{1}_d \otimes A). \end{aligned} \tag{22}$$

Taking the correspondence (21) into account, we obtain the following expression for $d\Gamma(A)$ in terms of creation/annihilation operators,

$$d\Gamma(A) = \sum_{i,j} A_{ij} a_i^\dagger a_j, \tag{23}$$

where $A_{ij} = \langle e_i|A|e_j\rangle$. In the physics literature, the operator $d\Gamma(A)$ is referred to as the *second quantization* of A.

In the present quantum-mechanical context, where the number of particles is kept fixed, we want to focus our attention on the operator $d\Gamma^{(k)}(A)$. One of the properties of this operator is that it preserves the symmetries of $\mathcal{H}^{(k)}$. Furthermore, the map $A \longrightarrow d\Gamma^{(k)}(A)$ allows us to study subalgebras of 1-particle observables.

The simplest example we can consider in order to illustrate entanglement issues for systems of identical particles is that of just two fermions. Let us then consider, as done in [25] and in [17], a 2-fermion system, where each fermion can be in a linear superposition of 4 basic states which, for the sake of physical interpretation, will be divided into internal and external degrees of freedom. So we describe 1-particle states in terms of a set of (fermionic) creation/annihilation operators $a_\lambda^{(\dagger)}, b_\lambda^{(\dagger)}$, where a stands for *"left"*, b for *"right"* (the external degrees of freedom) and $\lambda = 1, 2$ for spin up and down (the internal degrees of freedom). Hence, in this case we have $\mathcal{H}^{(1)} = \mathbb{C}^4$ and, for the 2-fermion space, $\mathcal{H}^{(2)} = \bigwedge^2 \mathbb{C}^4$. An orthonormal basis for $\mathcal{H}^{(2)}$ is given by the vectors

$$a_1^\dagger a_2^\dagger|0\rangle,\ \ b_1^\dagger b_2^\dagger|0\rangle,\ \ a_1^\dagger b_2^\dagger|0\rangle \text{ and } a_2^\dagger b_1^\dagger|0\rangle.$$

The two-particle algebra $\mathcal{A}$ of observables is thus isomorphic to the matrix algebra $M_6(\mathbb{C})$.

For $|\psi_\theta\rangle = (\cos\theta a_1^\dagger b_2^\dagger + \sin\theta a_2^\dagger b_1^\dagger)|\Omega\rangle$, the corresponding state ω_θ is given by $\omega_\theta(\alpha) = \langle\psi_\theta|\alpha|\psi_\theta\rangle$ for $\alpha \in \mathcal{A}$. We choose the subalgebra $\mathcal{A}_0$ to be the one generated by $\mathbb{1}_\mathcal{A}$, $n_{12} = a_1^\dagger a_1 a_2^\dagger a_2$, $N_a = a_1^\dagger a_1 + a_2^\dagger a_2$ and $T_{i=1,2,3} = (1/2)\ a_\lambda^\dagger(\sigma_i)^{\lambda\lambda'} a_{\lambda'}$. Physically, it corresponds to the subalgebra of one-particle observables corresponding to measurements at the left location.

Problem 2.14. *Consider the restriction of ω_θ to $\mathcal{A}_0$ and study the GNS representation corresponding to this choice. For $\theta = 0, \pi/2$ you should obtain vanishing entropy, in contrast to the result $S = \log_2 2$ obtained via partial trace for states with Slater rank one [17, 18, 23, 26].*

3. Spin chains

3.1. *The transverse Ising chain*

The Hamiltonian for the (quantum) Ising chain in a transverse field is given by

$$H = -\sum_{i=1}^{N-1} \sigma_i^x \sigma_{i+1}^x - \lambda \sum_{i=1}^{N} \sigma_i^z. \tag{24}$$

An important aspect when solving the model and studying its solution is the type of boundary conditions considered. We will be interested in both open as well as periodic boundary conditions. The first step in the solution of this model is the so-called Wigner-Jordan transformation, that allows us to express all spin operators in terms of *fermionic* creation/annhilation operators:

As can be easily checked, the operators defined by

$$\begin{aligned} a_i &= \sigma^z \otimes \cdots \otimes \sigma^z \otimes \sigma^+, \\ a_i^\dagger &= \sigma^z \otimes \cdots \otimes \sigma^z \otimes \sigma^-, \end{aligned} \tag{25}$$

where $\sigma^\pm = \frac{1}{2}(\sigma^x \pm i\sigma^y)$, are fermionic operators.

Problem 3.1. *Check that the operators defined above indeed satisfy the CAR algebra: $\{a_i, a_j^\dagger\} = \delta_{ij}$, $\{a_i, a_j\} = 0 = \{a_i^\dagger, a_j^\dagger\}$.*

The inverse transformation is given by

$$\begin{aligned}\sigma_i^z &= 1 - 2a_i^\dagger a_i,\\ \sigma_i^x &= \left(\prod_{m<i}(1-2a_m^\dagger a_m)\right)(a_i^\dagger + a_i),\\ \sigma_i^y &= i\left(\prod_{m<i}(1-2a_m^\dagger a_m)\right)(a_i^\dagger - a_i).\end{aligned} \tag{26}$$

Problem 3.2. *Check that (26) is the inverse transformation to (25).*

With this we obtain, for the interaction terms of the Hamiltonian,

$$\sigma_i^x \sigma_{i+1}^x = (a_i^\dagger - a_i)(a_{i+1}^\dagger + a_{i+1}),$$

and so H takes the form

$$H = -\sum_{i=1}^{N-1}(a_i^\dagger - a_i)(a_{i+1}^\dagger + a_{i+1}) - \lambda\sum_{i=1}^{N}(1-2a_i^\dagger a_i). \tag{27}$$

The term containing the external field λ is diagonal in this basis. The constant term $-\lambda N$ coming from the last sum is usually disregarded, because its only effect is to shift the energy spectrum. We will nevertheless keep all terms, in order to be able to compare with numerical solutions in the spin basis, for small values of N.

Expanding all terms in (27) and moving all creation operators to the left, we obtain (*cf.* Eq. (10.14) in [27]):

$$H = -\sum_{i=1}^{N-1}(a_i^\dagger a_{i+1} + a_{i+1}^\dagger a_i + a_i^\dagger a_{i+1}^\dagger - a_i a_{i+1}) + 2\lambda\left(\sum_{i=1}^{N} a_i^\dagger a_i\right) - \lambda N.$$

The point of using the Wigner-Jordan transformation for this model is that the Hamiltonian becomes "almost" diagonal. By this we mean that it is a sum of local, *quadratic* expressions in the creation and annihilation operators. As we will see, such models can be exactly solved. We now write H in a suggestive matrix notation. Arranging all creation and annihilation operators in rows and columns, we can write H as a kind of quadratic form. For example, we have, for $N = 2$,

$$(a_1^\dagger, a_2^\dagger)\begin{pmatrix}2\lambda & -1\\ -1 & 2\lambda\end{pmatrix}\begin{pmatrix}a_1\\ a_2\end{pmatrix} = -a_1^\dagger a_2 - a_2^\dagger a_1 + 2\lambda a_1^\dagger a_1 + 2\lambda a_2^\dagger a_2.$$

The expression corresponding to the same term for $N = 4$ is then

$$(a_1^\dagger, a_2^\dagger, a_3^\dagger, a_4^\dagger) \begin{pmatrix} 2\lambda & -1 & 0 & 0 \\ -1 & 2\lambda & -1 & 0 \\ 0 & -1 & 2\lambda & -1 \\ 0 & 0 & -1 & 2\lambda \end{pmatrix} \begin{pmatrix} a_1 \\ a_2 \\ a_3 \\ a_4 \end{pmatrix}$$

$$= -a_1^\dagger a_2 + a_2^\dagger a_1 - a_2^\dagger a_3 - a_3^\dagger a_2 - a_3^\dagger a_4 - a_4^\dagger a_3 + 2\lambda \sum_{i=1}^{4} a_i^\dagger a_i.$$

For arbitrary N, we may define the following $N \times N$ matrices:

$$A = \begin{pmatrix} 2\lambda & -1 & 0 & 0 & \cdots \\ -1 & 2\lambda & -1 & 0 & \\ 0 & -1 & 2\lambda & -1 & \\ 0 & 0 & -1 & 2\lambda & \cdots \\ \vdots & & & \vdots & \ddots \end{pmatrix}, \quad B = \begin{pmatrix} 0 & -1 & 0 & 0 & \cdots \\ 1 & 0 & -1 & 0 & \\ 0 & 1 & 0 & -1 & \\ 0 & 0 & 1 & 0 & \cdots \\ \vdots & & & \vdots & \ddots \end{pmatrix}. \tag{28}$$

The matrix elements of these two matrices can be written as follows:

$$A_{ij} = 2\lambda\delta_{i,j} - (\delta_{i+1,j} + \delta_{i,j+1}), \quad B_{ij} = -(\delta_{i+1,j} - \delta_{i,j+1}).$$

Hence, the Hamiltonian takes the following form:

$$H = \sum_{i,j=1}^{N} \left[a_i^\dagger A_{ij} a_j + \frac{1}{2} \left(a_i^\dagger B_{ij} a_j^\dagger - a_i B_{ij} a_j \right) \right] - \lambda N. \tag{29}$$

3.2. *Open boundary conditions*

For several reasons, including the study of edge states, it is instructive to explore the explicit solution of this model for *open* boundary conditions. We start with the Ising Hamiltonian written in the form (29). Any model that can be written as a quadratic form can be expressed in this way, the "only" difference being the explicit form of the matrices A and B (notice that, in order for H to be Hermitian, A has to be symmetric and B antisymmetric). In principle, thus, the method presented below (following the work of Lieb, Schultz and Mattis [28]) can be applied to any such model.

For periodic boundary conditions, it is usually more convenient to take into account translation invariance and hence to introduce Fourier transformed operators. But for open boundary conditions, translation invariance is "broken" and then it is a good idea to start right away with a Bogoliubov transformation, as explained below.

Recall that the operators a_i, $a_j^\dagger$ defined in (25) obey fermionic canonical anti-commutation (CAR) relations (*cf.* problem 3.1):

$$\{a_i, a_j^\dagger\} = \delta_{ij}, \quad \{a_i, a_j\} = 0 = \{a_i^\dagger, a_j^\dagger\}.$$

In general, a Bogoliubov transformation is a mapping induced by a change of basis on the one-particle Hilbert space (unitary transformation), its main effect being to provide a new set of creation/annihilation operators for which the Hamiltonian (29) becomes diagonal. Consider, then, a new set of operators given by

$$c_k = \sum_{i=1}^{N} \left(g_{ki} a_i + h_{ki} a_i^\dagger \right), \qquad c_k^\dagger = \sum_{i=1}^{N} \left(\bar{g}_{ki} a_i^\dagger + \bar{h}_{ki} a_i \right), \tag{30}$$

where g and h are $N \times N$ matrices to be chosen so that

(i) The new operators satisfy the same CAR algebra:

$$\{c_k, c_l^\dagger\} = \delta_{kl}, \quad \{c_k, c_l\} = 0 = \{c_k^\dagger, c_l^\dagger\}. \tag{31}$$

(ii) The Hamiltonian becomes *diagonal* in the new basis:

$$H = \sum_k \Lambda_k c_k^\dagger c_k + \mu \tag{32}$$

(with μ some constant).

Problem 3.3. *Show that the requirement (31) leads to the following conditions:*

$$\begin{aligned} gg^\dagger + hh^\dagger &= \mathbb{1}_N, \\ gh^t + hg^t &= 0. \end{aligned} \tag{33}$$

Problem 3.4. *Compute the trace of H in two different ways, in order to show that the constant term in (32) is given by*

$$\mu = \frac{1}{2} \left(TrA - \sum_k \Lambda_k \right) - \lambda N.$$

The second condition above, Eq. (32), will lead to an eigenvalue problem for g and h, the solution of which amounts -in principle- to the solution of the full problem.

Now we compute the commutator $[c_k, H]$ in two different ways, once using (29) and once using (32). This leads to the following set of equations:

$$g_{ki}\, \Lambda_k = \sum_{j=1}^{N} \left(g_{kj} A_{ji} - h_{kj} B_{ji} \right), \quad h_{ki}\, \Lambda_k = \sum_{j=1}^{N} \left(g_{kj} B_{ji} - h_{kj} A_{ji} \right). \tag{34}$$

In order to solve this eigenvalue problem, it proves convenient to introduce new matrices Φ and Ψ, as follows:

$$\Phi := g + h, \qquad \Psi := g - h.$$

If we now define for each k a vector $|\Phi_k\rangle$, the ith component of which is given by Φ_{ki}, and similarly for Ψ, we find that (34) can be written as follows:

$$(A - B)|\Psi_k\rangle = \Lambda_k|\Phi_k\rangle, \quad (A + B)|\Phi_k\rangle = \Lambda_k|\Psi_k\rangle,$$

or, equivalently, as

$$\begin{aligned}(A - B)(A + B)|\Phi_k\rangle &= \Lambda_k^2\ |\Phi_k\rangle \\ (A + B)(A - B)|\Psi_k\rangle &= \Lambda_k^2\ |\Psi_k\rangle.\end{aligned} \tag{35}$$

With this we have reduced our problem from the diagonalization of a $2^N \times 2^N$ matrix to that of diagonalizing two $N \times N$ ones. As mentioned before, this would be a very easy task if we would have chosen periodic boundary conditions. The reason being that for periodic boundary conditions the matrices $(A \pm B)(A \mp B)$ are *Toeplitz*. But for open boundary conditions the matrices A and B are given by (28), so that

$$\frac{1}{4}(A - B)(A + B) = \begin{pmatrix} \lambda^2 & -\lambda & & & & \\ -\lambda & 1+\lambda^2 & -\lambda & & & \\ & -\lambda & 1+\lambda^2 & \ddots & & \\ & & \ddots & \ddots & & \\ & & & & & -\lambda \\ & & & & -\lambda & 1+\lambda^2 \end{pmatrix} \tag{36}$$

and

$$\frac{1}{4}(A + B)(A - B) = \begin{pmatrix} 1+\lambda^2 & -\lambda & & & \\ -\lambda & 1+\lambda^2 & & & \\ & & \ddots & \ddots & \\ & & & & \\ & & \ddots & 1+\lambda^2 & -\lambda \\ & & & -\lambda & \lambda^2 \end{pmatrix}. \tag{37}$$

In order to solve (35), we propose the following ansatz:

$$\Psi_{kl} = \alpha e^{ikl} + \beta e^{-ikl}, \tag{38}$$

with α and β constants to be determined. The eigenvalue equation will then give 3 independent equations. The first one is obtained by equating the jth component of $(A+B)(A-B)|\Psi_k\rangle$ with $\Lambda_k^2\Psi_{kj}$, for $j=2,\ldots,N-1$. All these choices of j yield the same equation (by enforcing the vanishing of the coefficients of α and β), namely,

$$\left(\frac{\Lambda_k}{2}\right)^2 = \lambda^2 + 1 - 2\lambda\cos k. \tag{39}$$

This is (almost) the spectrum of our problem. We still need to find what are the allowed values of the label "k". This is done by considering the two other cases ($j=1$ and $j=N$), that give a system of equations for α and β:

$$\begin{aligned} 0 &= \quad \alpha \quad + \quad \beta, \\ 0 &= \left(\lambda e^{ik(N+1)} - e^{ikN}\right)\alpha + \left(\lambda e^{-ik(N+1)} - e^{-ikN}\right)\beta. \end{aligned}$$

The non-trivial solution $\alpha = -\beta$ is obtained provided the determinant of this matrix vanishes. This condition is equivalent to k being solution of the following transcendental equation, for which $\lambda \neq 0$ has to be assumed:

$$\sin kN = \lambda \sin k(N+1). \tag{40}$$

Notice that, since we must have N eigenvectors, we expect this equation to have N roots. The behavior of these roots as a function of λ is quite relevant; comparison with the periodic chain allows for the recognition of edge states.

3.3. *Periodic boundary conditions*

In the case of periodic boundary conditions, we extend the sums in (24) to $i=N$, adopting the convention that $\sigma_{N+1}^\alpha \equiv \sigma_1^\alpha$ ($\alpha = x,y,z$). This generates a boundary term that couples the first spin operator to the last one:

$$H = -\sum_{i=1}^{N-1} \sigma_i^x\sigma_{i+1}^x - \lambda\sum_{i=1}^{N}\sigma_i^z - \sigma_N^x\sigma_1^x. \tag{41}$$

We already know how to write the first two terms of this Hamiltonian in terms of creation and annihilation operators. Let us therefore consider the last term: $\sigma_N^x\sigma_1^x$. Here it is convenient to consider the *parity* operator, $e^{i\pi\mathcal{N}}$, where $\mathcal{N} = \sum_j a_j^\dagger a_j$ is the *number* operator. Using the identities

$$1 - 2a_j^\dagger a_j = e^{i\pi a_j^\dagger a_j}, \quad (e^{i\pi a_j^\dagger a_j})^2 = 1, \quad e^{i\pi a_j^\dagger a_j}(a_j^\dagger + a_j) = (a_j - a_j^\dagger),$$

and (26) we obtain:

$$\begin{aligned}\sigma_N^x \sigma_1^x &= \left(\prod_{m=1}^{N-1}(1-2a_m^\dagger a_m)\right)(a_N^\dagger + a_N)(a_1^\dagger + a_1)\\ &= \left(\prod_{m=1}^{N-1} e^{i\pi a_m^\dagger a_m}\right)(a_N^\dagger + a_N)(a_1^\dagger + a_1)\\ &= \left(\prod_{m=1}^{N} e^{i\pi a_m^\dagger a_m}\right) e^{i\pi a_N^\dagger a_N}(a_N^\dagger + a_N)(a_1^\dagger + a_1)\\ &= e^{i\pi\mathcal{N}}(a_N - a_N^\dagger)(a_1^\dagger + a_1)\\ &= (a_N - a_N^\dagger)(a_1^\dagger + a_1)e^{i\pi\mathcal{N}}.\end{aligned}$$

With this we can write H in terms of fermionic operators:

$$\begin{aligned}H = &-\sum_{i=1}^{N-1}(a_i^\dagger a_{i+1} + a_{i+1}^\dagger a_i + a_i^\dagger a_{i+1}^\dagger - a_i a_{i+1}) + 2\lambda\sum_{i=1}^{N} a_i^\dagger a_i\\ &-\lambda N + e^{i\pi\mathcal{N}}(a_N^\dagger a_1 + a_1^\dagger a_N + a_N^\dagger a_1^\dagger + a_1 a_N).\end{aligned} \tag{42}$$

An important fact (that can be easily checked) is that the parity operator commutes with H:

$$[H, e^{i\pi\mathcal{N}}] = 0.$$

Thus, it is possible to diagonalize the Hamiltonian separately in sectors of even and odd numbers of "particles". The eigenvalues of $e^{i\pi\mathcal{N}}$ are of the form $\sigma = \pm 1$, with the plus sign for states with an *even* number of particles and the minus sign for states with an *odd* number of particles. We can, as in the previous case, write H as a quadratic form in the fermion operators (*cf.* (29)), the only difference being the explicit form of the matrices A and B, for which we now get:

$$A = \begin{pmatrix} 2\lambda & -1 & 0 & \cdots & \sigma \\ -1 & 2\lambda & -1 & 0 & \cdots \\ 0 & -1 & 2\lambda & -1 & \\ \vdots & 0 & -1 & 2\lambda & \cdots \\ \sigma & \vdots & & \vdots & \ddots \end{pmatrix}, \quad B = \begin{pmatrix} 0 & -1 & 0 & \cdots & -\sigma \\ 1 & 0 & -1 & 0 & \cdots \\ 0 & 1 & 0 & -1 & \\ \vdots & 0 & 1 & 0 & \cdots \\ \sigma & \vdots & & \vdots & \ddots \end{pmatrix}. \tag{43}$$

In contrast to (35) or (37), the matrix $(A \pm B)(A \mp B)$ is now Toeplitz:

$$\frac{1}{4}(A \pm B)(A \mp B) = \begin{pmatrix} 1+\lambda^2 & -\lambda & 0 & \cdots & \sigma\lambda \\ -\lambda & 1+\lambda^2 & -\lambda & 0 & \cdots \\ 0 & -\lambda & 1+\lambda^2 & \ddots & \\ \vdots & 0 & \ddots & \ddots & -\lambda \\ \sigma\lambda & \vdots & & -\lambda & 1+\lambda^2 \end{pmatrix}. \tag{44}$$

The eigenvalue problem (35) can again be solved using the ansatz (38). The eigenvalues are again given by (39), with the difference that the allowed values of k can now be explicitly given[c], as follows from the following problem.

Problem 3.5. *Show that, in the even parity sector ($\sigma = 1$), the allowed values of k are given by*

$$k_m = \frac{(2m+1)}{N}\pi, \qquad (m = 0, 1, \ldots, N-1),$$

whereas in the odd parity sector ($\sigma = -1$) they are given by

$$k_m = \frac{2m}{N}\pi, \qquad (m = 0, 1, \ldots, N-1).$$

Problem 3.6. *Explain how the k's can be made to take positive and negative values and how then we obtain an explicit solution for the matrix Φ, of the form*

$$\Phi_{k,l} \sim \begin{cases} \sin(kl), \ k > 0 \\ \cos(kl), \ k < 0. \end{cases}$$

It should by now be clear that the solution of the eigenvalue problem is much easier in the periodic case. We have chosen to discuss the open chain mainly because of the role that boundary effects play in connection to symmetry breaking and also for studies of surface effects [29, 30]. Now we are going to take advantage of the translational symmetry in the periodic case. As discussed below, if instead of applying a Bogoliubov transformation right after transforming the spin operators to fermion operators via the Jordan-Wigner transformation, we perform a Fourier transformation as an intermediate step, we quickly obtain a much more elegant solution, that can be easily generalized to treat the limiting case of an infinite chain. Here

[c]This stands in contrast to the case of open boundary conditions, where they are given by the solutions of the transcendental equation (40).

we will again make use of the fact that the Hamiltonian commutes with the parity operator.

Looking back at (42), we realize that it is possible to express H in a more compact form if we introduce boundary conditions of the fermion operators, depending on the parity sector we are interested in. In fact, notice that if we choose $\sigma = -1$, then the boundary term can be included in the first sum, just by setting $a_{N+1} \equiv a_1$. But if we want to do the same in the sector $\sigma = +1$, we will need $a_{N+1} \equiv -a_1$. In view of our previous discussion, it is clear that the choice $\sigma = -1$ is more convenient computationally (the matrices $(A \pm B)(A \mp B)$ are circulant in that case). But if we are interested in the *ground state*, then we must consider the other choice ($\sigma = 1$). This can be seen if we compare the ground state energies of the fermionic Hamiltonian in each sector. As can be easily checked, the lowest energy eigenvalue indeed comes from the sector $\sigma = 1$. The difference between them decreases with increasing N and eventually disappears in the thermodynamic limit. It is for this reason that the boundary term is usually disregarded. We nevertheless feel that the price to pay if we keep all terms is actually very low, and in return we can have complete control over the spectrum and its degeneracies [31].

Problem 3.7. *Making use of the solution of the eigenvalue problem for both periodic and antiperiodic boundary conditions, plot the lowest energy eigenvalues for $\sigma = \pm 1$ as functions of λ and N.*

We therefore define

$$a_{N+1} := -e^{i\pi\mathcal{N}} a_1.$$

In this way the Hamiltonian becomes

$$H = -\sum_{i=1}^{N} (a_i^\dagger a_{i+1} + a_{i+1}^\dagger a_i + a_i^\dagger a_{i+1}^\dagger - a_i a_{i+1} - 2\lambda a_i^\dagger a_i) - \lambda N. \tag{45}$$

Notice that the sum now goes over $1 \le i \le N$. So we see that we can describe *both* sectors using the *same* quadratic form, the choice of parity being now encoded in the boundary conditions for the fermion operators. It should be remarked that when we consider anti-periodic boundary conditions for the fermions we are still considering periodic boundary conditions for the *spin chain*. Anti-periodicity for the fermions in this context is only related to the choice of a negative eigenvalue for the parity operator.

Let us now introduce the Fourier transformed operators

$$d_k := \frac{1}{\sqrt{N}} \sum_{l=1}^{N} a_l e^{-i\phi_k l}. \tag{46}$$

The choice of the phases ϕ_k must be made in such a way that the CAR algebra is preserved, but in addition it has to imply the boundary condition $a_{l+N} = -e^{i\pi\mathcal{N}} a_l$. Therefore, the phases ϕ_k will also depend on the parity sector.

Problem 3.8. *Show that in the* even *parity sector* ($\sigma = 1, a_{l+N} = -a_l$ *), the phases are given by*

$$\phi_k = \left(\frac{2k+1}{N}\right)\pi,$$

whereas in the odd *parity sector* ($\sigma = -1, a_{l+N} = a_l$ *) they are given by*

$$\phi_k = \left(\frac{2k+2}{N}\right)\pi.$$

A word of caution is perhaps in order: In principle, k is an integer that takes values in the range $1 \leq k \leq N$. Nevertheless, due to the translational symmetry, k actually belongs to a "Brillouin zone", meaning with this that there is no difference between k and $k \pm N$. So, for instance, for $\sigma = 1$ we have that $e^{i(\phi_k + \phi_{k'})} = 1$ whenever $2\pi(k + k' + 1)/N$ is an integer or, in other words, when

$$k + k' + 1 = 0 \pmod{N}. \tag{47}$$

For this reason, even if the allowed values of k are originally in the range $\{1, \ldots, N\}$, we will allow k to take any integer value, provided we interpret this (for $\sigma = 1$) in the sense of (47). With this convention we then obtain, for example:

$$\phi_{-k-1} = -\phi_k.$$

A similar remark applies to the case $\sigma = -1$. For the remaining part of this section we will confine ourselves to the even parity sector and so $\sigma = 1$ will be tacitly assumed.

We can now go back to the Hamiltonian (45) and insert there the new operators defined by (46). As a result, we obtain the following form of the Hamiltonian:

$$H = \sum_{k=1}^{N} \left[2(\lambda - \cos\phi_k) d_k^\dagger d_k - i \sin\phi_k (d_k^\dagger d_{-k-1}^\dagger + d_k d_{-k-1}) - \lambda \right]. \tag{48}$$

This form of the Hamiltonian is indeed very convenient as it suggests that a diagonalization by means of a Bogoliubov transformation that only mixes $d_k^{(\dagger)}$ with $d_{-k-1}^{(\dagger)}$ should be possible. For this reason we propose the following transformation:

$$c_k = \alpha_k d_k + \beta_k d^\dagger_{-k-1}, \tag{49}$$

with α_k, β_k (possibly complex) coefficients to be found. Imposing the condition $\{c_k, c^\dagger_{k'}\} = \delta_{k,k'}$ we readily obtain $|\alpha_k|^2 + |\beta_k|^2 = 1$. Furthermore, it is easy to check that the choices

$$\alpha_{-k-1} = \alpha_k, \qquad \beta_{-k-1} = -\beta_k,$$

enforce $\{c_k, c_{k'}\} = 0$. Inverting (49) and inserting the result in (48), we obtain:

$$\begin{aligned} H = &\sum_{k=1}^{N} \Big(c_k^\dagger c^\dagger_{-k-1} \Big[2(\cos\phi_k - \lambda)\alpha_k\beta_k - i\sin\phi_k(\alpha_k^2 + \beta_k^2) \Big] + \text{h.c.} \Big) \\ &+ 2\sum_{k=1}^{N} c_k^\dagger c_k \Big[(\lambda - \cos\phi_k)(|\alpha_k|^2 - |\beta_k|^2) + i\sin\phi_k(\bar{\alpha}_k\beta_k - \alpha_k\bar{\beta}_k) \Big] \\ &+ \sum_{k=1}^{N} \Big(2(\lambda - \cos\phi_k)|\beta_k|^2 + i\sin\phi_k(\alpha_k\bar{\beta}_k - \bar{\alpha}_k\beta_k) - \lambda \Big). \end{aligned} \tag{50}$$

Vanishing of the coefficients in the first sum leads to

$$\frac{\sin\phi_k}{\lambda - \cos\phi_k} = \frac{2i\alpha_k\beta_k}{\alpha_k^2 + \beta_k^2}. \tag{51}$$

Recalling that $|\alpha_k|^2 + |\beta_k|^2 = 1$, we see that the right hand side of this expression simplifies if we parameterize α_k and β_k as follows:

$$\alpha_k = \cos\frac{\theta_k}{2}, \quad \beta_k = -i\sin\frac{\theta_k}{2}.$$

In fact, with this choice (51) simplifies to

$$\tan\theta_k = \frac{\sin\phi_k}{\lambda - \cos\phi_k}.$$

We can now introduce a normalization factor Λ_k[d] by means of

$$\sin\theta_k = \frac{\sin\phi_k}{(\Lambda_k/2)}, \qquad \cos\theta_k = \frac{\lambda - \cos\phi_k}{(\Lambda_k/2)}.$$

---

[d]In section 3.5 we will multiply H by a factor $1/2$, in order to simplify some expressions.

This leads to

$$\Lambda_k = 2\sqrt{(\lambda - \cos\phi_k)^2 + \sin^2\phi_k}.$$

With these definitions, the remaining terms in (50) simplify further yielding the desired diagonal form for the Hamiltonian:

$$H = \sum_{k=1}^{N} \Lambda_k \Big(c_k^\dagger c_k - \frac{1}{2} \Big). \tag{52}$$

Problem 3.9. *Verify all the computations leading from (45) to (52).*

The Ground State. Having brought the Hamiltonian to the diagonal form (52), we now proceed to find an expression for the ground state. Since Λ_k has been chosen to be positive (or zero) for all values of k, the ground state $|\Omega(\lambda)\rangle$ will be given by the condition $c_k|\Omega(\lambda)\rangle = 0, \ \ \forall k \in \{1, \ldots, N\}$. Let $|0\rangle$ be the "vacuum" state for the operators d_k, defined by the condition $d_k|0\rangle = 0$. Observe now that if we define

$$B_k := \cos\frac{\phi_k}{2} + i\sin\frac{\phi_k}{2} d_k^\dagger d_{-k-1}^\dagger,$$

then we get $c_k B_k|0\rangle = 0$ for all k. Since $[B_k, B_k'] = 0$, one would think that $\prod_{k=1}^{N} B_k|0\rangle$ does the job, but we must take into account that $B_{k'} = B_k$ whenever $k' + k + 1 = 0 \pmod N$. If N is even, this means that *each* B_k appears *twice* in $\prod_{k=1}^{N} B_k$ and then an easy calculation shows that $c_{k'} \prod_{k=1}^{N} B_k|0\rangle$ does not vanish. Hence, we must be careful to include in the product only one instance of each B_k.

To do this, we distinguish two cases: N even and N odd. Restricting the domain of ϕ_k to the interval $(-\pi, \pi)$ we get, for N even, $0 < \phi_k < \pi$ for $1 \leq k \leq N/2 - 1$ (also for $k = N$) and $-\pi < \phi_k < 0$ for $N/2 \leq k \leq N - 1$. Now, for each k such that $0 < \phi_k < \pi$, there is *exactly* one k' with $-\pi < \phi_{k'} < 0$ and such that $B_k = B_{k'}$. In fact, let k be such that $0 < \phi_k < \pi$ and put $k' = N - k - 1$. It follows that $-\pi < \phi_{k'} < 0$ and $k' + k + 1 = 0 \ (\mathrm{mod} N)$, so that $B_k = B_{k'}$. We therefore see that, *for even* N, the ground state is given by

$$|\Omega(\lambda)\rangle = \prod_{0<\phi_k<\pi} B_k|0\rangle.$$

Writing $N = 2M$, we may as well consider k to be such that $-M \leq k \leq M - 1$. The ground state, then, takes the form $|\Omega(\lambda)\rangle = \prod_{k=0}^{M-1} B_k|0\rangle$.

For the case N odd, write $N = 2M + 1$. In this case all the B_k appear twice in $\prod_{k=1}^{N} B_k$, *except* when $k = M$. When $k = M$ we have $\phi_k = \pi$.

This, together with the fact that $d_M^\dagger \equiv d_{-M-1}^\dagger$, implies that $B_M \equiv 0$, so we must exclude it from the product. In analogy to the previous case, for each k for which $0 < \phi_k < \pi$ we can find exactly one k' such that $-\pi < \phi_{k'} < 0$ and $B_k = B_{k'}$, namely $k' = 2M - k$. Now, for $0 \leq K \leq M - 1$, we have $0 < \phi_k < \pi$, whereas for $M + 1 \leq k \leq 2M$, we have $\phi_k \in (-\pi, 0)$. The result is that, *for odd* N, the ground state is given by

$$|\Omega(\lambda)\rangle = \prod_{0<\phi_k<\pi} B_k d_M^\dagger |0\rangle. \tag{53}$$

The operator $d_M^\dagger$ has to be included in (53) in order to ensure that $c_M|\Omega(\lambda)\rangle = 0$ (recall that in this case $c_M = -id_M^\dagger$).

3.4. *Entanglement properties of the ground state*

Having obtained the exact spectrum of the model, as well as the ground state both for open and periodic boundary conditions, it is now possible -in principle- to obtain all relevant correlation functions and relate them to physical observables like magnetization, susceptibilities, and so on. As is well-known, the quantum Ising chain does not present any phase transition at finite temperature, but it certainly is one of the paradigmatic examples of a system displaying a quantum critical point. In fact, in the thermodynamic limit there is a quantum phase transition occurring when the external field λ approaches the critical value $\lambda_c = 1$. In this notes we will review an approach to the study of this quantum phase transition from the point of view of *entanglement*. We will also be interested in certain geometric and topological properties of the ground state that are relevant for this quantum phase transition.

Let us start by considering the model (24) for an open, two-site chain ($N = 2$). Putting $\alpha = \sqrt{1 + 4\lambda^2}$, we find that the ground state is given in the spin basis by

$$|\Omega(\lambda)\rangle = \frac{1}{\sqrt{2\alpha(\alpha - 2\lambda)}} \left(|++\rangle + (\alpha - 2\lambda)|--\rangle\right). \tag{54}$$

Taking the partial trace with respect to one of the sites, we obtain

$$\rho \equiv \mathrm{Tr}_1|\Omega(\lambda)\rangle\langle\Omega(\lambda)| = \frac{1}{2\alpha(\alpha - 2\lambda)} \left(|+\rangle\langle+| + (\alpha - 2\lambda)^2|-\rangle\langle-|\right),$$

which in terms of the basis $\{|+\rangle, |-\rangle\}$ takes the following simple matrix form:

$$\rho = \begin{pmatrix} \frac{1}{2} + \frac{\lambda}{\alpha} & 0 \\ 0 & \frac{1}{2} - \frac{\lambda}{\alpha} \end{pmatrix}. \tag{55}$$

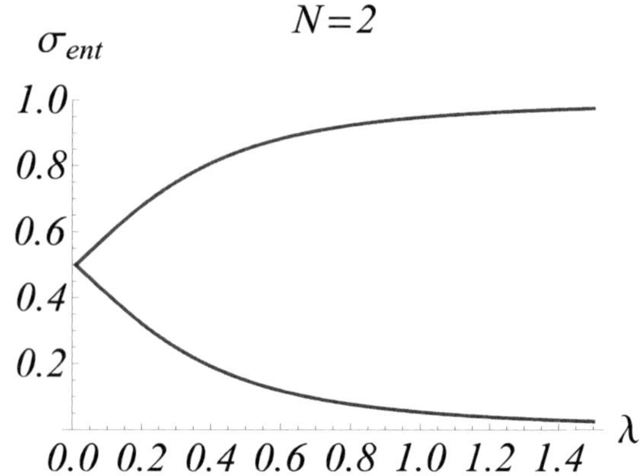

Fig. 2: Entanglement spectrum for two-site chain as a function of the external field λ

Figure 2 displays a plot of the eigenvalues of ρ as a function of λ. As we shall see, the behavior of this *entanglement spectrum* as the size of the chain increases will provide important information regarding the quantum phase transition in this model.

Problem 3.10. *Follow the steps outlined in section 3.2 to show that for $N = 2$ the matrices g and h introduced in (30) are given by*

$$g = \frac{1}{\sqrt{8\alpha}} \begin{pmatrix} \frac{2\lambda+\alpha-1}{\sqrt{\alpha-1}} & \frac{2\lambda+\alpha-1}{\sqrt{\alpha-1}} \\ \frac{-2\lambda-\alpha-1}{\sqrt{\alpha+1}} & \frac{2\lambda+\alpha+1}{\sqrt{\alpha+1}} \end{pmatrix}, \qquad h = \frac{1}{\sqrt{8\alpha}} \begin{pmatrix} \frac{2\lambda-\alpha+1}{\sqrt{\alpha-1}} & \frac{\alpha-1-2\lambda}{\sqrt{\alpha-1}} \\ \frac{-2\lambda+\alpha+1}{\sqrt{\alpha+1}} & \frac{\alpha+1-2\lambda}{\sqrt{\alpha+1}} \end{pmatrix}.$$

Use this in order to show that the condition $c_k|\Omega(\lambda)\rangle \stackrel{!}{=} 0 \ \ (k = 1, 2)$ leads exactly to (54).

As we have seen, in this simple case ($N = 2$) the density matrix for the reduced state can be easily computed. In spite of the fact that the model can be exactly solved, the computation of correlation functions (as those involved in the computation of the reduced density matrix) for arbitrary values of N is a non-trivial task. We will therefore present, following [32], a method that allows us to obtain the spectrum of ρ for arbitrary values of N and that can be used for any quadratic Hamiltonian. As a preparation for the general case, we first explain the idea using our very simple example of a two-site chain, where all calculations can be explicitly carried out.

Let $\mathcal{A}$ be the CAR algebra generated by the two fermionic operators a_1, a_2 and the unity $\mathbb{1}$. We consider the "Fock representation" where $a_i^\dagger$

acts on the state $|0\rangle \equiv |++\rangle$. This state is cyclic and therefore we can write the ground state (54) in terms of it (*cf.* problem 3.10):

$$|\Omega(\lambda)\rangle = \frac{1}{\sqrt{2\alpha(\alpha - 2\lambda)}} \left(|0\rangle + (\alpha - 2\lambda) a_1^\dagger a_2^\dagger |0\rangle \right). \tag{56}$$

As discussed in section 2.1, we can use it to define a state ω_λ in the *algebraic sense*:

$$\begin{aligned} \omega_\lambda : \mathcal{A} &\longrightarrow \mathbb{C} \\ A &\longmapsto \omega_\lambda(A) := \langle \Omega(\lambda)| A |\Omega(\lambda)\rangle. \end{aligned}$$

This is, of course, a pure state. Nevertheless, if we *restrict* it to the subalgebra $\mathcal{A}_1$ that is generated by $\mathbb{1}$ and a_1 ("a half-chain"), then we find that in general the resulting state $\omega_\lambda|_{\mathcal{A}_1}$ will be a *mixed* state. Let us consider this restriction in more detail. Defining $\omega_{\lambda,1} := \omega_\lambda|_{\mathcal{A}_1}$ we expect that for any $A \in \mathcal{A}_1$

$$\omega_{\lambda,1}(A) = \mathrm{Tr}(\rho A) \tag{57}$$

will hold, with

$$\rho = \frac{e^{-\varepsilon_1 a_1^\dagger a_1}}{Z},$$

$Z = \mathrm{Tr} e^{-\varepsilon_1 a_1^\dagger a_1}$, both traces being taken on the Hilbert space corresponding to the half-chain (which in this case has a basis given by $\{|0\rangle, a_1^\dagger|0\rangle\}$). Finding ε_1 amounts to finding the spectrum of ρ and this, in turn, can be used to compute the entanglement entropy. The specific form of ρ as the exponential of a quadratic form is to be expected in view of the fact that the ground state is Gaussian. In order to find ε_1, we notice that it is possible to compute the left hand side of (57). The only relevant case is $A = a_1^\dagger a_1$, so let us define $C_1 := \omega_{\lambda,1}(a_1^\dagger a_1)$. Using (56), we get:

$$\begin{aligned} C_1 &= \omega_{\lambda,1}(a_1^\dagger a_1) = \omega_\lambda(a_1^\dagger a_1) \\ &= \langle \Omega(\lambda)| a_1^\dagger a_1 |\Omega(\lambda)\rangle \\ &= \frac{1}{2} - \frac{\lambda}{\alpha}, \end{aligned}$$

where we still keep the notation $\alpha = \sqrt{1 + 4\lambda^2}$. On the other hand, we have

$$\mathrm{Tr}(\rho a_1^\dagger a_1) = \frac{1}{1 + e^{\varepsilon_1}}.$$

It follows that

$$1 - 2C_1 = \frac{2\lambda}{\alpha} = \tanh \frac{\varepsilon_1}{2}$$

or, equivalently,

$$\varepsilon_1 = \ln\left(\frac{1/2 + \lambda/\alpha}{1/2 - \lambda/\alpha}\right).$$

Since ρ is already diagonal in the basis $\{|0\rangle, a_1^\dagger|0\rangle\}$, we obtain:

$$\rho = \begin{pmatrix} \frac{1}{1+e^{-\varepsilon_1}} & 0 \\ 0 & \frac{1}{1+e^{\varepsilon_1}} \end{pmatrix} = \begin{pmatrix} \frac{1}{2} + \frac{\lambda}{\alpha} & 0 \\ 0 & \frac{1}{2} - \frac{\lambda}{\alpha} \end{pmatrix},$$

which coincides with (55). Notice, in passing, that C_1 can also be obtained directly from h. In fact, a small computation shows that $C_1 = (h^t h)_{11}$.

This last remark gives a hint towards the solution for the general case: As we shall see, the spectrum of ρ in the general case can be extracted from an eigenvalue problem that uses the matrices h and g as input. Let us, then, start by considering the algebra $\mathcal{A}$ generated by $\mathbb{1}$ and by all operators a_i, with $i = 1, \ldots, N$. Again, the ground state $|\Omega(\lambda)\rangle$ gives rise to a state $\omega_\lambda : \mathcal{A} \to \mathbb{C}$. We are interested in the restriction of this state to the subalgebra $\mathcal{A}_L$ generated by $\{\mathbb{1}, a_1, \ldots, a_L\}$. By the same argument as before, we expect this new state $\omega_{\lambda,L} := \omega_\lambda|_{\mathcal{A}_L}$ to be of the form $\omega_{\lambda,L}(A) = \mathrm{Tr}(\rho A)$ $(A \in \mathcal{A}_L)$, with

$$\rho = \frac{e^{-H_\rho}}{Z}, \qquad Z = \mathrm{Tr} e^{-H_\rho}$$

and

$$H_\rho = \sum_{i,j=1}^{L} \left(a_i^\dagger K_{ij} a_j + \frac{1}{2}(a_i^\dagger M_{ij} a_j - a_i M_{ij} a_j)\right).$$

For the Ising model, the matrices K and M (yet to be determined) will be real, so that K has to be symmetric, and M antisymmetric. Now we define [32], for $i, j = 1, \ldots, L$ the correlation functions

$$C_{ij} := \omega_{\lambda,L}(a_i^\dagger a_j), \quad F_{ij} := \omega_{\lambda,L}(a_i^\dagger a_j^\dagger). \tag{58}$$

Using (30) one shows that $C_{ij} = (h^t h)_{ij}$ and $F_{ij} = (h^t g)_{ij}$. Let us first consider the simpler case $M = 0$. In this case H_ρ can be diagonalized by a simple transformation of the form

$$b_l = \sum_{i=1}^{L} \gamma_{li} a_i, \quad l = 1, \ldots, L, \tag{59}$$

where the matrix γ is such that $\gamma^t\gamma = \mathbb{1}_L$. Solving the eigenvalue problem

$$K|\varphi_l\rangle = \varepsilon_l|\varphi_l\rangle$$

and introducing the convention $|i\rangle = (0, 0, \ldots, 1, \ldots, 0)$ (1 in the ith entry and 0 everywhere else), we set $\gamma_{li} := \langle i|\varphi_l\rangle$, *i.e.*, the rows of γ are the eigenvectors $|\varphi_l\rangle$. We then obtain

$$H_\rho = \sum_{l=1}^{L} \varepsilon_l b_l^\dagger b_l.$$

It follows that $Z = \prod_{l=1}^{L}(1 + e^{-\varepsilon_l})$, so that $\omega_{\lambda,L}(b_i^\dagger b_j) = \delta_{ij}(1 + e^{\varepsilon_i})^{-1}$. Using (59) we then obtain $C_{ij} = \sum_{l=1}^{L} \gamma_{li}\gamma_{lj}(1+e^{\varepsilon_l})^{-1}$, which is equivalent to the matrix identity $C = (1 + e^K)^{-1}$. This can also be written in the following form:

$$1 - 2C = \tanh\frac{K}{2}.$$

If $M \neq 0$, we replace (59) by

$$b_l = \sum_{i=1}^{L}(\gamma_{li}a_i + \eta_{li}a_i^\dagger),$$

where now we require (*cf.* problem 3.3):

$$\gamma^t\gamma + \eta^t\eta = \mathbb{1}_L, \quad \gamma^t\eta + \eta^t\gamma = 0.$$

Once again, the matrices γ and η should be such that ρ becomes diagonal, *i.e.*, $H_\rho = \sum_{l=1}^{L} \varepsilon_l b_l^\dagger b_l$ should hold. The eigenvalue problem to be solved is

$$\begin{aligned}(K - M)|\psi_l\rangle &= \varepsilon_l|\varphi_l\rangle,\\ (K + M)|\varphi_l\rangle &= \varepsilon_l|\psi_l\rangle.\end{aligned} \tag{60}$$

Using these eigenvectors as rows for the matrices ψ and φ and putting

$$\gamma = \frac{1}{2}(\varphi + \psi), \quad \eta = \frac{1}{2}(\varphi - \psi),$$

we obtain the desired diagonal form for H_ρ. As in the previous case, we can now write all correlation functions that involve the operators a_i in terms of correlation functions that only involve the operators b_l.

Problem 3.11. *Obtain the following two relations:*

$$C_{ij} = \sum_{l=1}^{L}\left(\frac{\gamma_{li}\gamma_{lj}}{1 + e^{\varepsilon_l}} + \frac{\eta_{li}\eta_{lj}}{1 + e^{-\varepsilon_l}}\right),$$

$$F_{ij} = \sum_{l=1}^{L}\left(\frac{\gamma_{li}\eta_{lj}}{1 + e^{\varepsilon_l}} + \frac{\gamma_{lj}\eta_{li}}{1 + e^{-\varepsilon_l}}\right),$$

and use them in order to prove the following identity:

$$\frac{1}{1+e^{(K\pm M)}} - \frac{1}{1+e^{-(K\pm M)}} = 2C - \mathbb{1}_L \pm 2F,$$

where C and F are the $L \times L$ matrices with components given by (58).

From this problem we obtain, making use of (60),

$$(2C - \mathbb{1}_L - 2F)(2C - \mathbb{1}_L + 2F)|\varphi_l\rangle = \tanh^2 \frac{\varepsilon_l}{2} |\varphi_l\rangle. \tag{61}$$

We will now make use of this expression in order to obtain the entanglement spectrum and the entanglement entropy of the ground state of the Ising chain when $L = N/2$, for various values of N and λ. In Fig. 3 we plot the entanglement spectrum (in logarithmic scale) for a chain of $N = 10$ sites, when the ground state is restricted to a half-chain. We notice that, in contrast to the case $N = 2$, the gap between the first two eigenvalues is now closed for almost all values of λ below $\lambda_c = 1$.

In Fig. 4 we compare the entanglement spectrum for an Ising chain of $N = 10$ sites against the spectrum of a chain of $N = 100$ sites. It is apparent that for $N = 100$ the gap near λ_c has almost closed.

This gap, when evaluated at the critical value λ_c, bears the name of *Schmidt gap.* We will use the notation Δ_S for the Schmidt gap. De Chiara and collaborators have shown, using finite-size scaling, that the Schmidt gap can be interpreted as an *order parameter* [33]. In fact, from a finite size scaling analysis, one can in fact obtain numbers μ_1 and μ_2 such that when we plot $\Delta_S N^{\mu_1}$ versus $|\lambda - \lambda_c| N^{\mu_2}$, all points collapse to a single curve,

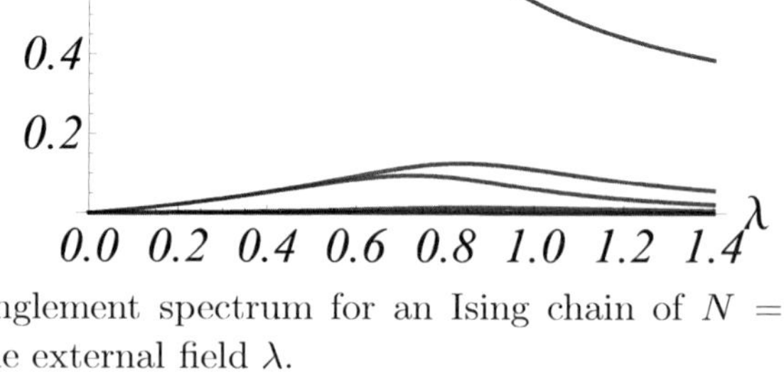

Fig. 3: Entanglement spectrum for an Ising chain of $N = 10$ sites as a function of the external field λ.

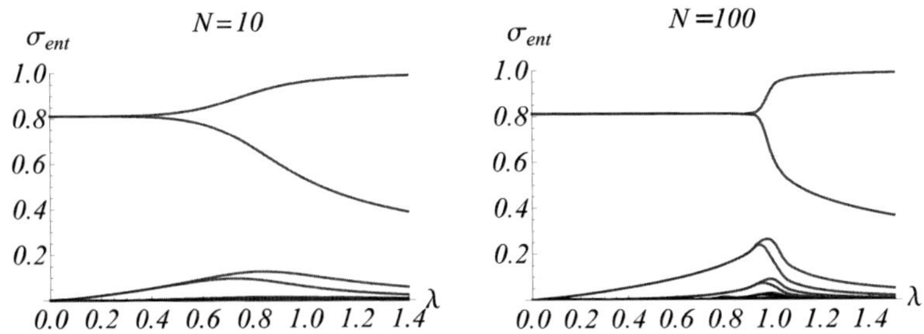

Fig. 4: Entanglement spectrum for an Ising chains of $N = 10$ and $N = 100$, showing how the gap at $\lambda = 1$ closes as the number of sites increases.

irrespective of the value of N chosen. This result was obtained in [33] using a DMRG algorithm. Figure 5, obtained by C. Rivera using the same DMRG technique [34] depicts the resulting curve, for which points corresponding to different lengths of the chain (ranging from $N \approx 700$ to $N = 8000$) are seen to collapse to a single curve. The values of μ_1 and μ_2 for which this is achieved give values for the critical exponents which are very close to the actual values ($\nu = 1$ and $\beta = 1/8$). The entanglement spectrum can also be

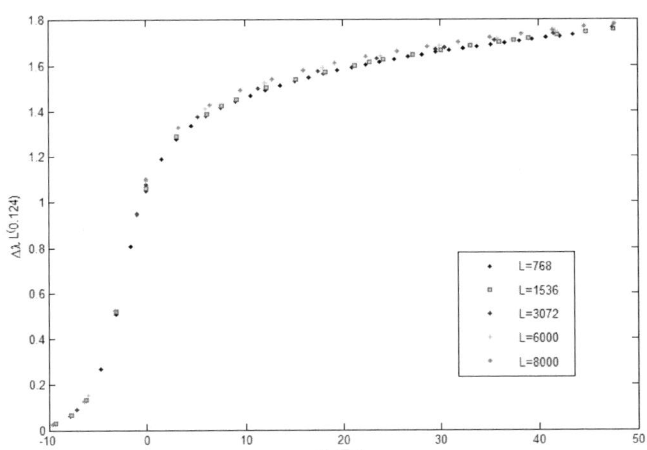

Fig. 5: Finite-size scaling

used to compute the entanglement entropy and its dependence on the size of the subsystem chosen. A numerical evaluation of this entropy gives the result expected from conformal field theory, that is, at the critical point, the entanglement entropy follows a logarithmic behavior, of the form

$$S(l) \sim \frac{c}{3} \log l,$$

where c denotes the central charge of the corresponding CFT. In this case (Ising chain), numerical evaluation of the entropy for different chain sizes gives a value of $c = 1/2$, in accordance with the predictions of CFT.

3.5. *The XY model*

The XY model is a generalization of the Ising model. The model also represents a spin chain with nearest-neighbor interactions. Its Hamiltonian is given by

$$H = -\frac{1}{2} \sum_{j=1}^{N} \left(\frac{1+\gamma}{2} \sigma_j^x \sigma_{j+1}^x + \frac{1-\gamma}{2} \sigma_j^y \sigma_{j+1}^y + \lambda \sigma_j^z \right).$$

There are two parameters in this model. As in the Ising case, λ represents an external field. The parameter γ is an anisotropy parameter. Notice that if $\gamma = 1$, we recover, up to an overall factor of 1/2, the Ising Hamiltonian (24). Our interest in this model lies in the fact that according to the values of the parameters γ and λ, we may study critical regions corresponding to *different* universality classes: If $\gamma > 0$, there is a critical line determined by $\lambda = 1$ that corresponds to the Ising universality class, with central charge $c = 1/2$. On the other hand, for $\gamma = 0$ the whole line $0 < \lambda < 1$ corresponds to the XX universality class, with central charge $c = 1$.

3.6. *Solution of the model*

The solution of this model can be obtained in the same fashion as we did in the case of the Ising model. The first step, then, consists in using the Wigner-Jordan transformation to bring the Hamiltonian to a quadratic form in fermionic operators. The result one obtains is

$$H = -\frac{1}{2} \sum_j (a_j^\dagger a_{j+1} + a_{j+1}^\dagger a_j + \gamma a_j^\dagger a_{j+1}^\dagger - \gamma a_j a_{j+1}) \\ + \lambda \left(\mathcal{N} - \frac{N}{2} \right), \tag{62}$$

with $\mathcal{N} \equiv \sum_j a_j^\dagger a_j$, the fermionic number operator. Since we are interested in the ground state, we consider the even parity sector and proceed, as with the Ising model, with a Fourier transformation, followed by a Bogoliubov transformation. For odd N $(= 2M + 1)$ and with

$$d_k = \frac{1}{\sqrt{N}} \sum_{l=1}^{N} a_l e^{-i\phi_k l}, \quad \phi_k := \left(\frac{2k+1}{N}\right)\pi, \quad -M \leq k \leq M,$$

we obtain:

$$H = \sum_k ((\lambda - \cos\phi_k) d_k^\dagger d_k - \frac{i\gamma}{2} \sin\phi_k (d_k^\dagger d_{-k-1}^\dagger + d_k d_{-k-1})) - \frac{\lambda N}{2}.$$

We can now consider a Bogoliubov transformation, as in (49),

$$c_k = \cos\frac{\theta_k}{2} d_k - i \sin\frac{\theta_k}{2} d_{-k-1}^\dagger, \quad \tan\theta_k = \frac{\gamma \sin\phi_k}{\lambda - \cos\phi_k},$$

and obtain the following diagonal form for the Hamiltonian:

$$H = \sum_{k=-M}^{M} \Lambda_k \left(c_k^\dagger c_k - \frac{1}{2}\right),$$

with

$$\Lambda_k = \sqrt{(\lambda - \cos\phi_k)^2 + \gamma^2 \sin^2\phi_k}.$$

Defining $B_k := \cos\frac{\phi_k}{2} + i \sin\frac{\phi_k}{2} d_k^\dagger d_{-k-1}^\dagger$ we obtain the following explicit form for the ground state $|\Omega(\lambda, \gamma)\rangle$:

$$|\Omega(\lambda, \gamma)\rangle = \left(\prod_{0 \leq k < M} B_k\right) d_M^\dagger |0\rangle,$$

where $|0\rangle$ is the vacuum state with respect to the operators d_k.

3.7. *Criticality*

The XY model has a rich phase diagram, and this provides an opportunity to check the behavior of entanglement entropy at the critical point and its relation to the central charge. In the previous section we focused on a finite-size scaling analysis for the Ising model, which in turn brought up the difficulty of having to compute correlation functions for finite size chains. In this section, we will only be interested in properties in the thermodynamic limit, *i.e.*, we will discuss the behavior of the entanglement entropy of a sub-chain of length L, considered as a subsystem of a chain of infinite length, or the geometric phase of the ground state associated to closed loops around the critical point. Calculations in the thermodynamic limit are greatly simplified because discrete sums become integrals that can be readily evaluated. Let us now very briefly describe how to obtain the entanglement entropy of a sub-chain of size L in the thermodynamic limit. We will follow the approach of Vidal *et al.* [35], which amounts to the real (*e.g.* Majorana) version of the method discussed in section 3.4. We begin by replacing the fermion operators (a_i) appearing in (62) by new, Majorana fermion operators $(\tilde{c}_i)$, defined as follows:

$$\begin{aligned}\tilde{c}_{2n-1} &:= (a_n + a_n^\dagger)\\ \tilde{c}_{2n} &:= i(a_n^\dagger - a_n).\end{aligned}$$

These Majorana operators satisfy commutation relations of the form $\{\tilde{c}_m, \tilde{c}_n\} = 2\delta_{mn}$. Now, consider the correlation matrix $\langle \tilde{c}_m \tilde{c}_n \rangle = \delta_{mn} + i\Gamma_{mn}$, where the expectation value is taken with respect to the ground state. The matrix Γ can be expressed as a block matrix, with the block (i, j) being given by the 2×2 matrix Π_{j-i} (i and j range from 0 to $N-1$, N being the size of the full chain), where

$$\Pi_l := \begin{pmatrix} 0 & g_l \\ -g_l & 0 \end{pmatrix}.$$

In the thermodynamic limit ($N \to \infty$) the coefficient function g_l takes the following compact form:

$$g_l = \frac{1}{2\pi} \int_0^{2\pi} d\phi e^{-il\phi} \frac{(\cos\phi - \lambda - i\gamma \sin\phi)}{|\cos\phi - \lambda - i\gamma\sin\phi|}.$$

The restriction of the correlation matrix to the sites belonging to a sub-chain of size L is then a $L \times L$ matrix Γ_L which is obtained just by deleting the entries of Γ that do not correspond to sites in the sub-chain.

Problem 3.12. *Following the method of section 3.4 as well as Vidal et al. [35], show that the entanglement entropy corresponding to the restriction of the full chain to a sub-chain of L sites is given by the formula*

$$S(L) = \sum_{m=0}^{L-1} H\left(\frac{1+\nu_m}{2}\right), \tag{63}$$

where H denotes the Shannon entropy as defined in (18) and $\{\nu_m\}_m$ *are the positive eigenvalues of* $i\Gamma_L$.

Figure 6 displays the behavior of the entropy as a function of L. One of the curves (dashed) shows the results for the XX chain, which corresponds to $\gamma = 0$. From the logarithmic dependence of the entropy we extract the

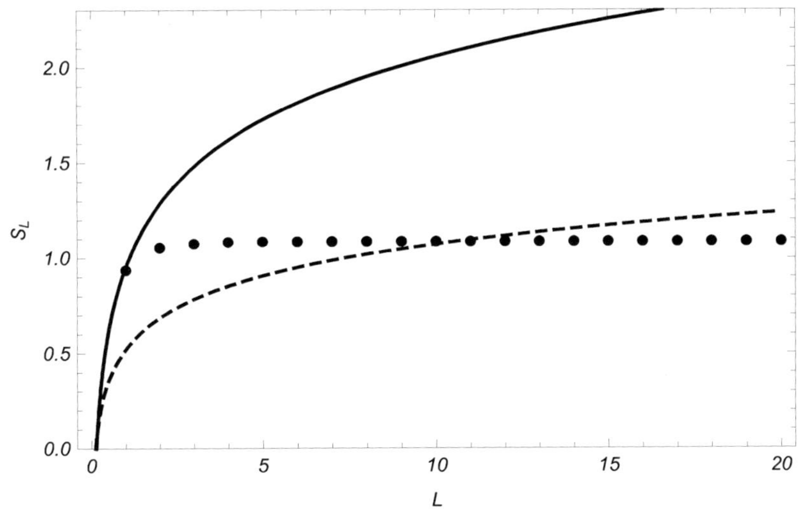

Fig. 6: Entanglement entropy as a function of the sub-chain size. L is represents the size of the sub-chain. For model parameters corresponding to critical phases, we observe a logarithmic behavior for the entanglement entropy. Outside the critical point, the entropy saturates.

value $c = 1$ for the central charge in this model. The continuous curve shows the entropy for the critical case $\lambda = 1, \gamma = 0.5$, which belongs to the Ising universality class. In this case, the value obtained for the central charge is $c = 1/2$. The dots correspond to the non-critical case $\lambda = 0.5, \gamma = 0.5$. We can see that in this case the entropy saturates with L. Thus, the results confirm the expected behavior of the entropy as predicted by CFT.

4. Quantum fields

4.1. *The scalar field*

In section 2 we mentioned that the *canonical commutation relations* (CCR) lead to a C^*-algebra, whose representation theory plays a very important role in quantum field theory. Let us recall that the operators $U(a)$ and $V(b)$ defined in (13) satisfy the so-called Weyl form of the CCR (*cf.* problem 2.2). The first thing we want to do is to recognize that any implementation of the CCR in their Weyl form corresponds in fact to the construction of a representation of a certain group (the Heisenberg group). The Heisenberg group of $\mathbb{R}^n$, whose underlying set is $(\mathbb{R}^n \times \mathbb{R}^n \times \mathbb{R})$, is the group defined by the following operation:

$$(\vec{a}_1, \vec{b}_1, r_1) \cdot (\vec{a}_2, \vec{b}_2, r_2) := (\vec{a}_1 + \vec{a}_2, \vec{b}_1 + \vec{b}_2, r_1 + r_2 + \frac{1}{2}(\vec{a}_1 \cdot \vec{b}_2 - \vec{a}_2 \cdot \vec{b}_1)).$$

Problem 4.1. *Study the Lie algebra of the Heisenberg group and explain how this Lie algebra is related to the CCR.*

Problem 4.2. *Show that the operators $\mathcal{R}(\vec{a}, \vec{b}, r)$ defined through*

$$\mathcal{R}(\vec{a}, \vec{b}, r) := V(\vec{b})U(\vec{a})e^{i\hbar(r + \frac{1}{2}\vec{a}\cdot\vec{b})},$$

with U and V as in (13), furnish a representation of the Heisenberg group.

The key point regarding these commutation relations is actually their relation to the (canonical) symplectic structure of the underlying classical phase space, $T^*\mathbb{R}^n$. In fact, if for $u = (\alpha, \beta) \in T^*\mathbb{R}^n$ we define

$$W(\alpha, \beta) := e^{-i(\alpha\hat{q} + \beta\hat{p})},$$

we obtain the following identity:

$$W(u)W(v) = e^{-\frac{i}{2}\sigma(u,v)}W(u+v),$$

where u and v denote elements of the *symplectic vector space* $T^*\mathbb{R}^n$, and σ the standard symplectic form. Notice that we have the following relations between all these operators:

$$W(\alpha, \beta) = e^{i\frac{\hbar}{2}\alpha\beta}V(\alpha)U(\beta) = \mathcal{R}(\beta, \alpha, 0).$$

Definition 4.1. (*cf.* [36]) Let V be a real vector space and $\sigma : V \times V \to \mathbb{R}$ a symplectic form (*i.e.* a bilinear, skew-symmetric, non-degenerate form). A $*$-algebra $\mathcal{W}(V, \sigma)$ is called a Weyl $*$-algebra of (V, σ) if there is a family $\{W(u)\}_{u\in V}$ of "generators" such that

(*i*) $W(u)W(v) = e^{-\frac{i}{2}\sigma(u,v)}W(u+v), \; W(u)^* = W(-u), \quad u, v \in V.$
(*ii*) $\mathcal{W}(V, \sigma)$ is *generated* by the family $\{W(u)\}_{u\in V}$, *i.e.*, it is the span of finite linear combinations of finite product of the $W(u)$.

In [36] it is proved that every symplectic vector space (V, σ) determines uniquely a Weyl $*$-algebra, up to $*$-isomorphism. Also, if $\phi : V_1 \to V_2$ is a symplectomorphism between two symplectic vector spaces, then the corresponding Weyl $*$-algebras are isomorphic. A most important fact is that $\mathcal{W}(V, \sigma)$ can be completed to a C^*-algebra, the Weyl C^*-algebra.

It is a fundamental result, due to Stone and von Neumann, that when the dimension of V is finite (and hence necessarily even), there is essentially only one representation of the CCR, which can be taken to be the standard Schrödinger representation [36]. But in infinite dimensions uniqueness is lost, and so *inequivalent* representations do exist. This fact is closely related to the non-uniqueness of a vacuum state for a free quantum field in a curved spacetime background. The following example illustrates how the construction of a (free, bosonic) quantum field runs in parallel to the previous discussion of CCR, the "only" (!) difference here being that the symplectic vector space we are dealing with is infinite dimensional.

Example 4.1 (The Klein-Gordon field, *cf.* [37]). *Let (M, g) denote a globally hyperbolic spacetime, and consider the Klein-Gordon equation on that background:*

$$(\Box_g + m^2)\varphi = 0.$$

Let Σ_0 denote a fixed Cauchy surface, to which the initial data for the solutions of the Klein-Gordon equation will be referred and consider the following vector space, a space of solutions of the K-G equation*:*

$$\mathcal{S} := \{\varphi \in C^\infty(M, \mathbb{R}) : \ (\Box_g + m^2)\varphi = 0 \ and \ \varphi|_{\Sigma_0} \in C_0^\infty(\Sigma_0)\}.$$

It is a most remarkable fact that the classical field equation comes equipped with a natural symplectic structure. In fact, if Σ is any space-like (Cauchy) hypersurface, and n denotes the unit normal, then on the vector space $\mathcal{S}$ defined above we can define the following symplectic form:

$$\sigma(\varphi_1, \varphi_2) := \int_\Sigma (\varphi_1 \nabla_\mu \varphi_2 - \varphi_2 \nabla_\mu \varphi_1) n_\mu dvol_g.$$

For $f \in C_0^\infty(M)$, let A and R denote the advanced and retarded fundamental solutions of the K-G equation. That is, for $f \in C_0^\infty(M)$ we have

$$(\Box_g + m^2)Af = f, \ \ (\Box_g + m^2)Rf = f,$$

where the support of Af lies in the causal past of the support of f (and analogously for Rf). Then $E := A - R$ gives a map

$$E : C_0^\infty(M) \to \mathcal{S}.$$

The CCR for the quantum field (in its smeared form*) then take the form*

$$[\hat{\varphi}(f), \hat{\varphi}(g)] = -i\sigma(Ef, Eg).$$

Just as in non-relativistic quantum mechanics, the operators obeying CCR are unbounded, and so it is convenient to go to the Weyl form, which leads us back to the Weyl C^*-algebra, this time the one corresponding to the infinite dimensional vector space $(\mathcal{S}, \sigma)$.

4.2. *Fermionic second quantization and Clifford algebras*

Representation spaces for the Weyl algebra of the previous example are bosonic Fock spaces. For the classical Dirac field, the transition to the corresponding quantum description can be performed in a completely analogous way, only that in this case, instead of an antisymmetric form, a symmetric one is used which, again, comes "for free" as part of the structure of the (vector space) of solutions of the classical field equation. It is an interesting fact that the existence of a natural bilinear antisymmetric form in the scalar field case (and of a symmetric one in the case of the Dirac field) leads to quantum fields that obey the physically correct relation between spin and statistics.

In very general terms, quantization of free fields can be formulated as follows. Suppose we are given a real vector space V together with a non degenerate (anti-)symmetric bilinear form $s : V \times V \to \mathbb{R}$. If in addition we choose a complex structure J which is compatible with s (meaning $s(u, v) = s(Ju, Jv)$ holds), we can complexify and obtain a complex vector space V_J. The bilinear form s can then be used to define an inner product on V_J, making it a Hilbert space. According to whether the bilinear form is symmetric or antisymmetric, one then considers the antisymmetric or the symmetric subspace of the tensor algebra of V_J, call it $\mathcal{F}_J$. Endomorphisms of V_J may be lifted to $\mathrm{End}(\mathcal{F}_J)$ in a natural (*i.e.* functorial) way. These liftings are then interpreted as the "second quantization" of observables (self-adjoint operators in V_J) and symmetries (unitary operators).

In this section we will focus on the fermionic case, following the beautiful treatment of the subject presented in [11], which makes strong use of Clifford algebras and spin group representations[e]. In the next section we will explore the connection between second quantization (as presented here) and quantum criticality in spin chains (as discussed previously). Let us start by first considering the simple case where $\dim V < \infty$. If V is a finite dimensional real vector space, and g a positive definite symmetric bilinear form on it, we can construct the corresponding Clifford algebra $Cl(V, g)$. As shown in [11], it is possible to obtain a concrete realization of the complexified Clifford algebra $\mathbb{C}l(V)$ as a subalgebra of $\mathrm{End}(\Lambda^\bullet V^{\mathbb{C}})$, but the corresponding representation is not irreducible. The situation changes if we assume V has even dimension, say $2m$. Then we may choose an orthogonal complex structure, that is, a linear operator $J \in \mathrm{End}_{\mathbb{R}}(V)$ such

[e]We urge the reader to consult [11] for details and explicit computations.

that $J^2 = -1$ and $g(Ju, Jv) = g(u, v)$ for all $u, v \in V$, and regard V as a complex vector space V_J of dimension m. The scalar product on V_J is given by $\langle u|v\rangle_J := g(u, v) + ig(Ju, v)$. It then turns out that the exterior algebra $\Lambda^\bullet V_J$ is an irreducible Clifford module. We call this module *Fock space* and use the notation $\mathcal{F}_J(V)$ for it. The action of $\mathbb{C}l(V)$ on $\mathcal{F}_J(V)$ is obtained by defining "creation" and "annihilation" operators $a_J^\dagger$, $a_J : V \to \mathrm{End}(\mathcal{F}_J(V))$ acting on $\mathcal{F}_J(V)$ as follows:

$$a_J(v)(u_1 \wedge \ldots \wedge u_k) := \sum_{j=1}^{k} (-1)^{j-1} \langle v|u_j\rangle_J u_1 \wedge \ldots \wedge \widehat{u}_j \wedge \ldots \wedge u_k,$$

$$a_J^\dagger(v)(u_1 \wedge \ldots \wedge u_k) := v \wedge u_1 \wedge \ldots \wedge u_k\,, \qquad a_J(v)(1) := 0.$$

From these definitions, we obtain the anti-commutation relations

$$\{a_J^\dagger(u), a_J(v)\} = \langle u|v\rangle_J, \quad \{a_J(u), a_J(v)\} = 0 = \{a_J^\dagger(u), a_J^\dagger(v)\},$$

from which $(a_J^\dagger(v) + a_J(v))^2 = \langle v|v\rangle_J$ follows. Hence, the map

$$\begin{aligned} \pi_J : V &\longrightarrow \mathrm{End}(\Lambda^\bullet V_J) \\ v &\longmapsto \pi_J(v) := a_J(v) + a_J^\dagger(v) \end{aligned}$$

may be extended to $\mathbb{C}l(V)$ by making use of the universal property of the Clifford algebra. The scalar product $\langle\ |\ \rangle_J$ on V_J induces one on $\mathcal{F}_J(V)$, which on basis elements is given by

$$\langle u_1 \wedge \ldots \wedge u_k | v_1 \wedge \ldots \wedge v_{k'}\rangle = \delta_{kk'} \det\langle u_i|v_j\rangle_J, \tag{64}$$

thus making $\mathcal{F}_J(V)$ a (finite dimensional) Hilbert space. It may be checked that $a_J^\dagger(v)$ is really the adjoint of $a_J(v)$, with respect to this scalar product. Our interest is to lift (or quantize) operators $A \in \mathrm{End}(V_J)$ to operators acting on Fock space. In the present context, this is done by means of maps Γ, $d\Gamma : \mathrm{End}(V_J) \to \mathrm{End}(\mathcal{F}_J(V))$ defined as the graded operators obtained from

$$\Gamma^{(k)} A(u_1 \wedge \ldots \wedge u_k) := Au_1 \wedge \ldots \wedge Au_k$$

and

$$d\Gamma^{(k)} A(u_1 \wedge \ldots \wedge u_k) := \sum_{j=1}^{k} u_1 \wedge \ldots \wedge u_{j-1} \wedge (Au_j) \wedge u_{j+1} \wedge \ldots \wedge u_k.$$

Problem 4.3. *Check the correspondence between the definition of $d\Gamma$ given here and the one given by (22) and (23).*

Problem 4.4 (*cf.* [11]). *Derive the following identity:* $[d\Gamma(A), a^\dagger(v)] = a^\dagger(Av)$.

Given any $x \in V$ and $v \in V^{\mathbb{C}}$ any *unitary* vector, we see that $\phi(v)(x) := \chi(v)\cdot x \cdot v = x - 2g(v,x)v$, *i.e.* the twisted conjugation by v using the grading automorphism χ is a reflection on the hyperplane perpendicular to v. But reflections generate the orthogonal group $O(V,g)$ so that the map ϕ, being defined for all invertible elements of the Clifford algebra, when restricted to the subgroup generated by all products of an even number of unitary vectors ($\mathrm{Spin}^c(V)$), reduces to a homomorphism onto $SO(V)$. This homomorphism may be restricted to $\mathrm{Spin}(V)$.

Let us recall that there is a canonical vector space isomorphism between the Clifford algebra and the exterior algebra, given explicitly in terms of an orthonormal basis by:

$$\begin{aligned} Q: \quad \Lambda V^{\mathbb{C}} \quad &\longrightarrow \mathbb{C}l(V) \\ v_1 \wedge \ldots \wedge v_k &\longmapsto v_1 \cdot v_2 \cdots v_k. \end{aligned} \tag{65}$$

Direct computation then shows that $\mathfrak{spin}(V)$ coincides with $Q(\Lambda^2 V)$ and hence $Q(\Lambda^2 V) \cong \mathfrak{so}(V)$, the isomorphism being given by

$$\begin{aligned} \mathrm{ad}: Q(\Lambda^2 V) &\longrightarrow \mathfrak{so}(V) \\ a \quad &\longmapsto [a, \cdot\,] \end{aligned}$$

Let $c: \mathbb{C}l(V) \to \mathrm{End}(S)$ be an irreducible representation of $\mathbb{C}l(V)$. Since $\mathfrak{spin}(V)$ is realized as a subspace of $\mathbb{C}l(V)$, we can compose c with the isomorphism ad^{-1} to obtain a representation, called the *infinitesimal spin representation*, of $\mathfrak{so}(V)$ on S. Setting $\widetilde{B} := ad^{-1}(B)$:

$$\begin{aligned} \dot{\mu}: \mathfrak{so}(V) \to Q(\Lambda^2 V) &\hookrightarrow \mathbb{C}l(V) \to \mathrm{End(S)} \\ B \;\mapsto\; \widetilde{B} \quad &\mapsto\; \widetilde{B} \;\;\mapsto\; c(\widetilde{B}), \end{aligned}$$

i.e.,

$$\dot{\mu}(B) := c(\widetilde{B}), \qquad B \in \mathfrak{so}(V).$$

The infinitesimal spin representation satisfies the following basic relation:

$$[\dot{\mu}(B), c(v)] = c(Bv). \tag{66}$$

Comparing (66) with the identity from problem 4.4, one realizes that $d\Gamma$ and $\dot{\mu}$ must be closely related. In order to express this relation, note that any real linear operator R on V can be written as $R = R_+ + R_-$, with R_+ linear in V_J and R_- antilinear in V_J. For an element $B \in \mathfrak{so}(V)$, B_+ is skewadjoint and B_- antilinear and skewsymmetric (in $(V_J, \langle\ |\ \rangle_J)$). Then, it can be shown that for any $B \in \mathfrak{so}(V)$, the following relation holds:

$$\dot{\mu}(B) = d\Gamma(B_+) + \frac{1}{2}(a^\dagger B_- a^\dagger - aB_- a) - \frac{1}{2}\mathrm{Tr} B_+, \tag{67}$$

where $a^\dagger B_- a^\dagger := \sum_{k,l} \langle u_k | B_- u_l \rangle_J a_k^\dagger a_l^\dagger$ and $aTa := \sum_{k,l} \langle T u_l | u_k \rangle_J a_l a_k$ (with $\{u_k\}_k$ any orthonormal basis in V_J) [11, 38]. This equation plays an important role for the definition of the quantization map in the infinite dimensional case, discussed below.

In the infinite dimensional context, we begin with a separable real vector space V on which a positive definite symmetric bilinear form g is defined. It is assumed that V is complete in the metric induced by g, so that $(V^{\mathbb{C}}, \langle\langle\ |\ \rangle\rangle\ )$ is a separable Hilbert space, where $\langle\langle u|v \rangle\rangle := 2g(\bar{u}, v)$. The Clifford algebra is constructed as follows. Consider the algebra $\mathbb{C}l_{\mathrm{fin}}(V)$ obtained from the union of all algebras $\mathbb{C}l(W)$, where W runs through all finite dimensional subspaces of V. There is a unique trace τ on $\mathbb{C}l_{\mathrm{fin}}(V)$, inherited from the traces on each $\mathbb{C}l(W)$. The scalar product induced by the trace makes $\mathbb{C}l_{\mathrm{fin}}(V)$ a prehilbert space. Its completion $\mathcal{H}_\tau$ allows one to represent $\mathbb{C}l_{\mathrm{fin}}(V)$ as a subalgebra of $\mathcal{B}(\mathcal{H}_\tau)$ via the GNS construction. The closure of this algebra is then a C^*-algebra, which is defined to be the Clifford algebra $\mathbb{C}l(V)$. The universal property remains valid in this context and in particular an orthogonal map $h \in O(V, g)$ extends to a C^*-algebra automorphism θ_h of $\mathbb{C}l(V)$ (Bogoliubov automorphsim).

In order to construct a representation of the Clifford algebra on Fock space we need, as before, an orthogonal complex structure J. The Fock space $\mathcal{F}_J(V)$ is now defined as the completion of the exterior algebra $\Lambda^\bullet V_J$ with respect to the scalar product (64). The action of $\mathbb{C}l(V)$ on the exterior algebra is defined as in the finite dimensional case, its extension giving rise to an irreducible representation $\pi_J : \mathbb{C}l(V) \to \mathrm{End}(\mathcal{F}_J(V))$. As in the finite dimensional case, the action of the orthogonal group $O(V)$ on the set of orthogonal complex structures is transitive, for given two complex structures J and K, there is a unitary map $h : V_J \to V_K$ such that $K = hJh^{-1}$. As can be shown, if the Bogoliubov automorphism θ_h is implementable by a unitary operator on Fock space, then π_J and π_K are equivalent. In general however, they will be inequivalent. The converse also holds: Implementability of θ_h follows from equivalence of the representations. Necessary and sufficient conditions on h for θ_h to be implementable are afforded by the Shale-Stinespring theorem (*cf.* [11], thm. 6.16), namely that the antilinear part $\frac{1}{2}(h + JhJ)$ of h be Hilbert-Schmidt. In terms of the representations, it says that π_J and π_K are unitarily equivalent if and only if $(J - K)$ is Hilbert-Schmidt.

Problem 4.5. *Provide a physical interpretation of the Shale-Stinespring theorem. Hint: Try relating the vacua associated to two complex structures J and K.*

For the construction of quantized currents (like the charge or the number operator) we need a map taking self-adjoint operators on $\mathcal{H} := V_{\mathbb{C}}$ to self-adjoint operators on $\mathcal{F}_J(V)$. The explicit construction of this map, which can be found in [11], will not be presented here. Instead, we will try to describe some of the motivations behind the construction.

Recall that $O(V)$ acts transitively on the set of orthogonal complex structures, $\mathcal{J}(V)$. Since we have chosen a fixed complex structure J to construct the representation π_J on Fock space, then any other complex structure will be given in terms of J and of an element $h \in O(V)$ $(J' = hJh^{-1})$. Recall also that an element of the unitary group $U_J(V)$ is just an orthogonal map $h \in O(V)$ that commutes with J. A unitary transformation $U \in U_J(V)$ is implemented in Fock space by a map Γ_J such that:

$$\Gamma_J(U)\pi_J(v)\Gamma_J(U)^{-1} = \pi_J(Uv).$$

Now, the observables of the 1-particle theory correspond to elements of the Lie algebra $\mathfrak{o}_J(V)$ of the restricted orthogonal group, which is defined as $O_J(V) := \{h \in O(V) \mid [h, J] \text{ is Hilbert-Schmidt}\}$. If an element $X \in \mathfrak{o}_J(V)$ is such that $[X, J] = 0$, then it follows that JX is self-adjoint on V_J and it can be quantized by means of $d\Gamma_J$. The question of (unitary) equivalence of the quantization obtained by means of J on one side and by means of $J' = hJh^{-1}$ on the other, may be formulated as follows. The complex structure J' is determined by h. The map $v \mapsto \pi_J(hv)$ extends to an automorphism of the Clifford algebra and the question is then whether this automorphism is implementable on Fock space. So, given $h \in O(V)$, we are looking for a unitary operator $\mu(h)$ on Fock space such that:

$$\mu(h)\pi_J(v)\mu(h)^{-1} = \pi_J(hv).$$

In view of the inclusion $U_J(V) \subseteq O(V)$, we see that we are looking for an extension of Γ_J to $O(V)$. If $[J, h]$ is Hilbert-Schmidt, that is, if $h \in O_J(V)$, then the map μ can be constructed (it is the pin representation of $O_J(V)$ on Fock space). Its restriction to $SO_J(V)$ is the spin representation, from which an infinitesimal version $\dot{\mu}$ can be defined on a dense domain of Fock space and turns out to be given by (67) without the trace term and, therefore, it is not a Lie algebra homomorphism. This gives rise to the so-called *Schwinger term*, defined as

$$\alpha(A, B) := [\dot{\mu}(A), \dot{\mu}(B)] - \dot{\mu}([A, B]).$$

The Schwinger term is a cocycle that gives an obstruction for $\dot{\mu}$ to be a Lie algebra homomorphism. As explained in [11], the Schwinger term is related to anomalies in quantum field theory and also to noncommutative geometry. It appears naturally when studying the Virasoro algebra, and therefore is also related to the notion of *central charge* in conformal field theory. As we have seen, central charges do also appear when we study entanglement entropy in spin chains that display critical behavior. Our aim in the next section is to give an idea of how all these concepts are interrelated. This will allow us to close the circle by providing a new topological interpretation of quantum criticality, for which the underlying geometry is not anymore the geometry of the space of external parameters, but a geometry in the sense of noncommutative geometry. This geometric interpretation of anomalies, central charges, Schwinger cocycles *etc.*, has been known for several decades (to the best of my knowledge the first person to realize this was Araki [39, 40]). But the connection to *quantum critical phenomena* does not seem to have been exploited.

4.3. *Cyclic cocycles and quantum criticality*

Previously in these notes we discussed the Gelfand-Naimark theorem, which serves as a motivation for the notion of a noncommutative space. It turns out that the Schwinger term described in the previous section also has a nice geometric interpretation, in the context of noncommutative geometry [11, 41]. In order to give at least a glimpse of how this interpretation appears we will present, following [5], a very basic (yet instructive) example, namely the Gauss-Bonnet theorem, in the language of noncommutative geometry. But before that we quickly review some aspects of the geometry of surfaces, following the beautiful presentation of the subject by Pressley [42]. We thus start with a given 2-dimensional surface, described locally through a parametrization

$$\begin{aligned} \sigma : U \subseteq \mathbb{R}^2 &\longrightarrow \quad \mathbb{R}^3 \\ (u, v) &\longmapsto \sigma(u, v) = (x(u, v), y(u, v), z(u, v)). \end{aligned}$$

At each point, the tangent plane is generated by $\sigma_u \equiv \frac{\partial\sigma}{\partial u}$ and $\sigma_v \equiv \frac{\partial\sigma}{\partial v}$ and we can define the normal unit vector as

$$n := \frac{\sigma_u \times \sigma_v}{\|\sigma_u \times \sigma_v\|}.$$

The infinitesimal length element on the surface is given by

$$ds^2 = Edu^2 + 2Fdudv + Gdv^2,$$

where

$$E = (\sigma_u, \sigma_u), \; F = (\sigma_u, \sigma_v), \; G = (\sigma_v, \sigma_v)$$

are the components of the *first fundamental form*:

$$\mathcal{F}_I = \begin{pmatrix} E & F \\ F & G \end{pmatrix}.$$

Recall that the first and second fundamental forms enter the definition of Gaussian curvature of a surface, which we now recall, starting from the curvature of a curve. For a plane curve, parametrized by $\gamma(t)$ with unit speed ($\|\dot\gamma(t)\| = 1$), let $n(t)$ denote the unit normal at $\gamma(t)$. Then the (signed) curvature κ is defined through the relation $\ddot\gamma = \kappa n$. The curvature gives us an idea of how much the curve deviates from a straight line [42]. In fact, we may compute

$$(\gamma(t+\Delta t) - \gamma(t)) \cdot n \approx (\dot\gamma(t)\Delta t + \frac{\ddot\gamma(t)}{2}\Delta t^2 + \cdots) \cdot n = \frac{1}{2}\kappa \Delta t^2.$$

Now we repeat the same reasoning, but with a surface. That is, we want to compute $(\sigma(u+\Delta u, v+\Delta v) - \sigma(u,v)) \cdot n$, where now n denotes the unit normal to the surface. From a second order Taylor expansion and using the fact that $\sigma_u \cdot n = 0 = \sigma_v \cdot n$, we get

$$\begin{aligned} &\sigma(u+\Delta u, v+\Delta v) - \sigma(u,v) \approx \\ &\quad \approx \frac{1}{2}(L\Delta u^2 + 2M\Delta u\Delta v + N\Delta v^2) = \frac{1}{2}(\Delta u, \Delta v)\begin{pmatrix} L & M \\ M & N \end{pmatrix}\begin{pmatrix} \Delta u \\ \Delta v \end{pmatrix}, \end{aligned}$$

where $L = \sigma_{uu} \cdot n$, $M = \sigma_{uv} \cdot n$ and $N = \sigma_{vv} \cdot n$ are the coefficients of the *second fundamental form*:

$$\mathcal{F}_{II} = \begin{pmatrix} L & M \\ M & N \end{pmatrix}.$$

Now, for any smooth curve γ we have a decomposition of the form

$$\ddot\gamma = \kappa_n n + \kappa_g n \times \dot\gamma,$$

where κ_n and κ_g denote, respectively, the *normal* and *geodesic* curvatures of γ. To define the curvature of the surface we need to consider only curves with $\kappa_g = 0$, for which the normal curvature takes the form

$$\kappa_n = \ddot{\gamma} \cdot n = L\dot{u}^2 + 2M\dot{u}\dot{v} + N\dot{v}^2.$$

The value of κ_n changes depending on the direction of the curve at the evaluation point, and it is clear it must attain maximum and minimum values. They are the *principal curvatures*, denoted as κ_1 and κ_2. The Gaussian curvature is defined as $K := \kappa_1\kappa_2$. We can obtain several useful formulas for the curvature using the fundamental forms. The principal curvatures κ_1 and κ_2 are obtained from an extremization problem, where we look for extremal values of the principal curvature as we change the direction of the curve -given by $(\dot{u}, \dot{v})$- subject to the unit-speed restriction $\|\dot{\gamma}\| = 1$. So, with $x = (x_1, x_2)$, $x_1 = \dot{u}, x_2 = \dot{v}$, we want to extremize

$$\begin{aligned} Q : \mathbb{R}^2 &\longmapsto \mathbb{R} \\ x &\longmapsto x^t \mathcal{F}_{II} x, \end{aligned}$$

subject to the restriction $\|\dot{\gamma}\|^2 = 1$. But we also have

$$\|\dot{\gamma}\|^2 = E\dot{u}^2 + 2F\dot{u}\dot{v} + G\dot{v}^2 = x^t \mathcal{F}_I x.$$

Problem 4.6. *Solve the extremization problem posed above, subject to the constraint $\|\dot{\gamma}\|^2 = 1$, and show that the solution is given by the solution of the eigenvalue problem*

$$(\mathcal{F}_I^{-1} \mathcal{F}_{II})\, x = \lambda x,$$

where the eigenvalue λ is the Lagrange multiplier of the problem. Show furthermore that, for a given solution (x, λ), the corresponding normal curvature coincides precisely with λ and conclude that the Gaussian curvature can be expressed as follows:

$$K = \kappa_1\kappa_2 = \det\left(\mathcal{F}_I^{-1} \mathcal{F}_{II}\right) = \frac{LN - M^2}{EG - F^2}.$$

There are other ways we can characterize K. One of them uses the Gauss map: For any point p in the surface, consider the unit normal, n_p. Regarding this vector as a point in the unit sphere, we obtain a map $p \mapsto n_p$ from the surface to the unit sphere, the Gauss map. As we go around a small loop centered at p, the Gauss map produces an image loop on the unit sphere. The quotient of the areas enclosed by the loops converges, in the limit where the original loop shrinks to a point, to the Gaussian curvature.

Yet another way to obtain the Gaussian curvature is by means of the *Weingarten map*. First notice that we have $n_u \cdot \sigma_u = -L$, $n_u \cdot \sigma_v = n_v \cdot \sigma_u = -M$ and $n_v \cdot \sigma_v = -N$. This can be easily checked; for instance, the first relation is obtained by taking the partial derivative of $\sigma_u \cdot n = 0$ with respect to u. The others are obtained in the same way.

The Weingarten map $\mathcal{W}$ is most easily defined in the basis $\{\sigma_u, \sigma_v\}$, where it has a matrix given by $\mathcal{F}_I^{-1}\mathcal{F}_{II}$. In fact, if we define

$$\mathcal{W}(\sigma_u) = -n_u,\ \mathcal{W}(\sigma_v) = -n_v$$

and write

$$\begin{aligned} -n_u &= \lambda_{11}\sigma_u + \lambda_{12}\sigma_v \\ -n_v &= \lambda_{21}\sigma_u + \lambda_{22}\sigma_v, \end{aligned}$$

then it follows that

$$\Lambda := \begin{pmatrix} \lambda_{11} & \lambda_{12} \\ \lambda_{21} & \lambda_{22} \end{pmatrix} = \mathcal{F}_I^{-1}\mathcal{F}_{II}.$$

Now we compute

$$\begin{aligned} n_u \times n_v &= (\det \Lambda)\, \sigma_u \times \sigma_v \\ &= (\det \mathcal{F}_I^{-1}\mathcal{F}_{II})\, \sigma_u \times \sigma_v, \end{aligned}$$

from which the following identity is obtained:

$$n \cdot (n_u \times n_v) = K \|\sigma_u \times \sigma_v\|.$$

Now, as the components of the unit normal $n = (n_1, n_2, n_3)$ are smooth functions of (u, v), we can compute their exterior derivatives

$$dn_i = (\partial_u n_i)du + (\partial_v n_i)dv,$$

in order to obtain

$$\varepsilon_{ijk} n_k dn_i \wedge dn_j = 2n \cdot (n_u \times n_v) du \wedge dv.$$

This leads to the following formula for the Euler characteristic:

$$\chi(\Sigma) = \frac{1}{4\pi} \int \varepsilon_{ijk} n_k dn_i \wedge dn_j.$$

This formula is particularly well suited to explain the noncommutative viewpoint, to which we now turn, following the example presented by Connes in [5]. Let us consider an algebra $\mathcal{A}$ (without assuming other properties, for the moment being) together with a trilinear map

$$\tau : \mathcal{A} \times \mathcal{A} \times \mathcal{A} \to \mathbb{C}$$

such that

$$\tau(a_0, a_1, a_2) = \tau(a_2, a_0, a_1) \tag{68}$$

and

$$\tau(a_0a_1, a_2, a_3) - \tau(a_0, a_1a_2, a_3) + \tau(a_0, a_1, a_2a_3) - \tau(a_3a_0, a_1, a_2) = 0. \tag{69}$$

Now let $t \mapsto e_t$ be a (continuous) family of idempotents, $e_t^2 = e_t$. Then it can be shown that the quantity $\tau(e_t, e_t, e_t)$ does not depend on t, providing an "invariant" of the algebra. A quick way to see what is going on is to assume that we can differentiate with respect to t. In this case we have

$$\dot{e}_t = [x_t, e_t],$$

with $x_t = [\dot{e}_t, e_t]$. Using this fact and (68), we obtain

$$\frac{d}{dt}\tau(e_t, e_t, e_t) = 3\tau([x_t, e_t], e_t, e_t).$$

But the last expression vanishes exactly. To see this, we put $a_0 = x_t$ and $a_1 = a_2 = a_3 = e_t$ in (69) and use $e_t^2 = e_t$. Now, coming back to Gauss-Bonnet, if we take the algebra to be $\mathcal{A} = C^\infty(\Sigma)$, we see that the trilinear map $(f_0, f_1, f_2) \mapsto \int_\Sigma f_0 df_1 \wedge df_2$ satisfies (68) and (69) above, provided $\partial\Sigma = \emptyset$. But, as explained in [5], in this algebra there are no interesting idempotents. This can be remedied by adding a "bit of noncommutativity" to this algebra, that is, by replacing it by the following one: $\mathcal{A} = C^\infty(\Sigma) \otimes M_2(\mathbb{C})$. Now, recalling the Gauss map, we realize that we can use the normal to the surface in order to obtain an idempotent, namely $e = 1/2\,(\mathbb{1}_2 + n \cdot \sigma)$, where $\sigma = (\sigma_1, \sigma_2, \sigma_3)$, a "vector" whose components are the Pauli matrices. Then, if we define -following [5]-

$$\tau(f_0 \otimes M_0, f_1 \otimes M_1, f_2 \otimes M_2) := \left(\int_\Sigma f_0 df_1 \wedge df_2\right) \mathrm{Tr}(M_0M_1M_2),$$

we obtain (an algebraic version of) the Euler characteristic.

Problem 4.7. *Prove the last statement, i.e., compute $\tau(e, e, e)$ and show that it is (up to a constant factor) precisely $\chi(\Sigma)$.*

What we learn from the previous example is that it seems plausible to encode geometry and topology in algebraic structures related to a manifold M. This is precisely the way noncommutative geometry works. After establishing such algebraic and operator-theoretic characterization of certain geometric or topological structures, one can try to keep the algebraic version, but now dropping the commutativity assumption. Such is the case with differential calculus. Let us then quickly list the main properties of de Rham cohomology and see whether they can be formulated for more general algebras. Given a compact manifold M, the space of 1-forms $\Omega^1(M)$ consists of elements of the form fdg, where f and g are smooth functions on M, and $dg = \sum_i(\partial_i g)dx_i$ is the exterior derivative of g. Differential forms of order k, $\Omega^k(M)$, are generated locally by expressions of the form $dx_I \equiv dx_{i_1} \wedge \ldots \wedge dx_{i_k}$ so that, in general, the exterior derivative $d : \Omega^k(M) \to \Omega^{k+1}(M)$ can be defined through $d(fdx_I) := df \wedge dx_I$. The exterior derivative satisfies the Leibniz rule

$$d(\omega \wedge \eta) = d\omega \wedge \eta + (-1)^{|\omega|}\omega \wedge \eta$$

and fulfills the basic identity $d^2 = 0$, from which we obtain the (de Rham) cohomology "groups":

$$H^k(M) = \text{Ker } d^{k+1}/\text{Im } d^k,$$

that contain important information about the topology of M. Finally, we also have Stoke's theorem:

$$\int_{\partial M} \omega = \int_M d\omega.$$

Now suppose we are given a unital algebra $\mathcal{A}$. Thinking of it as given by some space of functions on a "virtual" (*i.e.* noncommutative) space, we may try to define a differential calculus satisfying properties as close as possible to the ones recalled above. One possible construction goes as follows [11]:

In order to define the space of 1-forms, we consider the map $d : \mathcal{A} \to \mathcal{A} \otimes \mathcal{A}$ defined as $da := 1 \otimes a - a \otimes 1$. Its usefulness comes from the fact that it acts as a derivation:

$$d(ab) = (da)b + a(db).$$

But $\mathcal{A} \otimes \mathcal{A}$ is "too big" to be regarded as a space of 1-forms. We want to have a closer resemblance to $\Omega^1(M)$. Recalling that 1-forms can always be put in the form $\theta = fdg$, we consider the restriction to the subspace of $\mathcal{A} \otimes \mathcal{A}$ generated by elements of the form $a\,db$. Noticing that

$$a\,db = a(1 \otimes b - b \otimes 1) = a \otimes b - ab \otimes 1,$$

we arrive at the following definition:

$$\Omega^1\mathcal{A} := \text{Ker}(m) \cong \mathcal{A} \otimes \overline{\mathcal{A}} \qquad (\overline{\mathcal{A}} \equiv \mathcal{A}/\mathbb{C}),$$

where $m : \mathcal{A} \otimes \mathcal{A} \to \mathcal{A}$ is the multiplication map $m(a \otimes b) = ab$. Then, starting from $\Omega^1\mathcal{A} = \mathcal{A} \otimes \overline{\mathcal{A}}$ we can go on and define $\Omega^k\mathcal{A} := \mathcal{A} \otimes \overline{\mathcal{A}}^{\otimes k}$ for

$k \geq 1$. The exterior derivative, then, is simply given by

$$d(a_0 \otimes \bar{a}_1 \otimes \cdots \otimes \bar{a}_k) := 1 \otimes \bar{a}_0 \otimes \bar{a}_1 \otimes \cdots \otimes \bar{a}_k.$$

From these definitions we get, as can be easily checked:

(*i*) $d^2 = 0$ and Leibniz rule: $d(\omega_k \, \eta) = d\omega_k \, \eta + (-1)^k \omega_k \, d\eta$,
(*ii*) $\Omega^k \mathcal{A} \cdot \Omega^l \mathcal{A} \subseteq \Omega^{k+l} \mathcal{A}$.

This means that in any complex algebra we can find an analog of the space of differential forms (*i.e.*, a graded differential algebra). The importance of this construction lies in the fact that it satisfies a universal property, as any derivation of $\mathcal{A}$ into a bimodule factors through a bimodule morphism from $\Omega^l \mathcal{A}$.

We also want to have a notion of integration. In this context this is achieved through the introduction of *cycles*. An n-dimensional cycle is a complex graded differential algebra $(\Omega^\bullet = \bigoplus_{k \leq n} \Omega^k, d)$, along with an "integral", *i.e.* a linear map $\int : \Omega^\bullet \to \mathbb{C}$, such that:

$$\int \omega_k = 0 \text{ (for } k < n), \quad \int \omega_k \omega_l = (-1)^{kl} \int \omega_l \omega_k \quad \text{and} \quad \int d\omega_{n-1} = 0.$$

The third requirement is tantamount to Stokes' theorem for the case of spaces without boundary. Now, given an algebra $\mathcal{A}$, a *cycle* over $\mathcal{A}$ is a cycle $(\Omega^\bullet, d, \int)$ together with a homomorphism $\rho : \mathcal{A} \to \Omega^0$ [5, 11]. A basic example is afforded by the de Rham complex on a compact manifold without boundary.

Given a cycle over $\mathcal{A}$ we may now define, for $a_0, a_1, \ldots, a_n \in \mathcal{A}$,

$$\tau(a_0, \ldots, a_n) := \int \rho(a_0) d(\rho(a_1)) \cdots d(\rho(a_n)).$$

As shown in [5], this map satisfies (the generalization to n of) the relations (68) and (69), *i.e.*, it is a *cyclic cocycle.*

Definition 4.2. Given a complex algebra $\mathcal{A}$, an *odd* Fredholm module over $\mathcal{A}$ is given by a representation $\sigma : \mathcal{A} \to \mathcal{B}(\mathcal{H})$ on the algebra of bounded operators on a Hilbert space, together with a self-adjoint operator $F : \mathcal{H} \to \mathcal{H}$, such that $F^2 = \mathbb{1}$ and such that $[F, \sigma(a)]$ is a compact operator for each a in $\mathcal{A}$. An *even* Fredholm module contains, in addition to the structure of an odd Fredholm module, an additional grading of $\mathcal{H}$, say γ, such that $\gamma F = -\gamma F$ and $[\gamma, \sigma(a)] = 0, a \in \mathcal{A}$.

On a Fredholm module one can introduce the structure of a graded differential algebra defining the differential as

$$d(\sigma(a)) := i[F, \sigma(a)].$$

Letting Ω^1 be spanned by elements of the form $a_0 da_1$, and so on, we arrive at the formula $d\omega = i[F, \omega]_\pm$ as a suitable definition for d, where the bracket denotes a supercommutator (depending on the degree of ω).

Example 4.2 (Quasi-free Representations). *An important class of examples for cyclic-cocycles is afforded by the so-called Schwinger terms (as defined in the previous section) in quantum field theory. In the context of charged fermionic fields, we start with a real vector space V (given as a certain space of solutions to the classical field equation), a symmetric bilinear form g on V and a compatible (orthogonal) complex structure J. Physically, J is determined by the spectrum of the Hamiltonian. From the data (V, J, g) one then defines a complex Hilbert space which is the complexification of V using J as the complex structure. So we get a Hilbert space $\mathcal{H} = (V_J, \langle\cdot|\cdot\rangle_J)$, where the inner product is given by*

$$\langle u|v\rangle_J := g(u, v) + ig(Ju, v).$$

We also have a representation of the field algebra on Fock space,

$$\hat{\pi}_J : \mathbb{C}l(V) \to End(F_J(V)).$$

Now, the point is that the Schwinger term

$$\alpha(A, B) = [\dot{\mu}(A), \dot{\mu}(B)] - \dot{\mu}[A, B] = \frac{i}{8} Tr\left(J[J, A][J, B]\right), \tag{70}$$

defines a cyclic 1-cocycle. A geometric interpretation of this cocycle is then made possible in the context of non-commutative geometry described above.

When discussing the entanglement properties of spin chains, we found that the entanglement entropy of a sub-chain of length L, at the critical point, scales logarithmically with L. We mentioned that the constant in front of the logarithm was related to *central charge* of the corresponding conformal field theory. These central charges can also be understood as anomalies appearing in the quantization process of the effective field theories describing the spin chains. As an example, let us consider the Ising chain:

$$H = -\frac{1}{2}\sum_{n=1}^{N}\left(\sigma_n^x \otimes \sigma_{n+1}^x + \lambda\sigma_n^z\right).$$

Using the Wigner-Jordan transformation and then defining the (Majorana) operators [43]

$$\chi_1(n) = (a_n^\dagger + a_n), \quad \chi_2(n) = i(a_n^\dagger - a_n),$$

we obtain

$$H = \frac{i}{2} \sum_n \big(\chi_2(n)\chi_1(n+1) + \lambda\chi_1(n)\chi_2(n)\big).$$

We can go to the continuum limit by assuming the sites in the chain are separated by a unit length l_0 and defining $x_n = l_0 n$. Then, to first order in l_0, we have:

$$\chi_\alpha(x_{n\pm1}) \approx \chi_\alpha(x_n) \pm l_0 \partial_x \chi_\alpha(x_n).$$

Defining the "chiral" components $\chi_\pm = (\chi_1 \pm \chi_2)/2$ and the corresponding spinor

$$\psi = \begin{pmatrix} \chi_+ \\ \chi_- \end{pmatrix},$$

one obtains, using Heisenberg's equations of motion,

$$(i\gamma^\mu \partial_\mu - m)\,\psi = 0, \tag{71}$$

where $m = m(\lambda)$ is such that $m(\lambda_c) = 0$.

Problem 4.8. *Defining light-cone coordinates $x_\pm = x^0 \pm x^1$, show that, at the critical point ($m = 0$), the Dirac equation (71) is equivalent to $\partial_+\chi_- = 0 = \partial_-\chi_+$.*

Problem 4.9. *In the massless case, (71) can be obtained from the Lagrangian $\mathcal{L} = i\overline{\psi}\gamma^\mu\partial_\mu\psi$. Considering the spatial coordinate x_1 to be defined on S^1 (which amounts to imposing periodic boundary conditions on the spin chain), expand the chiral fields in Fourier modes and then express the Hamiltonian density $\mathcal{H} = \dot{\psi}\frac{\partial\mathcal{L}}{\partial\dot{\psi}} - \mathcal{L}$, and the energy function $H = \int_{S^1} dx_1 \mathcal{H}(x_0, x_1)$ in terms of the Fourier modes.*

As the previous calculations show, when $m = 0$ the two chiral fields are completely decoupled from each other. Thus, we may study them separately. Let us consider χ_+, for which the mode expansion must be of the form

$$\chi_+(t, x) = \sum_{n\in\mathbb{Z}} a_n e^{in(t+x)}.$$

Imposing canonical anticommutation relations to the field (which is real), we obtain:

$$\{a_n, a_m\} = \delta_{n+m,0}, \quad a_n^* = a_{-n}.$$

From the previous problem, we see that the contribution of χ_+ to the (second quantized) Hamiltonian will be

$$H_+ = \sum_{n\in\mathbb{Z}} n : a_{-n}a_n : ,$$

where normal ordering has been introduced. Now, how do we define normal ordering depends on what we regard as the "vacuum state". This has important consequences for the commutation relations of the (Fourier components of the) stress-energy tensor.

Problem 4.10. *Decomposing the stress-energy tensor*

$$T_{\mu\nu} = \frac{1}{4}(\overline{\psi}\gamma_\mu \overleftrightarrow{\partial}_\nu \psi + \overline{\psi}\gamma_\nu \overleftrightarrow{\partial}_\mu \psi)$$

in light-cone components T_{++}, T_{--}, T_{-+} *and* T_{+-}, *show that* $T_{+-} = T_{-+} = 0$, *and that the Fourier components* (L_n) *of* T_{++} *satisfy the following commutation relations (Virasoro algebra):*

$$[L_m, L_n] = (m-n)L_{m+n} + \frac{c}{12}(m^3 - m)\delta_{m+n,0},$$

with $c = 1/2$.

The correct solution of the previous problem depends crucially in a careful handling of the normal ordering prescription. This in turn is related to the "filling-up of the Dirac sea" which, in mathematical terms, can be reduced to the problem of quantizing a linear system with an appropriately chosen complex structure. In fact, writing the Hamiltonian density in terms of the spinor ψ, we obtain

$$\mathcal{H} = -i\psi^\dagger \sigma_3 \partial_x \psi.$$

This means that the 1-particle Hamiltonian (which is defined on the Hilbert space $L^2(S^1) \otimes \mathbb{C}^2$) is given by $H^{(1)} = -i\sigma_3\partial_x$. As we have seen, the splitting of the 1-particle Hilbert space into positive and negative energy states gives rise to a complex structure, which is the one that should be used for quantization. When this is done, one obtains the anomalous term of the Virasoro algebra as an anomaly (Schwinger term) given precisely as a cyclic 1-cocyle of the form (70). Now, the relevance of this fact in the context of quantum phase transitions is that there are already some geometric characterizations of the critical point in terms of, *e.g.*, Berry phases [44, 45]. This has also been related to the behavior of certain Chern numbers associated to the parameter space of the spin chain [46]. Now, using Araki's self-dual formalism, it should also be possible to compute the cocycle outside the critical point. The behavior of the cocycle as a function of the model's external parameters may provide a new geometric characterization of the critical point.

Acknowledgement

First of all, I would like to thank all students who enthusiastically took part in the lectures on which these notes are based. Many of them contributed with useful criticisms and suggestions. I also want to warmly thank A.P. Balachandran, Andrés Vargas and Nicolás Escobar, for their very careful reading of the manuscript. Their feedback helped to improve the presentation of this document in a substantial way. Financial support from the Faculty of Science and the Vice Rectorate for Research of Universidad de los Andes, through project No. P13.700022.005, is gratefully acknowledged.

References

[1] H. Maassen, Quantum probability and quantum information theory, in *Quantum Information, Computation and Cryptography*, eds. F. Benatti, M. Fannes, R. Floreanini and D. Petritis, Lecture Notes in Physics, Vol. 808 (Springer Berlin Heidelberg, 2010) pp. 65–108.

[2] A. Einstein, B. Podolsky and N. Rosen, Can quantum-mechanical description of physical reality be considered complete?, *Phys. Rev.* **47**, 777 (May 1935).

[3] D A Guzmán, L J Uribe, A Valencia, F J Rodríguez, and L Quiroga, Contrasting classical probability concepts with quantum mechanical behavior in the undergraduate laboratory, *European Journal of Physics* **36** (2015), no. 5, 055039.

[4] E. T. Jaynes, *Probability theory: The logic of science* (Cambridge university press, 2003).

[5] A. Connes, *Noncommutative geometry* (Academic press, 1995).

[6] D. Werner, *Funktionalanalysis*, 3 edn. (Springer, 2000).

[7] R. Haag, *Local Quantum Physics.* Texts and Monographs in Physics, Texts and Monographs in Physics, 2nd edn. (Springer Verlag, 1996).

[8] O. Bratteli, *Operator Algebras and Quantum Statistical Mechan-*

ics: Equilibrium States. Models in Quantum Statistical Mechanics (Springer, 1997).

[9] W. Rudin, *Functional Analysis*, 2nd edn. (McGraw-Hill, 1991).

[10] I. Gelfand and M. Neumark, On the imbedding of normed rings into the ring of operators in hilbert space., *Rec. Math. [Mat. Sbornik] N.S.* **12**, 197 (1943).

[11] J. Gracia-Bondía, J. Várilly and H. Figueroa, *Elements of Noncommutative Geometry* (Birkhäuser, 2001).

[12] M. Paschke, Von nichtkommutativen Geometrien, ihren Symmetrien und etwas Hochenergiephysik, PhD thesis, Institut für Physik, (Universität Mainz, 2001).

[13] N. Papadopoulos, M. Paschke, A. Reyes and F. Scheck, The spin-statistics relation in nonrelativistic quantum mechanics and projective modules, *Annales Mathématiques Blaise-Pascal* **11**, 205 (2004).

[14] N. Papadopoulos and A. Reyes-Lega, On the geometry of the Berry-Robbins approach to spin-statistics, *Found. Phys.* **40**, p. 829-851 (2010).

[15] A. Reyes-Lega, On the geometry of quantum indistinguishability, *Journal of Physics A: Mathematical and Theoretical* **44**, p. 325308 (2011).

[16] N. E. Wegge-Olsen, *K-theory and C*-algebras: A Friendly Approach* (Oxford University Press, 1994).

[17] A. P. Balachandran, T. R. Govindarajan, A. R. de Queiroz and A. F. Reyes-Lega, Entanglement and particle identity: A unifying approach, *Phys. Rev. Lett.* **110**, p. 080503 (Feb 2013).

[18] A. Balachandran, T. Govindarajan, A. R. de Queiroz and A. Reyes-Lega, Algebraic approach to entanglement and entropy, *Physical Review A* **88**, p. 022301 (2013).

[19] M. A. Nielsen and I. L. Chuang, *Quantum computation and quantum information* (Cambridge university press, 2010).

[20] M. Keyl, Fundamentals of quantum information theory, *Physics Reports* **369**, 431 (2002).

[21] A. S. Holevo, *Probabilistic and statistical aspects of quantum theory* (Springer, 2011).

[22] A. Balachandran, T. Govindarajan, A. R. De Queiroz and A. Reyes-Lega, Algebraic theory of entanglement, *Il Nuovo Cimento C* **36**, 27 (2013).

[23] M. C. Tichy, F. Mintert and A. Buchleitner, Essential entanglement for atomic and molecular physics, *Journal of Physics B: Atomic, Molecular and Optical Physics* **44**, p. 192001 (2011).

[24] F. Benatti, R. Floreanini and U. Marzolino, Bipartite entanglement in systems of identical particles: The partial transposition criterion, *Annals of Physics* **327**, 1304 (2012).

[25] K. Eckert, J. Schliemann, D. Bruss and M. Lewenstein, Quantum correlations in systems of indistinguishable particles, *Annals of Physics* **299**, 88 (2002).

[26] G. Ghirardi and L. Marinatto, General criterion for the entanglement of two indistinguishable particles, *Phys. Rev. A* **70**, p. 012109 (2004).

[27] S. Sachdev, *Quantum Phase Transitions*, 2nd edn. (Cambridge, 2011).

[28] E. Lieb, T. Schultz and D. Mattis, Two soluble models of an antiferromagnetic chain, *Annals of Physics* **16**, 407 (1961).

[29] M. Henkel, *Conformal invariance and critical phenomena* (Springer, 1999).

[30] D. Karevski, Surface and bulk critical behaviour of the XY chain in a transverse field, *Journal of Physics A: Mathematical and General* **33**, p. L313 (2000).

[31] U. Schmidt, The excitation spectrum of the one-dimensional XY-model in a magnetic field, *Zeitschrift für Physik* **267**, 271 (1974).

[32] I. Peschel and V. Eisler, Reduced density matrices and entanglement entropy in free lattice models, *Journal of Physics A: Mathematical and Theoretical* **42**, p. 504003 (2009).

[33] G. De Chiara, L. Lepori, M. Lewenstein and A. Sanpera, Entanglement spectrum, critical exponents, and order parameters in quantum spin chains, *Physical review letters* **109**, p. 237208 (2012).

[34] C. Rivera, DMRG algorithm and quantum phase transitions, Master's thesis, Universidad de los Andes (July 2014).

[35] G. Vidal, J. I. Latorre, E. Rico and A. Kitaev, Entanglement in quantum critical phenomena, *Phys. Rev. Lett.* **90**, p. 227902 (Jun 2003).

[36] V. Moretti, *Spectral Theory and Quantum Mechanics with an introduction to the Algebraic Formulation* (Springer-Verlag, 2012).

[37] R. M. Wald, *Quantum field theory in curved spacetime and black hole thermodynamics* (University of Chicago Press, 1994).

[38] J. M. Gracia-Bondía and J. C. Várilly, Quantum electrodynamics in external fields from the spin representation, *Journal of Mathematical Physics* **35**, 3340 (1994).

[39] H. Araki, Bogoliubov automorphisms and fock representations of canonical anticommutation relations, *Contemp. Math* **62**, 23 (1987).

[40] H. Araki, Schwinger terms and cyclic cohomology, in *Quantum Theories and Geometry*, (Springer, 1988) pp. 1–22.

[41] E. Langmann, *Geometric Methods for Quantum Field Theory: Proceedings of the Summer School: Villa de Leyva, Colombia, 12-30 July 1999* (World Scientific, 2001), ch. Quantum Theory of Fermion Systems: Topics between Physics and Mathematics, pp. 170–237.

[42] A. N. Pressley, *Elementary differential geometry* (Springer, 2010).

[43] E. Fradkin, *Field theories of condensed matter physics* (Cambridge University Press, 2013).

[44] A. C. M. Carollo and J. K. Pachos, Geometric phases and criticality in spin-chain systems, *Phys. Rev. Lett.* **95**, p. 157203 (Oct 2005).

[45] S.-L. Zhu, Scaling of Geometric Phases Close to the Quantum Phase Transition in the XY Spin Chain, *Phys. Rev. Lett.* **96**, p. 077206 (Feb 2006).

[46] H. Contreras and A. Reyes-Lega, Berry phases, quantum phase transitions and Chern numbers, *Physica B: Condensed Matter* **403**, 1301 (2008), Proceedings of the International Conference on Strongly Correlated Electron Systems.

K3 EN ROUTE FROM GEOMETRY TO CONFORMAL FIELD THEORY

Katrin Wendland

Mathematisches Institut,
Abteilung Reine Mathematik,
Albert-Ludwigs-Universität Freiburg,
Freiburg, Germany

To pave the way for the journey from geometry to conformal field theory (CFT), these notes present the background for some basic CFT constructions from Calabi-Yau geometry. Topics include the complex and Kähler geometry of Calabi-Yau manifolds and their classification in low dimensions. I furthermore discuss CFT constructions for the simplest known examples that are based in Calabi-Yau geometry, namely for the toroidal superconformal field theories and their $\mathbb{Z}_2$-orbifolds. En route from geometry to CFT, I offer a discussion of K3 surfaces as the simplest class of Calabi-Yau manifolds where non-linear sigma model constructions bear mysteries to the very day. The elliptic genus in CFT and in geometry is recalled as an instructional piece of evidence in favor of a deep connection between geometry and conformal field theory.

Keywords: Conformal Field Theory; Calabi-Yau geometry; K3 surfaces.

Introduction

These lecture notes aim to make a contribution to paving the way from geometry to conformal field theory.

While two-dimensional conformal field theories can be defined abstractly in mathematics without reference to the algebraic geometry of complex manifolds of dimension greater than one, some of the most interesting examples are expected to arise geometrically from Calabi-Yau manifolds of complex dimension two or higher. Indeed, Calabi-Yau manifolds are the candidates for consistent geometric backgrounds for superstrings. Roughly, the string dynamics in such a geometric background are governed by so-called NON-LINEAR SIGMA MODELS, whose equations of motion imply conformal invariance. The resulting quantum field theory on the world-sheet is thus a superconformal field theory. From a mathematical point of view, this looks encouraging, since conformal field theory allows an independent mathematical approach, while string theory as a whole is not comprehensible to a mathematically satisfactory degree, as yet. However, non-linear sigma model constructions are far from well understood, mathematically. The only examples of (compact) Calabi-Yau manifolds where explicit constructions of a non-linear sigma model are known are the complex tori and their orbifolds. It seems that we are jumping out of the frying pan into the fire: By focusing on conformal field theory instead of string theory, we trade one mathematically dissatisfactory approach by a concept whose abstract definition prevents it from being immediately applied. Though string theory in its full generality seems to allow non-geometric phases, it is still crucially connected to geometry. To advance the subject, I am convinced that it is a *Sine qua non* to get a better understanding of the precise relation between geometry and conformal field theory, beginning with those cases where constructions are explicitly known.

These lecture notes should be viewed as an invitation to this journey from geometry to conformal field theory. They give a lightning introduction to the subject, as do many other excellent sources, so I try to keep the exposition somewhat complementary to existing works. As key examples, which are both sufficiently simple and mysterious, K3 surfaces play a special role en route from geometry to conformal field theory.

These notes are structured as follows:

Section 1 is devoted to some background in Calabi-Yau geometry. In Section 1.1, I recall basic mathematical concepts leading to the definition of Calabi-Yau manifolds, and some of their fundamental properties. Section 1.2 introduces a number of topological invariants of Calabi-Yau manifolds and culminates in a summary of mathematical arguments that yield the classification of Calabi-Yau manifolds of complex dimension $D \leq 2$. The classification naturally leads to the definition of K3 surfaces. An important topic whose discussion would naturally follow is the structure of the moduli spaces of complex structures, Kähler structures and Hyperkähler structures on K3 surfaces. However, since these topics have already been discussed elsewhere, even by myself [NW01, Wen07], I omit them in these notes. Instead, Section 1.3 gives a detailed summary of the Kummer construction as a classical example of a geometric orbifold procedure. In particular, I argue that the Kummer surfaces constitute a class of K3 surfaces which are very well understood, because their geometric properties are entirely governed by much simpler manifolds, namely the underlying complex two-tori.

Section 2 discusses explicit examples of superconformal field theories that are obtained from non-linear sigma model constructions. Since I have done so elsewhere, in these notes I do not offer a proposal for the defining mathematical properties of conformal field theory. Instead, the recent review [Wen15] should be viewed as the conformal field theory companion to these lecture notes, while the present exposition, in turn, can be understood as the geometric companion of [Wen15]. Concretely, Section 2.1 gives a brief review of toroidal superconformal field theories. Lifting the Kummer construction of Section 1.3 to the level of conformal field theory, Section 2.2 addresses $\mathbb{Z}_2$-orbifold constructions of toroidal superconformal field theories. To substantiate the expectation that these orbifold conformal field theories are correctly interpreted as non-linear sigma models on Kummer surfaces, I include a brief discussion of elliptic genera: I introduce the conformal field theoretic elliptic genus as a counter part of the geometric elliptic genus. I then show that the geometric elliptic genus of K3 surfaces agrees with the conformal field theoretic elliptic genus of the $\mathbb{Z}_2$-orbifold conformal field theory obtained from a toroidal superconformal field theory on a complex two-torus, recalling the known proof [EOTY89].

The final Section 3 places K3 en route from geometry to conformal field theory: I motivate and discuss the definition of K3 theories, which is formulated purely within representation theory. While this may be mathematically satisfying, it entails a consideration of K3 theories against non-linear sigma models on K3, for which explicit direct constructions on smooth K3 surfaces are lacking. I recall the role of the chiral de Rham complex in the context of elliptic genera of Calabi-Yau manifolds and conformal field theories, respectively. I conclude with a few speculations on the vertex algebra which can be obtained from the cohomology of the chiral de Rham complex, as candidate for a recovery of some of the conformal field theory structure from purely geometric ingredients.

Acknowledgements

My sincere thanks go to the organizers Alexander Cardona, Sylvie Paycha, Andrés Reyes, Hernán Ocampo and the participants of the tremendously inspiring 8th Summer School on “Geometric, Algebraic and Topological Methods for Quantum Field Theory 2013” at Villa de Leyva, Colombia, where the lectures leading to these notes were presented.

I am grateful to the organizers of the 2nd COST MP2010 Meeting “The String Theory Universe”, particularly to the local organizer Gabriele Honecker, for inviting me to this stimulating event at the Mainz Institute for Theoretical Physics (MITP). My thanks go to the MITP for its hospitality and its partial support during the completion of these lecture notes.

This work was in part supported by my ERC Starting Independent Researcher Grant StG No. 204757-TQFT "The geometry of topological quantum field theories".

1. Calabi-Yau geometry

By definition, a CALABI-YAU MANIFOLD is a compact[a] Kähler manifold with trivial canonical bundle. This section gives an introduction to the mathematical ingredients of this definition and illustrates its consequences.

Examples of Calabi-Yau manifolds are the complex tori, which also furnish the only examples where a non-linear sigma model construction immediately yields an associated conformal field theory (see Section 2.1). In Section 1.2, we will see that in complex dimension one and two, apart from tori, the only Calabi-Yau manifolds are the K3 SURFACES. The final subsection is devoted to the KUMMER CONSTRUCTION, yielding a special class of K3 surfaces which are almost as well under control from a conformal field theoretic point of view as are the complex tori (see Section 2.2).

There are a number of excellent books on the topics presented in this section, see for example [BHPvdV04, GH78, Huy05, Joy00, Wel73]. In particular, if not stated otherwise, then proofs of the classical results below can be found in these references.

1.1. *The Calabi-Yau condition*

In the following, familiarity with the concepts of differential geometry and complex analysis is assumed. The definition of topological and Riemannian manifolds over $\mathbb{R}$ and of vector bundles is crucial and can be found in textbooks like [dC92, Lee09, Mor01, O'N83]. The concept of holomorphic functions and their special properties is the foundation of complex analysis, where good textbooks include [FB05, Lan99, Rem91].

Recall that a D-DIMENSIONAL COMPLEX MANIFOLD is a differentiable real $2D$-dimensional manifold Y together with a holomorphic atlas $(U_\alpha, \varphi_\alpha)_{\alpha\in A}$, *i.e.* an open covering $\{U_\alpha \mid \alpha \in A\}$ of Y with diffeomorphisms $\varphi_\alpha\colon U_\alpha \longrightarrow V_\alpha$, $V_\alpha \subset \mathbb{C}^D$ open for all $\alpha \in A$, such that all coordinate

[a]By COMPACT, I mean bounded and closed; one also frequently finds definitions of non-compact Calabi-Yau manifolds, but those will not be of interest in these lectures.

changes $\varphi_\alpha \circ \varphi_\beta^{-1} \colon \varphi_\beta(U_\alpha \cap U_\beta) \longrightarrow \varphi_\alpha(U_\alpha \cap U_\beta)$ are holomorphic. Such a holomorphic atlas is said to define a COMPLEX STRUCTURE on the real manifold Y. There is an alternative description, which sometimes is more convenient, and which is motivated by the following observations:

Consider $y \in U_\alpha$ and note that due to $\varphi_\alpha(U_\alpha) = V_\alpha \subset \mathbb{C}^D$, for the tangent space $T_{\varphi_\alpha(y)}V_\alpha$ we have a natural identification $\mathbb{C}^D = T_{\varphi_\alpha(y)}V_\alpha$, such that multiplication by $i = \sqrt{-1}$ on $\mathbb{C}^D$ yields an endomorphism of $T_{\varphi_\alpha(y)}V_\alpha$. On the tangent space T_yY this induces an endomorphism $I \in \mathrm{End}(T_yY)$ with $D\varphi_\alpha(Iv) = iD\varphi_\alpha(v) \in \mathbb{C}^D$ for all $v \in T_yY$. The endomorphism I is checked to be independent of the choice of (holomorphic!) coordinates. Its $\mathbb{C}$-linear extension to the complexification $T^{\mathbb{C}}Y := TY \otimes_{\mathbb{R}} \mathbb{C}$ of the tangent bundle TY is also denoted by I. By construction, $I^2 = -\mathbb{I}$, such that we have an eigenspace decomposition $T^{\mathbb{C}}Y = T^{1,0}Y \oplus T^{0,1}Y$ for I, where I acts on the fibers of $T^{1,0}Y$ by multiplication by i, and on those of $T^{0,1}Y$ by multiplication by $-i$. In fact, $T^{1,0}Y$ is a holomorphic vector bundle whose cocycles can be given by the Jacobians of the coordinate changes in any holomorphic atlas of Y, and $T^{0,1}Y = \overline{T^{1,0}Y}$ with the complex conjugation $\overline{v \otimes \lambda} := v \otimes \overline{\lambda}$ for $y \in Y$, $v \in T_yY$, $\lambda \in \mathbb{C}$. The induced decomposition of the cotangent bundle yields a decomposition $d = \partial + \overline{\partial}$ of the exterior differential d, where on $\mathcal{A}^{p,q}(Y)$, the space of (p,q)-forms on Y, we have

$$\partial \colon \mathcal{A}^{p,q}(Y) \longrightarrow \mathcal{A}^{p+1,q}(Y) \quad \text{and} \quad \overline{\partial} \colon \mathcal{A}^{p,q}(Y) \longrightarrow \mathcal{A}^{p,q+1}(Y).$$

Since $T^{1,0}Y$ is a holomorphic vector bundle, it is closed under the Lie bracket (see *e.g.* [O'N83, §1 and App. B] for a discussion of the Lie bracket).

Vice versa, on a differentiable real $2D$-dimensional manifold Y, a fiber-wise endomorphism $I \in \mathrm{End}(TY)$ with $I^2 = -\mathbb{I}$ is called an ALMOST COMPLEX STRUCTURE. Given an almost complex structure, the fiber-wise eigenspace decomposition $T^{\mathbb{C}}Y = T^{1,0}Y \oplus T^{0,1}Y$ for I holds as above, with complex vector bundles $T^{\mathbb{C}}Y$, $T^{1,0}Y$ and $T^{0,1}Y$. The almost complex structure I is called INTEGRABLE if the bundle $T^{1,0}Y$ closes under the Lie bracket. This condition is equivalent to $d = \partial + \overline{\partial}$ on each $\mathcal{A}^{p,q}(Y)$ as above, where on $\mathcal{A}^{p,q}(Y)$, the operators $\partial, \overline{\partial}$ are obtained by composing the operator d with the projector to $\mathcal{A}^{p+1,q}(Y)$ and $\mathcal{A}^{p,q+1}(Y)$, respectively. By the celebrated Newlander-Nirenberg theorem [NN75], every integrable almost complex structure on Y is induced by a unique complex structure.

In summary, the choice of a complex structure on Y is tantamount to the choice of an endomorphism I which defines an integrable almost complex structure on Y.

Next, Riemannian geometry comes into play: We choose a Riemannian metric g on our complex manifold Y, viewed as a real manifold, and discuss meaningful additional compatibility conditions. Let g also denote the sesquilinear continuation of g to $T^{\mathbb{C}}Y$, where I use the convention

$$\forall y \in Y, \;\; u, v \in T_y^{\mathbb{C}}Y, \;\; \lambda, \mu \in \mathbb{C}: \quad g(\lambda u, \mu v) = \lambda\overline{\mu} g(u, v).$$

With respect to local complex coordinates $(z^1, \ldots, z^D)$ on $U \subset Y$, and restricting to real tangent vectors,

$$\begin{aligned} g_{|TU \times TU} = \tfrac{1}{2} \sum_{j,k} \Big(& g_{j,\overline{k}} dz^j \otimes d\overline{z}^k + g_{k,\overline{j}} d\overline{z}^j \otimes dz^k \\ & + g_{j,k} dz^j \otimes dz^k + \overline{g_{j,k}} d\overline{z}^j \otimes d\overline{z}^k \Big), \end{aligned}$$

$$\text{where } g_{j,\overline{k}} := 2g\left(\tfrac{\partial}{\partial z^j}, \tfrac{\partial}{\partial z^k}\right), \;\; g_{j,k} := 2g\left(\tfrac{\partial}{\partial z^j}, \tfrac{\partial}{\partial \overline{z}^k}\right),$$

with a Hermitean matrix $(g_{j,\overline{k}})_{j,k \in \{1,\ldots,D\}}$. The complex structure on Y is said to be COMPATIBLE with the metric g, if the corresponding almost complex structure $I \in \mathrm{End}(TY)$ is orthogonal with respect to g, in other words if

$$\forall y \in Y, \;\; u, v \in T_yY: \quad g(Iu, Iv) = g(u, v).$$

Then $g_{j,k} = 0$ for all $j, k \in \{1, \ldots, D\}$, so

$$g_{|TU \times TU} = \tfrac{1}{2} \sum_{j,k} \left(g_{j,\overline{k}} dz^j \otimes d\overline{z}^k + g_{k,\overline{j}} d\overline{z}^j \otimes dz^k \right).$$

One checks directly that this compatibility condition implies that with respect to g, we have $T_y^{1,0}Y \perp T_y^{0,1}Y$ for all $y \in Y$.

If I is compatible with g, then the $\mathbb{C}$-linear extension ω of the bilinear form

$$\forall y \in Y, \;\; u, v \in T_yY: \quad \omega(u, v) := g(Iu, v)$$

to $T^{\mathbb{C}}Y$ is called the KÄHLER FORM. By construction, ω is real, and by a direct calculation one checks that compatibility of g with I implies that ω is antisymmetric. Denoting by $\mathcal{A}^k_{\mathbb{R}}(Y)$ the space of real k-forms on Y, and with respect to local complex coordinates $(z^1, \ldots, z^D)$ on $U \subset Y$ as above, we have

$$\omega \in \mathcal{A}^{1,1}(Y) \cap \mathcal{A}^2_{\mathbb{R}}(Y), \quad \omega_{|U} = \tfrac{i}{2} \sum_{j,k} g_{j,\overline{k}} dz^j \wedge d\overline{z}^k.$$

It is important to keep in mind that g is extended to $T^{\mathbb{C}}Y$ as a HERMITEAN SESQUILINEAR FORM, while ω is ANTISYMMETRIC AND $\mathbb{C}$-BILINEAR on $T^{\mathbb{C}}Y$. Note that any two ingredients of the triple (I, g, ω) determine the third one.

The following additional condition on the metric turns out to have far-reaching consequences [Käh33]:

Definition 1.1.1. Consider a Riemannian manifold (Y, g), equipped with a compatible complex structure. Then the metric g is called KÄHLER METRIC if and only if the Kähler form ω is a closed differential form: $d\omega = 0$.

A complex manifold Y is called KÄHLER MANIFOLD if a Kähler metric exists on Y.

A harbinger of the fact that this condition is a very natural and interesting one is the following lemma, which states as many as six equivalent formulations of the Kähler condition, each emphasizing a different geometric aspect:

Lemma 1.1.2. *Consider a Riemannian manifold (Y, g) with compatible complex structure I, such that Y is a complex D-dimensional manifold and ω its Kähler form. Then the following are equivalent:*

(1) The metric g is a Kähler metric.
(2) For every $y \in Y$, there is an open neighborhood $U \subset Y$ of y and a smooth function $f\colon U \to \mathbb{R}$, such that $\omega_{|U} = i\partial\overline{\partial} f$.
(3) The Levi-Civita connection for g, that is, the unique torsion-free metric connection, agrees with the Chern (or holomorphic) connection for I, i.e. with the unique metric connection whose $(0,1)$ part is $\overline{\partial}$.
(4) The almost complex structure I of Y is parallel with respect to the Levi-Civita connection for g.
(5) With respect to arbitrary holomorphic coordinates $(z^1, \dots, z^D)$ on $U \subset Y$, the coefficients of $\omega_{|U} = \frac{i}{2}\sum_{j,k} g_{j,\overline{k}} dz^j \wedge d\overline{z}^k$ obey $\frac{\partial g_{k,\overline{l}}}{\partial z^m} = \frac{\partial g_{m,\overline{l}}}{\partial z^k}$ for all k, l, $m \in \{1, \dots, D\}$, or equivalently $\frac{\partial g_{k,\overline{l}}}{\partial \overline{z}^m} = \frac{\partial g_{k,\overline{m}}}{\partial \overline{z}^l}$ for all k, l, $m \in \{1, \dots, D\}$.
(6) The metric OSCULATES TO SECOND ORDER *with the standard Euclidean metric, that is, for every $y \in Y$, there are holomorphic coordinates $(z^1, \dots, z^D)$ around y such that $z(y) = (z^1, \dots, z^D)(y) = 0$ and*

$$g\left(\tfrac{\partial}{\partial z^j}, \tfrac{\partial}{\partial z^k}\right) = \delta_{j,k} + \mathcal{O}(|z|^2) \quad \forall j,\, k \in \{1, \dots, D\}.$$

(7) The holonomy representation of Y is unitary on each tangent space.

In short, Kähler metrics are very similar to the Euclidean metric.

The Kähler condition has a number of important consequences. For example, the standard Laplace operators on a Kähler manifold obey a very simple relation: $\Delta_d = 2\Delta_\partial = 2\Delta_{\overline{\partial}}$, where $\Delta_d = dd^* + d^*d$, $\Delta_\partial = \partial\partial^* + \partial^*\partial$, and $\Delta_{\overline{\partial}} = \overline{\partial}\,\overline{\partial}^* + \overline{\partial}^*\overline{\partial}$. Here, d^*, ∂^* and $\overline{\partial}^*$ can be viewed as the L^2-duals of d, ∂ and $\overline{\partial}$ if Y is compact. One furthermore has

Lemma 1.1.3. *Consider a compact complex Kähler manifold Y, and denote its de Rham and Dolbeault cohomology groups, respectively, by*

$$H^k(Y,\mathbb{C}) = \ker(d : \mathcal{A}^k(Y) \longrightarrow \mathcal{A}^{k+1}(Y))/\mathrm{im}(d : \mathcal{A}^{k-1}(Y) \longrightarrow \mathcal{A}^k(Y)),$$
$$H^{p,q}(Y,\mathbb{C}) = \ker(\overline{\partial} : \mathcal{A}^{p,q}(Y) \longrightarrow \mathcal{A}^{p,q+1}(Y))/\mathrm{im}(\overline{\partial} : \mathcal{A}^{p,q-1}(Y) \longrightarrow \mathcal{A}^{p,q}(Y))$$

with $\mathcal{A}^k(Y)$ the space of k-forms on Y. Then for all k, p, $q \in \mathbb{N}$,

$$H^k(Y,\mathbb{C}) = \bigoplus_{r+s=k} H^{r,s}(Y,\mathbb{C}), \quad H^{p,q}(Y,\mathbb{C}) = \overline{H^{q,p}(Y,\mathbb{C})}.$$

Apart from the Euclidean spaces and complex tori, important examples of Kähler manifolds are the complex projective spaces $\mathbb{P}^D$. Indeed, the FUBINI-STUDY METRIC defines a Kähler metric on $\mathbb{P}^D$; with respect to homogeneous coordinates $(z^0 : \cdots : z^D)$, its Kähler form on the standard coordinate neighborhood $U_j := \{(z^0 : \cdots : z^D) \in \mathbb{P}^D \mid z^j \neq 0\}$ is

$$\omega_{|U_j} = \tfrac{i}{2\pi}\partial\overline{\partial}\log\left(\sum_{k=0}^{D}\left|\tfrac{z^k}{z^j}\right|^2\right). \tag{1}$$

Since the restriction of a Kähler metric on a complex manifold Y to a complex submanifold yields a Kähler metric on the submanifold, this implies that all closed algebraic manifolds are compact Kähler manifolds.

Finally, we are ready to state

Definition 1.1.4. A manifold Y is called CALABI-YAU MANIFOLD if Y is a compact Kähler manifold with trivial canonical bundle K_Y.

Recall that the canonical bundle K_Y of a complex D-manifold Y is the determinant line bundle of the holomorphic cotangent bundle $\Omega_Y := (T^{1,0}Y)^*$, *i.e.* $K_Y = \Lambda^D\Omega_Y$. Hence a Kähler manifold is Calabi-Yau if and only if $h^{D,0}(Y) := \dim_{\mathbb{C}} H^{D,0}(Y,\mathbb{C}) = 1$, or equivalently if the holonomy representation of Y is special unitary on each tangent space.

Examples of Calabi-Yau manifolds include all complex tori $\mathbb{C}^D/L$ ($L \subset \mathbb{C}^D$ a lattice of rank $2D$) and the degree $D+2$ hypersurfaces in $\mathbb{P}^{D+1}$.

It is natural to investigate whether a given Calabi-Yau manifold Y possesses any Kähler metrics with particularly appealing properties. Indeed, the search within KÄHLER CLASSES turns out to be fruitful. Here, I make use of the fact that the Kähler form ω of a Kähler metric on Y is a real, closed

$(1,1)$-form. Thus, ω represents a class $[\omega] \in H^{1,1}(Y,\mathbb{C}) \cap H^2(Y,\mathbb{R})$, which in fact is non-zero if Y is compact. Another Kähler metric on Y is said to BELONG TO THE SAME KÄHLER CLASS AS g, if its Kähler form represents the same class $[\omega] \in H^{1,1}(Y,\mathbb{C}) \cap H^2(Y,\mathbb{R})$. One now has the seminal

Theorem 1.1.5 (Calabi-Yau Theorem [Cal54, Yau78]). *Consider a Calabi-Yau manifold Y with Kähler metric g. Then there exists a unique* RICCI-FLAT KÄHLER METRIC *in the Kähler class of g.*

The proof of this theorem is non-constructive [Yau78]; with the exception of complex tori, explicit forms of Ricci-flat Kähler metrics on smooth Calabi-Yau manifolds are not known.

1.2. *Classifying Calabi-Yau manifolds*

A classification of Calabi-Yau D-manifolds turns out to be possible in dimension $D \leq 2$, as I shall explain in this section. In the following, if not stated otherwise, let Y denote a connected Calabi-Yau D-manifold. I first recall some basic topological invariants of Y.

The theory of elliptic differential operators ensures that the cohomology of Y is finite dimensional, $h^{p,q}(Y) := \dim_{\mathbb{C}} H^{p,q}(Y,\mathbb{C}) < \infty$ for all p, $q \in \mathbb{N}$. These so-called HODGE NUMBERS are topological invariants which enjoy a number of constraints for our connected Calabi-Yau D-manifold Y:

$$\begin{aligned} h^{0,0}(Y) = h^{D,D}(Y) = \quad & h^{D,0}(Y) \quad = h^{0,D}(Y) = 1, \\ h^{p,q}(Y) = \quad h^{q,p}(Y) \quad & = h^{D-p,D-q}(Y) \quad \forall p,\, q \in \{0,\dots,D\}. \end{aligned} \tag{2}$$

Here, $h^{p,q}(Y) = h^{q,p}(Y)$ is an immediate consequence of Lemma 1.1.3, while $h^{p,q}(Y) = h^{D-p,D-q}(Y)$ is a consequence of the SERRE DUALITY [Ser55]. $h^{0,0}(Y) = h^{D,D}(Y) = 1$ is true by assumption, since Y is compact and connected, while $h^{D,0}(Y) = h^{0,D}(Y) = 1$ follows from the triviality of the canonical bundle.

Using the Hodge numbers, one obtains the following classical topological invariants:

Definition 1.2.1. For a compact Kähler manifold Y of dimension D, the EULER CHARACTERISTIC $\chi(Y)$, HOLOMORPHIC EULER CHARACTERISTIC $\chi(\mathcal{O}_Y)$ and SIGNATURE $\sigma(Y)$ are defined by

$$\chi(Y) := \sum_{p,q=0}^{D} (-1)^{p+q} h^{p,q}(Y), \quad \chi(\mathcal{O}_Y) := \sum_{q=0}^{D} (-1)^q h^{0,q}(Y),$$

$$\sigma(Y) := \sum_{p,q=0}^{D} (-1)^q h^{p,q}(Y).$$

For any holomorphic vector bundle $E \to Y$,

$$\chi(E) := \sum_{q=0}^{D} (-1)^q \dim H^q(Y, E)$$

is the HOLOMORPHIC EULER CHARACTERISTIC of E.

By (2), the signature $\sigma(Y)$ vanishes if the complex dimension D of Y is odd; I have thus trivially extended the traditional definition of the signature of oriented compact manifolds whose real dimension is divisible by 4 to all compact complex Kähler manifolds.

We are now in a position to state a first classification result:

Theorem 1.2.2. *Every connected Calabi-Yau one-manifold is biholomorphic to a torus $\mathbb{C}/L$ with $L \subset \mathbb{C}$ a lattice of rank* 2.

Proof (sketch):
Since $D = 1$, the constraints (2) already fix all Hodge numbers. Hence by Definition 1.2.1 we have $\chi(Y) = 0$. Therefore, the claim follows from the classification of compact Riemann surfaces. A discussion of this deep and fundamental classification result can be found in textbooks on differential topology, for example [For81, Hir76]. In particular, one needs to use the fact that $\chi(Y)$ agrees with the Euler-Poincaré characteristic of Y known from topology. This in turn is deeply linked to the relation between de Rham cohomology, Čech cohomology and singular homology, which is explained in depth in [Mor01]. □

Another type of topological invariants are the CHARACTERISTIC CLASSES, all of which can be traced back to CHERN CLASSES for our Calabi-Yau manifold Y. Recall that by definition, a complex vector bundle $E \to Y$ of rank r has Chern classes $c_k(E) \in H^{2k}(Y, \mathbb{Z})$, $k \in \{0, \dots, D\}$, which are uniquely determined by the following four conditions (a)-(d):

(a) $c_0(E) = [1]$.
(b) For the dual $\mathcal{O}(1) \to \mathbb{P}^D$ of the tautological line bundle, $c_1(\mathcal{O}(1))$ is the Kähler class of the Fubini-Study metric on $\mathbb{P}^D$ with Kähler form (1).
(c) For smooth $f\colon X \longrightarrow Y$ and $k \in \{0, \dots, D\}$, one has $c_k(f^*E) = f^*c_k(E)$.
(d) The TOTAL CHERN CLASS $c(E) := \sum\limits_{k=0}^{D} c_k(E) = \sum\limits_{k=0}^{r} c_k(E)$ for line bundles $L_1, \dots, L_r$ on Y obeys

$$c(L_1 \oplus \cdots \oplus L_r) = c(L_1) \wedge \cdots \wedge c(L_r).$$

It follows that the total Chern class of every trivial bundle is $[1] \in H^0(Y, \mathbb{Z})$. By convention, the Chern classes $c_k(Y) \in H^{2k}(Y, \mathbb{Z})$ of a complex manifold Y are the Chern classes of its holomorphic tangent bundle $T := T^{1,0}Y$. One checks that $c_1(Y) = -c_1(K_Y)$, where as above, K_Y is the canonical bundle of Y. Hence every Calabi-Yau manifold Y has vanishing first Chern class $c_1(Y) = 0$. Vice versa, using the exponential sheaf sequence one finds: If Y is a compact Kähler manifold with $c_1(Y) = 0$ and vanishing first Betti number, $b_1(Y) = \dim_{\mathbb{C}} H^1(Y, \mathbb{C}) = 0$, then Y is Calabi-Yau[b].

If $E \to Y$ is a complex vector bundle of rank r, then by the so-called SPLITTING PRINCIPLE, we may work with FORMAL CHERN ROOTS $e_1, \ldots, e_r$, such that $c(E) = \prod_{j=1}^r (1 + e_j)$, where the e_j are elements of a ring extension of $H^*(Y, \mathbb{R})$. It then follows that $c_k(Y) = \sigma_k(e_1, \ldots, e_r)$ with the elementary symmetric polynomials σ_k. More generally, for an analytic function f on a neighborhood of 0 in $\mathbb{C}^r$, by $f(e_1, \ldots, e_r)$ one denotes the power series expansion of f about the origin with insertions $(e_1, \ldots, e_r)$. This allows the definition of further topological invariants:

Definition 1.2.3. Consider a compact complex D-manifold Y with holomorphic tangent bundle $T := T^{1,0}Y$, and a complex vector bundle $E \to Y$ of rank r. Let $y_1, \ldots, y_D$ and $e_1, \ldots, e_r$ denote the formal Chern roots, such that $c(Y) = \prod_{j=1}^D (1 + y_j)$ and $c(E) = \prod_{j=1}^r (1 + e_j)$.
Then the TODD GENUS of Y is given by

$$\mathrm{Td}(Y) := \prod_{j=1}^{D} \frac{y_j}{1 - \exp(-y_j)} = c_0(Y) + \tfrac{1}{2} c_1(Y) + \tfrac{1}{12}(c_1(Y)^2 + c_2(Y)) + \cdots$$

The CHERN CHARACTER of the bundle E is

$$\mathrm{ch}(E) := \sum_{j=1}^{r} \exp(e_j) = r\, c_0(Y) + c_1(E) + \tfrac{1}{2}(c_1(E)^2 - 2c_2(E)) + \cdots$$

On first sight, the topological invariants that are obtained from the Hodge numbers in Definition 1.2.1 and those that are obtained from the Chern classes in Definition 1.2.3 have very different flavors: While the former are integers, the latter are cohomology classes. However, evaluation of

[b]Instead of Definition 1.1.4, some authors give a more restrictive definition of Calabi-Yau manifolds: A complex D-manifold Y is then called Calabi-Yau, if and only if Y is a compact Kähler manifold with $h^{1,0}(Y) = 0$ and $c_1(Y) = 0$ or, equivalently, such that the holonomy representation on every tangent space is an irreducible representation of $SU(D)$. Hence complex tori are not Calabi-Yau according to this definition, whereas I view them as the simplest examples of Calabi-Yau manifolds.

such cohomology classes on the fundamental cycle of Y, denoted by $\int_Y$, yields integers from integral cohomology classes. Here, it is understood that $\int_Y \alpha = 0$ if $\alpha \in H^k(Y,\mathbb{Z})$ with $k \neq 2D$. The ATIYAH-SINGER INDEX THEOREM [AS63] governs the deep relationship between the two types of invariants; in the context of complex manifolds and holomorphic bundles, which is relevant to our discussion, a precursor of this theorem is the following seminal

Theorem 1.2.4 (Hirzebruch-Riemann-Roch Formula [Hir54]). *Let $E \to Y$ denote a holomorphic vector bundle on a compact complex D-manifold Y. With notations as in Definitions 1.2.1 and 1.2.3,*

$$\chi(E) = \int_Y Td(Y) ch(E).$$

To appreciate this theorem, recall that representatives of the Chern classes $c_k(Y)$, $c_k(E)$ can be obtained in terms of the curvature forms of arbitrary Hermitean metrics on Y and E, respectively (see *e.g.* [LM89, BGV92] for a detailed discussion). The theorem thus yields the topological invariant $\chi(E)$ in terms of the integral of a local curvature expression. The essence of this theorem is the fact that $\not{D}^E = \partial_E + \overline{\partial}_E$ is a Dirac operator on $E = E^+ \oplus E^-$, $\not{D}^E : \Gamma(E^\pm) \to \Gamma(E^\mp)$, whose index $\dim\ker(\not{D}^E_{|E^+}) - \dim\ker(\not{D}^E_{|E^-}) = \chi(E)$ can therefore alternatively be calculated by heat kernel methods in terms of an integral over local curvature data.

To recover the invariants of Definition 1.2.1, for the holomorphic Euler characteristic $\chi(\mathcal{O}_Y)$ one uses the trivial bundle E, so $\not{D}^E = \partial + \overline{\partial}$ and $\mathrm{Td}(Y)\mathrm{ch}(E) = \mathrm{Td}(Y)$ with expansion into Chern classes as in Definition 1.2.3. The usual Euler characteristic $\chi(Y)$ arises for the virtual bundle $E = E^+ - E^-$ with $E^+ = \oplus_{p\equiv 0(2)} \Lambda^p T^*$, $E^- = \oplus_{p\equiv 1(2)} \Lambda^p T^*$ and $T = T^{1,0}Y$ as before. With $\chi(E) := \chi(E^+) - \chi(E^-) = \chi(Y)$, and with notations as in Definition 1.2.3,

$$\begin{aligned}
\mathrm{ch}(E) &= \mathrm{ch}(E^+) - \mathrm{ch}(E^-) = \prod_{j=1}^{D} (1 - \exp(-y_j)), \\
\mathrm{Td}(Y)\mathrm{ch}(E) &= \prod_{j=1}^{D} y_j = c_D(Y).
\end{aligned}$$

Thus the Hirzebruch-Riemann-Roch Formula 1.2.4 yields

$$\begin{aligned}\chi(\mathcal{O}_Y) &= \int_Y \left(c_0(Y) + \tfrac{1}{2}c_1(Y) + \tfrac{1}{12}(c_1(Y)^2 + c_2(Y)) + \cdots\right),\\ \chi(Y) &= \int_Y c(Y).\end{aligned} \tag{3}$$

To obtain the signature $\sigma(Y)$ from the Hirzebruch-Riemann-Roch Formula, one uses the same total bundle $E^+ \oplus E^- = \oplus_p \Lambda^p T^*$ with different $\mathbb{Z}_2$-grading $\oplus_p \Lambda^p T^* = \tilde{E}^+ \oplus \tilde{E}^-$ and $E = \tilde{E}^+ - \tilde{E}^-$.

There is another relation between the Hodge numbers of Y and its top Chern class, due to the interpretation of $\chi(Y)$ in terms of the POINCARÉ-HOPF INDEX THEOREM [Poi85, Hop26]. One combines this classical result from differential topology, which is discussed in textbooks like [GP74, Mor01], with the celebrated Weitzenböck formula from differential geometry, see *e.g.* the textbook [Jos95]. Indeed, if $\chi(Y) \neq 0$, then the Poincaré-Hopf Index Theorem implies that every holomorphic one-form on Y has at least one zero. On the other hand, using a Ricci-flat metric on Y, which exists by the Calabi-Yau Theorem 1.1.5, the Weitzenböck formula implies that every holomorphic one-form on Y has constant norm. In other words:

Lemma 1.2.5. *If a Calabi-Yau D-manifold Y has non-vanishing Euler characteristic, $\chi(Y) \neq 0$, then $h^{1,0}(Y) = 0$.*

We are now ready to classify Calabi-Yau two-manifolds; the resulting Theorem 1.2.8 was first proved in [Kod64], but with more recent results the proof can be simplified. First note

Lemma 1.2.6. *Let Y denote a connected Calabi-Yau two-manifold. Then Y has Hodge numbers*

$$h^{1,0}(Y) = 2,\ h^{1,1}(Y) = 4 \quad \text{or} \quad h^{1,0}(Y) = 0,\ h^{1,1}(Y) = 20.$$

This fixes all Hodge numbers of Y.

Proof (sketch):
Since $D = 2$, using (2) it is clear that the values of $h^{1,0}(Y)$ and $h^{1,1}(Y)$ determine all the Hodge numbers of Y.

Generalizing Lemma 1.2.5, if a holomorphic k-form on Y has a zero, then it vanishes identically on Y [Bog74]. This implies $h^{1,0}(Y) \leq 2$. Since Y is Kähler, such that $h^{1,1}(Y) \neq 0$, one deduces that $\chi(Y) = 0$ can only hold if $h^{1,0}(Y) = 2,\ h^{1,1}(Y) = 4$.

On the other hand, if $\chi(Y) \neq 0$, then Lemma 1.2.5 implies $h^{1,0}(Y) = 0$, so by (2) we have $h^{1,1}(Y) = \chi(Y) - 4$. Moreover, by (2) and Definition 1.2.1, Y has holomorphic Euler characteristic $\chi(\mathcal{O}_Y) = 2$. Hence

$$\chi(Y) \overset{(3)}{=} \int_Y c_2(Y) \overset{(3),\, c_1(Y)=0}{=} 12\chi(\mathcal{O}_Y) = 24$$

and therefore $h^{1,1}(Y) = \chi(Y) - 4 = 20$. □

If $\chi(Y) = 0$ for a Calabi-Yau two-fold Y, then one can use the Calabi-Yau Theorem 1.1.5 and a result by Bochner and Yano [YB53] to prove that Y is a torus $\mathbb{C}^2/L$ with a lattice $L \subset \mathbb{C}^2$ of rank 4. If $\chi(Y) \neq 0$, then Lemma 1.2.6 implies that Y is a K3 SURFACE, according to

Definition 1.2.7. A connected, compact complex surface Y with trivial canonical bundle and $b_1(Y) = \dim_{\mathbb{C}} H^1(Y, \mathbb{C}) = 0$ is called a K3 SURFACE.

It was conjectured independently by Andreotti and Weil [Wei58] and proved by Siu [Siu83] that all K3 surfaces are Kähler and thus Calabi-Yau[c]. Using Lemma 1.2.6, I have thus summarized a derivation of the first claim of

Theorem 1.2.8 ([Kod64]). *If Y is a connected Calabi-Yau two-manifold, then Y is either a complex two-torus or a K3 surface. Viewed as real four-manifolds, all complex two-tori are diffeomorphic to one another, and all K3 surfaces are diffeomorphic to one another.*

In summary, one main ingredient to the proof of the classification result Theorem 1.2.8 is the Atiyah-Singer Index Theorem in the form (3), which for Calabi-Yau two-manifolds Y implies $\chi(Y) = 12\chi(\mathcal{O}_Y)$. For Calabi-Yau three-manifolds Y, the corresponding formula does not suffice to fix the topological type of Y. Indeed, the problem of classifying all Calabi-Yau three-manifolds is wide open – the naive observation that there are more independent Chern classes to keep under control is precisely the source of the problem.

[c]It is not clear to me who introduced the name "K3 surface", and when. The standard explanation and first mention in writing that I am aware of is due to Weil, who in [Wei58], declares "Dans la seconde partie de mon rapport, il s'agit des variétés Kähleriennes dites K3, ainsi nommées en l'honneur de Kummer, Kodaira, Kähler et de la belle montagne K2 au Cachemire." The explanation of course relies on the Kähler property for all K3 surfaces, which at the time was only conjectural. It probably also relies on the fact that among mountaineers, K2 is often understood as the most challenging summit to the very day, which however had been vanquished only a few years before Weil completed his report [Wei58], namely in 1954.

1.3. *The Kummer construction*

In the previous section, I stated that all K3 surfaces are diffeomorphic to one and the same real four-manifold X [Kod64]. The choice of a complex structure and Kähler class on X, of course, greatly influences its geometric properties, for example the symmetries of the K3 surface. The KUMMER CONSTRUCTION, which shall be discussed in the present section, amounts to a special choice of complex structure and (degenerate) Kähler class, governed by the geometry of an underlying complex two-torus T_L:

Definition 1.3.1. Let $T_L = \mathbb{C}^2/L$ denote a complex two-torus, where $L \subset \mathbb{C}^2$ is a lattice of rank 4, and T_L carries the complex structure and Kähler metric induced from the standard complex structure and Euclidean metric on $\mathbb{C}^2$. This Calabi-Yau two-manifold enjoys a biholomorphic isometry $\kappa \in \mathrm{Aut}(T_L)$ of order 2 which is induced by $z \mapsto -z$ on $\mathbb{C}^2$.

The quotient $T/\mathbb{Z}_2$ of T_L by the group $\{\mathbb{I}, \kappa\} \cong \mathbb{Z}_2$ is called the SINGULAR KUMMER SURFACE with underlying torus T_L.

The singular Kummer surface $T_L/\mathbb{Z}_2$ is indeed singular: We choose $\epsilon > 0$ and denote by $B_\epsilon(0) \subset \mathbb{C}^3$ the open ball of radius ϵ with respect to the standard Euclidean metric on $\mathbb{C}^3$. Then the map

$$\mathfrak{u} : \mathbb{C}^2 \longrightarrow \mathbb{C}^3, \quad \mathfrak{u}(z^1, z^2) := \left((z^1)^2,\, (z^2)^2,\, z^1 z^2\right)$$

descends to an open neighborhood $U_\epsilon \subset T_L/\mathbb{Z}_2$ of 0 in the singular Kummer surface, where it is also denoted $\mathfrak{u}$, such that $\mathfrak{u}$ bijectively maps U_ϵ to

$$\mathfrak{u}(U_\epsilon) = \{u = (u^1,\, u^2,\, u^3) \in B_\epsilon(0) \mid u^1 u^2 = (u^3)^2\}.$$

The map $\mathfrak{u}$ is biholomorphic upon restriction to

$$U_\epsilon \setminus \{0\} \stackrel{\mathfrak{u}}{\longrightarrow} \mathfrak{u}(U_\epsilon) \setminus \{0\}.$$

The equation $u^1 u^2 = (u^3)^2$ of $\mathfrak{u}(U_\epsilon)$ in $B_\epsilon(0)$ immediately shows that $\mathfrak{u}(U_\epsilon)$ is a double cone with an isolated singularity at $u = 0$. Let us define the MINIMAL RESOLUTION of this singularity:

Definition 1.3.2. With notations as above, the point $0 \in U_\epsilon$, and equivalently its image $0 \in \mathfrak{u}(U_\epsilon)$, is called a SINGULARITY OF TYPE A_1. Let $\mathcal{K}_\epsilon := \mathfrak{u}(\overline{U_\epsilon})\setminus\{0\}$, and with homogeneous coordinates $v = (v^1 : v^2 : v^3) \in \mathbb{P}^2$,

$$W_\epsilon := \left\{(u, v) \in \mathcal{K}_\epsilon \times \mathbb{P}^2 \mid u^j v^k = u^k v^j \ \ \forall j,\, k \in \{1, 2, 3\}\right\}.$$

Moreover, let V_ϵ denote the interior of the closure $\overline{W}_\epsilon$ of $W_\epsilon \subset \mathbb{C}^3 \times \mathbb{P}^2$. Then

$$\sigma \colon V_\epsilon \longrightarrow \mathfrak{u}(U_\epsilon), \quad \sigma(u, v) := u$$

is called the BLOW-UP of the singularity $0 \in \mathfrak{u}(U_\epsilon)$ of type A_1, and $E := \sigma^{-1}(0)$ is its EXCEPTIONAL DIVISOR.

If $(u, v) \in W_\epsilon$, then $u \neq 0$, and the defining equations of W_ϵ imply $u^1u^2 = (u^3)^2$ and $(v^1 : v^2 : v^3) = (u^1 : u^2 : u^3)$. Denote this by $v = [u] \in \mathbb{P}^2$ and observe that $v^1v^2 = (v^3)^2$ follows. Then for $t \in \mathbb{R}$ close to $t = 0$, say $t \in (0, \delta)$ for an appropriate $\delta > 0$, the map $t \mapsto \gamma_u(t) := (tu, [u]) \in W_\epsilon$ yields a smooth curve with $\lim_{t\to 0} \gamma_u(t) = (0, [u]) \in E$. In fact,

$$E = \{(u, v) \in \mathbb{C}^3 \times \mathbb{P}^2 \mid u = 0,\ v^1v^2 = (v^3)^2\},$$

so the exceptional divisor E is biholomorphic to $\mathbb{P}^1$,

$$\mathbb{P}^1 \xrightarrow{\cong} E \quad \text{under} \quad (t^1 : t^2) \mapsto \left(0,\ ((t^1)^2 : (t^2)^2 : t^1t^2) \right).$$

The most important properties of the blow-up of a singularity of type A_1 are summarized in the following

Proposition 1.3.3. *The resolution $\sigma\colon V_\epsilon \longrightarrow \mathfrak{u}(U_\epsilon)$ of the singularity $0 \in U_\epsilon$ of type A_1 yields a smooth complex two-manifold V_ϵ. The restriction $\sigma_{|V_\epsilon \setminus E}\colon V_\epsilon \setminus E \longrightarrow \mathfrak{u}(U_\epsilon) \setminus \{0\}$ is biholomorphic, and the exceptional divisor $E \subset V_\epsilon$ is biholomorphic to $\mathbb{P}^1$. Moreover, V_ϵ has trivial canonical bundle.*

Proof (sketch):
The proof of Proposition 1.3.3 can be performed by a direct calculation.

For example, smoothness of V_ϵ close to the point $p_1 := (0, (1 : 0 : 0)) \in E$ can be checked in the chart $U_1 := \{(u, v) \in V_\epsilon \mid v^1 \neq 0\}$. Indeed, $(u^1, v^3) \mapsto ((u^1, u^1(v^3)^2, u^1v^3), (1 : (v^3)^2 : v^3))$ yields a smooth parametrization of U_1 near p_1. Analogously, one obtains smoothness everywhere, and holomorphicity of changes of coordinates is immediate, as are the claims about biholomorphicity of $\sigma_{|V_\epsilon \setminus E}$ and $E \cong \mathbb{P}^1$ by what was said above. The triviality of the canonical bundle of V_ϵ follows since $\eta_{(z^1,z^2)} = dz^1 \wedge dz^2$ on $\mathbb{C}^2$ descends to a section of the canonical bundle of V_ϵ over $V_\epsilon \setminus E$. On U_1 and with respect to coordinates (u^1, v^3) as above, we have

$$\eta_{(u^1,v^3)} = \frac{du^1 \wedge du^3}{2u^1} = \tfrac{1}{2} du^1 \wedge dv^3$$

as long as $u^1 \neq 0$. However, this implies that η has a holomorphic continuation to all of U_1, which never vanishes. By proceeding similarly for other coordinate neighborhoods, η can be continued to a nowhere vanishing global section of the canonical bundle of V_ϵ, which thus is trivial. □

The KUMMER CONSTRUCTION is now summarized in

Theorem 1.3.4 (Kummer construction). *The singular Kummer surface $T_L/\mathbb{Z}_2$ of Definition 1.3.1 has 16 singularities of type A_1, situated at $L/2L \subset T_L/\mathbb{Z}_2$. The complex surface X obtained by blowing up each of these singularities is a K3 surface.*

Proof (sketch):
One immediately checks that the singular points of $T_L/\mathbb{Z}_2$ are precisely the images of the fixed points of $\mathbb{Z}_2$ in T_L under the quotient $T_L \longrightarrow T_L/\mathbb{Z}_2$. If $[z]$ denotes the image of $z \in \mathbb{C}^2$ under the natural projection $\mathbb{C}^2 \longrightarrow T_L$, then $[z]$ is fixed under $\mathbb{Z}_2$ if and only if $[z] = [-z]$, that is, if and only if $2z \in L$. Since the lattice L has rank 4, we find that $L/2L \cong \mathbb{F}_2^4$ contains 16 points. That each of these singularities is of type A_1 is also immediate – one translates the coordinates induced from $\mathbb{C}^2$ on an open neighborhood of $y \in L/2L \subset T_L$ by $-y$ and finds that a punctured neighborhood of the origin is then biholomorphically mapped to $U_\epsilon \setminus \{0\}$ as in Definition 1.3.2.

Now X is obtained from the singular Kummer surface by blowing up all the singularities, that is, by replacing a neighborhood U_ϵ of each of the singular points by a copy of the blow-up V_ϵ. Since $U_\epsilon \setminus \{0\}$ is biholomorphic to $V_\epsilon \setminus \{0\}$ and according to Proposition 1.3.3, X is a smooth complex surface with trivial canonical bundle. Moreover, X is compact and connected by construction. I claim that $b_1(X) = 0$. Indeed, $\kappa \in \mathbb{Z}_2$ acts by multiplication by -1 on $H^1(T_L, \mathbb{C})$, such that none of the classes in $H^1(T_L, \mathbb{C})$ can descend to $H^1(X, \mathbb{C})$. Furthermore, according to Proposition 1.3.3, the exceptional divisor of each blow-up is biholomorphic to $\mathbb{P}^1$ with $h^{p,q}(\mathbb{P}^D) = \delta_{p,q}$, and thus the blow-ups only contribute to the cohomology of X in even degree.

In summary, X is a connected, compact complex surface with trivial canonical bundle and $b_1(X) = 0$. By Definition 1.2.7, I have shown that X is a K3 surface. □

Note that the standard Kähler metric on $\mathbb{C}^2$, which is the Euclidean one, induces a DEGENERATE Kähler metric on X. Indeed, the induced metric assigns vanishing volume to each irreducible component $E \cong \mathbb{P}^1$ of the exceptional divisor and thus does not correspond to a smooth, Riemannian metric on X.

Definition 1.3.5. The K3 surface X obtained from a complex torus $T_L = \mathbb{C}^2/L$ by the Kummer construction of Theorem 1.3.4 is called a KUMMER SURFACE. In other words, a Kummer surface carries the complex structure and (degenerate) Kähler structure induced from the respective structures inherited by T_L from the standard ones on $\mathbb{C}^2$.

By [Nik75], every K3 surface X which is obtained from some singular surface by the minimal resolution of 16 distinct singularities of type A_1 is biholomorphic to a Kummer surface. In other words, there exists a rank 4 lattice $L \subset \mathbb{C}^2$ such that X is biholomorphic to the minimal resolution of the singular Kummer surface $T_L/\mathbb{Z}_2$. There are many examples of singular quartic hypersurfaces in $\mathbb{P}^3$ with 16 singularities of type A_1, *e.g.* the one defined by

$$\left\{ z = (z^0 : \cdots : z^3) \in \mathbb{P}^3 \;\Big|\; \sum_{j=0}^{3} (z^j)^4 - 4\prod_{j=0}^{3} z^j \right\}.$$

Such singular Kummer surfaces were first studied by Kummer [Kum64]. The construction has become a classical one by now and is described in detail, for example, in [BHPvdV04].

2. Conformal field theory

As argued in the introduction, the Calabi-Yau geometry which the previous section was devoted to plays a crucial role in string theory. Indeed, a so-called NON-LINEAR SIGMA MODEL CONSTRUCTION is predicted to yield a SUPERCONFORMAL FIELD THEORY as the world-sheet theory for superstrings in a Calabi-Yau target geometry.

Unfortunately, for a generic Calabi-Yau manifold it is still hopeless to attempt a non-linear sigma model construction explicitly. The only exceptions, in general, are the complex tori, which carry flat metrics, such that non-linear sigma models are only little more complicated than free field theories. Furthermore, orbifold constructions, which can be viewed as generalizations of the Kummer construction of Section 1.3, are reasonably well understood.

Non-linear sigma models on K3 surfaces, which will be discussed in Section 3, furnish an intermediate case in that many K3 THEORIES are accessible through orbifold procedures, and in that the moduli space of these theories has been determined globally under a few additional assumptions [Sei88, Cec90, AM94, NW01]. That superconformal field theories allow an independent, mathematical approach is a crucial ingredient to that result. The current section therefore gives an overview on aspects of conformal field theory related to the particular Calabi-Yau geometries that the previous section was focused on, namely the non-linear sigma models on complex tori (Section 2.1) and their $\mathbb{Z}_2$-orbifolds (Section 2.2). Due to restrictions of space and time, I content myself with giving an overview and pointing to some relevant literature.

2.1. *Toroidal superconformal field theories*

In this treatise, I do not attempt to give a definition of conformal field theory (CFT), though hopefully these notes are useful also for the CFT-novice, in that they discuss some of the basic ingredients to CFT. I have summarized my own view on a definition of CFT elsewhere, see *e.g.* [Wen10]. To make best use of the restricted amount of space, I refer the reader to the recent review [Wen15] as a companion paper to the present lecture notes. Indeed, while in [Wen15] the main emphasis lies on CFT aspects, here I focus on the geometric point of view. In particular, for the notions of holomorphic fields, operator product expansions (OPEs) and normal ordered products, following [LW78, FK81, Bor86, Kac98, FBZ04], see [Wen15, Sect. 2.1], and for a summary of a definition of conformal and superconformal field theories, see [Wen15, Sect. 2.2]. As in this reference, in what follows, I solely discuss TWO-DIMENSIONAL EUCLIDEAN UNITARY CONFORMAL FIELD THEORIES. All superconformal field theories (SCFTs) are assumed to be NON-CHIRAL and to enjoy SPACE-TIME SUPERSYMMETRY as well as $N = (2,2)$ WORLD-SHEET SUPERSYMMETRY.

An important ingredient to SCFT is the representation theory of the (super-) Virasoro algebra. Indeed, the SPACE OF STATES of any of our SCFTs is a $\mathbb{Z}_2 \times \mathbb{Z}_2$-graded complex vector space

$$\begin{aligned} \mathbb{H} \;=\; \mathbb{H}^{NS} \oplus \mathbb{H}^{R}, \qquad \text{where} \qquad & \mathbb{H}^{NS} = \mathbb{H}_b^{NS} \oplus \mathbb{H}_f^{NS},\ \mathbb{H}^{R} = \mathbb{H}_b^{R} \oplus \mathbb{H}_f^{R}, \\ & \mathbb{H}_b := \mathbb{H}_b^{NS} \oplus \mathbb{H}_b^{R}, \quad \mathbb{H}_f := \mathbb{H}_f^{NS} \oplus \mathbb{H}_f^{R}. \end{aligned}$$

The subspaces $\mathbb{H}^{NS}$ and $\mathbb{H}^{R}$ are called NEVEU-SCHWARZ (NS) and RAMOND (R) SECTORS, while $\mathbb{H}_b$ and $\mathbb{H}_f$ are the subspaces of BOSONIC and FERMIONIC states, respectively. The space $\mathbb{H}$ carries a representation of two copies of a super-Virasoro algebra at central charges $(c, \overline{c})$, where for all examples discussed here we have $c = \overline{c}$. As usual, the zero-modes of left- and right-moving Virasoro fields and $U(1)$-currents are denoted L_0, $\overline{L}_0$ and J_0, $\overline{J}_0$, respectively[d]. Then space-time supersymmetry implies that $(-1)^F := e^{\pi i (J_0 - \overline{J}_0)}$ acts as identity operator $\mathbb{I}$ on $\mathbb{H}_b$ and as $-\mathbb{I}$ on $\mathbb{H}_f$.

---

[d]As in [Wen15, Sect. 2.2], I assume a COMPACTNESS CONDITION for our SCFTs, namely that the operators L_0, $\overline{L}_0$, J_0, $\overline{J}_0$ are simultaneously diagonalizable, and that simultaneous eigenspaces of L_0 and $\overline{L}_0$ are finite dimensional.

Every SCFT possesses a modular invariant PARTITION FUNCTION $Z(\tau, z)$ in two complex variables $\tau,\ z \in \mathbb{C}$ with $\Im(\tau) > 0$,

$$\text{with } q := e^{2\pi i\tau},\ y := e^{2\pi iz}: \quad Z(\tau, z) = \mathrm{tr}_{\mathbb{H}_b}\left(q^{L_0 - c/24} y^{J_0} \overline{q}^{\overline{L}_0 - \overline{c}/24} \overline{y}^{\overline{J}_0}\right).$$

Space-time supersymmetry is tantamount to a decomposition of the partition function into four sectors,

$$Z(\tau, z) = \tfrac{1}{2}\left(Z_{NS}(\tau, z) + Z_{\widetilde{NS}}(\tau, z) + Z_R(\tau, z) + Z_{\widetilde{R}}(\tau, z)\right),$$

for $S \in \{NS,\ R\}$:

$$\begin{aligned} Z_S(\tau, z) &:= \mathrm{tr}_{\mathbb{H}^S}\left(q^{L_0 - c/24} y^{J_0} \overline{q}^{\overline{L}_0 - \overline{c}/24} \overline{y}^{\overline{J}_0}\right), \\ &\quad \text{with } Z_R(\tau, z) = (q\overline{q})^{c/24}(y\overline{y})^{c/6} Z_{NS}(\tau, z + \tfrac{\tau}{2}), \end{aligned} \tag{4}$$

$$Z_{\widetilde{S}}(\tau, z) := \mathrm{tr}_{\mathbb{H}^S}\left((-1)^{J_0 - \overline{J}_0} q^{L_0 - c/24} y^{J_0} \overline{q}^{\overline{L}_0 - \overline{c}/24} \overline{y}^{\overline{J}_0}\right) = Z_S(\tau, z + \tfrac{1}{2}).$$

Equation (4) reflects an isomorphism $\mathbb{H}^{NS} \cong \mathbb{H}^R$ of representations of the $N = (2, 2)$ superconformal algebra, known as SPECTRAL FLOW, which holds due to our assumption of space-time supersymmetry.

The simplest example of such an SCFT is a TOROIDAL $N = (2, 2)$ SUPERCONFORMAL FIELD THEORY, see [Wen15, Def. 5] for a more detailed account. At central charges $(c, \overline{c}) = (3D, 3D)$, $D \in \mathbb{N}$, such a theory is in particular characterized by the fact that the chiral algebras on the left and on the right contain a $\mathfrak{u}(1)^{2D}$-current algebra each, along with D left-moving and D right-moving Dirac fermions, the superpartners of the $\mathfrak{u}(1)$-currents. The corresponding fields $j^k(z)$, $k \in \{1, \ldots, 2D\}$ and $\psi_l^\pm(z)$, $l \in \{1, \ldots, D\}$, whose only non-vanishing OPEs are

$$\begin{aligned} \forall k, l \in \{1, \ldots, 2D\}: \quad j^k(z) j^l(w) &\sim \frac{\delta^{k,l}}{(z - w)^2}, \\ \forall k, l \in \{1, \ldots, D\}: \quad \psi_k^+(z) \psi_l^-(w) &\sim \frac{\delta^{k,l}}{(z - w)}, \end{aligned} \tag{5}$$

along with their right-moving analogues are called the BASIC FIELDS. They are examples of so-called FREE FIELDS (see *e.g.* [Kac98]).

The space of states $\mathbb{H}$ of a toroidal SCFT arises as FOCK SPACE REPRESENTATION of the modes of these basic fields, that is, of the super-Lie algebra generated by $\mathbb{I}$ and a_n^k, $(\psi_l^\pm)_m \in \mathrm{End}(\mathbb{H}^S)$, $k \in \{1, \ldots, 2D\}$, $l \in \{1, \ldots, D\}$, $S \in \{NS, R\}$, where $n \in \mathbb{Z}$, and on the NS-sector $\mathbb{H}^{NS}$,

$m \in \mathbb{Z} + \frac{1}{2}$, while on the R-sector $\mathbb{H}^R$, $m \in \mathbb{Z}$,

$$[a^j_m, a^k_n] := a^j_m a^k_n - a^k_n a^j_m \quad = m\delta^{j,k}\delta_{m+n,0}\mathbb{I},$$
$$\{\psi^j_m, \psi^k_n\} := \psi^j_m \psi^k_n + \psi^k_n \psi^j_m = \delta^{j,k}\delta_{m+n,0}\mathbb{I} \quad \forall j,\, k,\, m,\, n,$$

and analogously for the right-movers. All other (super-)commutators between the a^k_n, $(\psi^\pm_l)_m$ and their right-moving analogues vanish. One then has $\mathbb{H}^{NS} = \oplus_{\gamma\in\Gamma}\mathbb{H}_\gamma$ with Γ the so-called CHARGE LATTICE, and $\mathbb{H}_\gamma$ a lowest weight representation of the above super-Lie algebra of modes, with lowest weight vector $|\gamma\rangle \in \mathbb{H}_\gamma$ such that

$$\forall j \in \{1,\dots,2D\}: \quad a^j_0|\gamma\rangle = \gamma^j|\gamma\rangle,$$
$$\forall k \in \{1,\dots,D\}: \; a^j_m|\gamma\rangle = \psi^k_{m+\frac{1}{2}}|\gamma\rangle = 0 \;\; \forall m \in \mathbb{Z} \text{ with } m < 0,$$

and analogously for the right-moving modes. Here, $\Gamma \subset \mathbb{R}^{2D,2D}$, by which I mean $\mathbb{R}^{2D,2D} = \mathbb{R}^{2D} \oplus \mathbb{R}^{2D}$ and

$$\forall\gamma \in \Gamma: \quad \gamma = (\gamma_L, \gamma_R) \text{ with } \gamma_L, \gamma_R \in \mathbb{R}^{2D}, \;\; \langle\gamma,\gamma\rangle = \gamma_L\cdot\gamma_L - \gamma_R\cdot\gamma_R$$

with the standard Euclidean scalar product $\cdot$ on $\mathbb{R}^{2D}$ and $\gamma_L = (\gamma^1,\dots,\gamma^{2D})^T$, $\gamma_R = (\gamma^{2D+1},\dots,\gamma^{4D})^T$. The representation $\mathbb{H}^R$ is obtained from $\mathbb{H}^{NS}$ by spectral flow, as was mentioned in the discussion of equation (4). A counting argument then shows

$$\mathrm{tr}_{\mathbb{H}_\gamma}\left(q^{L_0-c/24}y^{J_0}\overline{q}^{\overline{L}_0-\overline{c}/24}\overline{y}^{\overline{J}_0}\right) = \frac{q^{\frac{1}{2}\gamma_L\cdot\gamma_L}\overline{q}^{\frac{1}{2}\gamma_R\cdot\gamma_R}}{|\eta(\tau)|^{4D}} \cdot \left|\frac{\vartheta_3(\tau,z)}{\eta(\tau)}\right|^{2D}$$

with the Dedekind eta function $\eta(\tau)$ and the standard Jacobi theta functions $\vartheta_k(\tau,z)$, $k \in \{1,\dots,4\}$. Here, the charge lattice Γ is an even self-dual lattice of signature $(2D,2D)$, given in terms of an embedding into $\mathbb{R}^{2D,2D}$, which is specified by the decomposition $\gamma = (\gamma_L,\gamma_R)$ for every $\gamma \in \Gamma$ as above. Generalizing Kac's holomorphic lattice algebras [Kac98] to the non-holomorphic case, in [KO03] the lattice vertex operator algebra corresponding to such an indefinite lattice is described. One has:

Proposition 2.1.1 ([CENT85, Nar86]). *A toroidal superconformal field theory is uniquely characterized by its charge lattice $\Gamma \subset \mathbb{R}^{2D,2D}$. For a theory with charge lattice Γ, setting*

$$Z_\Gamma(\tau) := \sum_{\gamma=(\gamma_L,\gamma_R)\in\Gamma} \frac{q^{\frac{1}{2}\gamma_L\cdot\gamma_L}\overline{q}^{\frac{1}{2}\gamma_R\cdot\gamma_R}}{|\eta(\tau)|^{4D}},$$

the four sectors of the partition function $Z(\tau,z)$ are

$$Z_{NS}(\tau,z)=Z_\Gamma(\tau)\cdot\left|\tfrac{\vartheta_3(\tau,z)}{\eta(\tau)}\right|^{2D},\; Z_{\widetilde{NS}}(\tau,z)=Z_\Gamma(\tau)\cdot\left|\tfrac{\vartheta_4(\tau,z)}{\eta(\tau)}\right|^{2D},$$
$$Z_{R}(\tau,z)=Z_\Gamma(\tau)\cdot\left|\tfrac{\vartheta_2(\tau,z)}{\eta(\tau)}\right|^{2D},\quad Z_{\widetilde{R}}(\tau,z)=Z_\Gamma(\tau)\cdot\left|\tfrac{\vartheta_1(\tau,z)}{\eta(\tau)}\right|^{2D},$$

hence

$$Z(\tau,z)=Z_\Gamma(\tau)\cdot Z_f(\tau,z)\quad \textit{with}\quad Z_f(\tau,z):=\tfrac{1}{2}\sum_{k=0}^{4}\left|\tfrac{\vartheta_k(\tau,z)}{\eta(\tau)}\right|^{2D}.$$

From Proposition 2.1.1 one also reads the moduli space of such theories [CENT85, Nar86], see [Wen15, Thm. 1].

As mentioned above, the complex tori are the unique Calabi-Yau manifolds for which a non-linear sigma model construction can be performed directly and explicitly; in fact, for a torus $T_L=\mathbb{C}^D/L$ with $L\subset\mathbb{C}^D$ a lattice of rank $2D$, the non-linear sigma model construction yields a toroidal SCFT as discussed above. Geometrically, each $\mathfrak{u}(1)$-current $j^k(z)$, $k\in\{1,\ldots,2D\}$, as in (5), arises as the field corresponding to a parallel tangent vector field, given by a standard Euclidean coordinate vector field in $\mathbb{R}^{2D}\cong\mathbb{C}^D$. The real and imaginary parts of its fermionic superpartners are the fields corresponding to the dual cotangent vector fields. The resulting non-linear sigma model also depends on the choice of a B-FIELD, which in this setting can be described by a constant, skew-symmetric endomorphism of $\mathbb{R}^{2D}$. Identifying $\mathbb{R}^{2D}\cong\mathbb{C}^D$ with its dual by means of the standard Euclidean scalar product, the DUAL LATTICE of L is given by

$$L^*:=\left\{\alpha\in\mathbb{R}^{2D}\mid \alpha\cdot\lambda\in\mathbb{Z}\ \forall\lambda\in L\right\},$$

and the charge lattice of the resulting toroidal SCFT is

$$\Gamma=\left\{\tfrac{1}{\sqrt{2}}(\mu-B\lambda+\lambda,\mu-B\lambda-\lambda)\mid\lambda\in L,\ \mu\in L^*\right\}.$$

Vice versa, every toroidal SCFT allows a GEOMETRIC INTERPRETATION in terms of a non-linear sigma model, see *e.g.* [NW01]:

Definition 2.1.2. Consider a toroidal $N=(2,2)$ SCFT at central charges $(c,\overline{c})=(3D,3D)$ with charge lattice $\Gamma\subset\mathbb{R}^{2D,2D}$. A GEOMETRIC INTERPRETATION of the theory is any choice of complementary $2D$-dimensional subspaces Y, $Y^0\subset\mathbb{R}^{2D,2D}$, such that $\mathbb{R}^{2D,2D}=Y\oplus Y^0$, both Y and Y^0 are null, and both $\Gamma\cap Y$ and $\Gamma\cap Y^0$ are rank $2D$ lattices.

The terminology deserves explanation:

Given a geometric interpretation $\mathbb{R}^{2D,2D} = Y \oplus Y^0$ of a theory with charge lattice $\Gamma \subset \mathbb{R}^{2D,2D}$, without loss of generality we can set

$$\Gamma \cap Y = \left\{ \tfrac{1}{\sqrt{2}}(\mu, \mu) \mid \mu \in L^* \right\}, \quad \Gamma \cap Y^0 = \left\{ \tfrac{1}{\sqrt{2}}(\lambda - B\lambda, -\lambda - B\lambda) \mid \lambda \in L \right\}$$

for some rank $2D$ lattice $L \subset \mathbb{R}^{2D} \cong \mathbb{C}^D$, $L^* \subset \mathbb{R}^{2D}$ its dual, and B some skew-symmetric linear endomorphism of $\mathbb{R}^{2D}$. By Proposition 2.1.1 and the explanations preceding Definition 2.1.2, our theory then agrees with a non-linear sigma model on $T_L = \mathbb{C}^D/L$ with B-field B.

Given such a geometric interpretation of a toroidal SCFT, it is now also clear how certain GEOMETRIC SYMMETRIES of the torus T_L may induce symmetries of a toroidal SCFT on T_L: If a symmetry of T_L is given in terms of $A \in \mathrm{End}(\mathbb{C}^D)$, then A has to act as lattice automorphism of L. Hence $A \in O(2D)$ and A also acts as lattice automorphism of L^*. If in addition, $AB = BA$, then A acts as lattice automorphism on Γ which respects the embedding $\Gamma \subset \mathbb{R}^{2D,2D}$. One checks that the induced action on the space of states $\mathbb{H}$ yields a symmetry of the toroidal SCFT.

2.2. $\mathbb{Z}_2$-*orbifold conformal field theories*

In Section 1.3, I presented the classical Kummer construction, which yields a K3 surface from the much simpler complex two-torus $T_L = \mathbb{C}^2/L$ by $\mathbb{Z}_2$-ORBIFOLDING. The construction begins with the projection to $T_L/\mathbb{Z}_2$, where $\mathbb{Z}_2$ is generated by the symmetry κ of T_L which is induced from $-\mathbb{I}$ on $\mathbb{C}^2$.

By the discussion at the end of the previous section, κ also acts as symmetry on the non-linear sigma model constructed on T_L with an arbitrary B-field B. The present section is devoted to the lift of this $\mathbb{Z}_2$-orbifold procedure to the level of CFT, to construct a new superconformal field theory from the much simpler toroidal one. This string theory procedure was inspired by the techniques that were originally developed in the context of MONSTROUS MOONSHINE [FLM84] for holomorphic vertex algebras, and it can be carried out in much greater generality [DHVW85, DHVW86], that is, in arbitrary dimensions and with more general orbifolding-groups. Indeed, already the Kummer construction can be generalized to obtain K3 surfaces by orbifolding an underlying complex two-torus $T_L = \mathbb{C}^2/L$ with the appropriate symmetry by groups $\mathbb{Z}_3$, $\mathbb{Z}_4$ or $\mathbb{Z}_6$, or even by certain examples of non-Abelian groups, see [Wen01] and references therein. In the following I provide an overview of the basic ideas behind the orbifold procedure in CFT. For the sake of brevity, I content myself with the discussion of $\mathbb{Z}_2$-orbifolds as an instructive example.

If not stated otherwise, in the following I consider a toroidal superconformal field theory at central charges $(c, \overline{c}) = (3D, 3D)$ with geometric interpretation on a complex torus $T_L = \mathbb{C}^D/L$ and with some B-field B as discussed in Section 2.1. As before, the space of states of the SCFT is denoted by $\mathbb{H} = \mathbb{H}_b \oplus \mathbb{H}_f$, and κ denotes the symmetry of order 2 of this theory which is induced by $\mathbb{C}^D \longrightarrow \mathbb{C}^D$, $z \mapsto -z$. It acts by multiplication by -1 on the basic fields (5) and on the charge lattice Γ, and therefore as linear involution on the bosonic and fermionic spaces of states $\mathbb{H}_b$ and $\mathbb{H}_f$, respectively.

Let us investigate the κ-invariant subsector $\mathbb{H}_b^{\mathbb{Z}_2}$ of the bosonic space of states. With notations as in Proposition 2.1.1, a counting argument shows

$$\begin{aligned}\mathrm{tr}_{\mathbb{H}_b^{\mathbb{Z}_2}}\left(q^{L_0-c/24}y^{J_0}\overline{q}^{\overline{L}_0-\overline{c}/24}\overline{y}^{\overline{J}_0}\right) &= \tfrac{1}{2}\left(Z_\Gamma(\tau)\cdot Z_f(\tau,z) + Z_{-\mathbb{I}}(\tau,z)\right)\\ \text{with}\quad Z_{-\mathbb{I}}(\tau,z) &= \left|\frac{2\eta(\tau)}{\vartheta_2(\tau,0)}\right|^{2D}\cdot Z_f(\tau,z).\end{aligned}$$

This expression is not modular invariant, since $Z_{-\mathbb{I}}(\tau,z)$ isn't, while $Z_\Gamma(\tau)$ and $Z_f(\tau,z)$ are. Hence $\mathbb{H}_b^{\mathbb{Z}_2}$ is not the space of bosonic states of a full-fledged superconformal field theory. However, one can construct a TWISTED SECTOR $\mathbb{H}_b^{tw}$ such that $\mathbb{H}_b^{\mathbb{Z}_2} \oplus \mathbb{H}_b^{tw}$ has this property. To obtain information about such a would-be twisted sector, we observe that $Z_{-\mathbb{I}}(\tau,z)$ has a natural modular invariant completion

$$Z_{-\mathbb{I}}(\tau,z) + Z_{-\mathbb{I}}\left(-\tfrac{1}{\tau},\tfrac{z}{\tau}\right) + Z_{-\mathbb{I}}\left(-\tfrac{1}{\tau+1},\tfrac{z}{\tau+1}\right).$$

On the basis of a path integral interpretation of each summand, see *e.g.* [Gin88, §8.3], modular invariance of this expression is expected; it is shown to hold by a direct calculation. The key idea behind the orbifolding procedure is to interpret the three summands of the above expression as follows:

$$\begin{aligned}Z_{-\mathbb{I}}(\tau,z) &= \mathrm{tr}_{\mathbb{H}_b}\left(\kappa q^{L_0-c/24}y^{J_0}\overline{q}^{\overline{L}_0-\overline{c}/24}\overline{y}^{\overline{J}_0}\right),\\ Z_{-\mathbb{I}}\left(-\tfrac{1}{\tau},\tfrac{z}{\tau}\right) &= \mathrm{tr}_{\widetilde{\mathbb{H}}_b^{tw}}\left(q^{L_0-c/24}y^{J_0}\overline{q}^{\overline{L}_0-\overline{c}/24}\overline{y}^{\overline{J}_0}\right),\\ Z_{-\mathbb{I}}\left(-\tfrac{1}{\tau+1},\tfrac{z}{\tau+1}\right) &= \mathrm{tr}_{\widetilde{\mathbb{H}}_b^{tw}}\left(\kappa q^{L_0-c/24}y^{J_0}\overline{q}^{\overline{L}_0-\overline{c}/24}\overline{y}^{\overline{J}_0}\right).\end{aligned}$$

Here, $\widetilde{\mathbb{H}}_b^{tw}$ needs to be constructed as some representation of the super-Virasoro algebras that enjoys a symmetry κ of order 2 such that $\mathbb{H}_b^{tw}$ is the κ-invariant subspace of $\widetilde{\mathbb{H}}_b^{tw}$. Indeed, in the present example an appropriate space $\widetilde{\mathbb{H}}_b^{tw}$ can be found, that is, such that $\mathbb{H}_b^{\mathbb{Z}_2} \oplus \mathbb{H}_b^{tw}$ can be interpreted as bosonic space of states of a full-fledged superconformal field theory. The details of the construction are quite technical but have been worked out, see [FFRS10] and references therein[e].

Proposition 2.2.1. *Consider a toroidal $N = (2,2)$ superconformal field theory at central charges $(c,\bar{c}) = (3D, 3D)$ with charge lattice $\Gamma \subset \mathbb{R}^{2D,2D}$ and space of states $\mathbb{H}$. Let κ denote the symmetry of order 2 which acts by multiplication by -1 on Γ and on the basic fields (5).*

Then there exists a $\mathbb{Z}_2$-ORBIFOLD CONFORMAL FIELD THEORY of this toroidal theory, whose space of states is $\mathbb{H}^{\mathbb{Z}_2} \oplus \mathbb{H}^{tw}$, where $\mathbb{H}^{\mathbb{Z}_2} \subset \mathbb{H}$ denotes the κ-invariant subspace. The four sectors of the partition function $Z^{orb}(\tau, z)$ of this theory are obtained from its $\widetilde{R}$-sector by application of the spectral flow formulas (4). With notations as in Proposition 2.1.1 we have

$$Z^{orb}_{\widetilde{R}}(\tau, z) = \tfrac{1}{2} \cdot \left(Z_\Gamma(\tau) \cdot \left| \frac{\vartheta_1(\tau, z)}{\eta(\tau)} \right|^{2D} + \left| \frac{2\vartheta_2(\tau, z)}{\vartheta_2(\tau, 0)} \right|^{2D} + \left| \frac{2\vartheta_4(\tau, z)}{\vartheta_4(\tau, 0)} \right|^{2D} + \left| \frac{2\vartheta_3(\tau, z)}{\vartheta_3(\tau, 0)} \right|^{2D} \right).$$

If the $\mathbb{Z}_2$-orbifold CFT of Proposition 2.2.1 is obtained from a toroidal SCFT with geometric interpretation on $T_L = \mathbb{C}^D/L$ with some B-field B, then it is believed that the orbifold theory can be obtained by a non-linear sigma model construction from the orbifold limit of the Calabi-Yau D-manifold obtained from blowing up all singularities in $T_L/\mathbb{Z}_2$. To support this belief, one checks from the partition function $Z^{orb}(\tau, z)$ given in Proposition 2.2.1 that the dimension of the space of TWISTED GROUND STATES in $\mathbb{H}^{tw}$ is 2^{2D}. This is precisely the number of singular points in $T_L/\mathbb{Z}_2$. For more complicated orbifolding-groups, each singular point corresponds to a higher dimensional subspace of twisted ground states, depending on the order of the stabilizer group. This supports the idea that, at least naively[f], the introduction of twisted sectors corresponds to the resolution of the singular points in the non-linear sigma model on $T_L/\mathbb{Z}_2$.

If $D = 2$, then by Theorem 1.3.4 the resolution of all singularities in $T_L/\mathbb{Z}_2$ yields a K3 surface. We should hence expect the $\mathbb{Z}_2$-orbifold conformal field theory of Proposition 2.2.1 to allow a non-linear sigma model

[e] Also see Yi-Zhi Huang's blog [Hua14] and the references therein to appreciate the mathematical problems that the introduction of a SCFT on $\mathbb{H}_b^{\mathbb{Z}_2} \oplus \mathbb{H}_b^{tw}$ poses.

[f] Note however, at least in the $D = 2$-dimensional case, a subtlety concerning the B-field on the orbifold [NW01], which turns out to be non-zero on every irreducible component of the exceptional divisor in the blow-up.

interpretation on a Kummer surface in this case. In [NW01], a map between the respective moduli spaces of conformal field theories is constructed which is compatible with this expectation. Earlier evidence in favor of the prediction arises from a calculation of the so-called CONFORMAL FIELD THEORETIC ELLIPTIC GENUS [EOTY89], according to

Definition 2.2.2. Consider the $\widetilde{R}$-sector $Z_{\widetilde{R}}$ of the partition function of an $N = (2,2)$-superconformal field theory, viewed as a function of four complex variables $Z_{\widetilde{R}} = Z_{\widetilde{R}}(\tau, z; \overline{\tau}, \overline{z})$. The CONFORMAL FIELD THEORETIC ELLIPTIC GENUS of the theory is

$$\mathcal{E}(\tau, z) := Z_{\widetilde{R}}(\tau, z; \overline{\tau}, \overline{z} = 0).$$

For a CFT that arises as non-linear sigma model on some Calabi-Yau D-manifold Y, it is expected that the conformal field theoretic elliptic genus agrees with the GEOMETRIC ELLIPTIC GENUS of Y. The latter can be defined as the holomorphic Euler characteristic of a formal vector bundle $\mathbb{E}_{q,-y}$ on Y, more precisely a formal power series in $q = e^{2\pi i\tau}$ and $y = e^{2\pi iz}$ whose coefficients are holomorphic vector bundles on Y. Using the notations $\Lambda_x E$, $S_x E$ for any vector bundle $E \to Y$ and a formal variable x with

$$\Lambda_x E := \bigoplus_{p=0}^{\infty} x^p \Lambda^p E, \quad S_x E := \bigoplus_{p=0}^{\infty} x^p S^p E,$$

where $\Lambda^p E$, $S^p E$ denote the pth exterior and symmetric powers of E, and with $T := T^{1,0}Y$ the holomorphic tangent bundle of Y,

$$\mathbb{E}_{q,-y} = y^{-D/2} \bigotimes_{n=1}^{\infty} \left(\Lambda_{-yq^{n-1}} T^* \otimes \Lambda_{-y^{-1}q^n} T \otimes S_{q^n} T^* \otimes S_{q^n} T \right). \quad (6)$$

The holomorphic Euler characteristic of Definition 1.2.1 is naturally extended to formal power series with coefficients in holomorphic vector bundles on Y,

$$\chi\left(\sum_{n=0}^{\infty} x^n F_n \right) := \sum_{n=0}^{\infty} x^n \chi(F_n).$$

We then have:

Definition 2.2.3. For a Calabi-Yau D-manifold Y with holomorphic tangent bundle $T := T^{1,0}Y$, the GEOMETRIC ELLIPTIC GENUS $\mathcal{E}_Y(\tau, z)$ is the holomorphic Euler characteristic of the bundle $\mathbb{E}_{q,-y}$ introduced in (6),

$$\mathcal{E}_Y(\tau, z) := \chi(\mathbb{E}_{q,-y}).$$

Following [Wit87], the geometric elliptic genus can be interpreted as a regularized version of a $U(1)$-equivariant index of a Dirac operator on the loop space of Y. It is a topological invariant of Y with many beautiful mathematical properties. In particular, its modular transformation properties agree with those of the conformal field theoretic elliptic genus of Definition 2.2.2 for a theory at central charges $c = \overline{c} = 3D$, $D \in \mathbb{N}$: Both are WEAK JACOBI FORMS of weight 0 and index $\frac{D}{2}$ (with a character, if D is odd) [Hir88, Wit88, EOTY89, Kri90, DY93, Wit94, BL00]. See Section 3 and [Wen15, Sect. 2.4] for a more detailed discussion of the two versions (geometric vs. conformal field theoretic) of the elliptic genus and their expected relationship, and for further references.

From Propositions 2.1.1 and 2.2.1 one immediately finds the conformal field theoretic elliptic genera of the non-linear sigma model on a complex two-torus $T_L = \mathbb{C}^2/L$ and of its $\mathbb{Z}_2$-orbifold CFT:

$$\begin{aligned} \mathcal{E}_{T_L}(\tau, z) &= 0, \\ \mathcal{E}_{K3}(\tau, z) &= 8\left(\frac{\vartheta_2(\tau,z)}{\vartheta_2(\tau,0)}\right)^2 + 8\left(\frac{\vartheta_3(\tau,z)}{\vartheta_3(\tau,0)}\right)^2 + 8\left(\frac{\vartheta_4(\tau,z)}{\vartheta_4(\tau,0)}\right)^2. \end{aligned} \tag{7}$$

These functions agree with the known geometric elliptic genera of T_L and a Kummer surface, respectively, and thereby of all complex two-tori resp. K3 surfaces [EOTY89]:

Proposition 2.2.4. *Consider a toroidal $N = (2,2)$ superconformal field theory at central charges $c = \overline{c} = 6$ with geometric interpretation on a complex two-torus $T_L = \mathbb{C}^2/L$ with some B-field B. Then its conformal field theoretic elliptic genus agrees with the geometric elliptic genus of T_L, and thereby of every complex two-torus. The conformal field theoretic elliptic genus of the $\mathbb{Z}_2$-orbifold CFT of this toroidal theory agrees with the geometric elliptic genus of any (and thereby every) K3 surface.*

3. Outlook: Towards superconformal field theory on K3 and beyond

The discussion in the previous section foreshadows a rather subtle relation between geometry and conformal field theory. This final section of these lecture notes gives a rough overview and outlook on attempts to get a better understanding of this relation. Special attention is paid to K3 surfaces and conformal field theories associated to them, because as we have seen, the K3 surfaces furnish the simplest examples where non-linear sigma model constructions are not fully understood.

From a mathematical point of view, an abstract approach to CFT, based in representation theory, is desirable. However, then the route back to geometry is not immediate[g]. As mentioned before, direct constructions of superconformal field theories from Calabi-Yau data are sparse – they are essentially restricted to toroidal SCFTs and their orbifolds.

However, if the existence of a non-linear sigma model on a Calabi-Yau D-manifold Y is assumed, then a number of additional properties are known for the resulting SCFT. First, it enjoys $N = (2, 2)$ (world-sheet) supersymmetry at central charges $c = \overline{c} = 3D$ as well as space-time supersymmetry. Second, all eigenvalues of the linear operators J_0 and $\overline{J}_0$ on the space of states $\mathbb{H} = \mathbb{H}^{NS} \oplus \mathbb{H}^R$ are integral in the Neveu-Schwarz sector[h] $\mathbb{H}^{NS}$.

At small central charges, more precisely at $c = \overline{c} = 3$ and $c = \overline{c} = 6$, these conditions severely restrict the types of theories that can occur. One first observes that the conformal field theoretic elliptic genus of Definition 2.2.2 either vanishes, or it agrees with the geometric elliptic genus $\mathcal{E}_{K3}(\tau, z)$ of K3 surfaces[i] as in (7): For $c = \overline{c} = 3$ this follows immediately, since the space of weak Jacobi forms of weight 0 and index $\frac{1}{2}$ is trivial, for $c = \overline{c} = 6$ see [Wen15, Prop. 2(1)]. This reproduces the classification of Calabi-Yau D-manifolds with $D \leq 2$ according to Theorems 1.2.2 and 1.2.8 on the level of the elliptic genus. If the conformal field theoretic elliptic genus vanishes for a theory at $c = \overline{c} = 3$ or $c = \overline{c} = 6$ which obeys all the above assumptions on supersymmetry and on the $J_0, \overline{J}_0$-eigenvalues, then one actually finds that the underlying theory is toroidal (see [Wen15, Prop. 2(2)] for the case $c = \overline{c} = 6$ – the case $c = \overline{c} = 3$ is in fact simpler and can be treated analogously). This motivates the following definition, see also [Wen15, Def. 8]:

Definition 3.0.1. A superconformal field theory is called a K3 THEORY, if the following conditions hold: The CFT is an $N = (2, 2)$ superconformal field theory at central charges $c = \overline{c} = 6$ with space-time supersymmetry, all the eigenvalues of J_0 and of $\overline{J}_0$ are integral, and the conformal field theoretic elliptic genus of the theory agrees with the geometric elliptic genus of K3 surfaces $\mathcal{E}_{\mathrm{K3}}(\tau, z)$ as in (7).

---

[g] Specifically concerning the discussion of the "geometry of CFTs", see also Yi-Zhi Huang's blog [Hua14] and the references therein.

[h] Since space-time supersymmetry implies an equivalence of representations of the $N = (2, 2)$ superconformal algebra $\mathbb{H}^{NS} \cong \mathbb{H}^R$ under spectral flow, this integrality condition implies that $J_0 - \overline{J}_0$ has only integral eigenvalues on all of $\mathbb{H}$, and that the eigenvalues of J_0 and $\overline{J}_0$ in the Ramond sector $\mathbb{H}^R$ lie in $\frac{D}{2} + \mathbb{Z}$.

[i] This can only happen if $c = \overline{c} = 6$.

By Proposition 2.2.4, all $\mathbb{Z}_2$-orbifold conformal field theories obtained from toroidal superconformal ones at $c = \overline{c} = 6$ are examples of K3 theories.

One needs to appreciate that Definition 3.0.1 does not make use of non-linear sigma model assumptions. This also means that one needs to carefully distinguish between K3 THEORIES and CONFORMAL FIELD THEORIES ON K3. It is wide open, yet interesting and important, whether all K3 theories are theories on K3 in the sense that they can be constructed as non-linear sigma models. At least to our knowledge, no counter example is known.

There are various geometric properties of K3 surfaces that can be recovered from abstractly defined K3 theories. Let us only mention that under few additional assumptions, listed in [Sei88, Cec90, AM94, NW01], each connected component of the moduli space of K3 theories agrees with the moduli space expected for non-linear sigma models on K3. From this identification one obtains the notion of GEOMETRIC INTERPRETATION for K3 theories in the spirit of Definition 2.1.2, see [AM94]. For the so-called GEPNER MODELS, this allows to interpret certain symmetries of these theories in terms of geometric symmetries of an underlying K3 surface, see *e.g.* [EOTY89, NW01, Wen02, Wen07, Wen06]. The precise role of geometric symmetries for the properties of the elliptic genus is central to more recent discussions [TW13a, TW13c, TW13b] of the MATHIEU MOONSHINE phenomena, in search of a geometric explanation for the seminal observations of [EOT11, Gan12], see [Wen15, Sect. 4] for a summary. Using mirror symmetry, one can also explicitly construct a family of K3 theories with a geometric interpretation on a family of smooth K3 surfaces, which are expected to provide examples of non-linear sigma models on such smooth manifolds [Wen06].

It is natural, but much more subtle, to reconstruct aspects of the vertex algebra which underlies a superconformal field theory from geometric data, if the theory is expected to have a non-linear sigma model interpretation. From a string theory perspective, one should introduce free fields on local coordinate patches of Y, analogously to the fields $j^k(z)$, $\psi_l^{\pm}(z)$ of (5), however it is unclear how to relate these "locally defined" fields to the ones that are used in abstract conformal field theories.

A more "rigid" approach, inspired by and closely related to TOPOLOGICAL FIELD THEORY, uses $bc{-}\beta\gamma$ systems on holomorphic coordinate patches of a compact, connected complex manifold Y. Roughly, one replaces the local holomorphic coordinate functions and holomorphic tangent vector fields by free bosonic fields of scaling dimensions zero and one, respectively, (the *bc*-FIELDS), while cotangent vector fields correspond to fermionic fields (the $\beta\gamma$-FIELDS). For these fields, transition functions between holomorphic coordinate patches can be defined according to the known geometric transformation rules. By this construction, one arrives at a sheaf of conformal vertex algebras on Y, which allows the definition of a natural $\mathbb{Z}$-grading and a differential of degree one. The resulting complex is known as the CHIRAL DE RHAM COMPLEX [MSV99] because it is quasi-isomorphic to the classical (Dolbeault-) de Rham complex on Y.

The construction of the chiral de Rham complex is "rigid" or topological in that an underlying conformal covariance manifests itself in a global section of the chiral de Rham complex, which yields a Virasoro field at central charge zero. One can perform a topological twist to a traditional $N = 2$ superconformal structure at central charge $c = 3D$ if Y is a Calabi-Yau D-manifold. This structure is found to descend to the sheaf cohomology of the chiral de Rham complex, which carries the structure of a superconformal vertex operator algebra [Bor01, BL00]. Thereby, the above-mentioned problem is solved, that the geometrically motivated fields are only defined locally on coordinate patches, while abstractly defined conformal field theories do not exhibit such "locally defined" fields.

Armed with this success, one would naturally expect the vertex algebra that is obtained from the chiral de Rham complex to be closely related to the one which underlies a non-linear sigma model on Y. This expectation is reinforced by the fact that almost by construction, the GRADED EULER CHARACTERISTIC of the chiral de Rham complex on a Calabi-Yau D-manifold Y yields the geometric elliptic genus $\mathcal{E}_Y(\tau, z)$ of Definition 2.2.3 [Bor01, BL00]. However, the precise relation between the relevant vertex algebras turns out to be more subtle. In particular, the construction of the chiral de Rham complex does not depend on the choice of the Kähler class on Y, in contrast to what one expects for the vertex algebras obtained in non-linear sigma models. According to [Kap05], this problem can be solved by performing a large volume limit to identify the BRST-cohomology of a topologically half twisted non-linear sigma model on Y with the sheaf cohomology of the chiral de Rham complex.

Returning to K3 theories, where the notion of geometric interpretations is well understood, one may hope for more concrete results. Indeed, for generic Calabi-Yau D-manifolds it is notoriously hard to calculate the sheaf cohomology of the chiral de Rham complex and its superconformal vertex algebra structure. On the other hand, for toroidal SCFTs and their $\mathbb{Z}_2$-

orbifolds at $c = \overline{c} = 6$, I expect that a direct calculation should show that the vertex algebra obtained from the sheaf cohomology of the chiral de Rham complex in fact agrees with the one obtained by a topological half-twist from the respective superconformal field theories [CGW15]. Intriguingly, our calculations seem to indicate that although the construction of the chiral de Rham complex crucially depends on the choice of complex structure on Y, the resulting vertex algebras show no dependence on the choice of complex structure (nor Kähler structure) within these two classes of examples.

In short, there are many intriguing mathematical and physical mysteries left when it comes to the journey from geometry to conformal field theory, even en route at the potentially simpler K3 geometry and K3 theories.

References

[AM94] P.S. Aspinwall and D.R. Morrison, *String theory on K3 surfaces*, in: Mirror symmetry II, B. Greene and S.T. Yau, eds., AMS, 1994, pp. 703–716; hep-th/9404151.

[AS63] M.F. Atiyah and I.M. Singer, *The index of elliptic operators on compact manifolds*, Bulletin American Mathematical Society **69** (1963), 322–433.

[BGV92] N. Berline, E. Getzler, and M. Vergne, *Heat kernels and Dirac operators*, Springer-Verlag, Berlin Heidelberg New York, 1992.

[BHPvdV04] W. Barth, K. Hulek, C. Peters, and A. van de Ven, *Compact Complex Surfaces*, Second enlarged ed., Springer-Verlag, Berlin Heidelberg, 2004.

[BL00] L.A. Borisov and A. Libgober, *Elliptic genera of toric varieties and applications to mirror symmetry*, Invent. Math. **140** (2000), no. 2, 453–485.

[Bog74] F.A. Bogomolov, *Kähler manifolds with trivial canonical class*, Izv. Akad. Nauk SSSR Ser. Mat. **38** (1974), 11–21.

[Bor86] R.E. Borcherds, *Vertex algebras, Kac–Moody algebras and the monster*, Proc. Nat. Acad. Sci. U.S.A. **83** (1986), 3068–3071.

[Bor01] L.A. Borisov, *Vertex algebras and mirror symmetry*, Commun. Math. Phys. **215** (2001), no. 2, 517–557; math.AG/9809094.

[Cal54] E. Calabi, *The space of Kähler metrics*, Proceedings of the International Congress of Mathematicians, Amsterdam, 1954, vol. II, Groningen-Amsterdam, pp. 206–207.

[Cec90] S. Cecotti, *$N = 2$ supergravity, type IIB superstrings and algebraic geometry*, Commun. Math. Phys. **131** (1990), 517–536.

[CENT85] A. Casher, F. Englert, H. Nicolai, and A. Taormina, *Consistent superstrings as solutions of the $D = 26$ bosonic string theory*, Phys. Lett. **B162** (1985), 121–126.

[CGW15] S. Carnahan, F. Grimm, and K. Wendland, *work in progress.*

[dC92] M.P. do Carmo, *Riemannian geometry*, Mathematics: Theory & Applications, Birkhäuser Boston, Inc., Boston, MA, 1992, Translated from the second Portuguese edition by Francis Flaherty.

[DHVW85] L.J. Dixon, J. Harvey, C. Vafa, and E. Witten, *Strings on orbifolds*, Nucl. Phys. **B261** (1985), 678–686.

[DHVW86] ——, *Strings on orbifolds II*, Nucl. Phys. **B274** (1986), 285–314.

[DY93] P. Di Francesco and S. Yankielowicz, *Ramond sector characters and $N = 2$ Landau-Ginzburg models*, Nucl. Phys. **B409** (1993), 186–210; `hep-th/9305037`.

[EOT11] T. Eguchi, H. Ooguri, and Y. Tachikawa, *Notes on the $K3$ surface and the Mathieu group M_{24}*, Exp. Math. **20** (2011), no. 1, 91–96; `arXiv:1004.0956 [hep-th]`.

[EOTY89] T. Eguchi, H. Ooguri, A. Taormina, and S.-K. Yang, *Superconformal algebras and string compactification on manifolds with $SU(n)$ holonomy*, Nucl. Phys. **B315** (1989), 193–221.

[FB05] E. Freitag and R. Busam, *Complex analysis*, Universitext, Springer-Verlag, Berlin, 2005, Translated from the 2005 German edition by Dan Fulea.

[FBZ04] E. Frenkel and D. Ben-Zvi, *Vertex algebras and algebraic curves*, second ed., Mathematical Surveys and Monographs, vol. 88, American Mathematical Society, Providence, RI, 2004.

[FFRS10] J. Fröhlich, J. Fuchs, I. Runkel, and Ch. Schweigert, *Defect lines, dualities and generalised orbifolds*, in: XVIth International Congress on Mathematical Physics, World Sci. Publ., Hackensack, NJ, 2010, pp. 608–

613; `arXiv:0909.5013 [math-ph]`.

[FK81] I.B. Frenkel and V.G. Kac, *Basic representations of affine Lie algebras and dual resonance models*, Invent. Math. **62** (1980/81), no. 1, 23–66.

[FLM84] I. Frenkel, J. Lepowsky, and A. Meurman, *A natural representation of the Fischer-Griess Monster with the modular function J as character*, Proc. Nat. Acad. Sci. U.S.A. **81** (1984), no. 10, Phys. Sci., 3256–3260.

[For81] O. Forster, *Lectures on Riemann surfaces*, Graduate Texts in Mathematics, vol. 81, Springer-Verlag, New York-Berlin, 1981, Translated from the German by Bruce Gilligan.

[Gan12] T. Gannon, *Much ado about Mathieu*; `arXiv:1211.5531 [math.RT]`.

[GH78] P. Griffiths and J. Harris, *Principles of algebraic geometry*, John Wiley & Sons, New York, 1978.

[Gin88] P. Ginsparg, *Applied conformal field theory*, Lectures given at the Les Houches Summer School in Theoretical Physics 1988 (Les Houches, France), pp. 1–168.

[GP74] V. Guillemin and A. Pollack, *Differential topology*, Prentice-Hall, Inc., Englewood Cliffs, N.J., 1974.

[Hir54] F. Hirzebruch, *Arithmetic genera and the theorem of Riemann-Roch for algebraic varieties.*, Proc. Nat. Acad. Sci. U.S.A. **40** (1954), 110–114.

[Hir76] M.W. Hirsch, *Differential topology*, Springer-Verlag, New York-Heidelberg, 1976, Graduate Texts in Mathematics, No. 33.

[Hir88] F. Hirzebruch, *Elliptic genera of level N for complex manifolds*, in: Differential geometric methods in theoretical physics, K. Bleuler and M. Werner, eds., Kluwer Acad. Publ., 1988, pp. 37–63.

[Hop26] H. Hopf, *Vektorfelder in n-dimensionalen Mannigfaltigkeiten*, Math. Ann. **96** (1926), 225–250.

[Hua14] Y.-Z. Huang, *Conformal Field Theory*; `https://qcft.wordpress.com/2014/09/16/a-program-to-construct-and- study-conformal-field-theories/`.

[Huy05] D. Huybrechts, *Complex geometry*, Universitext, Springer-Verlag, Berlin, 2005.

[Jos95] J. Jost, *Riemannian geometry and geometric analysis*, Universitext, Springer-Verlag, Berlin, 1995.

[Joy00] D.D. Joyce, *Compact manifolds with special holonomy*, Oxford Mathematical Monographs, Oxford University Press, Oxford, 2000.

[Kac98] V.G. Kac, *Vertex algebras for beginners*, second ed., University Lecture Series, vol. 10, American Mathematical Society, Providence, RI, 1998.

[Käh33] E. Kähler, *Über eine bemerkenswerte Hermitesche Metrik*, Abh. math. Sem. Univ. Hamburg **9** (1933), 173–186.

[Kap05] A. Kapustin, *Chiral de Rham complex and the half-twisted sigma-model*; hep-th/0504074.

[KO03] A. Kapustin and D. Orlov, *Vertex algebras, mirror symmetry, and D-branes: the case of complex tori*, Commun. Math. Phys. **233** (2003), no. 1, 79–136; hep-th/0010293.

[Kod64] K. Kodaira, *On the structure of compact complex analytic surfaces*, I. Am. J. Math. **86** (1964), 751–798.

[Kri90] I. Krichever, *Generalized elliptic genera and Baker-Akhiezer functions*, Math. Notes **47** (1990), 132–142.

[Kum64] E.E. Kummer, *Über die Flächen vierten Grades mit sechzehn singulären Punkten*, in: Gesamtwerk, 1864, pp. 246–260, Monatsbericht der Königlich Preußischen Akademie der Wissenschaften zu Berlin aus dem Jahre 1864; 18. April. Sitzung der physikalisch-mathematischen Klasse.

[Lan99] S. Lang, *Complex analysis*, Graduate Texts in Mathematics, vol. 103, Springer-Verlag, New York, 1999.

[Lee09] J.M. Lee, *Manifolds and differential geometry*, Graduate Studies in Mathematics, vol. 107, American Mathematical Society, Providence, RI, 2009.

[LM89] H.B. Lawson and M.-L. Michelsohn, *Spin Geometry*, Princeton University Press, Princeton, New Jersey, 1989.

[LW78] J. Lepowsky and R.L. Wilson, *Construction of the affine Lie algebra* $A_1(1)$, Commun. Math. Phys. **62** (1978), no. 1, 43–53.

[Mor01] S. Morita, *Geometry of differential forms*, Translations of Mathematical Monographs, vol. 201, American Mathematical Society, Providence, RI, 2001, Translated from the two-volume Japanese original (1997, 1998) by Teruko Nagase and Katsumi Nomizu, Iwanami Series in Modern Mathematics.

[MSV99] F. Malikov, V. Schechtman, and A. Vaintrob, *Chiral de Rham complex*, Commun. Math. Phys. **204** (1999), no. 2,

439–473; math.AG/9803041.

[Nar86] K.S. Narain, *New heterotic string theories in uncompactified dimensions* < 10, Phys. Lett. **169B** (1986), 41–46.

[Nik75] V.V. Nikulin, *On Kummer surfaces*, Math. USSR Isv. **9** (1975), 261–275.

[NN75] A. Newlander and C. Nirenberg, *Complex analytic coordinates in almost complex manifolds*, Ann. Math. **65** (1975), 391–404.

[NW01] W. Nahm and K. Wendland, *A hiker's guide to K3 – Aspects of $N = (4,4)$ superconformal field theory with central charge $c = 6$*, Commun. Math. Phys. **216** (2001), 85–138; hep-th/9912067.

[O'N83] B. O'Neill, *Semi-Riemannian geometry with applications to relativity*, Pure and Applied Mathematics, vol. 103, Academic Press, Inc. (Harcourt Brace Jovanovich, Publishers), New York, 1983.

[Poi85] H. Poincaré, *Sur les courbes définies par les équations différentielles (3ème partie)*, Journal de mathématiques pures et appliquées **4** (1885), 167–244.

[Rem91] R. Remmert, *Theory of complex functions*, Graduate texts in mathematics, vol. 122, Springer-Verlag, 1991.

[Sei88] N. Seiberg, *Observations on the moduli space of superconformal field theories*, Nucl. Phys. **B303** (1988), 286–304.

[Ser55] J.-P. Serre, *Un théorème de dualité*, Comment. Math. Helv. **29** (1955), 9–26.

[Siu83] Y.T. Siu, *Every K3 surface is Kähler*, Invent. Math. **73** (1983), no. 1, 139–150.

[TW13a] A. Taormina and K. Wendland, *The overarching finite symmetry group of Kummer surfaces in the Mathieu group M_{24}*, JHEP **08** (2013), 125; arXiv:1107.3834 [hep-th].

[TW13b] ——, *Symmetry-surfing the moduli space of Kummer K3s*, accepted for publication in Proc. of the Conference String-Math 2012, PSPUM; arXiv:1303.2931 [hep-th].

[TW13c] ——, *A twist in the M_{24} moonshine story*, accepted for publication in Confluentes Mathematici; arXiv:1303.3221 [hep-th].

[Wei58] A. Weil, *Final report on contract AF 18(603)-57*, in: Oeuvres scientifiques - Collected Works, vol. II (1979), Springer Verlag, 1958, pp. 390–395.

[Wel73] R.O. WELLS JR., *Differential analysis on complex manifolds*, Prentice-Hall, Inc., Englewood Cliffs, N.J., 1973, Series in Modern Analysis.

[Wen01] K. WENDLAND, *Consistency of orbifold conformal field theories on $K3$*, Adv. Theor. Math. Phys. **5** (2001), no. 3, 429–456; `hep-th/0010281`.

[Wen02] ——, *Orbifold constructions of $K3$: A link between conformal field theory and geometry*, in: Orbifolds in Mathematics and Physics, vol. 310 of Contemporary Mathematics, AMS, Providence R.I., 2002, pp. 333–358; `hep-th/0112006`.

[Wen06] ——, *A family of SCFTs hosting all 'very attractive' relatives of the $(2)^4$ Gepner model*, JHEP **03** (2006), 102; `hep-th/0512223`.

[Wen07] ——, *On superconformal field theories associated to very attractive quartics*, Frontiers in Number Theory, Physics and Geometry II (P. Cartier, B. Julia, P. Moussa, and P. Vanhoeve, eds.), Springer 2007, pp. 223–244; `hep-th/0307066`.

[Wen10] ——, *On the geometry of singularities in quantum field theory*, Proceedings of the International Congress of Mathematicians, Hyderabad, August 19-27, 2010, Hindustan Book Agency, pp. 2144–2170.

[Wen15] ——, *Snapshots of conformal field theory*, in: Mathematical Aspects of Quantum Field Theories, D. Calaque and Th. Strobl, eds., Springer-Verlag, 2015, pp. 89–129; `arXiv:1404.3108 [hep-th]`.

[Wit87] E. WITTEN, *Elliptic genera and quantum field theory*, Commun. Math. Phys. **109** (1987), 525–536.

[Wit88] ——, *The index of the Dirac operator in loop space*, in: Elliptic curves and modular forms in algebraic geometry, P. Landweber, ed., Springer-Verlag, 1988, pp. 161–181.

[Wit94] ——, *On the Landau-Ginzburg description of $N = 2$ minimal models*, Int. J. Mod. Phys. **A9** (1994), 4783–4800; `hep-th/9304026`.

[Yau78] S.T. YAU, *On the Ricci curvature of a compact Kähler manifold and the complex Monge-Ampère equation, I*, Comm. Appl. Math. **31** (1978), 339–411.

[YB53] K. YANO AND S. BOCHNER, *Curvature and Betti numbers*, Annals of Mathematics Studies, No. 32, Princeton University Press, Princeton, N. J., 1953.

GEOMETRIC ASPECTS OF BLACK HOLES

Bruno Carneiro da Cunha
Departamento de Física, Universidade Federal de Pernambuco, 53901-970, Recife, Pernambuco, Brazil
E-mail: bcunha@df.ufpe.br

Black Holes in General Relativity and Supergravity are presented and their geometrical aspects discussed. The microcanonical description of the entropy of the black hole in the D1-D5 system in type IIB supergravity is sketched. These are the class notes of the 12-hour series of lectures presented during July 15 to 27, 2013 at Villa de Leyva, Colombia as part of the Summer School on Geometric, Algebraic and Topological Methods for Quantum Field Theory.

1. Introduction & Notation

Einstein's General Relativity theory associates gravitational phenomena to the curvature of space-time. The application of the theory to physical phenomena has evolved from corrections to solar system gravitational dynamics to the very foundation of many astrophysical and cosmological systems. All along its story, the search for exact solutions of Einstein's equations and the study of its geometric aspects has had great impact on the understanding of the theory and the associated physical applications. Mathematically, it is a very eclectic area, involving concepts from integrable systems, functional analysis, dynamical systems and of course differential and algebraic geometry [1, 2].

Of particular interest to both physics and mathematics stand the solutions known as black holes. These are solutions of General Relativity due to localized distributions of matter and display special geometric properties. These are interesting for physicists not only because of their applications to General Relativity and Supergravity but also of the renewed interest in black holes dynamics due to the gauge/gravity conjecture. From the mathematicians' point of view, black hole solutions provide non-trivial examples of metrics with special algebraic properties, for instance, Calabi-Yau [3], Einstein-Sasaki [4], and more generally special holonomy manifolds [5].

The purpose of this series of lectures is to survey the geometric aspects of black holes, with an emphasis on their event horizon and global structure. We will take an approach and methods from physics, first pointing out that evolution of generic initial conditions lead to event horizons, and discussing how gravitational collapse seems to be the natural outcome of stellar evolution. We then move to a more geometrical approach, introducing the notion of a Killing horizon and talking about the rigidity theorems in the classical (electro-)vacuum solutions of 4-dimensional General Relativity. We will then talk about higher-dimensional black holes, in Supergravity and String Theory. We will end by describing the Strominger-Vafa microcanonical formulation for the Bañados-Teitelboim-Zanelli (BTZ) black hole in 3-dimensional spacetime.

In this series we will assume some knowledge of General Relativity, and the Riemannian geometry associated to it, up to the level of first part of [1]. We will suppose throughout that we are dealing with a smooth (analytic) manifold M equipped with a Lorentzian metric g_{ab} with "GR-styled" signature $\{-,+,\ldots\}$. The covariant derivative associated to g_{ab} will be denoted by ∇_a. The Christoffel symbols are defined as the local expression for the connection acting on vectors:

$$\nabla_\mu X^\nu = \partial_\mu X^\nu + \Gamma^\nu_{\mu\rho} X^\rho; \qquad \Gamma^\nu_{\mu\rho} = \frac{g^{\nu\sigma}}{2}(\partial_\mu g_{\sigma\rho} + \partial_\rho g_{\mu\sigma} - \partial_\sigma g_{\mu\rho}), \quad (1)$$

with ∂_μ the usual (Lie) derivative associated to a local coordinate system x^μ. Greek letters from the middle of the alphabet refer to spacetime indices of those coordinates. Finally, the curvature is defined as the anticommutator of the covariant derivative acting on 1-forms (covariant vectors) ω_a:

$$(\nabla_a \nabla_b - \nabla_b \nabla_a)\omega_c = R_{abc}{}^d \omega_d. \quad (2)$$

The reader may have noted that vectors are denoted by capital Latin letters, while 1-forms are denoted by greek letters for the end of the alphabet. We will denote the outer product of 1-forms either by the wedge notation or by antisymmetry of the indices: $\omega_{[a}\zeta_{b]} \equiv \omega \wedge \zeta$. We will try otherwise to follow the conventions in [1].

2. Historical Remarks

Let us start with some physical intuition. At first, black hole solutions were considered unphysical before the work of Chandrasekhar because of the extreme conditions necessary for their creation – we will revisit this point in the next section, but the minded reader is referred to [6]. In order to get an appraisal of the physical conditions involved in gravitational collapse, let us consider a neutral monoatomic gas subjected to Newtonian gravity. In a straightforward calculation, one can see that the pressure of a non-relativistic dense gas in equilibrium is given by:

$$P_F = (3\pi^2)^{1/3}/4\hbar c n^{4/3}, \text{ for } n >> m^3c^3/\hbar^3, \tag{3}$$

which can counter the gravity up to the point were the total mass of the cloud is ($c = 1$, G_N is Newton's constant):

$$M \propto \left(\frac{\hbar}{G_N}\right)^{3/2} \frac{1}{m_B^2} \sim M_\odot. \tag{4}$$

The inclusion of non-gravitational interactions can change this somewhat, but the lesson that stars owe their equilibria to the capacity of Fermi pressure to counter gravitational attraction is the most important.

In General Relativity, the metric is deformed by the presence of matter. Let us study a static, spherically symmetric distribution of matter, with metric (line element) given by:

$$ds^2 = -e^{2\phi(r)}dt^2 + h(r)dr^2 + r^2 d\Omega^2,$$

which fixes the geometry up to two functions $\phi(r)$ and $h(r)$. Assuming a perfect fluid (static) form for the stress energy-tensor,

$$T_{ab} = \rho u_a u_b + P(u_a u_b + g_{ab}), \qquad \text{where } u^a = e^{-\phi}\partial_t,$$

Calculating the components of Einstein's equations (G_{ab} is the Einstein tensor), we have:

$$\begin{aligned} G_{tt}: \quad h(r) = \left(1 - \frac{2m(r)}{r}\right)^{-1}, \qquad m(r) = 4\pi \int_0^r \rho(r')r'^2\, dr' \\ G_{rr}: \quad \frac{d\phi}{dr} = \frac{m(r) + 4\pi r^3 P}{r(r - 2m(r))}, \end{aligned} \tag{5}$$

and one notes the similarity of $h(r)$ with the Newtonian potential. The remaining (angular) components of the equations give the Tolman-Oppenheimer-Volkoff (TOV) equation:

$$\frac{dP}{dr} = -(P + \rho)\frac{m(r) + 4\pi r^3 P}{r(r - 2m(r))} \tag{6}$$

this equation is the starting point for stellar equilibrium. Even though it only applies to very specialized situations, one can draw generic conclusions from it. First and foremost, a greater $\rho(r)$ is needed for achieving equilibrium in General Relativity as opposed to Newtonian gravity. More quantitatively, one sees that, if one assumes constant density, for radii in the interval:

$$R \geq \sqrt{1/3\pi\rho} \quad \text{or} \quad R \leq 9M/4, \tag{7}$$

there can be no equilibrium. Thirdly, if one assumes a reasonable equation of state for the matter – where thermal equilibrium results in a dependence of the pressure with the energy density $P = P(\rho)$, an upper bound for the total mass M of the star can be proved. Lastly, one can also get an estimate for stellar mass equilibrium by assuming $\rho = \rho_0$ at about nuclear densities and a plausible behavior beyond that scale to arrive at:

$$M_{\text{max}} \sim (2-3)M_{\odot}. \tag{8}$$

In other words, after that mass, either the equilibrium of astronomic bodies cannot be explained in terms of known physics, or the body in question will undergo *gravitational collapse*: the other forces, and Fermi gas degeneracy will not be strong enough to counter the action of gravity.

Notwithstanding these general remarks, it is extremely hard to give mathematical proof of the existence of collapsing solutions. In fact, one of the milestones in the study of General Relativity during its first century came directly from this study. First we have the combination of analytical and numerical methods which allowed for studies of gravitational collapse and black holes in astrophysical settings – modeling successfully a myriad of phenomena such as quasars, gamma ray bursts and their influence on galactic evolution [7]. Second, there was the proof of existence of collapsing solutions – for spherically symmetric scalar massless fields, using standard analytical methods [8]. Albeit recent, the outline and the methods used in the proof were already influential in discovering black hole and even "vacuum space" instabilities. No doubt its impact will unfold more thoroughly in the following years.

3. The Schwarzschild Black Hole

So we gave some evidence that black holes are important from the physical point of view. Let us continue with the assumption of spherical symmetry. After the collapse has settled, Birkhoff's theorem assures that the exterior metric will be Schwarzschild's:

$$ds^2 = -\left(1 - \frac{2M}{r}\right) dt^2 + \frac{dr^2}{1 - \frac{2M}{r}} + r^2 d\Omega^2, \tag{9}$$

with $d\Omega^2$ being the line element of the unit 2−sphere. The metric was one of the first exact solutions obtained from Einstein's equations and its significance will be outlined in this section. As stated, the metric has a problem at the radial position $r = 2M$. At first physicists assumed that the region $r < 2M$ is unphysical, since, as seen in the preceding section one needs dramatic conditions to compress matter beyond this point. On the other hand, the mathematical notion of manifold explicitly leaves the possibility of having more than one chart (or system of coordinates) covering the full of spacetime. So the problem at $r = 2M$ may be either due to physical singularities, or to the failure of our coordinate system chosen.

As a preamble to illustrate the second point, let us consider Rindler's metric:

$$ds^2 = -x^2 dt^2 + dx^2, \tag{10}$$

which is a two-dimensional metric used to illustrate the problem. As Schwarzschild's metric at $r = 2M$, the Rindler metric has a problem at $x = 0$: the metric becomes non-invertible. However, a straightforward calculation shows that the Rindler metric is flat, and by changing coordinates:

$$T = x \sinh t, \quad X = x \cosh t \quad \Longrightarrow \quad ds^2 = -dT^2 + dX^2, \tag{11}$$

one recovers the two-dimensional Minkowski metric. As far as the full manifold is concerned, however, only the part with $X > 0$ and $X^2 > T^2$ is covered by the Rindler coordinates. The transformation also allows for a geometrical interpretation of the parameters of the Rindler metric: the curves with $x = x_0$ constant are actually constant proper acceleration time-like curves, so the Rindler metric is suited to the study of the effects of acceleration in the physics.

Coming back to the Schwarzschild case, the question that poses itself now is how to find a change a coordinates which "solve" the singular surface $r = 2M$. If this surface is "physical", we should see a breakage of description in terms of metric in any coordinate system we chose. If, however, it is not physical, the singularity should go away in any coordinate system with a direct physical interpretation.

A convenient way to implement such a coordinate system is to follow a particular congruence of null (light-like) geodesics (zero acceleration curves) as they travel inside the black hole. Null geodesics are described by the conditions:

$$u^a u_a = 0, \qquad u^b \nabla_b u^a = 0, \tag{12}$$

where the attentive reader may recognize the geodesic equation in its geometric form as the second condition. Using a coordinate patch $\{x^\mu\}$ to write it,

$$\frac{d^2x^\mu}{d\lambda^2} + \Gamma^\mu_{\nu\rho}\frac{dx^\nu}{d\lambda}\frac{dx^\rho}{d\lambda} = 0; \tag{13}$$

where λ is the affine parameter of the geodesic – the integral parameter of the curve.

In the Schwarschild case, the existence of local isometries – Killing Vector Fields (KVFs) – associated to time and angular translations:

$$t^a = \left(\frac{\partial}{\partial t}\right)^a, \qquad \phi^a = \left(\frac{\partial}{\partial\phi}\right)^a, \tag{14}$$

reduce the null vector condition $u^a u_a$ to a simple relation between $\dot{r}$ and $\dot{t}$. For zero angular momentum, this reduces to a simple differential equation for the radial position as a function of global time $r(t)$:

$$\left(\frac{dr}{dt}\right)^2 = 1 - \frac{2M}{r}, \tag{15}$$

which can be integrated to:

$$dt = \pm\frac{dr}{1-\frac{2M}{r}} \to x_\pm = t \pm \left[r + 2M\log\left(\frac{r-2M}{2M}\right)\right], \tag{16}$$

the new coordinates $x_\pm$ now correspond to the congruence of null ingoing x_+ and outgoing x_- null geodesics. This is illustrated in Figure 1. The function of r appearing above:

$$r_* = r + 2M\log\left(\frac{r-2M}{2M}\right), \tag{17}$$

is sometimes referred to as "tortoise coordinate".

If we write the line element in terms of the null coordinates $x_\pm$,

$$ds^2 = -\left(1-\frac{2M}{r}\right)dx_+dx_- + r^2d\Omega^2,\ 1-\frac{2M}{r} = \frac{e^{-r/2M}}{r}e^{(x_+-x_-)/4M}, \tag{18}$$

where the scale factor in front of dx_+dx_- is given implicitly as a function of $x_+ - x_-$. The problematic surface at $r = 2M$ corresponds then to $x_+ - x_- \to \infty$. One can simplify this divergence by a redefinition of the coordinates $X_+ = \exp(x_+/4M)$ and $X_- = -\exp(-x_-/4M)$, in terms of which the line element becomes:

$$\begin{aligned} ds^2 &= -\frac{16M^2e^{-r/2M}}{r}dX_+dX_- + r^2d\Omega^2 \\ &= -\frac{32M^2e^{-r/2M}}{r}(dT^2 + dX^2) + r^2d\Omega^2, \end{aligned} \tag{19}$$

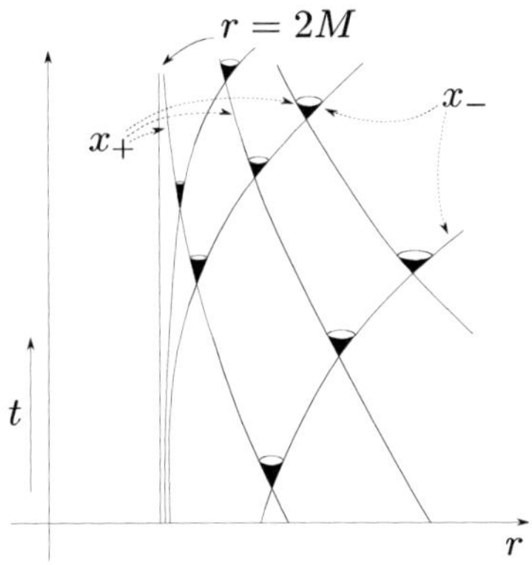

Fig. 1: The ingoing and outgoing null geodesics in the Schwarzschild metric. The surface $r = 2M$ can be seen to be a limiting value for those. Also depicted are the future cones at various points, showing their "compression" as one approaches the limiting surface.

which shows no problem at $r = 2M$ (or $T = X$). The metric now can be continued down to $r = 0$, where the divergence has an invariant meaning, since the scalar – "Kretschmann invariant":

$$R_{abcd}R^{abcd} = 48M^2/r^6 \tag{20}$$

blows up there. The new, non-problematic coordinates are written in terms of the Schwarzschild coordinates as

$$\begin{aligned} T &= \left(1 - \frac{2M}{r}\right)^{1/2} e^{r/4M} \sinh\left(\frac{t}{4M}\right), \\ X &= \left(1 - \frac{2M}{r}\right)^{1/2} e^{r/4M} \cosh\left(\frac{t}{4M}\right). \end{aligned} \tag{21}$$

So the problematic surface at $r = 2M$ was not problematic at all, and the manifold described by the Schwarzschild metric can be extended past it. The new coordinates T and X seem to cover the region $r < 2M$ up to the point $r = 0$, where a curvature singularity ensues. Mathematically, the extension will be outlined in the next subsection.

Physically, this failure of the Schwarzschild coordinates can be understood heuristically. The coordinate r in (9) has a definite geometric meaning: the spherical symmetry can be used to generate a 2-surface from a given point. The area of this surface is $4\pi r^2$, thus giving a definition of the function r and a definition of the normal one form $(dr)_a$. Note that this may not – and actually have not – anything to do with the notion of geodesic distance to the fixed point of rotations, which only hold in flat spacetime. The horizon is the surface where $(dr)_a$ changes from being space-like $r > 2M$, to time-like $r < 2M$. In physical terms, there's no fundamental principle forbidding this change, so we don't expect this surface to be of physical significance locally. Its significance is global, and will be described below. We will turn to the problem of describing precisely what is in the region $r \leq 2M$.

3.1. *The Kruskal Extension*

One global significance of the surface $r = 2M$ is that of a "horizon": the limiting surface as far as the t evolution of the ingoing null geodesics is concerned (see Fig. 1). An asymptotic observer sitting at $r = \infty$ does not have access to the region beyond this point, and it is not affected by it. Of course, this interpretation comes about by verifying that this observer measures t as proper time. Yet another characterization is that it is a surface of infinite redshift: the "horizon" is the place where the time translation symmetry $t^a = \partial_t$ becomes null: $g_{tt} = -(1 - 2M/r) = 0$. In order to describe the horizon structure, let us work with the proper coordinates $X_\pm$, in terms of which the metric is given by

$$ds^2 = -\frac{32M^3}{r} e^{r/2M} dX_+ dX_- + r^2 d\Omega_2. \tag{22}$$

In terms of these – the Kruskal coordinates – the normal 1-form $(dr)_a$ and $(dt)_a$ are given by:

$$dr = \frac{4M^2}{r} e^{-r/2M} (X_+ dX_- + X_- dX_+), \quad dt = 2M \left(\frac{dX_+}{X_+} - \frac{dX_-}{X_-} \right). \tag{23}$$

One sees that $(dt)_a$ diverges and $(dr)_a$ vanishes at the horizon $X_+ = 0$ or $X_- = 0$. The covariant derivative of t_a, however, has a clear limit:

$$\begin{aligned} \nabla_a t_b &= \frac{8M^3}{r^3} e^{-r/2M} \left(1 - \frac{2M}{r} \right) (dT)_{[a} (dX)_{b]} \\ &= \frac{16M^3}{r^3} e^{-r/2M} dX_+ \wedge dX_-. \end{aligned} \tag{24}$$

We will define the quantity κ, the norm of this derivative, computed at $r = 2M$:

$$\kappa^2 = -\frac{1}{2} (\nabla^a t^b)(\nabla_a t_b) = \left. \frac{M^2}{r^4} \right|_{r=2M} \longrightarrow \kappa = \frac{1}{4M}, \tag{25}$$

which is called *surface gravity* at the horizon. Numerically, it is equal to a force (per unit mass) exerted at infinity, to keep a particle at the point $r = 2M$. We will see below that this quantity is constant throughout the horizon for any black hole.

The surface gravity κ also takes its name and interpretation from more direct calculations. In order to compute the acceleration of the time translation Killing vector field ∂_t let us first write it in terms of the Kruskal coordinates:

$$\begin{aligned} t^a &= \frac{1}{4M}\left(X_+\frac{\partial}{\partial X_+} - X_-\frac{\partial}{\partial X_-}\right), \\ t_a &= -\frac{4M^2}{r}e^{-r/2M}(X_-dX_+ - X_+dX_-). \end{aligned} \tag{26}$$

We now define the acceleration of the time translation by the usual formula:

$$\rho_b = t^a\nabla_a t_b = \frac{4M^3}{r^3}e^{-r/2M}(X_+dX_- + X_-dX_+), \tag{27}$$

with norm $\rho^b\rho_b = 2M^3/r^5 e^{-r/2M}X_-X_+$, a space-like vector in the region $r > 2M$ $(X_+ > 0)$. The surface gravity can also be defined by the value of this norm, divided by the norm of t^a, $\kappa^2 = -\rho^b\rho_b/t^c t_c$, so $\kappa = M/r^2$. At the horizon $(X_+ = 0)$, however, ρ_c becomes null:

$$\rho_a = \frac{1}{2M}X_-dX_+ = -\kappa t_a = -\frac{1}{4M}2Me^{-r/2M}X_-dX_+, \tag{28}$$

and actually parallel to t_c. Integral curves of t^a at the horizon are thus null geodesics, although not affinely parametrized, so that $t^b\nabla_b t^a = -\kappa t^a$. The name surface gravity for κ is now fully justified.

Turning now to the other problem, one sees from (22) that the region $r > 2M$ corresponds to part of the T, X plane:

$$X^2 - T^2 = \left(\frac{r}{2M} - 1\right)e^{r/2M}, \qquad \frac{t}{4M} = \operatorname{arc\,tanh}\left(\frac{T}{X}\right), \tag{29}$$

specifically, $r > 2M$ only covers $X > T$, $X > 0$. What are the other regions?

To better illustrate them, let us remark that the extension of null geodesics carried out by the Kruskal procedure has a direct interpretation in terms of domains of causality. By extending null geodesics to the future, we are extending lightcones and thus the causal structure. We are describing all of spacetime which can be causally affected by the region $r > 2M$. Likewise, by extending null geodesics to the past, we are describing all of spacetime which causally affects the region $r > 2M$. The causal structure of a spacetime has an extra symmetry: it is invariant by (proper) conformal transformation. Indeed, the trajectory of null geodesics is unchanged by a (local) scale transformation of the metric:

$$g_{ab} \to \tilde{g}_{ab} = e^{2\psi} g_{ab}, \qquad k^a \nabla_a k^b = 0 \to k^a \tilde{\nabla}_a k^b = (2k^c \nabla_c \psi) k^b, \tag{30}$$

so the causal structure is the same. Through the scale transformation, we can study the infinity structure. For spaces with a large number of symmetries, one can choose the scale transformation judiciously to describe the causal structure in terms of a finite diagram – called causal – or Penrose – diagram. As an example, consider the standard Minkowski metric in spherical coordinates. By a straightforward change of coordinates, we have:

$$ds^2 = -dt^2 + d\mathbf{r}^2 = \frac{-dpdq + \frac{1}{4}\sin^2(p-q)d\Omega^2}{\cos p \cos q}, \tag{31}$$

where p and q can be seen to be lightcone coordinates. The spacetime in these coordinates is only defined for a finite interval of these coordinates $p, q \in (-\pi/2, \pi/2)$. The scaled region is represented in Figure 2. One sees that the past region $t \to -\infty$ is causally divided between that can be achieved by following lightrays to the past $\mathcal{I}^-$ – called null past infinity, and the region which cannot called timelike past infinity i^-. The future is equally described by a future null $\mathcal{I}^+$ and timelike infinity i^+. In between the null infinities lies the region only achieved by spacelike curves, called spacelike infinity i^0. In the Figure some of these degenerate to a point, but in fact this is just some artifact of the singularity of the scale transformation there.

From this, it is now a straightforward exercise to work out the causal structure for Schwarzschild spacetime in Kruskal coordinates. From (22), we see that all four regions:

$$\mathrm{I}: X, T > 0; \;\; \mathrm{II}: X < 0, T > 0; \;\; \mathrm{III}: X > 0, T < 0; \;\; \mathrm{IV}: X, T < 0, \tag{32}$$

represented in the Figure 3. Each of these four regions can be covered by a coordinate system involving the integral parameter of the Killing vector field t, and the coordinate r described in terms of area as above. We will leave this proof as an exercise to the reader. The regions I and IV display the same asymptotic structure of Minkowski space, with two copies of future and past null infinities, future and past timelike infinities and spacelike infinities. Spaces with the same asympotic structure of Minkowski spacetime are called asympotically flat.

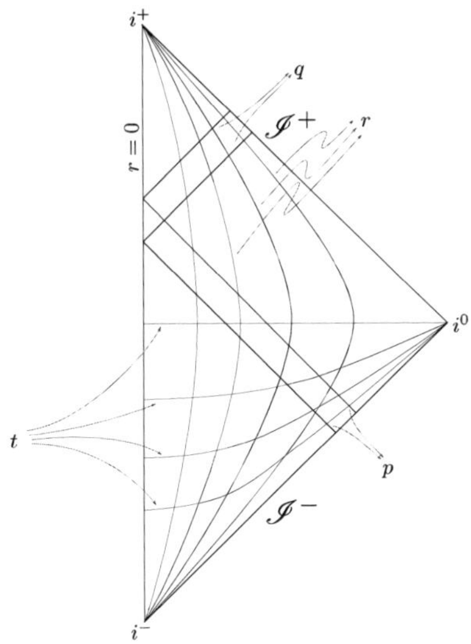

Fig. 2: The causal structure of Minkowski space. The straight lines correspond to p or q constant, physically lightrays leaving the null past $\mathcal{I}^-$ and bouncing off of the origin $r = 0$ to escape to future infinity $\mathcal{I}^+$. Also represented are the curves of constant t (horizontal lines), and constant r (vertical lines).

Problem 3.1. *Manipulate the relative signs in* (29) *in order to find that the metric in each region is given by the standard Schwarzschild form* (9), *albeit for regions II and III we are mapping* $r < 2M$.

On the other hand, the region $r = 0$ continues to be a true boundary of the spacetime, given that the curvature blows up there in an invariant manner. It is a region where curvature, and as such gravitational effects, becomes strong. As we can see from the extension, it is also a spacelike region, and as such it displays a curious feature of Schwarzschild spacetime: in order to describe local physics in it, even in the region I, one should prescribe data in a spacelike (Cauchy) surface. But these would include the singularities if one pushes these surfaces to include all of the past. One way around this is to only define physical processes as those which begin in one past region and end in one future region, like in region I. the boundary of this region, the surface $r = 2M$ is not only an event horizon, but also a Cauchy horizon: one cannot describe the region behind it with physical processes alone. Although in this example they are the same, they could be

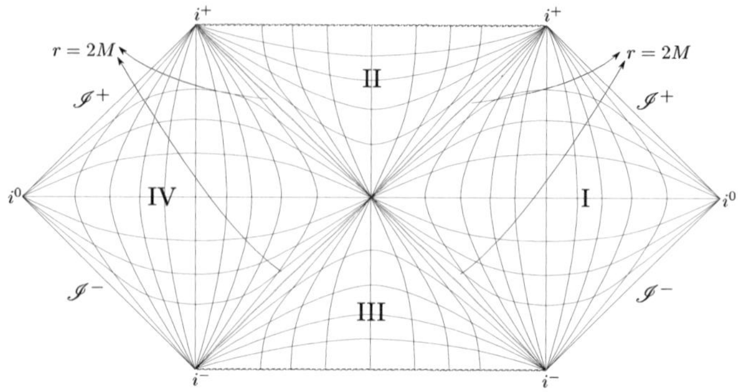

Fig. 3: The causal structure for the Schwarzschild solution in Kruskal coordinates.

in principle different. We will move now from the very singular example of Schwarschild to more generic cases and see which lessons can be generalized.

4. The Reissner-Nordström and Kerr-Newman Black Holes

Shortly after Schwarzschild's solution (9), a generalization for electrically charged, spherically symmetric sources was obtained – independently – by Reissner and Nordström. In the static gauge, the only non-trivial component of the gauge field is the Couloumbian electric potential $A_0 = \phi = -e/r$. The new metric, however, is deformed by the (static) electric field density energy. For a source with mass M and electric charge e, we have:

$$ds^2 = -\left(1 - \frac{2M}{r} + \frac{e^2}{r^2}\right) dt^2 + \frac{dr^2}{1 - \frac{2M}{r} + \frac{e^2}{r^2}} + r^2 d\Omega^2, \tag{33}$$

which displays a similar problem as Schwarzschild's metric for $g_{tt} = 0$, or:

$$r = r_\pm, \qquad r_\pm = M \pm \sqrt{M^2 - e^2}, \tag{34}$$

which are real roots for $e^2 < M^2$. We can now use the same methods as before, and introduce null coordinates $x_\pm = t \pm r_*$ where

$$\begin{aligned} r_* &= \int \frac{dr}{1 - \frac{2M}{r} + \frac{e^2}{r^2}} \\ &= r + \frac{r_+^2}{r_+ - r_-} \log(r - r_+) - \frac{r_-^2}{r_+ - r_-} \log(r - r_-), \quad e^2 < m^2, \end{aligned} \tag{35}$$

where one can see that the positions of the horizon r_+ is mapped to $r_* = -\infty$, and thus only the interval $r \in (r_+, \infty)$, the exterior of the black hole is mapped. Following the discussion above, we define new coordinates

$$X_{\pm} = \pm \exp\left(\pm \frac{r_+ - r_-}{4r_+^2} x_{\pm}\right), \tag{36}$$

which bring the horizon to a finite value $X_+ = 0$ or $X_- = 0$. Now, written in terms of $X_{\pm}$, the metric is

$$ds^2 = \frac{4r_+^4}{(r_+ - r_-)^2} \frac{(r - r_-)^{r_-^2/r_+^2} e^{-\frac{r_+ - r_-}{r_+^2} r}}{r^2} dX_+ dX_- + r^2 d\Omega^2, \tag{37}$$

where, as above, r is defined implicitly:

$$X_- X_+ = -\exp\left(\frac{r_+ - r_-}{r_+^2} r_*\right), \tag{38}$$

and one can see that, $X_{\pm}$ can be continued beyond $r = r_+$ ($X_+ = 0$ or $X_- = 0$). Another, similar, coordinate transformation can be used to continue past $r = r_-$ up to $r = 0$, we will leave the details as an exercise. Once again, the singularity at $r = 0$ is physical:

$$R_{abcd} R^{abcd} = \frac{8(6M^2 r^2 - 12 M r e^2 + 7e^4)}{r^8}. \tag{39}$$

The horizon is the limit value of the r coordinates for geodesics parametrized by time t as counted by an asymptotic observer. The Killing vector field associated to time translations there ∂_t becomes null at $r = r_+$. However, the covariant derivative of t_a:

$$\nabla_a t_b = -\frac{2}{r^3}(Mr - e^2)(dt)_{[a}(dr)_{b]}, \tag{40}$$

has a well defined limit at the horizon. By the formulas presented in the preceding section, this results a constant surface gravity κ at the horizon $r = r_+$:

$$\kappa^2 = \left.\frac{(Mr - e^2)^2}{r^6}\right|_{r=r_+} \longrightarrow \kappa = \frac{(M^2 - e^2)^{1/2}}{2M[M + (M^2 - e^2)^{1/2}] - e^2} = \frac{r_+ - r_-}{2r_+^2} \tag{41}$$

Problem 4.1. *Verify that $\chi^a \nabla_a \chi^b = \kappa \chi^b$ at the horizon and find κ.*

The causal structure is obtained by the same methods as above. One sees, however, that now there will be an infinite number of asymptotic regions, depicted in Fig. 4. The first type, called I, has a regular horizon at $r = r_+$ and null, timelike and spacelike infinities just like Schwarzschild. The point where two regions type I meet is called a *bifurcation point*, and one can check that the Killing vector field ∂_t vanishes there – it is then a fixed point of the isometry. The region between the horizons r_- and r_+ is labeled II in the Figure. The inner horizon at r_- is a Cauchy horizon, and the region III inside it hides the singularity at $r = 0$, which is now timelike.

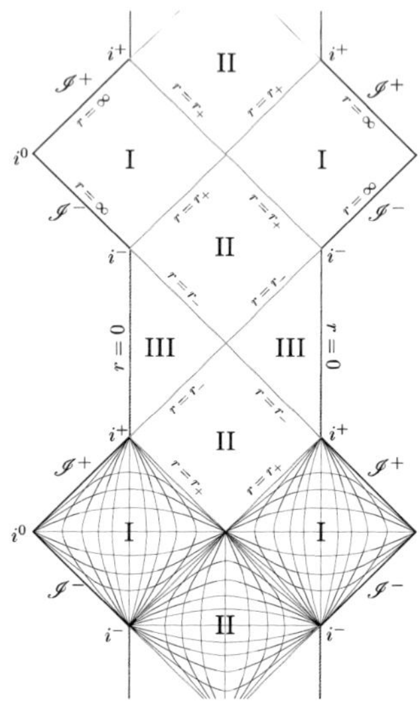

Fig. 4: The causal structure of the extension of the Reissner-Nordström space-time, assuming $r_+ > r_-$. Along with the asymptotically flat regions, which extend indefinitely into the future and the past, we see that the curvature singularity regions are now time-like.

4.1. *Near the Horizon*

Reissner-Nordström metric is simple enough to allow for exact calculations, but at the same time interesting enough to illustrate another general fact about black holes which Schwarzschild case did not allow. The positions of the horizons at $r = r_\pm$ are now tunable, whereas in Schwarzschild they where fixed at $r = 2M$ and $r = 0$ – which was not amenable to studies since it coincided with the position of the singularity. We can now study the local physics near $r = r_+$ and see how it changes with the position of the inner horizon r_-. Making $r = r_+ + y$, $y \ll 1$, the $r - t$ part of the metric changes to:

$$ds^2 \simeq -\frac{r_+ - r_-}{r_+^2} y dt^2 + \frac{r_+^2}{r_+ - r_-} \frac{dy^2}{y} = -\frac{r_+ - r_-}{4r_+^2} z^2 dt^2 + dz^2, \tag{42}$$

where we have made $y = (r_+ - r_-)(z/2r_+)^2$ in the last equation. One recognizes the Rindler metric with the form given by (10) if we rescale time:

$$\tau = \frac{r_+ - r_-}{2r_+^2} t = \kappa t, \tag{43}$$

so that $ds^2 = -z^2 d\tau^2 + dz^2$. As showed above, this metric is now part of the Minkowski metric, but, due to the time rescaling, the curves of constant z now correspond to observers of constant proper acceleration given by κ. The structure is sketched in the left side of Fig. 5.

In the above we assumed $r_+ > r_-$. The case $r_+ = r_-$ is quite different and it is known as the extremal Reissner-Nordström. Its causal structure is different, since there is no region II and no asymptotically flat regions I to the left of the diagram.

Problem 4.2. *Sketch the causal diagram of the extremal $r_+ = r_-$ ($e^2 = M^2$) Reissner-Nordström geometry.*

The near horizon behavior is also very different: making $r = r_+ + x$ in the metric (33), we have, near $x = 0$,

$$ds^2 \simeq -\frac{x^2}{r_+^2} dt^2 + \frac{r_+^2}{x^2} dx^2 = r_+^2 \frac{-dt^2 + dz^2}{z^2} \tag{44}$$

where $z = r_+^2/x$. This is actually a space of constant negative (scalar) curvature, called AdS_2. Its isometries include a "dilation":

$$\xi^a = t\frac{\partial}{\partial t} + z\frac{\partial}{\partial z}, \tag{45}$$

and as such, there is no scale associated to this near horizon structure, as opposed to the non-extremal case, where the surface gravity κ still played a role near the horizon.

The causal structure of AdS_2 is quite different from that of Rindler. It can be obtained through a series of coordinate changes. By assigning

$$X_0 = \frac{r_+}{2z}\left[z^2 - t^2 + 1\right], \quad X_1 = r_+\frac{t}{z}, \quad X_2 = \frac{r_+}{2z}\left[z^2 - t^2 - 1\right], \tag{46}$$

one arrives at the isometric embedding of AdS_2 in $\mathbb{R}^{1,2}$:

$$ds^2 = -dX_0^2 + dX_1^2 - dX_2^2 \tag{47}$$

With $X_0^2 - X_1^2 + X_2^2 = r_+^2$. Now, if we make:

$$X_0 = r_+ \cosh r \, \cos T, \quad X_1 = r_+ \sinh r, \quad X_2 = r_+ \cosh r \, \sin T. \tag{48}$$

We will arrive at the global metric of anti-de Sitter ($\cosh r = 1/\sin X$):

$$ds^2 = -\cosh^2 r \; dT^2 + dr^2 = \frac{-dT^2 + dX^2}{\sin^2 X}, \tag{49}$$

where only the patch depicted in the right hand side of Fig. 5 is covered by the original coordinates t, z. This is sometimes refereed to as the Poincaré patch. In the new coordinates, one can see that it can be now continued to all T, but the coordinate X is restricted to be in the interval $(0, \pi)$. The causal structure is the similar to a strip in Minkowski space, with spacelike infinity represented by the line $X = 0, \pi$, and null and timelike infinity collapsed to a point. As far as lightrays are concerned, the causal structure has the peculiar property that they arrive at $X = \pi$ at finite global time. All other lightlike excitations, however, are confined to a finite region of AdS_2. This has two very important consequences, which can be generalized to higher dimensions:

- The dynamics of massless particles – or fields – in Anti-de Sitter spacetimes have to be supplemented by boundary conditions at spacelike infinity. Most string-theoretical applications choose reflecting boundary conditions;
- The massive excitations of particles – or fields – in the near-horizon limit of extremal black holes are causally disconnected from the asymptotic infinity region – asymptotically flat in the Reissner-Nordström case. One can then choose values for the fields in the near horizon independently from those values at infinity. This fact is sometimes referred to as an "attractor mechanism" [11].

Problem 4.3. *Study the dynamics of time-like geodesics in* AdS_2 *and show that these remain in a finite portion of space for a finite value of the energy.*

4.2. *The Black Hole with Charge and Angular Momentum*

Generalizing the Schwarzschild and the Reissner-Nordström metrics, the Kerr-Newman metric includes angular momentum in the source:

$$\begin{aligned} ds^2 = &-\left(\frac{\Delta - a^2 \sin^2\theta}{\Sigma}\right) dt^2 - \frac{2a\sin^2\theta(r^2 + a^2 - \Delta)}{\Sigma} dt d\phi + \\ &\left[\frac{(r^2+a^2)^2 - \Delta a^2 \sin^2\theta}{\Sigma}\right] \sin^2\theta \, d\phi^2 + \frac{\Sigma}{\Delta} dr^2 + \Sigma \, d\theta^2, \end{aligned} \tag{50}$$

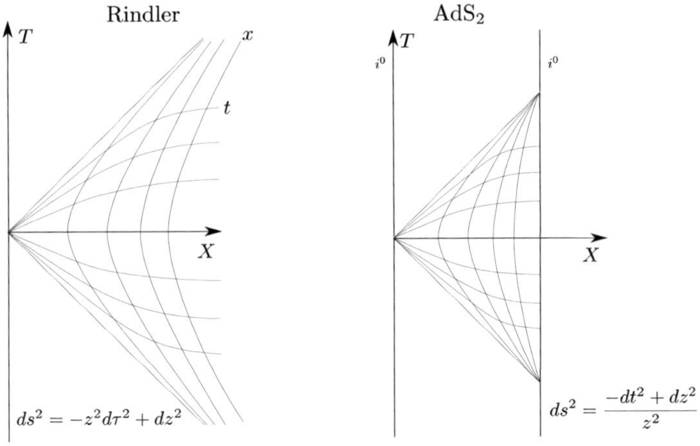

Fig. 5: The near horizon structure for Reissner-Nordström. To the left, the generic case $r_+ \neq r_-$. Vertical lines correspond to constant proper acceleration given by κ. To the right, the extremal case $r_+ = r_-$. Note that light rays reach spacelike infinity i^0 at a finite (global) time.

the electromagnetic potential is now $A_a = -\frac{er}{\Sigma}[(dt)_a - a\sin^2\theta(d\phi)_a]$, and the parameters are given by

$$\Sigma = r^2 + a^2\cos^2\theta, \qquad \Delta = r^2 + a^2 + e^2 - 2Mr, \tag{51}$$

with $a = J/M$ is a measure of the angular momentum of the black hole.

The Kerr, and Kerr-Newman solution have a rich history and profound impact in the development of General Relativity [9]. We will focus on its geometric properties. As above, there are horizons at $r = r_\pm = M \pm (M^2 - a^2 - e^2)^{1/2}$. The surface gravity at the horizon is computed by the same procedure as above:

$$\kappa = \frac{(M^2 - a^2 - e^2)^{1/2}}{2M[M + (M^2 - a^2 - e^2)^{1/2}] - e^2}, \tag{52}$$

which is constant at the horizon, We now have the angular velocity:

$$\Omega_H = \frac{a}{r_+^2 + a^2}, \tag{53}$$

also a constant. The causal structure is complicated, but with essentially the same structure as Reissner-Nordström: asymptotically flat regions, separate event and Cauchy horizons and bifurcation points. It is sketched in

Fig. 6. The physical (curvature) singularity is now at

$$\Sigma = r^2 + a^2 \cos^2\theta = 0, \tag{54}$$

which requires not only $r = 0$ but also $\theta = \pi/2$ to vanish. This can be checked to correspond to a ring topology singularity S^1.

It should be pointed out that the Killing vector field that becomes null at $r = r_+$ is now a combination of time translation and angular rotation:

$$\chi^a = \frac{\partial}{\partial t} + \Omega_H \frac{\partial}{\partial \phi}, \tag{55}$$

and this fact results in some geometric properties of the horizon, which we will explore in the next section below, as well as physical implications. Perhaps the most interesting of the latter is the fact that now there is a region where $g_{tt} < 0$ but still χ^a is spacelike, which means that one cannot have static curves with finite acceleration, but one can still escape the blackhole. This region is called the *ergosphere* and the absence of static observers there has serious implications for astrophysical models relying on black holes. We refer to [9] for a more thorough discussion.

Problem 4.4. *Work out the near horizon geometry for chargless $e = 0$ Kerr-Newman in the extremal case $a = M$. See [13] for help.*

5. The Laws of Black Holes Thermodynamics

With the previous examples in mind, we can try and single out some generic features of black hole solutions. All of the solutions above correspond to non-geodesically complete spacetimes, with the obvious patch only covering an asymptotically flat region, up to the outer horizon at $r = r_+$. For reasonable physical range of the parameters (positive M, etc.), the outer horizon is accompany of an inner horizon, at $r = r_-$, which hides a true curvature singularity at $r = 0$ in Schwarzschild and Reissner-Nordström case and $\Sigma = 0$ in the Kerr-Newman case. The inner horizon is then a Cauchy horizon, but the metric can be geodesically extended to define an infinite number of asymptotically flat regions and horizons.

The horizons themselves can be seen as regions where an isometry generator becomes null. This creates the notion of a *Killing horizon*, which we will explore here. Albeit of mathematical interest, the reader must be warned that this notion may or may not be useful in physical applications. At any rate, there are very good arguments [10] which leads one to expect a Killing horizon in stationary black holes. In more generic situations, the

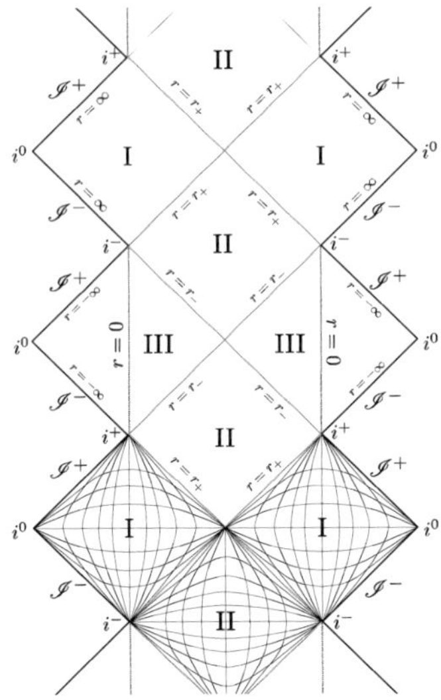

Fig. 6: The causal structure of Kerr(-Newman) black hole in the non-extremal case. Note that, since now the singularity has a ring topology, one could in principle avoid it to hit another asymptotically flat region at $r = -\infty$.

definition of marginally trapped surface – explained below – is much more useful, for instance in numerical simulations. Given the integrability implications of the existence of a Killing horizon, we will explore them here. A similar construction can be found in [10] and it is usually referred to as the *rigidity theorems.*

Suppose that our (four-dimensional) spacetime allows for two *commuting* Killing vector fields ξ^a, ϕ^a. In physical terms, we will think of one of them to be time translation and the other as axial rotations, both seen by an asymptotic observer. Our spacetime will In these coordinates $\xi^a = \partial_t$ and $\phi^a = \partial_\varphi$. Another way of phrasing this fact is by Frobenius theorem: the action of the two Killing vector fields on any point of spacetime will generate a 2-dimensional submanifold which is integrable, and their coordinates are given by t and ϕ. Such submanifolds are sometimes called distributions, and the condition for this to happen can be phrased as below:

$$d\xi \wedge \xi \wedge \phi = d\phi \wedge \xi \wedge \phi = 0 \tag{56}$$

On the other hand, the associated 1-forms to ξ^a and ϕ^a depend on terms of the metric:

$$\xi_a = -V(dt)_a + W(d\varphi)_a, \qquad \phi_a = W(dt)_a + X(d\varphi)_a, \tag{57}$$

where $V = -\xi_a\xi^a$, $W = \xi^a\phi_a$ and $X = \phi^a\phi_a$. The horizon is the region where $\rho^2 = VX + W^2$ vanishes. Now define $\Omega = -W/X$ and consider

$$X^2 d\Omega = -XdW + WdX. \tag{58}$$

A little algebra shows that

$$d\phi \wedge \phi = (WdX - XdW) \wedge d\varphi \wedge dt = -\frac{1}{\rho^2} X^2 d\Omega \wedge \xi \wedge \phi, \tag{59}$$

so that, when $\rho^2 = 0$, since the left hand side has a smooth limit, and the gradient of Ω lies in the plane spanned by ξ and ϕ. But this is not possible, since ξ and ϕ are Killing vector fields and Ω is a function constructed out of the metric. Therefore, $d\Omega$ has to vanish and then it is constant throughout the horizon. The limit of Ω as $\rho \to 0$ is called the *angular velocity*, Ω_H, of the horizon.

Now we define the Killing vector field

$$\chi = \frac{\partial}{\partial t} + \Omega_H \frac{\partial}{\partial \varphi}, \tag{60}$$

which becomes null at the horizon. By construction, then, it is also orthogonal to the horizon and hence hypersurface orthogonal $\chi \wedge d\chi = 0$. The surface gravity, κ, defined as:

$$\kappa^2 = -\frac{1}{2}(\nabla^a \chi^b)(\nabla_a \chi_b) \tag{61}$$

is also constant at the horizon (see, for instance, [12]). This comes about by noting that, on the horizon, $\chi^a \nabla_a \chi^b = \kappa \chi^b$, which results from the fact that χ^a is the only null vector normal to the horizon – itself a null surface. A lengthy series of manipulations (eqs. (12.5.24) to (12.5.30) in [1]) shows that:

$$\chi_{[a}\nabla_{b]}\kappa = -\chi_{[a}R_{b]}{}^c\chi_c, \tag{62}$$

which, for Einstein spaces, imply that $\nabla_a\kappa$ is parallel to χ_c and then constant over the horizon.

Problem 5.1. *Show that* $\chi^a\nabla_a\chi^b = \kappa\chi^b$ *on the horizon for the Schwarzschild, Reissner-Nordstrom and Kerr-Newman metrics.*

Along with the constancy of the surface gravity and the angular velocity over the horizon, one can show that the outer horizon itself is (linearly) stable under small metric perturbations. Due to the very specialized algebraic structure behind it – so far our working definition of the horizon stems from the vanishing of g_{tt}, a second order polynomial – one could think that, in more realistic settings, the horizon properties would be quite different. This does not seem so: at least perturbatively, the surface of the horizon does not seem to collapse or to bifurcate under small perturbations of the Kerr-Newman metric. One now tries to extend this result to generic black hole horizons, not necessarily Killing horizons, with limited success, and to higher dimensions, with greater success. This stability is the first indication of the thermodynamical nature of the horizon, in which the constancy of these "intrinsic quantities" have the status of the zeroth law:

Zeroth Law	The surface gravity κ and the angular velocity Ω_H are constant over a black hole horizon

One should at this point remember that the geometry of spacetime near the horizon is that of Rindler space – or AdS_2 in the extremal case. This feature also seems universal – in the case of non-zero angular momentum the $r-t$ portion of spacetime is fibered in terms of the other directions, see [13]. The near-horizon geometry presents then a higher degree of isometries if compared to the generic region. For instance, in the Kerr-Newman case we have two Killing vector fields, whereas in the near horizon limit we find 4: the 3 isometries of Rindler – secretly flat space – or AdS_2 – secretly a two-dimensional maximally symmetric space; along with the angular rotation. This "enhancement" seems to be generic: even in higher dimensional black holes we have such structure [14].

5.1. *The First Law*

The analogy between black hole horizons and thermodynamics can be furthered by considering quasi-static processes where matter is deposited in the black hole. In such processes, the perturbations of the stress-energy tensor are translated to small changes in the metric, which alter the values of the asymptotic charges of the black hole.

So our first task is to define global charges. This can be done for any covariant theory following a Nöther procedure [16]. We will perform the procedure for the Einstein-Hilbert Lagrangian density [15]. Upon a metric variation, the density changes by:

$$\delta(R\epsilon) = -2(R^{ab} - \frac{1}{2}Rg^{ab})\delta g_{ab}\epsilon + \nabla^a(\nabla^b\delta g_{ab} - g^{cd}\nabla_a\delta g_{cd}) \cdot \epsilon, \tag{63}$$

where ϵ is the invariant spacetime volume form and the $\cdot$ means contraction with the first index of ϵ. By Nöther's procedure, only the last term contributes to the conserved charge, since, on shell, the first term vanishes. The second term is actually a boundary term, by Stokes theorem:

$$\int_\Sigma d(v \cdot \epsilon) = \int_{\partial\Sigma} v \cdot \epsilon, \tag{64}$$

with $v_a = (\nabla^b\delta g_{ab} - g^{cd}\nabla_a\delta g_{cd})$. Also, one can verify that $(\nabla_a v^a)\epsilon = d(v\cdot\epsilon)$. Now, by assuming "on shell" perturbations, generated by a vector field ξ^a: $\delta g_{ab} = \mathcal{L}_\xi g_{ab} = 2\nabla_{(a}\xi_{b)}$, and the charge is actually defined in a 2-dimensional submanifold, which we will take to be an sphere with large radius approaching spacelike infinity S^2_∞:

$$Q[\xi] = -\frac{1}{8\pi}\int_{S^2_\infty} *d\xi, \tag{65}$$

where the numerical value was inserted for convenience, and $*$ denotes the Hodge star – contraction with the volume form ϵ. Although the final form for the conserved charge is quite simple, the procedure is very powerful and yields results in virtually any theory of gravitation.

Problem 5.2. *Show that this gives the right value for the mass of a Schwarzschild and Reissner-Nordström black holes if we take* $\xi = t_a$ *and* S^2_∞ *to be an "sphere at infinity"* $r \to \infty$.

In the Kerr Black hole case, the conserved quantities depend on both Killing vector fields ξ and ϕ. The expressions for the derivatives of ξ_a and ϕ_a can be worked from (57):

$$\begin{aligned} \nabla_a\xi_b &= -(dV)_{[a}(dt)_{b]} + (dW)_{[a}(d\varphi)_{b]}, \\ \nabla_a\phi_b &= (dW)_{[a}(dt)_{b]} + (dX)_{[a}(d\varphi)_{b]}, \end{aligned} \tag{66}$$

which, after substitution of V, W and X from the metric (50) results in a rather long expression. For the purposes of calculating the charge (65), however, we can cut some corners. The domain of integration will be a topological sphere, which in our coordinates is parametrized by θ and φ. Only the part of the integrand proportional to $d\theta\wedge d\varphi$ will then contribute. These come from the $dr \wedge dt$ term of the derivative of the Killing vector field, because of the contraction of the volume form. A straightforward but lengthy calculation shows:

$$\nabla_a \xi_b = -\frac{2M(r^2 - a^2\cos^2\theta)}{\Sigma^2} dt \wedge dr + \dots,$$
$$\nabla_a \phi_b = \frac{2aM(r^2 - a^2\cos^2\theta)\sin^2\theta}{\Sigma^2} dt \wedge dr + \dots, \tag{67}$$

which will yield the expected charges after integration.

The charges are important by themselves, but one can go further [15]. Stokes theorem states that the difference of the integral in (65) computed at two different surfaces is the integral of the exterior derivative of the integrand over the 3-volume between them:

$$\int_{S^2_\infty} *d\chi - \int_{S^2_i} *d\chi = \int_{\Sigma^*} d * d\chi \tag{68}$$

the right hand side is generically difficult to evaluate, but one can show that its *variation* vanishes, provided the equations of motion holds. Therefore:

$$\delta \int_{S^2_\infty} *d\chi = \delta \int_{S^2_i} *d\chi. \tag{69}$$

One now considers χ to be the Killing vector field which becomes null at the horizon, and $S^2_i = S^2_{\mathcal{H}}$ to the the Horizon itself. By the zeroth law, we have $\nabla_a \chi_b = \kappa \epsilon_{ab}$, where ϵ_{ab} is the normalized bivector normal to the horizon – this actually can be worked out from (67). Thus, the right hand side gives a term proportional to the area of the horizon, whereas the left hand side gives the charges at infinity, the total mass and angular momentum. Taking the variation, and using a "Gibbs relation" (see, for instance, [17]), one arrives at:

$$\delta M - \Omega_H \delta J = \frac{\kappa}{8\pi} \delta A, \tag{70}$$

which is now known as the first law of black hole thermodynamics.

First Law	In quasi static processes, energy (mass) and angular momentum transfers obey conservation of energy, if we interpret $\kappa\,\delta A/8\pi$ as the heat transferred to the black hole.

5.2. *The Second Law*

This results strongly hint that we can interpret the surface gravity κ as a measure of temperature and the area of the horizon as a measure of entropy. The latter is enforced from Raychaudhuri's equation. This equation deals with the evolution of a congruence of geodesics. Let us consider then such a congruence, but of null geodesics k^a tangent to the horizon. Being a geodesic, and being null, one can see that $\nabla_a k_b$ is normal to the horizon. One can therefore decompose the derivative in its "trace part", the antisymmetric part, and the traceless and symmetric part:

$$\nabla_a k_b \equiv \theta \hat{h}_{ab} + \hat{\omega}_{ab} + \hat{\sigma}_{ab}, \tag{71}$$

where $\equiv$ means that we are omitting terms parallel to k_a – being null, the horizon has the property that k^a is both parallel *and* normal to it. We introduced the induced metric $\hat{h}_{ab}$ on the normal subspace not parallel to k_a. Now, each term in (71) has an interpretation in terms of parallel transport: θ will be responsible for contraction of normal vectors, $\hat{\omega}_{ab}$ will rotate them, and $\hat{\sigma}_{ab}$ will shear them, transform infinitesimal circles normal to k^a into ellipses while maintaining the area.

Raychaudhuri's equation is now a statement on the derivative of the expansion factor θ as a function of the affine parameter of the geodesic:

$$\frac{d\theta}{d\lambda} = -\frac{1}{2}\theta^2 - \hat{\sigma}_{ab}\hat{\sigma}^{ab} + \hat{\omega}_{ab}\hat{\omega}^{ab} - R_{ab}k^a k^b, \tag{72}$$

which can be worked out from (71). One notes that, if we chose k^a to be an outward-directed null vector at the horizon, then it will be parallel to the Killing vector field χ^a and as such its expansion factor θ will vanish. The horizon is an example of what is called a marginally trapped surface. If, however, any regular matter passes through the horizon, then we can have $R_{ab}k^a k^b \neq 0$, and the horizon will move.

Matter satisfying the *null energy condition* will obey $R_{ab}k^a k^b \geq 0$, so, now k^a, which previously belonged to the horizon now is trapped: it will be inside the black hole. The black hole region then increases, and so its area. Therefore, under processes in which matter satisfying the null energy condition, the horizon area cannot decrease.

Problem 5.3. *Work out the constraint the null energy condition $R_{ab}k^a k^b \geq 0$ places on the stress energy tensor and on the pressure and energy density, assuming the latter describing a perfect fluid $T_{ab} = \rho u_a u_b + P(g_{ab} + u_a u_b)$, where u_a is normalized $u_a u^a = -1$.*

A more careful consideration of processes, for instance the near horizon behavior of quantum fields, ascribe exactly $\kappa/2\pi$ as the temperature of the black hole. This can be understood from the fact that the near horizon metric is Rindler (10), with "time" given by $\tau = \kappa t$ (43). Since τ translations are isometries of Rindler, then the scalar propagator in Rindler space will only depend on the relative τ-positions. In order to have a well-defined, analytical propagator, we have to take care about conical singularities at $\tau \simeq 0$ as we continue to imaginary τ – the analytically continued Rindler metric is now that of a cone. The conical singularity will not be there, and then the propagator will be analytic, if τ has period $2\pi i$, this means that t has now period $2\pi i/\kappa$. But imaginary periodicity of the propagator with respect to time has the interpretation of a temperature background. So we can think of $2\pi/\kappa$ as the inverse temperature. Now, from (70), if $\kappa/2\pi$ is the temperature, one can interpret $A/4$ as the entropy of the black hole. We now arrive at:

Second Law	Under generic processes involving matter satisfying the null energy condition, the area of the black hole cannot decrease.

5.3. *Consequences*

One can actually check that this assignment of the entropy and temperature is consistent with both classical processes and semiclassical processes like particle emission [18–20]. The recognition of the "Laws of black hole mechanics" [21] as principles governing the dynamics and interaction of black holes with matter is one of the pinnacles of the so-called "golden era of general relativity".

The laws, however, leaves us with a connundrum: black hole dynamics seem to be very rigid, and black hole systems are likewise very simple – simple enough so that we can compute exact solutions and draw generic conclusions based purely on our models. However, the attribution of the entropy as area (we will momentarily recover the SI units):

$$S = \frac{c^3}{\hbar}\frac{A}{4G_N}, \tag{73}$$

is at odds with the picture of entropy in statistical mechanics as the number of microscopic states available to a system where macroscopic (thermodynamical) variables is set. To put it bluntly: where are these states?

One could think that the degrees of freedom counted by (73) are gravitational. There are strong hints for that: the formula is universal, independent of the coupling between the black hole and other fields. However, there are uniqueness theorems (see [22] for a review) which state, that for generic conditions, the Kerr (or Kerr-Newman) metric is unique, with the only observables given by the total mass M, the angular momentum J and the electric charge e, assuming the existence of a regular event horizon – which requires $e^2 + a^2 \leq M^2$. This is the so-called "no-hair theorem". if the degrees of freedom are gravitational, therefore, they do not seem to be counted by general relativity. The situation in more than 4 dimensions is less stabilished, with more requirements necessary in order to prove uniqueness of black hole solutions. But one generically encounters the situation where a metric corresponding uniquely to a set of observed charges – including the total mass and angular momentum – is associated to a non-trivial entropy.

Perhaps the most dramatic effect from this interpretation is the Hawking radiation. Given the attribution of entropy to the black hole and the temperature as $\kappa/2\pi$, one can show that, by considering the behavior of quantum fields in a black hole background, an observer sitting at infinity should see a thermal flux of particles consistent to a black body at temperature $\kappa/2\pi$. So, while classically a black hole is defined as the region where not even lightrays can escape, this is not so if one considers quantum mechanics. Note that up to this point one still considers gravity to be classical, with the matter fields behaving quantum mechanically.

Problem 5.4. *Assume Stefan-Boltzmann law $\dot{M} \propto T^4$ and the expression of the temperature of the black hole as a function of its mass to show that the black hole evaporates in a finite time. Calculate this time for a solar-mass black hole.*

The fact that the evaporation process takes a finite, albeit very long, time poses all the questions about unitarity and consistency of time evolution in quantum gravity. There's the so-called "information paradox": how the uniqueness of the matter state giving rise to the black hole – one can think of an unbounded number of different initial conditions which will undergo gravitational collapse – will result in a generic, "thermal" natured Hawking radiation. The question could be in principle be answered if we had access to higher cumulants of the statistical distribution, but these depend on higher point functions and cannot be computed without a more complete quantum description of the black hole. This is sometimes stated in a very dramatic way by posing the question of the "black hole relic", depicted in Fig. 7.

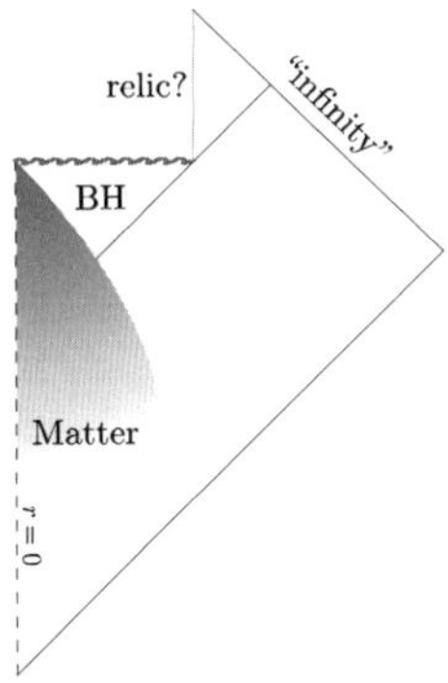

Fig. 7: The collapse and posterior evaporation depicted as a causal diagram. An observer sitting at "infinity" will at some time face with the question: "what remains after a black hole evaporates?"

6. Black Holes in Supergravity

Supersymmetric theories consider an equivalence between the number of fermionic and bosonic excitations of the system. Because of the spin-statistics theorem [23], such equivalence will mean that the field content of the theory will ge grouped in sets with different spins, called supersymmmetric multiplets [24]. These theories are usually justified from the physical point of view because the supersymmetric extension of the Poincaré group (isometries of Minkowski space) are the only possible extension of the symmetries of the scattering matrix (save from the global charges) [25], and the better renormalization properties for the theory – see for instance [26] and [27]. Mathematically, supersymmetric theories also relate the theory of representation theory and the topological properties of manifolds in a neat and deep way [28].

It will bring us far from the purpose of these lectures to attempt at any reasonable review of supersymmetric theories or supergravity (SUGRA) [29]. The latter includes in a generic setting all theories where the field content include a spin 2 field (gravity multiplet). They can be shown to descend via Kaluza-Klein compactification from an unique, 11-dimensional, maximally supersymmetric theory, 11d-SUGRA – see [30] and [31] for a review on 11d-SUGRA and Kaluza Klein compactification. By "maximally supersymmetric" one means that the gravity multiplet is formed with all fields with spin ≤ 2. They are also shown to describe the low energy limit of superstring theory and M-theory, so they are interesting to study quantum gravity. Our purpose here will be to describe one such classical background which corresponds to a black hole, but in which an microcanonical description of sorts is available.

The backgrouns of SUGRA are usually parametrized by non-zero values for fields in the bosonic sector. Turning on the fermionic sector usually leaves us with a less symmetric state. The bosonic sector of SUGRA generalizes Einstein-Maxwell: instead of a single vector field A_a coupled to gravity, one now considers generic $(p+2)$-form fields $F^{(p+2)}$ – the reason for the strange dimensional parametrization will become clear below. These fields couple with "charges" associated with dimension p objects, called p-branes, as depicted in Fig. 8:

$$S_{\text{int}} = \int_{\Sigma^{(p+1)}} A^{(p+1)}. \tag{74}$$

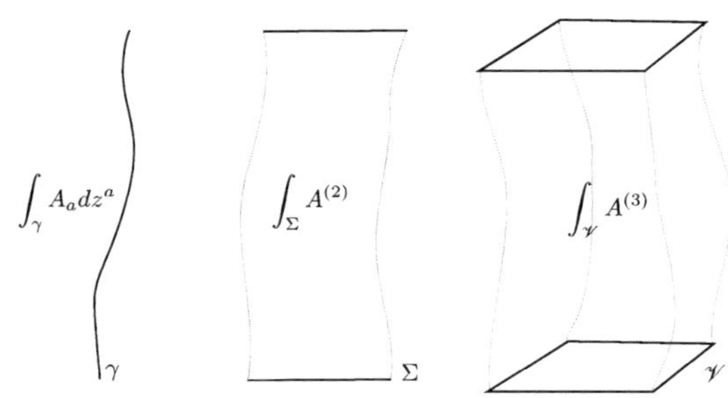

Fig. 8: The world line of p-dimensional "branes" spanning a $p+1$-dimensional trajectory in spacetime $\gamma, \Sigma, \mathcal{V}$, for $p = 0, 1, 2$, respectively. In each case, the brane serves as a source for a $p+1$ form potential $A^{(p+1)}$, with coupling displayed, and each potential is in turn associated to a $p+2$ form field $F^{(p+2)}$.

The form field associated to this potential is computed using the exterior derivative

$$F^{(p+2)} = dA^{(p+1)} \equiv \partial_{[\mu_1} A^{(p+1)}_{\mu_2 \ldots \mu_{p+2}]} dx^{\mu_1} \wedge \ldots \wedge dx^{\mu_{p+2}}, \tag{75}$$

and from this the Lagrangian density for the $(p+2)$-form fields $F^{(p+2)} \wedge$

${}^*F^{(p+2)}$ – see below. The $(p+2)$-form fields then interact with gravity much the same way the Maxwell field does, and in this sense generalize it for the case where there are extended charges. For instance, with the Lagrangian density as above, one can show that the analogue of the Coulomb field of a static, uniformly charged p-brane with volume element ϵ in D-dimensional Minkowski space is:

$$F^{p+2} = \frac{N}{r^{D-p-2}} dt \wedge dr \wedge \epsilon, \tag{76}$$

where r is the distance perpendicular to the p-brane. One can recognize in this expression essentially the solution of the delta-sourced Laplacian equation in $D-p$ dimensions. Likewise, much in the same way as in usual, 4-dimensional electrodynamics localized distributions of charge will have finite energy, in this generalization static solutions have finite energy as long as $p < (D-3)/2$.

The value of the charge N in (76) can be computed from an analogue of Gauss's law, sketched in Fig. 9. In most applications from string theory, the charge and mass (tension) of the brane are equal as per the BPS condition. These branes are not only sources for the $(p+2)$-form fields, but also string field theory semiclassical solutions where open string can end, the so-called D-branes. In the weakly coupled string regime, stability of such semiclassical solutions require some remaining supersymmetry, which in turn require the balance between charge and tension.

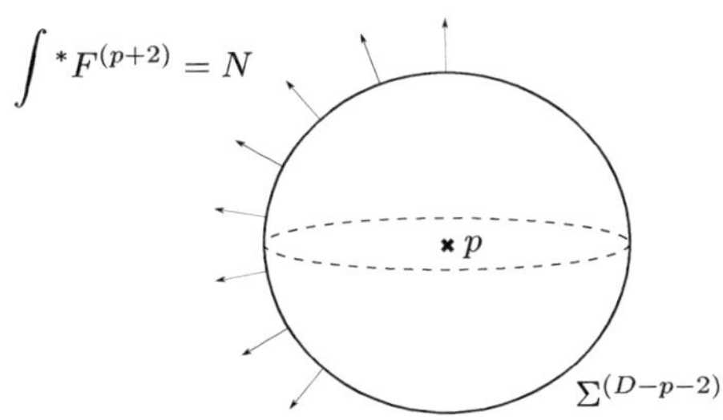

Fig. 9: The integral over a Gaussian hypersurface representing the charge of the p-brane. The brane itself is represented by a point, and we are depicting the transverse directions

One last ingredient for our application are the "magnetic dual" for p-branes. These are defined much in the same way as the electric charges, as a flux integral:

$$\int F^{(D-p-2)} = N_m, \tag{77}$$

but now involving the field itself and not the Hodge dual. A well-known quantum-mechanical argument from Dirac about the univalence of the wavefunction of the electric charge in the presence of a magnetic monopole implies a quantization condition for the charges, so N and N_m are thought of as integers. As far as the description of the magnetic charges are concerned, these $F^{(D-p-2)}$ form fields come from a $D-p-3$-form potential $F^{(D-p-2)} = dA^{(D-p-3)}$, and these in turn couple in the way depicted by Fig. 8 with $D-p-4$-branes. For our string-theoretical applications, $D = 10$ and we will think of the p-branes and the $(6-p)$-branes as magnetically dual to each other.

Like stated above, such a short introduction cannot have the aspiration to serve as a intelligible guide to supergravity. The interested reader should refer to the literature for all the details. For our purposes, which is the study of the geometry of a static p-brane solution, let us now consider the generic form of the action from SUGRA, involving the metric g_{ab}, a scalar field called the dilaton ϕ and the $(p+2)$-form field:

$$\begin{aligned} S_{\text{SUGRA}} &= \\ &= \frac{1}{16\pi G_{10}} \int d^{10}x\sqrt{-g}\left[e^{-2\phi}(R + 4\nabla_a\phi\nabla^a\phi) - \frac{1}{2}(F^{(p+2)})^2\right], \end{aligned} \tag{78}$$

where G_{10} is the 10-dimensional Newton's constant. Following the Euler-Lagrange procedure, we associate to this action the equations of motion:

$$\begin{aligned} R_{ab} = -4\nabla_a\phi\nabla_b\phi + \frac{e^{2\phi}(p+2)}{2}\left(F_{a\dots}F_b^{\ \dots} - \frac{p-1}{4(p+2)}F^2 g_{ab}\right) \\ \nabla_a F^{a\dots} = 0 \\ \nabla^2\phi - (\nabla\phi)^2 + \frac{1}{4}R = 0. \end{aligned} \tag{79}$$

In order to find static solutions, we begin with a metric Ansatz which is invariant by symmetries over the brane and by rotations in the transverse directions:

$$ds^2 = \underbrace{-e^{2A(r)}dt^2 + e^{2B(r)}dx_i dx^i}_{(p+1)\ \parallel} + \underbrace{e^{2C(r)}dr^2 + e^{2D(r)}r^2 d\Omega^2_{8-p}}_{(9-p)\ \perp}, \tag{80}$$

where $\parallel$ denotes the directions parallel to the p-brane – including time, and $\perp$ denotes the perpendicular directions. The form factors $A(r)$, $B(r)$, $C(r)$,

$D(r)$ will now determine the geometry. For the purely electric $N_m = 0$ case, the solution is:

$$\begin{aligned} ds^2 &= -H_+(r)H_-(r)^{-1/2}dt^2 + H_-(r)^{1/2}dx^i dx^i \\ &\quad + H_+(r)^{-1}H_-(r)^{\gamma}dr^2 + r^2 H_-(r)^{\gamma+1}d\Omega_{8-p}, \\ \gamma &= -\frac{1}{2} - \frac{5-p}{7-p}, \quad H_\pm(r) = 1 - \left(\frac{r_\pm}{r}\right)^{7-p} \end{aligned} \tag{81}$$

and

$$e^\phi = H_-(r)^{(p-3)/4}, \qquad {}^*F_{p+2} = N\omega_{8-p} \tag{82}$$

for the dilaton and the form field. The total tension and charge are:

$$T = \frac{\Omega_{8-p}}{2G_N}\left[(8-p)r_+^{7-p} - r_-^{7-p}\right], \qquad N = \frac{7-p}{2}(r_+ r_-)^{(7-p)/2}, \tag{83}$$

where Ω_{8-p} is the volume of an unit $8-p$-dimensional sphere. From these expressions one can recognize the same generic features as the 4-dimensional solutions considered in the previous Sections: there are two horizons, at $r = r_\pm$, and a metric singularity at $r = 0$. The value of the dilaton above controls the "effective Newton's constant", since it multiplies the Ricci scalar in the action (78). A self-consistent string description can only be achieved if the dilaton does not diverge, which can only happen in the case considered here if $p = 3$, where it is a constant. Note that in this case, the 3-branes can be both electric and magnetic sources for a 5-form field. As it happens, the reduction from type IIB string theory which delivers this type of supergravity, the 5-form field is self-dual, with equal values for the electric and magnetic charge, and in this case $F_5 = N(\omega_5 + {}^*\omega_5)$ may be self-dual. The BPS condition translates into $r_+ = r_-$, and in this case, the metric simplifies markedly:

$$\begin{aligned} ds^2 &= \left[1 + \left(\frac{r_0}{r}\right)^{7-p}\right]^{-1/2} (-dt^2 + dx_i dx^i) \\ &\quad + \left[1 + \left(\frac{r_0}{r}\right)^{7-p}\right]^{1/2} (dr^2 + r^2 d\Omega_{8-p}). \end{aligned} \tag{84}$$

In the near horizon limit, close to r_0, the metric reduces to a higher dimensional version of AdS_2:

$$ds^2 \simeq (r/r_0)^2(-dt^2 + dx_i dx^i) + (r_0/r)^2 dr^2 + r_0^2 d\Omega_5, \tag{85}$$

which corresponds to the geometry of $AdS_5 \times S^5$. This system can be shown to be an exact background of (type IIB) string theory and invaluable to many applications. We sketch the space in Fig. 10. The near-horizon region share the property of AdS_2 that time-like (positive mass squared) excitations that are generated in the near-horizon region remain there, so the system near the horizon decouples in a sense from the physics in the large r region. Thus $AdS_5 \times S^5$ can be seen – which adoption of proper boundary conditions for light-like excitations – as a background of string theory in its own right.

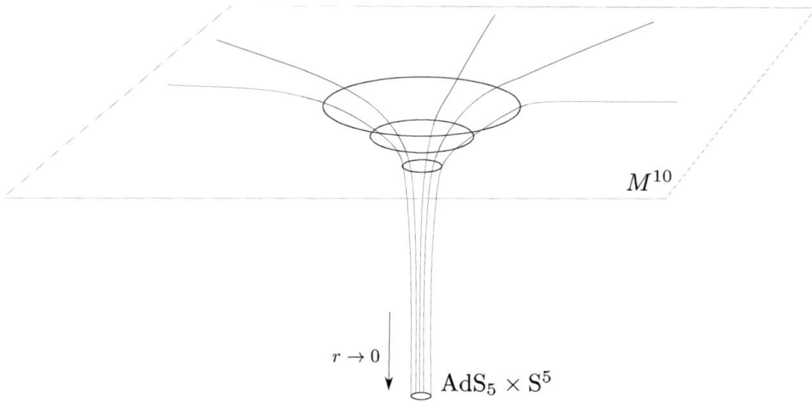

Fig. 10: The near horizon limit of the metric, depicted as the vertical "throat" $AdS_5 \times S^5$ region, and the large r, asymptotically flat region M^{10}.

The important message is to remember that this non-trivial metric is the result of the gravitational backreaction to the $D3$-brane system. So, there are two regimes of the same system where one can study an effective theory of perturbations of the system. In one of them, there is no gravity, and the excitations are described as fluctuations of the positions of the branes, which can be seen as open string excitations – gauge fields – between the brane. The amount of supersymmetry of the system constrains heavily the gauge field dynamics: in this particular example, it is the maximally supersymmetric $\mathcal{N} = 4$, four dimensional Yang-Mills theory based on SU(N). The other description involves gravity, and is based on fluctuations of the $AdS_5 \times S^5$ metric found above. The consistency between the two seemingly independent descriptions can be tested from quantities which do not depend on the gravity coupling. The statement that the two descriptions for generic brane systems are consistent is called the gauge-gravity correspondence.

7. The BTZ Metric and the D1-D5 System

The consistency between the two descriptions, one involving gravity and the other not, opens up the window for a microcanonical description of a black hole. We will now outline such a description for the D1-D5 brane system, where one can not only construct a brane system which, in the near-horizon limit, has a black hole metric, but also whose non-gravitational description can have the entropy measured. Moreover, we will see in the end that the entropy is counting geometric degrees of freedom. The latter feature is very special for the D1-D5 system, but nowadays a microcanonical description for the black hole entropy arising from brane systems can be obtained – and furthered – for a variety of scenarios.

Let us begin with the geometric part. The black hole discovered by Bañados, Teitelboim and Zanelli (BTZ) is an asymptotically anti-de Sitter metric in 3 dimensions. It is reviewed in [32]. The metric is given by:

$$ds^2 = -\frac{(r^2 - r_+^2)(r^2 - r_-^2)}{l^2 r^2} dt^2 + \frac{l^2 r^2}{(r^2 - r_+^2)(r^2 - r_-^2)} dr^2 + r^2 \left(d\phi - \frac{r_+ r_-}{l r^2} dt \right)^2 , \tag{86}$$

where ℓ is the AdS_3 radius. Its mass and angular momentum are parametrized by the positions of the inner and outer horizons $r_\pm$:

$$M = \frac{r_+^2 + r_-^2}{l^2}, \qquad J = \frac{2 r_+ r_-}{l}. \tag{87}$$

One can indeed check that the radial positions $r_\pm$ correspond to event horizons as defined in the preceeding lectures of this series.

One of the great surprises in finding the metric (86) was that the black hole properties in four dimensions seem intrinsically tied to the fact that the four dimensional geometry is not determined by Einstein's equations alone: indeed, the Einstein tensor – and then the Ricci tensor – has 10 independent components, whereas the Riemann tensor has 20. For instance, the Schwarzschild metric is a non-trivial solution to Einstein equations exactly because the non-vanishing components of the Riemann tensor are those which do not contribute to the Ricci tensor.

The BTZ metric, however, has the property that is maximally symmetric: its Riemann tensor is completely defined by the Ricci scalar, which is constant. Being maximally symmetric means that the metric can be locally put in the "pure AdS_3" form.

Problem 7.1. *Verify this claim and take* (86) *with* $r_- = 0$. *Verify that* *the coordinate transformation $t \to \alpha t$, $r \to r/\alpha$ and $\phi \to \alpha\phi$ changes M to $\alpha^2 M$!*

The black hole nature of (86), therefore, must derive from a global condition. We will show below that coordinate transformations such as the one above are not "true gauge transformations" in the sense that they change the value of the global charges. This fact, pointed out in [33] is fundamental for the microcanonical description of this black hole.

Before delving further into this question, let us consider the embedding of the BTZ metric into the brane system. This is accomplished by the so-called D1-D5 brane background in type IIB string theory. In this system have Q_1 D1-branes supplying electric charge for a F_3 form field and Q_5 D5-branes, supplying magnetic charges. The D5-branes are compactified in a 4-torus T^4 (sometimes another Calabi Yau space $K3$ is used, but the distinction will not be of relevance here). Finally, we include momentum for the branes on the direction x^5 given by p_5. The whole construction can be reviewed in [34] As per considerations from last section, the metric and the dilaton are:

$$\begin{aligned} ds^2 &= Z_1^{-1/2} Z_5^{-1/2} \left[\eta_{\mu\nu} dx^\mu dx^\nu + (Z_n - 1)(dt + dx_5)^2\right] \\ &\qquad + Z_1^{1/2} Z_5^{1/2} dx_i dx^i + Z_1^{1/2} Z_5^{-1/2} dx_m dx^m; \\ e^{-2\phi} &= Z_5/Z_1 = (r^2 + r_5^2)/(r^2 + r_1^2), \end{aligned} \tag{88}$$

where μ, ν span time, the x^5 and the radial directions to the system. The indices i, m span the other transverse directions to the $D1$ branes, with m parallel to the $D5$-system. The functions and parameters are given in terms of the charges and the momentum as

$$\begin{gathered} Z_1 = 1 + \frac{r_1^2}{r^2}, \quad r_1^2 = (2\pi)^4 g Q_1 \alpha'^3 / V_4; \qquad Z_5 = 1 + \frac{r_5^2}{r^2}, \quad r_5^2 = g Q_5 \alpha'; \\ Z_n = 1 + \frac{r_n^2}{r^2}, \quad r_n^2 = (2\pi)^5 g^2 p_5 \alpha'^4 / L V_4. \end{gathered} \tag{89}$$

In the near-horizon limit, $r \to 0$, metric decomposes into $\mathrm{AdS}_3 \times \mathrm{S}^3 \times \mathrm{T}^4$. In order to compute the Bekenstein-Hawking entropy we need to work in the Einstein metric – the one whose action is given by the usual Einstein-Hilbert term. As one can sort out from (78), the Einstein metric is related to the "string metric" appearing in the SUGRA action by a conformal transformation $g_E = e^{-\phi/2} g$. The area of the horizon at $r = 0$ is given in terms of the charges as:

$$S = A/4 = 2\pi (Q_1 Q_5 n_5)^{1/2}. \tag{90}$$

In terms of the brane system, the light excitations of the D1-D5 system correspond to strings stretching between the branes. For zero separation between the branes, these have zero mass. As one pulls one of the branes out, a superpotential for the D1-D5 system is generated:

$$V = \frac{1}{2\pi\alpha'}|X_i\chi - \chi Y_i|^2 + \frac{g_1^2}{4}D_1^A D_1^A + \frac{g_5^2}{4V_4}D_5^A D_5^A \tag{91}$$

where X_i, Y_i and χ are matricial degrees of freedom representing an open string stretching between any two of the Q_1 D1-branes, any two of the Q_5 branes and finally between a D1 and a D5-brane, respectively. These transverse bosonic D1-D5 strings are in number $4Q_1Q_5$ – the factor of 4 arising from the supersymmetric multiplet. Because of the large gauge symmetry $U(Q_1)$ and $U(Q_5)$ from the superposition of the branes, these D1-D5 strings are the only gauge independent part of the low lying spectrum. Therefore they span the gauge free part of the "moduli", the manifold of light excitations. This picture of excitations is muddled at finite coupling because gravity effects begin to interfere, but the important aspect of degeneracy is protected by non-renormalization theorems, themselves stemming from the SUSY structure.

7.1. *Counting States*

These low-lying excitations are two-dimensional in essence, spanning the x^5 dimension common to both the D1 and the D5 branes, and time. Along with this "spacetime picture", however, they are described in string theory from a worldsheet point of view in terms of a σ-model. The σ-model is model for string embeddings into the "target-space" metric given by the D1-D5 system. Since the latter is a solution of SUGRA, the RG flow of the σ-model is trivial and its "central charge" does not change over the flow. At weak coupling, each degree of freedom is well-modeled by a two-dimensional free bosonic system:

$$S = \frac{1}{4\pi}\int d^2z\ \partial\phi\bar{\partial}\phi, \tag{92}$$

which can be quantized from usual methods. In order to recover the entropy of the black hole, we will have to find the degeneracy of states at a given energy level. The stress energy tensor of this free bosonic system is $T_{\mu\nu}$ is traceless and the component zz:

$$T_{zz} = T(z) = \frac{1}{2} : \partial\phi(z)\partial\phi(z) : \tag{93}$$

is holomorphic by the equations of motion – likewise, the component $T_{\bar{z}\bar{z}}$ is antiholomorphic. The holomorphic current which can expanded in a Fourier series, with each term then corresponding to a conserved current:

$$L_n = \oint \frac{dz}{2\pi i} z^{n+1} T(z), \qquad T(z) = \sum_{n\in\mathbb{Z}} L_n z^{-n-2}. \tag{94}$$

Now, the L_n are the generators of conformal transformations in field space. Their algebra is then universal:

$$[L_n, L_m] = (n-m)L_{n+m} + \tfrac{1}{12}cn(n^2-1)\delta_{n+m}, \tag{95}$$

and called the Virasoro algebra. The parameter c controls the algebra and is called the central charge. For the free boson one can check that canonical commutations relations set $c = 1$, and for the free (chiral) fermion we have $c = 1/2$. Since the stress energy tensor for non-interacting fields is the sum of each constituent, it follows that the central charge is an additive quantity. For our D1-D5 system, we have 4 bosonic and 4 fermionic degrees of freedom from each string, so the total central charge is $6Q_1Q_5$.

Now, the spectrum of the theory in general is determined from a highest-weight construction from a "vacuum" state $|0\rangle$ which is annihilated by L_n, $n \geq 0$. The states are then constructed by acting on the vacuum with the "creation" operators L_n, $n \geq 0$. Because of the commutation relations, these states will have integer values for the energy L_0. Consider the many ways of forming states at level n:

$$L_{-n}|0\rangle, L_{n-1}L_{-1}|0\rangle, \dots, (L_{-1})^n|0\rangle. \tag{96}$$

Problem 7.2. *Show, using the Virasoro algebra* (95)*, that all states in* (96) *are eigenstates of* L_0 *with the same eigenvalue.*

The degeneracy, the number of states in (96), is equal to the number of partitions of n, the number of different ways we can write n as a sum of positive integers. This problem is then a direct application of number theory. Let us consider $P(n)$ the number of partitions of n. One can show that its generating function is:

$$F(x) = \sum_{n\geq 1} P(n)x^n = \prod_{n\geq 1} \frac{1}{(1-x^n)}. \tag{97}$$

We are interested in an estimation of $P(n)$ for large values of n. From the form of the expansion, we see that these should dominate the essential singularity of $F(x)$ at $x = 1$. Working the product expansion:

$$F(x) = \exp\left[-\sum_{n\geq 1}\log(1-x^n)\right]$$
$$= \exp\left[\sum_{m,n\geq 1}\frac{x^{mn}}{m}\right] = \exp\left[\sum_{m\geq 1}\frac{x^m}{m(1-x^m)}\right]. \qquad (98)$$

We can find an approximation of $F(x)$ for x near $x = 1$:

$$F(x) \approx \exp\left[\frac{1}{1-x}\sum_{m\geq 1}\frac{1}{m^2}\right] = \exp\left[\frac{\pi^2}{6(1-x)}\right]. \qquad (99)$$

Now, we have c such partitions of unity, so we have to consider the expansion of $(F(x))^c$. The coefficient of x^n can be extracted by a contour integral:

$$P(n) = \frac{1}{2\pi i}\oint\frac{(F(x))^c}{x^{n+1}}dx \approx Cn^{-A}\exp\left[2\pi\sqrt{\frac{c(n+1)}{6}}\right]. \qquad (100)$$

Applying this formula to the D1-D5 system, with $c = 6Q_1Q_5$ and $n = n_5$, gives the entropy which agrees up to logarithm corrections to the Bekenstein-Hawking formula.

The agreement between an string and a gravitational system seem to be nothing more than a coincidence. The description of the D1-D5 system we used in the σ-model is after all without gravity. One can show, however, that SUSY prevents this number to get corrections as one turns the string coupling – and hence gravity – on. So, even when the gravitational backreaction of the branes is strong, the formula of the entropy should hold. But in this regime, the description is exactly that of the near-horizon $AdS_3 \times S^3$ geometry described above. At low enough energies, the excitations on the S^3 decouple by the usual Kaluza-Klein argument and the degrees of freedom are just counted by the Bekenstein-Hawking formula.

We have thus succeeded in providing a microcanonical description of the BTZ entropy in terms of string excitations. One can further show that even the corrections to the entropy from (100) match what is expected from general relativity. But one is still lacking a geometric interpretation of those states counted by the Bekenstein-Hawking formula. We now turn to this problem.

7.2. *3-dimensional Gravity*

As stated before, in three dimensions the Riemann tensor is completely determined from the Ricci tensor. All Einstein spaces are maximally symmetric. The Ricci and Riemann tensor are of the form:

$$R_{ab} = -\frac{2}{\ell^2} g_{ab}, \qquad R_{abcd} = -\frac{1}{\ell^2}(g_{ac}g_{bd} - g_{ad}g_{bc}). \tag{101}$$

Since pure gravity in 3-dimensions does not have local excitations – no gravitons – it helps to consider an action for pure gravity which enforces the Riemann equation. Let us introduce the dreibein

$$g_{ab} = \eta_{ij} e^i_a e^j_b, \qquad \eta_{ij} = \mathrm{diag}(+,-,+), \tag{102}$$

and rewrite the Riemann tensor in terms of it:

$$R_{ab}{}^{ij} = -\frac{1}{\ell^2} e^i_{[a} e^j_{b]}. \tag{103}$$

We now introduce the spin connection ${\omega^i}_j$ via the Maurer-Cartan equations:

$$de^i + {\omega^i}_j \wedge e^j = 0, \tag{104}$$

and consider the "Lorentz dual" $\omega^i = \frac{1}{2}\epsilon^{ij}{}_k {\omega^k}_j$. Now, the equation for (103) and the Maurer-Cartan equations can be written in a compact form:

$$de + \omega \wedge e + e \wedge \omega = 0; \qquad d\omega + \omega \wedge \omega = -\frac{1}{\ell^2} \tag{105}$$

with $e = e^i \gamma_i$ and $\omega = \omega^i \gamma_i$ with γ_i a proper choice of generators of $\mathrm{SL}(2,\mathbb{R})$.

Problem 7.3. *Choose*

$$\gamma_1 = \sigma_1 = \begin{pmatrix} 0 & 1 \\ 1 & 0 \end{pmatrix}, \quad \gamma_2 = i\sigma_2 = \begin{pmatrix} 0 & 1 \\ -1 & 0 \end{pmatrix}, \quad \gamma_3 = \sigma_3 = \begin{pmatrix} 1 & 0 \\ 0 & -1 \end{pmatrix} \tag{106}$$

and verify that (105) *imply the Maurer-Cartan equations and* (103).

By defining $A_\pm = \omega \pm \frac{1}{\ell} e$, one can write (105) as zero curvature $\mathrm{SL}(2,\mathbb{R})$ connections:

$$F_\pm = dA_\pm + A_\pm \wedge A_\pm = 0, \tag{107}$$

whose natural action is the Chern-Simons form:

$$\sum_{i\in\{+,-\}} \int_M \mathrm{Tr}\left(A_i \wedge dA_i + \frac{2}{3} A_i \wedge A_i \wedge A_i \right). \tag{108}$$

Now, unlike the four-dimensional case, the boundary conditions are fundamental for singling out the spaces we want to consider. Following [35], an asymptotically anti-de Sitter spacetime is characterized by the existence of spacial infinity as the limit of the level curves of a function Ω as one takes $\Omega \to 0$. If we choose lightcone coordinates u, v for spacial infinity, it is a nice result that the line element near spacial infinity is of the form:

$$ds^2 = \ell^2 \frac{d\Omega^2 - dudv}{\Omega^2} + \mathcal{O}(\Omega^0) = \ell^2(d\rho^2 - e^{2\rho}dudv) + \mathcal{O}(\rho^0), \qquad (109)$$

where we made $\Omega = e^\rho$. This condition can be translated to an asymptotic form of the connections $A_\pm$:

$$\ell A_\pm = \pm e^1\gamma_1 + (-e^3 \pm e^2)\gamma_2 + (-e^2 \pm e^3)\gamma_3. \qquad (110)$$

And therefore, to have an asymptotically anti-de Sitter space, one needs to impose the boundary conditions: $A_+^+ = A_-^- = 0$ and $A_\pm^1 = \pm\frac{1}{\ell}d\rho$.

Now, due to its topological character, the variation of the Chern-Simons action (108) is a boundary term:

$$\begin{aligned} \delta \int_M \mathrm{Tr}\left(A_\pm \wedge dA_\pm + \frac{2}{3}A_\pm \wedge A_\pm \wedge A_\pm\right) \\ = 2\int_M \mathrm{Tr}\,(\delta A_\pm \wedge F_\pm) + \int_{\partial M} \mathrm{Tr}\,(\delta A_\pm \wedge A_\pm)\,. \qquad (111)\end{aligned}$$

And the boundary term vanishes from the conditions (110) above. Further setting a gauge where $A_\pm^1 = \pm\frac{1}{\ell}d\rho$, one can rewrite the action as

$$\int_M \mathrm{Tr}\left(A_\pm^t \wedge \partial_\rho A_\pm^t \wedge d\rho\right) + 2\int_M \mathrm{Tr}\left(A_\rho \wedge F^t\right) - \int \mathrm{Tr}[d^t(A^t \wedge A_\rho)], \quad (112)$$

where A^t refers to the transverse part to ρ of the connection. Varying the action with respect to A_ρ, we have $F^t = 0$, or rather:

$$A^t = -d\mathcal{G}\mathcal{G}^{-1}; \qquad \mathcal{G} = e^{-\rho\gamma_1/\ell}g(u,v). \qquad (113)$$

Now, rewriting the action in terms of $g(u,v)$ yields a Wess-Zumino-Witten (WZW) action:

$$\begin{aligned} S[g] = -\frac{\ell}{8\pi G}\int_{\partial M} \mathrm{Tr}(g^{-1}\partial_u g g^{-1}\partial_v g)du\,dv \\ -\frac{\ell}{48\pi G}\int_M \mathrm{Tr}(g^{-1}dg \wedge g^{-1}dg \wedge g^{-1}dg) \qquad (114)\end{aligned}$$

with an analogue for A_-. We are still halfway into the gauge-fixing process. The expression for the connection is

$$A_\pm = \pm d\rho\gamma_1 - e^{\pm\rho\gamma_1/\ell}dg(u,v)g^{-1}(u,v)e^{\mp\rho\gamma_1/\ell}. \qquad (115)$$

Now, the vanishing of $A^{\pm}_{\pm}$ in (110) have to be imposed through a Lagrange multiplier. By a Gauss parametrization of SL(2, $\mathbb{R}$):

$$g(u,v) = e^{\chi\gamma_-}e^{2\phi\gamma_1}e^{\varphi\gamma_+}. \tag{116}$$

One arrives at the action:

$$S_+ = \int d^2x[\partial_u\phi\partial_v\phi + e^{2\phi}\partial_u\varphi\partial_v\chi + \theta(e^{2\phi}\partial_u\varphi \pm 1/\ell)], \tag{117}$$

which is purely two-dimensional. One can see this as the spacetime version of the theory of fluctuations of the D1-D5 system last section, in the strongly coupled regime. Of course, the form is completely set by covariance in three-dimensions, so one does not need string theory to arrive at it. By completing the gauge fixing procedure, enforcing the constraint and using the rest of the gauge symmetry to fix $\mu = 0$, we arrive at the free action for ϕ, just like (92). However, the field ϕ is not a free field, because of the dimensionful nature of the constraint. Generically, the change under conformal transformations is generated in field space by the stress-energy tensor (93), which now will have to be modified. The only other term with the right scaling dimensions can be added to the stress-energy:

$$T(u) = \frac{1}{2}\partial\phi\partial\phi + k\partial^2\phi, \tag{118}$$

and such modification can be expected to change the central charge. To see how, consider the current:

$$J_L = -\partial_u g g^{-1} du = \frac{1}{\ell}\gamma_+ du + \frac{1}{2k\ell}L(u)\gamma_- du, \tag{119}$$

one can see that it changes by a conformal transformation in the u plane:

$$g \to e^{\log\partial_u w\gamma_1}e^{\ell\partial_u \log\partial_u w\gamma_+}g \tag{120}$$

like

$$\tilde{L}(w) = (\partial_u w)^2[L(u) - \frac{k}{4}\{w;u\}], \tag{121}$$

and the last term yields $c = 3k = 3\ell/2G$. It should be stressed that, unlike the results obtained above for the entropy of the D1-D5 system, these considerations are purely classical. There is no justification at this point to consider the representation of the Virasoro algebra as creation/annihillation operators acting on a "vacuum state".

In the end, the full metric obtained is defined up to two functions $L(u)$ and $\bar{L}(v)$:

$$\ell^{-2}ds^2 = d\rho^2 + e^{2\rho}dudv + L(u)du^2 + \bar{L}(v)dv^2 + L(u)\bar{L}(v)e^{-2\rho}dudv. \tag{122}$$

This solution is singularity-free for constant L and $\bar{L}$. This is actually the BTZ metric written in another coordinate system.

Problem 7.4. *Work the coordinate transformation that brings* (86) *to the canonical BTZ form* (86).

With the procedure outlined, one can make wild guesses about the results obtained in the last section. For instance, the function $L(u)$ plays the role of the stress-energy tensor, and a constant value will yield L_0, and likewise for $\bar{L}(v)$. By blindingly applying Cardy's formula (100) one arrives to:

$$S = 2\pi\sqrt{\frac{cL_0}{6}} + 2\pi\sqrt{\frac{c\bar{L}_0}{6}}, \tag{123}$$

with $c = 3\ell/G$, $L_0 = M + J$ and $\bar{L}_0 = M - J$, one arrives at the right Bekenstein-Hawking entropy. But of course this is a classical system, and the "states" being counted by this entropy are nothing but improper conformal transformations (120) which change the values of observables. They have therefore a classical interpretation. But one needs an underlying unitary quantum theory – string theory in this case – to understand this entropy as microscopic states.

8. Closure

This series of lectures gave some examples of the classical puzzle posed by the laws of thermodynamics of black holes and one such attempt to understand the thermodynamics in a statistical mechanical (quantum) setting. In this particular setting, the entropy associated to the black hole counts geometries which are inequivalent in the sense of their asymptotic structure but still have the same global charges. One should note that in string theory, there are a number of such descriptions for a variety of black hole settings. In a sense the description is very accurate, yielding the right Bekenstein-Hawking value even for theories of gravitation not sharing the Einstein-Hilbert action, or including string corrections. Almost none of these cases, however, allow for a geometric description of the entropy – the proposal in [14] being a notable exception to the rule.

In the three-dimensional case, however, the theory has a geometric interpretation to the start: if one takes the Chern-Simons (108) action as a starting point, the observables are holonomies, and although quantization is not straightforward, there is a sense in which large gauge transformations – which changes the values of the holonomies – are observables for generic values of the couplings. String theory did give the input that one can consider a quantum unitary theory underlying it, but the question remains if one could not make sense of it without string theory. At any rate, the tools developed here could help in diagnosing these geometric degrees of freedom in more general cases. All in all, the problem of giving a consistent quantum description to the gravitational collapse and the Hawking radiation remains outstanding, for all our knowledge.

Acknowledgements

The author is indebted to the organizers of the Villa de Leyva school, notably A. Reyes-Lega and S. Paycha, and the whole audience of students, which were very active and interested.

References

[1] R. M. Wald, *General Relativity*, The University of Chicago Press, Chicago (1981).

[2] H. Stephani, D. Kramer, M. A. H. MacCallum, C. Hoenselaers and E. Herlt, *Exact Solutions of Einstein's Field Equations*, Cambridge University Press, Cambridge 2003.

[3] B. R. Greene, *"String theory on Calabi-Yau manifolds,"* [arXiv:hep-th/9702155].

[4] J. Sparks, *"Sasaki-Einstein Manifolds,"* Surveys Diff. Geom. **16**, 265 (2011) [arXiv:1004.2461 [math.DG]].

[5] S. S. Gubser, *"TASI lectures: Special holonomy in string theory and M theory,"* [arXiv:hep-th/0201114].

[6] S. Chandrasekhar, *An Introduction to the Study of Stellar Structure*, Dover Pub., (1967).

[7] J.-P. Lasota, E. Gourgoulhon, M. Abramowicz, A. Tchekhovskoy and R. Narayan, *"Extracting black-hole rotational energy: The generalized Penrose process,"* Phys. Rev. D **89**, no. 2, 024041 (2014) [arXiv:1310.7499 [gr-qc]].

[8] D. Christodoulou, *The Formation of Black Holes in General Relativity*,

[9] S. A. Teukolsky, *The Kerr Metric*, Class. Quant. Grav., **2015**, 32, 124006.

[10] B. Carter, *Black Holes Equilibrium States*, in *Black Holes - Les Astres Occlus* (C. DeWitt and B. DeWitt, eds.), Les Houches 1972. Gordon & Breach, New York, 1973.

[11] A. Sen, *"Black hole entropy function, attractors and precision count-*

ing of microstates", *Gen.Rel.Grav.* **40:2249-2431,2008** (Aug., 2007) [arXiv:0708.1207 [hep-th]].

[12] B. Carneiro da Cunha and A. de Queiroz, *"Killing Horizons and Spinors,"* Phys. Rev. D **89**, 104031 (2014) [arXiv:1312.4911 [hep-th]].

[13] J. M. Bardeen and G. T. Horowitz, *"The Extreme Kerr Throat Geometry: A Vacuum Analog of AdS(2) x S(2),"* Phys. Rev. D **60**, 104030 (1999), [arXiv:hep-th/9905099].

[14] M. Guica, T. Hartman, W. Song, and A. Strominger, *"The Kerr/CFT Correspondence"*, arXiv:0809.4266 [hep-th]

[15] R. M. Wald, *"Black hole entropy is the Noether charge,"* Phys. Rev. D **48**, 3427 (1993) [arXiv:gr-qc/9307038].

[16] V. Iyer and R. M. Wald, *"Some properties of Noether charge and a proposal for dynamical black hole entropy,"* Phys. Rev. D **50**, 846 (1994).

[17] B. C. da Cunha and A. R. de Queiroz, *Kerr-CFT From Black-Hole Thermodynamics*, *JHEP* **08** (2010) 076, [arXiv:1006.0510 [hep-th]].

[18] J. D. Bekenstein, *"Black holes and entropy,"* Phys. Rev. D **7**, 2333 (1973).

[19] J. D. Bekenstein, *"Generalized second law of thermodynamics in black hole physics,"* Phys. Rev. D **9**, 3292 (1974).

[20] S. W. Hawking, *"Particle Creation by Black Holes,"* Commun. Math. Phys. **43**, 199 (1975) [Commun. Math. Phys. **46**, 206 (1976)].

[21] J. M. Bardeen, B. Carter and S. W. Hawking, *"The Four laws of black hole mechanics,"* Commun. Math. Phys. **31**, 161 (1973).

[22] P. O. Mazur, *"Black hole uniqueness theorems,"* hep-th/0101012.

[23] R. F. Streater and A. S. Wightman, *"PCT, Spin and Statistics, and all that"*, Princeton University Press, Princeton (US) (2000).

[24] J. Wess and J. Bagger, *"Supersymmetry and Supergravity"*, Princeton University Press, Princeton (US) (1992).

[25] S. Coleman and J. Mandula, *"All possible symmetries of the S-matrix,'* Phys. Rev. **159** (1967) 1251.

[26] N. Seiberg, *"Naturalness versus supersymmetric nonrenormalization theorems,"* Phys. Lett. B **318**, 469 (1993) [arXiv:hep-ph/9309335].

[27] N. Seiberg, *"Supersymmetry and Nonperturbative beta Functions,"* Phys. Lett. B **206**, 75 (1988).

[28] E. Witten, *"Supersymmetry and Morse theory,"* J. Diff. Geom. **17**, Np. 4 (1982), 661-692.

[29] S. J. Gates, M. T. Grisaru, M. Rocek and W. Siegel, *"Superspace Or One Thousand and One Lessons in Supersymmetry,"* Front. Phys. **58**,

1 (1983) [hep-th/0108200].

[30] T. Appelquist, A. Chodos and P. G. O. Freund, *"Modern Kaluza-Klein Theories,"*, READING, USA: ADDISON-WESLEY (1987) 619 P. (FRONTIERS IN PHYSICS, 65).

[31] M. J. Duff, B. E. W. Nilsson and C. N. Pope, *"Kaluza-Klein Supergravity,"* Phys. Rept. **130**, 1 (1986).

[32] S. Carlip, *"The (2+1)-Dimensional black hole,"* Class. Quant. Grav. **12**, 2853 (1995) [arXiv:gr-qc/9506079].

[33] J. D. Brown and M. Henneaux, *"Central Charges in the Canonical Realization of Asymptotic Symmetries: An example from three-dimensional gravity"*, Commun. Math. Phys. **104**, 207-226 (1986).

[34] J. Polchinski, *"String theory. Vol. 2: Superstring theory and beyond,"* Cambridge University Press, Cambridge (UK) (1998).

[35] A. Ashtekar and A. Magnon, *"Asymptotically anti-de Sitter space-times,"* Class. Quant. Grav. **1**, L39 (1984).

REAL RELATIVISTIC FLUIDS IN HEAVY ION COLLISIONS

E. Calzetta

Departamento de Física, Facultad de Ciencias Exactas y Naturales, Universidad de Buenos Aires and IFIBA, CONICET, Ciudad Universitaria, Buenos Aires 1428, Argentina
E-mail: calzetta@df.uba.ar

The theory of real relativistic fluids is in the rather unique situation that there is a natural relativistic extension of the nonrelativistic theory, but it is physically untenable [1]. On the other hand, mounting evidence that matter created in relativistic heavy ion collisions behaves as a relativistic fluid with small but finite viscosity has given the quest for an alternative a definite goal [2]. We shall review different approaches to relativistic real fluids, their link to relativistic kinetic theory, and their application to the analysis of heavy ion collisions [3].

Keywords: Relativistic fluids; Navier-Stokes equations; heavy ion collisions.

1. Introduction

The goal of these lectures is to discuss the theory of relativistic real fluids under the excuse of its application to the description of relativistic heavy ion collisions [4, 5]. More concretely, our goal is to discuss, in the simplest possible terms, why the "Einstein elevator" concept is not useful in the development of this theory. Namely, we cannot build a theory of relativistic real fluids by just asking that it reduces to its nonrelativistic counterpart, namely the Navier–Stokes equations, in an inertial frame where the fluid is at rest. For this reason we shall begin with a "derivation" of the nonrelativistic theory of ideal fluids (those which flow with no entropy production) from thermodynamics [6]. For ideal fluids, the relativistic theory may be found as the covariant extension of the nonrelativistic theory enforced in the rest frame. We shall then show that already in this simple framework we may obtain a working picture of how a relativistic heavy ion collision works, yielding remarkably accurate predictions regarding several concrete observables.

Among these observables there is the so-called elliptic flow, namely the anisotropy of the particle yield in the transverse plane to the beam direction. If the matter deposited in the collision region behaves as a nearly ideal fluid, then elliptic flow arises as the simple transduction of pressure gradients in the original configuration into velocities in the asymptotic expanding state. This is an obvious prediction of the Euler equations which is surprisingly hard to reproduce in non-hydrodynamic models. The success in predicting that there must be an elliptic flow is one of the most compelling reasons to believe that matter in the collision region, from some very early time after the actual collision up to the break up in the individual hadrons which are eventually detected, behaves as a fluid. However, ideal hydrodynamics overestimates the amount of elliptic flow [4]. This suggests there must be a mechanism that tends to isotropize flow, and indeed viscosity acts in precisely that direction. Therefore, the next natural step is to consider hydrodynamic models based on real fluid hydrodynamics.

At this point we return to the formal theory, and generalize our previous "thermodynamic" derivation to the case of real fluids. The resulting non relativistic theory is given by the Navier–Stokes equations, which is satisfactory, but the relativistic theory emerging from the "Einstein elevator" is essentially flawed - if covariance is enforced, then the theory has no stable solutions [1]. How to get out of this cul de sac is the subject of the rest of the lectures.

Since the "thermodynamic" approach led us nowhere, we shall begin anew from a more fundamental point of view, that provided by relativistic kinetic theory. In this case, there is an agreed upon relativistic formulation to build on [7, 8]. The problem is how to reduce it to the hydrodynamic level. In the nonrelativistic case, the method of choice is the so-called Chapman–Enskog expansion, which leads to the Navier - Stokes equations [9]. In the relativistic case, it works likewise, and so it must be rejected. However, a second well known technique, the Grad expansion, is successful in building a relativistic hydrodynamics where linear perturbations of an equilibrium state evolve causally.

However, the application to relativistic heavy ion collisions forces us to face the relationship between kinetic and hydro descriptions a second time, because what is actually measured are individual hadrons. This means that there must be some space time "break up" or "freeze out" surface (or region) where the fluid nucleates into individual particles that fly away to the detectors (a process that is by no means simple, but does not belong to these lectures). We must be able to read the one particle distribution function of these particles out of the hydrodynamic state of the fluid just before break up. If we simple invert the Grad expansion, we are led to a one particle distribution function which is not nonnegative throughout phase space, a defect that contaminates the predictions of the theory regarding observables which are sensitive to high momenta.

This drawback of the Grad approach, and also the need to include nonlinear corrections to the linearized equations of motion without spoiling stability and covariance, prompted us to suggest that the Grad approach may be just the linear approximation to a more comprehensive view of the kinetic - hydro relationship [10]. The basic insight is that any realistic kinetic theory displays a whole hierarchy of relaxation times, from the relaxation times characteristic of large scale inhomogeneities to the much shorter relaxation times of hard modes. Fluctuations in the hydrodynamic modes relax on the longer time scales, and are perceived by the harder modes as externally imposed thermodynamic forces which prevent their relaxation to true equilibrium. In other words, the truly kinetic modes relax not to equilibrium but to a nonequilibrium steady state constrained by the instantaneous configuration of the hydrodynamic modes. It has been known from long ago that such steady s tates are the solutions to variational problems [11, 12]. Prigogine among others has proposed that they are the extrema of the entropy *production*, as opposed to the extrema of the entropy itself, which are the true equilibria [13]. Known proofs of the so-called "Prigogine theorem" are restricted to linear irreversible thermodynamics [14, 15]; we shall appeal to it on a heuristic rather than formal basis - the idea is that most kinetic modes are in the linear regime most of the time anyway, so a theory which is good in the linear regime is good enough to compute global observables such as stress tensor components, but we shall not attempt to formulate this insight in any rigorous way [16, 17].

We shall therefore conclude these lectures with a brief presentation of this "entropy production variational method", its relationship to positivity of the one particle distribution function, and its nonlinear generalization.

The lectures are organized as follows. In next section we present the theory of relativistic ideal fluids, as derived from thermodynamics, and its application to the description of relativistic heavy ion collisions (RHICs). The main success of the theory, namely the prediction of elliptic flow, is also its fatal drawback, because experimental data show that ideal hydrodynamics overestimates the flow anisotropy [4]. We are therefore motivated to extend the theory to include viscosity, whereby we hit on the stability problem [1, 18, 19]. We present the theory of real fluids in Section III; since the thermodynamic derivation fails, we base our discussion on relativistic kinetic theory. We must confront the *closure* problem, that is, to relate the one particle distribution function in the kinetic description to the hydrodynamic degrees of freedom. We present the Chapman-Enskog solution (which reproduces the results from the thermodynamic analysi s, and therefore is unsuitable) and the Grad solution, and finally identify the Grad solution of the closure problem as the linearized approximation to an approach based on the entropy production variational method (EPVM). We conclude with some brief final remarks.

If relativistic hydrodynamics may be founded on kinetic theory, kinetic theory itself comes from quantum field theory, the main subject matter of this school. We show the main lines of the derivation of kinetic theory from quantum field theory in the Appendix, which relies heavily on Ref. [3].

2. From Thermodynamics to Hydrodynamics

2.1. *Nonrelativistic Ideal Fluids*

Let us begin by reviewing non relativistic hydrodynamics, as "derived" from thermodynamics. The basic tenets of thermodynamics we need to keep in mind are the following: we have a (simple) system described by some intensive parameters (temperature T, chemical potential μ, pressure p, etc.) whose meaning we take for granted (*e.g.* we already know everything about the zeroth law) and extensive parameters, such as energy U, entropy S, volume V , and particle number N (we use particle number for concreteness, but any, or several, conserved charge(s) would serve just as well). In equilibrium, all these quantities are position independent. Their first deviations from equilibrium are related by the first law

$$TdS = dU + pdV - \mu dN.$$

Extensive quantities are homogeneous functions of each other, so we must have

$$TS = U + pV - \mu N.$$

From the differential of this second identity we obtain the Gibbs–Duhem relation

$$dp = sdT + nd\mu,$$

where $s = S/V$ and $n = N/V$ are the entropy and particle number densities, respectively. This means

$$\left.\frac{\partial p}{\partial T}\right|_{\mu} = s = \frac{\rho + p - \mu n}{T},$$

$$\left.\frac{\partial p}{\partial \mu}\right|_{T} = n,$$

where $\rho = U/V$ is the energy density.

Let us now consider a non equilibrium, inhomogeneous configuration. The above relations still hold for the densities of the conserved quantities energy, momentum (which we must include) and particle number. The dynamics is given by the conservation laws

$$\frac{\partial \rho}{\partial t} = -\nabla_j J_u^j, \quad \frac{\partial p^i}{\partial t} = -\nabla_j T^{ij}, \quad \frac{\partial n}{\partial t} = -\nabla_j J^j.$$

Here ρ, p^i and n are the energy, momentum and charge densities, J_u^j, T^{ij} and J^j the respective currents. T^{ij} must be symmetric [20]. The entropy density

$$s = \frac{1}{T}\left[\rho + p - v_i p^i - \mu n\right]$$

obeys the first Law

$$T ds = d\rho - v_i dp^i - \mu dn,$$

which implies the Gibbs–Duhem relation

$$dp = sdT + p^i dv_i + nd\mu.$$

In these formulae we have introduced the velocity v_i and the chemical potential μ. We therefore have

$$\begin{aligned}\frac{\partial s}{\partial t} &= \frac{1}{T}\left\{\frac{\partial \rho}{\partial t} - v_i \frac{\partial p^i}{\partial t} - \mu \frac{\partial n}{\partial t}\right\} \\ &= \frac{-1}{T}\left\{\nabla_j J_u^j - v_i \nabla_j T^{ij} - \mu \nabla_j J^j\right\}.\end{aligned}$$

We then have

$$\begin{aligned}\frac{\partial s}{\partial t} = &-\nabla_j\left\{\frac{1}{T}\left[J_u^j - v_i T^{ij} - \mu J^j\right]\right\} + \left[J_u^j - v_i T^{ij} - \mu J^j\right]\nabla_j\left(\frac{1}{T}\right) \\ &-\frac{1}{T} T^{ij}\nabla_j v_i - \frac{1}{T} J^j \nabla_j \mu.\end{aligned}$$

To proceed we write $J^j = nv^j + \chi^j$. In the terms involving the velocity we eliminate the gradient of the chemical potential by using the Gibbs–Duhem relation

$$\begin{aligned}\frac{\partial s}{\partial t} = &-\nabla_j\left\{\frac{1}{T}\left[J_u^j - v_i T^{ij} - \mu J^j\right]\right\} + \left[J_u^j - v_i T^{ij} - \mu J^j\right]\nabla_j\left(\frac{1}{T}\right) \\ &-\frac{1}{T} T^{ij}\nabla_j v_i - \frac{1}{T} v^j\left[\nabla_j p - s\nabla_j T - p^i \nabla_j v_i\right] - \frac{1}{T}\chi^j \nabla_j \mu;\end{aligned}$$

so far

$$\frac{\partial s}{\partial t} = -\nabla_j \left\{ \frac{1}{T} \left[J_u^j + pv^j - v_i T^{ij} - \mu J^j \right] \right\} + \left[J_u^j + pv^j - v_i T^{ij} - \mu J^j - Tsv^j \right] \nabla_j \left(\frac{1}{T} \right) - \frac{1}{T} \left[T^{ij} - p^i v^j - p\delta^{ij} \right] \nabla_j v_i - \frac{1}{T} \chi^j \nabla_j \mu.$$

Using again the second law, we may write

$$J_u^j + pv^j - v_i T^{ij} - \mu J^j - Tsv^j = J_u^j - (\rho + p)\, v^j - v_i \left(T^{ij} - p^i v^j - p\delta^{ij} \right) - \mu \chi^j,$$

and so

$$\frac{\partial s}{\partial t} = -\nabla_j \left\{ \frac{1}{T} \left[J_u^j + pv^j - v_i T^{ij} - \mu J^j \right] \right\} + \left[J_u^j - (\rho + p)\, v^j - v_i \left(T^{ij} - p^i v^j - p\delta^{ij} \right) \right] \nabla_j \left(\frac{1}{T} \right) - \frac{1}{T} \left[T^{ij} - p^i v^j - p\delta^{ij} \right] \nabla_j v_i - \chi^j \nabla_j \alpha, \tag{1}$$

where $\alpha = \mu/T$. Since we have allowed for the possibility of the energy and particle number flux not being collinear, we must adopt a convention about what is the fluid velocity. We shall adopt the so-called Landau–Lifshitz prescription, namely, that J_u^j and p^j vanish when $v^j = 0$ [21].

We define an ideal fluid as one that flows with no entropy production. Thus for an ideal fluid

$$T^{ij} = p^i v^j + p\delta^{ij}, \qquad J_u^j = (\rho + p)\, v^j, \qquad \chi^j = 0.$$

The symmetry of T^{ij} requires p^i to be proportional to v^i. The equations for an ideal fluid are thus

$$\frac{\partial n}{\partial t} + \nabla_j n v^j = 0,$$

$$\frac{\partial \rho}{\partial t} + \nabla_j (\rho + p)\, v^j = 0,$$

$$\frac{\partial p^i}{\partial t} + \nabla_j p^i v^j + \delta^{ij} \nabla_j p = 0.$$

2.2. *Relativistic Ideal Fluids*

We now generalize the above framework of thermodynamics to a relativistic fluid evolving in a spacetime with an arbitrary metric $g_{\mu\nu}$. All derivatives shall be covariant derivatives with respect to the Levi–Civita connection, so that $g_{\mu\nu;\rho} = 0$. We write the Minkowsky metric as $\eta^{00} = -1$, $\eta^{0i} = 0$ and $\eta^{ij} = \delta^{ij}$, where a 0 refers to coordinate $x^0 = ct$

To simplify matters we will describe the construction of a covariant theory in terms of a set of rules [8]:

(a) Intensive quantities (T, p, μ) are associated with scalars, which represent the value of the quantity at a given event, as measured by an observer at rest with respect to the fluid.

(b) Extensive quantities (S, V, N) are associated with vector currents S^μ, u^μ, N^μ. If a given observer measures a density x and a flux J^i_x for the quantity X, then $X^\mu = \left(x, J^i_x/c\right)$. The quantity u^μ associated with volume is the fluid 4-velocity; in Minkowsky space

$$u^\mu = \frac{1}{\sqrt{1 - \frac{v^2}{c^2}}} \left(1, \frac{v^i}{c}\right).$$

u^μ obeys the additional constraint $u^2 = -1$. We call density tout court the density measured by an observer comoving with the fluid, namely $x = -u_\mu X^\mu$.

If the quantity X is conserved, then $X^\mu_{;\mu} = 0$. If we consider the variation of the entropy content within some spatial region as a function of time, the second law demands that the increase in entropy should be higher than the entropy flow through the boundary . Thus the covariant statement of the second law is that entropy production must be positive, *i.e.* $S^\mu_{;\mu} \geq 0$.

(c) Energy and momentum are combined into a single extensive quantity described by an energy-momentum tensor $T^{\mu\nu}$ which is symmetric. Therefore any given observer will identify the energy density as T^{00}, the energy flux as cT^{0i}, the momentum density as T^{i0}/c and the momentum flux as T^{ij}. The symmetry of the energy momentum tensor implies that the momentum density is c^{-2} times the energy flux [21]. We define the rest frame of the fluid as the frame where the energy flux vanishes. As before, the energy density ρ is defined as the energy density in the rest frame. Both the fluid velocity and energy density may be found from the eigenvalue equation $T^{\mu\nu}u_\nu = -\rho u^\mu$

Let us now describe a relativistic ideal fluid, namely one that flows with no entropy production. We shall proceed by correspondence between the relativistic and non relativistic theories. To this end, we assume that in the rest frame we have the decomposition

$$N^\mu = (n, 0)\,, \qquad T^{\mu\nu} = \begin{pmatrix} \rho & 0 \\ 0 & p\delta^{ij} \end{pmatrix}.$$

So in an arbitrary frame we must have

$$N^\mu = nu^\mu,$$
$$T^{\mu\nu} = \rho u^\mu u^\nu + p\Delta^{\mu\nu}, \tag{2}$$

where $\Delta^{\mu\nu} = g^{\mu\nu} + u^\mu u^\nu$. The entropy current S^μ in the rest frame is given by $S^\mu = (s, 0)$, where $Ts = p + \rho - \mu n$. In an arbitrary frame $S^\mu = su^\mu$ or

$$S^\mu = -T^{\mu\nu}\beta_\nu + pu^\mu - \alpha N^\mu,$$

where the inverse temperature vector is$\beta^\mu = T^{-1}u^\mu$ and $\alpha = \mu/T$. Observe that $T^{-2} = -\beta_\mu\beta^\mu$; then

$$S^\mu_{;\mu} = -\beta_\nu T^{\mu\nu}_{;\mu} - \alpha N^\mu_{;\mu}. \tag{3}$$

This means that entropy production vanishes for an ideal fluid, provided the conservation laws of energy-momentum and particle number hold.

2.3. *Ideal Hydrodynamic Models of Heavy Ion Collisions*

The theory of relativistic real fluids has been brought to the limelight by the mounting evidence that one such system has been seen in actual experiments, namely, relativistic heavy ion collisions (RHICs). Indeed, one can get a working understanding of RHICs from the theory of ideal fluids we have just described. Nevertheless, precisely the demands of a better fit between theory and experiment will prompt us to develop a theory or real fluids, namely, to add viscosity to the flow equations. In this section we shall briefly describe RHICs, and discuss what can and cannot be reproduced by models based on ideal hydrodynamics.

RHICs have been produced in several experiments, namely the Super Proton Synchrotron (SPS) at CERN, the Relativistic Heavy-Ion Collider (RHIC) at Brookhaven and the Large Hadron Collider (LHC) once again at CERN. The RHIC experiments in particular are described in detail in the so-called 'white papers', which are possibly the most reliable source on the subject [22–25]. One of the goals of the RHIC program is to probe into possible new phases of nuclear matter at high energies. In such a high energy phase, matter is expected to form a plasma of gluons and (massless) quarks (quark-gluon plasma, QGP).

Virtually all theoretical analysis of RHICs assume a space-time picture of the collision provided by the Bjorken model [26]. The colliding nuclei are seen as slabs of quark and gluon matter. In the center of mass frame, both slabs approach each other at near light speed. Upon collision, the two slabs of matter will mostly go through each other, leaving behind a wake of hot plasma. We may then distinguish three different regions: the two fragmentation regions corresponding to the receding slabs, and the central region corresponding to the plasma in between. We are interested in phenomena in the central region.

At the time of crossing a number of hard scattering processes will occur, whose products will reach directly the detectors. These hard processes are unrelated to the nonequilibrium dynamics of the plasma; and may presumably be predicted on perturbative QCD grounds. In what follows, we will assume this hard component has been isolated despite great difficulty to achieve this in reality.

The hot plasma will expand and cool, and eventually fragment into ordinary particles in flight intercepted by the detectors. We wish to predict the number of particles of each species to be detected, as a function of the angle θ between the direction of flight and the direction z of the beam. It is remarkable that with this simple picture we can state a first observable prediction already.

Indeed, because of Lorentz contraction, we may think of the approaching slabs as infinitely thin in the direction of motion z, and in a first approach to the problem, as infinite and homogeneous in the transverse directions x and y. This picture is invariant under boost in the z direction, and so is the final distribution of particles. So if we parametrize the momentum of an out-going particle as $p^0 = E$, $p^3 = p$ and $\left(p^1, p^2\right) = p_\perp$, then the distribution of particles may depend only upon the transverse momentum and $E^2 - p^2 = m^2 + p_\perp^2$. In particular, it must be independent of θ, since $\cos\theta \sim p/E$ is not invariant. It is conventional to plot the yield of the collision in terms of the *rapidity* Y, defined by $p/E \equiv \tanh Y$, or rather the pseudorapidity $\eta = -\ln\tan\left[\theta/2\right]$, $\tanh\eta = p/\left|\mathbf{p}\right|$. Rapidity and pseudo-rapidity agree at momenta which are large compared to the mass of the particle. Then the prediction in this picture is that there is a *plateau* in the (pseudo) rapidity distribution, at least for small rapidity ($|\eta| \to \infty$ corresponds to the fragmentation rather than the central region). Although this prediction is not quantitatively borne out by the RHIC data [23], the experimental curve flattens enough at low rapidity that it may be accepted as a working rough approximation.

We may elaborate the Bjorken picture further. Let us assume that the plasma is formed on the plane $z = 0$ at the time $t = 0$ of the collision, and then expands along the z direction. A given plasma element will cool according to its own proper time τ. Eventually, at some given constant τ

surface, the plasma will be cold enough (and/or dilute enough) to break up into hadrons. Assuming that the product hadrons are thermally distributed, massless and at zero chemical potential, the Bose-Einstein distribution predicts that the energy per particle is $\epsilon/n = 2.7\ T$. Since temperature is constant on the break up surface, this means that in all collisions particles should have the same average energy. Indeed, it is observed that the energy per particle is about 0.8 GeV, regardless of center-of-mass energy and impact parameter.

To obtain a more quantitative description of the process, we describe the plasma as a relativistic ideal fluid. To close the hydrodynamic system of equations we must provide the equation of state. The central feature of this is the "softening" near the critical point, meaning that the speed of sound $c_s^2 = \partial p/\partial\epsilon \to 0$ as we approach the transition point. The softening of the equation of state affects the evolution of the fireball, which then becomes a signal of whether the transition point has been reached or not.

Since perfect fluids conserve entropy, the total entropy within the fireball remains constant, and T scales as $V^{-1/3}$. So, if the expansion is one-dimensional, and we consider the volume enclosed between two fixed rapidities, then $T \sim \tau^{-1/3}$, where τ is the proper time. In particular, the energy density scales as $\tau^{-4/3}$.

We now consider more closely the phenomenon of break-up [4, 5]. Assume this occurs on a 3-dimensional surface Σ defined by some equation $\Sigma(x^\mu) = 0$. If x_0 is a solution, then the normal vector at x_0 is $n^\mu = (\alpha)\,\Sigma^{,\mu}$, $\alpha = \left(-\Sigma_{,\mu}\Sigma^{,\mu}\right)^{-1/2}$. We shall assume that n^μ is timelike. The invariant measure on Σ is given by $d^3\sigma = d^4x\,\delta(\Sigma)\,\alpha^{-1}$.

Let us assume that right up to break-up we can describe matter as a perfect relativistic fluid, and as a noninteracting relativistic gas thereafter. Let $K_a = \partial/\partial x^a$ be the four Killing vectors of Minkowski space. Then Gauss' theorem shows that the quantities $n_\mu K_{a\nu} T^{\mu\nu}$ and $n_\mu N^\mu$ are continuous across the break-up surface (we shall consider only one conserved current, corresponding to, say, the baryon number). These conditions plus the equation of state of the hadronic phase define the energy density, pressure, baryon number density (or equivalently, the temperature and chemical potential) and the four-velocity of the hadrons at break-up. The detailed spectrum is found by assuming that the hadrons are thermally distributed.

The total number of emitted particles is

$$\int d^3\mathbf{x}\ K_{0\mu} N^\mu_{\text{had}},$$

where the integral is over some t = constant surface well to the future of the collision. Because of Gauss theorem, we may replace the integral by an integral over the break-up surface (we may have to complete this surface to get a Cauchy surface, but the particle density flux will vanish on these additions anyway). But then we may use the matching conditions to express this integral in terms of the particle current before break-up. We obtain the total number of emitted particles as

$$\int d^4x \, \delta(\Sigma) \, \Sigma_{,\mu} N^{\mu}_{\text{hydro}}.$$

In practice, we may wish to smear a little the position of the break-up surface, thus writing the total number of emitted particles as

$$\int d^4x \left[\frac{e^{-\Sigma^2/2(\Delta\Sigma)^2}}{\sqrt{2\pi}\,(\Delta\Sigma)} \right] \Sigma_{,\mu} N^{\mu}_{\text{hydro}}.$$

The total number of particles of species i with momentum p^{μ} is

$$g_i \int d^4x \left[\frac{e^{-\Sigma^2/2(\Delta\Sigma)^2}}{\sqrt{2\pi}\,(\Delta\Sigma)} \right] \delta\left(p_i^2 - m_i^2\right) \frac{\Sigma_{,\mu} p_i^{\mu}}{\left[\exp\left(-\beta_\nu p_i^\nu - \mu_i\right) - \varepsilon_i\right]},$$

where the temperature four vector and chemical potentials are read out from the (ideal) energy momentum tensor and charged currents of the fluid just before break up.

The two basic observables are the total number of particles with transverse (respect to the beam axis) momentum $p_\perp$, which is usually given in terms of the transverse mass $m_\perp^2 = m^2 + p_\perp^2$, and the elliptic flow coefficient v_2, which results from fitting the particle spectrum in the transverse plane to a second harmonic $(1 + 2v_2\,(p_\perp)\cos 2\phi)$, where ϕ is the angle measured from the reaction plane. This is equivalent to considering an elliptic fireball, in which case v_2 measures the eccentricity of the ellipse. The first harmonic is called directed flow, and would represent a shifted spherical fireball in the transverse plane.

The agreement of predictions from hydrodynamical simulations with experimental data is good, provided the simulation is started very early (earlier than 1 fm/c after the collision). If one believes that the validity of hydrodynamics demands (local) equilibration, this very short time is somewhat of a puzzle.

2.4. *Elliptic Flow*

Of the several predictions of hydrodynamic models, elliptic flow is one of the most compelling, because it is hard to match by alternative models. Let us show how elliptic flow arises in hydro models.

Let the velocity of the fluid (with respect to laboratory time) in the longitudinal direction be $dz/dt = v^z$. We may define a longitudinal rapidity

$$y_L = \frac{1}{2}\ln\frac{1+v^z}{1-v^z}.$$

Let us take a boost invariant flow in the beam direction as background solution. Then $v^z = z/t$ and the longitudinal rapidity becomes identical with the space time rapidity

$$y_L = \eta = \frac{1}{2}\ln\frac{t+z}{t-z}.$$

We introduce Milne time $\tau = \sqrt{t^2 - z^2}$ and Milne coordinates $\xi^\alpha = (\tau, \eta, x, y)$. Define $dx^\mu/d\tau = (t/\tau, z/\tau, \vec{v}_\perp)$. We find

$$-\eta_{\mu\nu}\frac{dx^\mu}{d\tau}\frac{dx^\nu}{d\tau} = 1 - v_\perp^2.$$

Introducing the transverse rapidity

$$y_\perp = \frac{1}{2}\ln\frac{1+v_\perp}{1-v_\perp},$$

so that $v_\perp = \tanh y_\perp$, the four velocity becomes

$$u^\mu = \cosh y_\perp \;(\cosh\eta, \sinh\eta, \vec{v}_\perp)\,;$$

or else, in Milne coordinates,

$$u^\alpha = \cosh y_\perp \;(1, 0, \vec{v}_\perp)\,.$$

The interval element in Milne coordinates is $ds^2 = -d\tau^2 + \tau^2 d\eta^2 + dx^2 + dy^2$, and so there are nontrivial Christoffel symbols

$$\Gamma^\tau_{\eta\eta} = \tau \qquad \text{and} \qquad \Gamma^\eta_{\tau\eta} = \frac{1}{\tau}.$$

In the strictly boost invariant case $y_\perp = 0$ we have $u_{\eta;\eta} = \tau$ and $u^\alpha_{;\alpha} = 1/\tau$, although $u^\beta u_{\alpha;\beta} = \Gamma^\tau_{\alpha\tau} = 0$. The covariant Euler equation is identically satisfied, $n \propto \tau^{-1}$, while the energy density $\rho \propto \tau^{-4/3}$ for conformal matter.

Obviously boost invariant expansion cannot create elliptic flow. The next step is to consider a nontrivial dynamics in the transverse plane. It is convenient to parametrize the transverse plane in terms of Lagrangian coordinates $\vec{q}$. Then the full coordinates are $\zeta^\rho = \left(\tau, \eta, q^1, q^2\right)$; the transformation back to Euler coordinates is given by the system

$$\frac{\partial x^i}{\partial \tau}|_{q^a} = v^i\left(\tau, x^i\right),$$

with initial condition $x^i(\tau = 0) = q^i$. Now write

$$\begin{aligned} ds^2 &= -d\tau^2 + \tau^2 d\xi^2 + \delta_{ij} dx^i dx^j \\ &= -d\tau^2 + \tau^2 d\xi^2 + \delta_{ij} \left(\frac{\partial x^i}{\partial q^a} dq^a + v^i d\tau \right) \left(\frac{\partial x^j}{\partial q^b} dq^b + v^j d\tau \right). \end{aligned}$$

In the new coordinates the metric reads

$$g_{\rho\sigma} = \begin{pmatrix} -\left(1 - g^{ab} N_a N_b\right) & 0 & N_b \\ 0 & \tau^2 & 0 \\ N_a & 0 & g_{ab} \end{pmatrix},$$

where

$$g_{ab} = \delta_{ij} \frac{\partial x^i}{\partial q^a} \frac{\partial x^j}{\partial q^b}, \qquad N_a = v_i \frac{\partial x^i}{\partial q^a}.$$

Observe that $g^{ab} N_a N_b = v_i v^i = \tanh^2 y_\perp$, so $g_{00} = -1/\cosh^2 y_\perp$. The lapse function is then $N = 1$ and the four metric determinant is $g = -\tau^2 g^{(2)}$, where $g^{(2)} = \det\, g^{ab}$. In Milne-Lagrange coordinates the four velocity is $u^\alpha = (\cosh y_\perp, \vec{0})$, so

$$u^\nu_{;\nu} = \frac{1}{\tau\sqrt{g^{(2)}}} \frac{\partial}{\partial \tau} \tau \sqrt{g^{(2)}} \cosh y_\perp,$$

and the ideal hydrodynamic equations

$$\begin{aligned} \frac{\dot{T}}{T} + \frac{1}{3} u^\nu_{,\nu} &= 0, \\ \dot{u}^\mu + \Delta^{\mu\nu} \frac{T_{,\nu}}{T} &= 0, \end{aligned}$$

give

$$T = \frac{T_0(\vec{q})}{\left[\tau \sqrt{g^{(2)}} \cosh y_\perp\right]^{1/3}}$$

and

$$\cosh^2 y_\perp \Gamma^a_{00} + \frac{1}{T}\left[N^a T_{,\tau} + g^{ab} T_{,b} - N^a N^b T_{,b}\right] = 0.$$

We shall consider only linear deviations from homogeneity, where

$$x^a = q^a + \xi^a(\vec{q}, \tau),$$

with $\xi^a(\vec{q}, 0) = 0$, and

$$N^a = v^a = \frac{\partial}{\partial \tau} \xi^a(\vec{q}, \tau), \qquad \sqrt{g^{(2)}} = 1 + \xi^a_{,a}, \qquad \Gamma^a_{00} = \frac{\partial^2}{\partial \tau^2} \xi^a(\vec{q}, \tau).$$

So, writing $T_0(\vec{q}) = T_0(1 + \delta(\vec{q}))$,

$$\frac{\partial^2 \xi_a}{\partial \tau^2} - \frac{1}{3\tau}\frac{\partial \xi_a}{\partial \tau} - \frac{1}{3}\xi^b_{,ba} = -\delta_{,a}.$$

Given our initial conditions, there are no transverse velocity components. Writing $\xi_a = \phi_{,a}$ we get

$$\frac{\partial^2 \phi}{\partial \tau^2} - \frac{1}{3\tau}\frac{\partial \phi}{\partial \tau} - \frac{1}{3}\boldsymbol{\Delta}\phi = -\delta.$$

Write

$$\delta = \int \frac{d^2k}{(2\pi)^2}\, e^{ikq}\delta_k.$$

Then,

$$\phi = \int \frac{d^2k}{(2\pi)^2}\, e^{ikq}\phi_k,$$

$$\frac{\partial^2 \phi_k}{\partial \tau^2} - \frac{1}{3\tau}\frac{\partial \phi_k}{\partial \tau} + \frac{k^2}{3}\phi_k = -\delta_k.$$

Writing $\dot{\phi_k} = \varphi_k$, we obtain

$$\frac{\partial^2 \varphi_k}{\partial \tau^2} - \frac{1}{3\tau}\frac{\partial \varphi_k}{\partial \tau} + \frac{1}{3}\left[k^2 + \frac{1}{\tau^2}\right]\varphi_k = 0.$$

The solution, which is regular at $\tau = 0$, is $\varphi_k = c_k \tau^{2/3} J_{1/3}\left(k\tau/\sqrt{3}\right)$. As $\tau \to 0$ we get $\varphi_k \approx c_k \tau \left(k/\sqrt{3}\right)^{1/3}$, $\phi_k \approx c_k \tau^2 \left(k/\sqrt{3}\right)^{1/3}/2$, so

$$c_k = \frac{-3}{2}\left(\frac{k}{\sqrt{3}}\right)^{-1/3}\delta_k.$$

We may now find the velocities

$$v_a = \frac{\partial}{\partial \tau}\phi_{,a} = \varphi_{,a} = \left(\frac{-3^{3/2}}{2}\right)\int \frac{d^2k}{(2\pi)^2}\, e^{ikq}\frac{ik_a}{k}\left(\frac{k\tau}{\sqrt{3}}\right)^{2/3}\delta_k J_{1/3}\left(k\tau/\sqrt{3}\right).$$

At early times $v \propto -\tau\nabla\delta$, and thus the flow transforms an initial density (or pressure) gradient into a flow anisotropy. This simple effect is remarkably hard to reproduce in parton models, under any realistic initial condition.

The important conclusion is that even in this simple model we get a qualitative picture of a relativistic heavy ion collision that serves as a framework for further analysis. In reality the flow is not really boost invariant, and it is certainly not homogeneous in the transverse plane. The hydrodynamic models offer a natural way to couple flow to pressure gradients in the transverse plane, and in this respect are superior to all other known alternatives. On the other hand, ideal hydrodynamic models overestimate the amount of anisotropy in the transverse flow [4]; this underlines the need for a relativistic theory of real fluids to really model relativistic heavy ion collisions.

3. Real Fluids

3.1. *Relativistic Real Fluids from Thermodynamics*

Let us return to our analysis of non relativistic fluids. Our starting point is Eq. (2), where now we allow for entropy production, which must be positive. This suggests modifying the constitutive relations for an ideal fluid. We must take into account that T^{ij}must be symmetric, and the relativistic constraint $p^i = J^i_u/c^2$. This suggests writing

$$T_{ij} = \frac{1}{2}\left(p^j v^i + v^j p^i\right) + p\delta^{ij} + \tau_{ij},$$

where

$$\tau_{ij} = -\eta\sigma_{ij} - \xi\delta_{ij}\nabla_k v^k,$$

$$\sigma_{ij} = \nabla_i v_j + \nabla_j v_i - \frac{2}{3}\delta_{ij}\nabla_k v^k,$$

and

$$J^j_u = (\rho + p)\, v^j + \frac{1}{2}\left(p^j v^2 - v^j\left(v_i p^i\right)\right) + v_i \tau^{ij},$$

$$\chi_j = -\kappa\nabla_j\alpha.$$

In what follows, we assume $\xi = 0$. With these constitutive relations we find

$$\frac{\partial s}{\partial t} = -\nabla_j S^j + \kappa\left(\nabla_j\alpha\right)^2 + \frac{\eta}{T}\sigma^{ij}\sigma_{ij} + \frac{\xi}{T}\left(\nabla_j v^j\right)^2 + \frac{1}{T}\left(\vec{p}\times\vec{v}\right)\nabla\times\vec{v}.$$

The last term vanishes in the rest frame and will not affect our argument. To find the relativistic generalization of this theory, we observe that in the rest frame

$$T^{\mu\nu} = \begin{pmatrix} \rho & 0 \\ 0 & p\delta^{ij} + \tau^{ij} \end{pmatrix}.$$

Thus, in any frame we may write

$$T^{\mu\nu} = T^{\mu\nu}_0 + \Pi^{\mu\nu},$$

where $T_0^{\mu\nu} = \rho u^\mu u^\nu + p\Delta^{\mu\nu}$ and $\Pi^{\mu\nu} = -\eta\sigma^{\mu\nu}$, with σ the shear tensor

$$\sigma_{\mu\nu} = \Delta_\mu^\lambda \Delta_\nu^\rho \left[u_{\rho,\lambda} + u_{\lambda,\rho} - \frac{2}{3}\Delta_{\lambda\rho} u^\alpha_{,\alpha} \right].$$

This is the so-called Chapman–Enskog prescription, and although it leads to a covariant generalization of the Navier-Stokes equations, it is untenable.

The conservation equations read

$$T^{\mu\nu}_{0,\nu} + \Pi^{\mu\nu}_{,\nu} = 0.$$

Now.

$$T^{\mu\nu}_{0,\nu} = \rho_{,\nu} u^\mu u^\nu + (\rho + p)\, u^\mu_{,\nu} u^\nu + (\rho + p)\, u^\mu u^\nu_{,\nu} + \Delta^{\mu\nu} p_{,\nu} = 0.$$

Therefore,

$$\dot{\rho} + (\rho + p)\, u^\nu_{,\nu} - u_\mu \Pi^{\mu\nu}_{,\nu} = 0,$$

$$\dot{u}^\mu + \Delta^{\mu\nu} \frac{p_{,\nu} + \Pi^\lambda_{\nu,\lambda}}{(\rho + p)} = 0.$$

Moreover, since $\Pi^{\mu\nu}$ is both traceless and transverse, we may write the first equation as

$$\dot{\rho} + (\rho + p)\, u^\nu_{,\nu} + \frac{1}{2}\Pi^{\mu\nu}\sigma_{\mu\nu} = 0.$$

3.2. *The Causality Problem*

Since in the Chapman–Enskog approach the viscous energy-momentum tensor (VEMT) is slaved to the four velocity, the only dynamic degrees of freedom are the components of the four velocity themselves. Let us consider the linear perturbations around a solution with $\beta^\mu =$ constant. Let us write $u^i = un^i$, $n^i n_i = 1$ and $u^0 = \sqrt{u^2 + 1}$. We seek solutions that depend on a single variable $k_\mu x^\mu = k_i x^i - \omega t$. We further decompose $k^i = Kn^i + k^i_\perp$, $x^i = xn^i + x^i_\perp$. Since the theory is covariant, we may always perform a Lorentz transformation to the rest frame, namely we introduce new time and space variables

$$t' = u^0 t - ux, \qquad x' = u^0 x - ut$$

and new frequency and wave number

$$\omega' = u^0\omega - uK, \qquad K' = u^0 K - u\omega,$$

with $x'^i_\perp = x^i_\perp$ and $k'^i_\perp = k^i_\perp$. We seek solutions where k^i is real and $\mathrm{Im}\,[\omega] \leq 0$.

In this frame $u'^i = 0$, $u'^0 = 1$ and $T' = T$. The linearized perturbations must have $\delta u'^0 = 0$. For the other variables we seek a solution

$$\delta' T' = T\delta' e^{-i\omega' t' + ik'_j x'^j}, \qquad \delta' u'^i = \delta'^i e^{-i\omega' t' + ik'_j x'^j}.$$

The equations of motion become

$$-i\omega'\delta' + ic^2 k'_j \delta'^j = 0,$$

$$-i\omega'\delta'^j + ik'^j\delta' + \gamma\left(\left(K'^2 + k_\perp^2\right)\delta'^j + \frac{1}{3}k'^j k'_k \delta'^k\right) = 0,$$

where $\gamma = \eta/\rho + p$. For a transverse perturbation $k'_k\delta'^k = 0$, $\delta' = 0$ we must have $\omega' = -i\gamma\left(K'^2 + k_\perp^2\right)$. We observe that because $u^0 > 0$, the imaginary parts of ω and ω' have the same sign, so we can check stability by checking the imaginary part of ω'. Under a Galilean transformation $K' = K$ and $\omega' = \omega - uK$, so the theory is stable in all frames. However, under a relativistic transformation $K' = (K - u\omega')/u^0$. If $u \neq 0$ we get a quadratic equation for ω'. For $k_\perp = 0$ this is

$$\omega'^2 - 2\left(\frac{K}{u} + \frac{iu^{02}}{2\gamma u^2}\right)\omega' + \frac{K^2}{u^2} = 0,$$

with solution

$$\omega' = \frac{K}{u}\left[1 + \frac{iu^{02}}{2\gamma K u} \pm \sqrt{\left(1 + \frac{iu^{02}}{2\gamma K u}\right)^2 - 1}\right].$$

When $u \to 0$, one solution is the expected one $\omega' = -i\gamma K^2$ but the second solution becomes $\omega' = iu^{02}/\gamma u^2$ and is unstable. For a longitudinal perturbation $\delta'^j = k'^j\delta''$ and the dispersion relation is

$$\omega'^2 - c^2\left(K'^2 + k_\perp^2\right) + \frac{4}{3}i\gamma\omega'\left(K'^2 + k_\perp^2\right) = 0.$$

The same analysis as before leads to a cubic equation. When $u \to 0$ two roots correspond to damped sound waves; the third root $\omega' \approx 3i/4\gamma u^2$ is unstable.

3.3. *Relativistic Kinetic Theory*

In the kinetic theory description [7] the transport equation reads

$$p^\mu \partial_\mu f = \frac{-1}{\tau}\operatorname{sign}\left(p^0\right) I_{\mathrm{col}},$$

where τ is the so-called relaxation time. The current is

$$J^\mu = e\int Dp\, p^\mu f,$$

and the EMT is

$$T^{\mu\nu} = \int Dp\, p^\mu p^\nu f,$$

where

$$Dp = \frac{2d^4p\delta\left(p^2\right)}{(2\pi)^3} = \frac{d^4p}{(2\pi)^3\, p}\left(\delta\left(p^0 - p\right) + \delta\left(p^0 + p\right)\right).$$

For simplicity we assume massless particles. We do not assume particle number conservation. To enforce energy-momentum conservation we require

$$\int Dp\, p^\mu \mathrm{sign}\left(p^0\right) I_{\mathrm{col}} = 0.$$

assuming for simplicity Maxwell–Boltzmann statistics, the equilibria are of the form

$$f_0 = \exp\left\{-\left|\beta_\mu p^\mu\right|\right\}.$$

For a given $T^{\mu\nu}$ we can always find a local equilibrium distribution f_0 such that the ideal fluid energy momentum tensor built from it,

$$T_0^{\mu\nu} = \int Dp\, p^\mu p^\nu f_0 = \rho u^\mu u^\nu + p\Delta^{\mu\nu},$$

obeys

$$T_0^{\mu\nu} u_\nu = -\rho u^\mu$$

with the same energy density and four velocity as from the Landau–Lifshitz prescription. We then define a temperature from

$$\rho = \sigma T^4,$$

where $\sigma = \pi^2/15 \approx 0.66$. Using the Maxwell–Juttner distribution rather than the Bose–Einstein one is equivalent to the approximation

$$\sum_{n=1} n^{-4} = \frac{\pi^4}{90} \approx 1,$$

whereby σ becomes $6/\pi^2 \approx 0.61$. We make this approximation from now on.

It follows that the viscous energy momentum tensor

$$\Pi^{\mu\nu} = T^{\mu\nu} - T_0^{\mu\nu},$$

is traceless and transverse:

$$\Pi^{\mu\nu} u_\nu = 0. \tag{4}$$

We parametrize

$$f = f_0 \left[1 + Z\right],$$

with $Z = 0$ at equilibrium. The transversality condition becomes

$$\int Dp\, p^\mu \left(-u_\nu p^\nu\right) f_0 Z = 0. \tag{5}$$

Assuming a simple Boltzmann type entropy flux [27]

$$S^\mu = -\int Dp\ \left(\mathrm{sign}\left(p^0\right)\right) p^\mu f \left[\ln \frac{f}{f_0} - 1\right],$$

we get the entropy production

$$S^\mu_{,\mu} = \frac{1}{\tau} \int Dp\, I_{\mathrm{col}} \ln\left[1 + Z\right].$$

3.4. *Chapman–Enskog and Grad*

The Chapman–Enskog procedure seeks a formal expansion for Z in powers of τ. To this end, one parametrizes Z as

$$Z = \tau Z_1 + \tau^2 Z_2 + \ldots.$$

The "spatial" derivatives $\Delta^{\mu\nu} T_{,\nu}$ and $\Delta^{\mu\nu} u^\lambda_{,\nu}$ are regarded as zeroth order quantities, while the "time" derivatives $\dot{T} = u^\mu T_{,\mu}$ and $\dot{u}^\lambda = u^\mu u^\lambda_{,\mu}$ are derived from energy-momentum conservation, which is also a necessary consistency condition.

To find Z_1 we only need the linearized collision integral. This must be a symmetric operator, it must obey the energy momentum conservation constraint, must lead to non negative entropy production and must admit thermal distributions as the only homogeneous distributions. To satisfy these requirements, we write

$$I_{\mathrm{col}}\left(p\right) = F\left[p\right] f_0\left(p\right) \left[Z\left(p\right) - \int Dp'\ K\left[p, p'\right] F\left[p'\right] f_0\left(p'\right) Z\left(p'\right)\right].$$

The second term is there to enforce the constraints; namely,

$$\int Dp\, \mathrm{sign}\left(p^0\right) p^\mu F f_0 Z = \int DpDp'\ \mathrm{sign}\left(p^0\right) p^\mu F f_0 K\left[p, p'\right] F' f'_0 Z'.$$

Since the kernel K is symmetric and these hold for any Z we must have

$$\int Dp'\ K\left[p, p'\right] \mathrm{sign}\left(p'^0\right) p'^\mu F' f'_0 = p^\mu \mathrm{sign}\left(p^0\right).$$

It also makes I_{col} to vanish when Z is just a variation in β_μ.

We adopt the Anderson–Witting prescription $F = |-u_\mu p^\mu|$ [28, 29]. Write

$$K[p,p'] = K_{\rho\sigma} p^\rho \mathrm{sign}\left(p^0\right) p'^\sigma \mathrm{sign}\left(p'^0\right).$$

Then,

$$K_{\rho\sigma} I^{\sigma\mu} = \delta^\mu_\rho,$$

where

$$I^{\sigma\mu} = u_\lambda \int Dp\ \mathrm{sign}\left(p^0\right) p^\sigma p^\mu p^\lambda f_0 = A_3 \left[u^\sigma u^\mu + \frac{1}{3}\Delta^{\sigma\mu}\right],$$

$$A_3 = \int Dp\ |-u_\mu p^\mu|^3 f_0 = \frac{24}{\pi^2} T^5.$$

This means that

$$K_{\rho\sigma} = A_3^{-1}\left[u_\rho u_\sigma + 3\Delta_{\rho\sigma}\right].$$

Finally the collision integral is

$$I_{\mathrm{col}}(p) = |-u_\mu p^\mu| f_0(p) \left[Z(p) - K_{\rho\sigma} p^\rho \mathrm{sign}\left(p^0\right) \int Dp'\ p'^\sigma \left(-u_\mu p'^\mu\right) f_0(p') Z(p')\right].$$

If Z satisfies the constraint Eq. (5), then the second term is zero and the entropy production is positive, provided $Z \geq -1$.

To first order

$$Z_1 = -\frac{p^\mu p^\nu}{|-p^\alpha u_\alpha|}\beta_{\nu,\mu}.$$

Now, in general,

$$\begin{aligned}\beta_{\nu,\mu} &= \frac{1}{T}\left[u_{\nu,\mu} - u_\nu \frac{T_{,\mu}}{T}\right] \\ &= \frac{1}{T}\left[u_\mu u_\nu \frac{\dot{T}}{T} + \frac{1}{2}\left[\sigma_{\mu\nu} + \frac{2}{3}\Delta_{\mu\nu} u^\lambda_{,\lambda}\right] - \frac{1}{2}\left[u_\nu \Delta^\lambda_\mu + u_\mu \Delta^\lambda_\nu\right]\left(\frac{T_{,\lambda}}{T} + \dot{u}_{,\lambda}\right)\right. \\ &\quad \left. -\frac{1}{2}\left[u_\nu \Delta^\lambda_\mu - u_\mu \Delta^\lambda_\nu\right]\left(\frac{T_{,\lambda}}{T} - \dot{u}_{,\lambda}\right) + \frac{1}{2}\Delta^\lambda_\nu \Delta^\rho_\mu \left[u_{\lambda,\rho} - u_{\rho,\lambda}\right]\right],\end{aligned}$$

so, using the zeroth order time derivatives and $c^2 = 1/3$

$$Z_1 = \frac{-1}{2T\,|p^\alpha u_\alpha|}\sigma_{\mu\nu} p^\mu p^\nu. \tag{6}$$

The viscous energy momentum tensor is

$$\Pi_1^{\mu\nu} = \frac{-\tau}{2T}\int Dp\ f_0 \frac{p^\mu p^\nu p^\rho p^\lambda}{|p^\alpha u_\alpha|}\sigma_{\rho\lambda} = -\eta\sigma^{\mu\nu},$$

where

$$\eta = \frac{\tau}{15T}\int Dp\ f_0 \left|-p^\alpha u_\alpha\right|^3 = \frac{8}{5\pi^2}\tau T^4.$$

The Chapman–Enskog procedure cannot generate a true dynamical equation for the viscous EMT, and thus does not solve the causality/stability problems. The Grad approach takes the different strategy of keeping the form Eq. (6), but replacing $\sigma_{\mu\nu}$ by a new tensor $C_{\mu\nu}$ regarded as a new independent variable, so that

$$Z_G = \frac{-1}{2T\left|-p^\alpha u_\alpha\right|}C_{\mu\nu}p^\mu p^\nu. \tag{7}$$

Since $C_{\mu\nu}$ is defined up to a multiple of $g_{\mu\nu}$, we may assume it is traceless, and because of the constraint Eq. (5) it must be transverse. The viscous energy-momentum tensor becomes

$$\Pi_G^{\mu\nu} = \frac{-1}{2T}\int Dp\ f_0 \frac{p^\mu p^\nu p^\rho p^\lambda}{\left|-p^\alpha u_\alpha\right|}C_{\rho\lambda} = -\frac{\eta}{\tau}C^{\mu\nu}.$$

The next step is to substitute this into the Boltzmann equation. The nonlocal term vanishes and we get

$$\frac{1}{\tau}\left(p^\nu u_\nu\right) f_0 Z = \frac{\partial}{\partial x^\mu}\left[p^\mu f_0\left(1+Z\right)\right].$$

We wish to extract from this an equation for $C^{\mu\nu}$. The time-honored procedure is to consider the moments of this equation. Since Z is even the zeroth moment vanishes, and the first moment gives back energy-momentum conservation.

$$\frac{\partial}{\partial x^\lambda}\int Dp\ p^\rho p^\lambda f_0\left(1+Z\right) = 0.$$

Therefore the first non-trivial choice is to use the second moments

$$\frac{\partial}{\partial x^\mu}\int Dp\,\mathrm{sign}\left(p^0\right) p^\rho p^\lambda p^\mu f_0\left(1+Z\right) = \frac{-1}{\tau}\int Dp\ p^\rho p^\lambda \left|-p^\nu u_\nu\right| f_0 Z.$$

To evaluate these equations, we use the following identities:

$$\int Dp\ p^\rho p^\lambda f_0 = A_2\left[u^\rho u^\lambda + \frac{1}{3}\Delta^{\rho\lambda}\right],$$

$$\int Dp\, \mathrm{sign}\left(p^0\right) p^\rho p^\lambda p^\mu f_0$$
$$= A_3 \left[u^\rho u^\lambda u^\mu + \frac{1}{3}\left(\Delta^{\rho\lambda} u^\mu + \Delta^{\mu\lambda} u^\rho + \Delta^{\mu\rho} u^\lambda\right)\right],$$
$$\int Dp\, p^\rho p^\lambda p^\mu p^\nu f_0$$
$$= A_4 \left\{ u^\rho u^\lambda u^\mu u^\nu + \frac{1}{3}\left(\Delta^{\rho\lambda} u^\mu u^\nu + ...\right) + \frac{1}{15}\left(\Delta^{\rho\lambda}\Delta^{\mu\nu} + ...\right)\right\},$$
$$\int Dp\, \mathrm{sign}\left(p^0\right) p^\rho p^\lambda p^\mu p^\nu p^\theta f_0$$
$$= A_5 \left\{ u^\rho u^\lambda u^\mu u^\nu u^\theta + \frac{1}{3}\left(\Delta^{\rho\lambda} u^\mu u^\nu u^\theta + ...\right) + \frac{1}{15}\left(\Delta^{\rho\lambda}\Delta^{\mu\nu} u^\theta + ...\right)\right\},$$
$$\int Dk\, k^\rho k^\lambda k^\mu k^\nu k^\theta k^\phi F\left[\left(k^\alpha u_\alpha\right)\right]$$
$$= A_6 \left\{ u^\rho u^\lambda u^\mu u^\nu u^\theta u^\phi + \frac{1}{3}\left(\Delta^{\rho\lambda} u^\mu u^\nu u^\theta u^\phi + ...\right)\right.$$
$$\left. + \frac{1}{15}\left(\Delta^{\rho\lambda}\Delta^{\mu\nu} u^\theta u^\phi + ...\right) + \frac{1}{105}\left(\Delta^{\rho\lambda}\Delta^{\mu\nu}\Delta^{\theta\phi} + ...\right)\right\}.$$

In the following, we shall adopt

$$A_a = \int Dp\, \left|-p^\alpha u_\alpha\right|^a f_0 = (a+1)!\frac{T^{a+2}}{\pi^2}.$$

The equation becomes

$$\frac{\partial}{\partial x^\mu} T^5 C^{\rho\lambda\mu} = \frac{T^5}{\tau} C^{\rho\lambda},$$

where

$$C^{\rho\lambda\mu} = \left\{3u^\rho u^\lambda u^\mu + \Delta^{\rho\lambda} u^\mu + \Delta^{\mu\lambda} u^\rho + \Delta^{\mu\rho} u^\lambda - u^\lambda C^{\rho\mu} - u^\rho C^{\mu\lambda} - u^\mu C^{\rho\lambda}\right\}.$$

It is not really possible to kill all second moments simultaneously. We shall be happy to kill the transverse traceless contribution, namely

$$\left[\Delta_{\tau\rho}\Delta_{\sigma\lambda} - \frac{1}{3}\Delta_{\tau\sigma}\Delta_{\rho\lambda}\right] \frac{\partial}{\partial x^\mu} T^5 C^{\rho\lambda\mu} = \frac{T^5}{\tau} C_{\tau\sigma},$$

which reduces to

$$\frac{-1}{T^5}\frac{\partial}{\partial x^\mu} T^5 u^\mu C_{\tau\sigma} - \Delta_{\sigma\lambda} C^{\rho\lambda\mu}\frac{\partial u_\tau u_\rho}{\partial x^\mu} - \Delta_{\tau\rho} C^{\rho\lambda\mu}\frac{\partial u_\sigma u_\lambda}{\partial x^\mu}$$
$$+\frac{1}{3}\Delta_{\tau\sigma} C^{\rho\lambda\mu}\frac{\partial u_\rho u_\lambda}{\partial x^\mu} + \frac{1}{3}\Delta_{\rho\lambda} C^{\rho\lambda\mu}\frac{\partial u_\sigma u_\tau}{\partial x^\mu} = \frac{1}{\tau} C_{\tau\sigma}. \quad (8)$$

Keeping only linear terms, this is an equation of Maxwell–Cattaneo type for $C_{\mu\nu}$ [30, 31]

$$\dot{C}_{\tau\sigma} + \frac{1}{\tau}C_{\tau\sigma} - \sigma_{\tau\sigma} + \text{ho} = 0.$$

Actually, we may get an expression for the time derivative of Z directly from the transport equation and then use it to find an equation for the viscous energy momentum tensor [32]. However, beyond the leading order this equation involves integrals which cannot be simple expressed in terms of the known moments of the distribution function. Therefore, to obtain a definite equation we must provide a *closure*, *i.e.*, an expression for Z allowing us to compute the integrals. For example, if we use the Grad ansatz Z_G we get an equation with the same structure as Eq. (8), though the coefficients may be different.

We wish to check that the Grad approach is consistent with causality and stability [1, 18, 19]. Introducing a new perturbation for C^{ij} we get the equations

$$\begin{aligned} -i\omega'\delta' + ic^2k'_j\delta'^j &= 0, \\ -i\omega'\delta'^i + ik'^i\delta' - i\frac{\gamma}{\tau}k'_jC'^{ij} &= 0, \\ -i\omega'C'^{ij} + \frac{1}{\tau}C'^{ij} - i\left(k'^i\delta'^j + k'^j\delta'^i - \frac{2}{3}\delta^{ij}k'_k\delta'^k\right) &= 0. \end{aligned}$$

We revert to the Chapman–Enskog equations with the replacement $\gamma \to \gamma/1 - i\omega'\tau$. For transverse perturbations we now get a quadratic equation

$$(1 - i\omega'\tau)\,\omega' + i\gamma K'^2 = 0.$$

It is easy to see that both roots are stable when $u = 0$. For general u stability obtains if $\tau \geq \gamma$. It is interesting to observe that with the expressions above $\tau \approx 5\gamma$.

We now consider longitudinal waves with $k_\perp = 0$. The dispersion relation is

$$(1 - i\omega'\tau)\left[\omega'^2 - \frac{c^2}{u_0^2}\left(K - u\omega'\right)^2\right] + \frac{4}{3u_0^2}i\gamma\omega'\left(K - u\omega'\right)^2 = 0.$$

We rearrange this as

$$\omega'\left[\omega'^2 - C^2\left(K - u\omega'\right)^2\right] + \frac{i}{\tau}\left[\omega'^2 - \frac{c^2}{u_0^2}\left(K - u\omega'\right)^2\right] = 0,$$

where

$$C^2 = \frac{c^2}{u_0^2}\left(1 + \frac{4\gamma}{3c^2\tau}\right).$$

This yields

$$\omega'\left(\omega'-\frac{CK}{(1+Cu)}\right)\left(\omega'+\frac{CK}{(1-Cu)}\right)$$
$$+\frac{i}{(1-C^2u^2)\,\tau}\left[\omega'^2-\frac{c^2}{u_0^2}\left(K-u\omega'\right)^2\right]=0.$$

In the formal limit $\tau\to\infty$ we find three real roots $\omega'=0,\pm\omega_\pm$, where $\omega_\pm=CK/1\pm Cu$. At large but finite τ, the solution that goes to zero behaves as

$$\omega_0=\frac{-i}{\tau}\frac{c^2}{u_0^2C^2}$$

and it is stable. The solutions that converge to $\pm\omega_\pm$ behave as

$$2CK\omega_\pm\left(\omega'-\omega_\pm\right)+\frac{4i\gamma}{3u_0^2\tau^2}\left(K\mp u\omega_\pm\right)^2=0,$$

so they are stable too. Of course, we also have solutions with $\delta=\delta^i=k'_jC^{ij}=0$ and $\omega'=-i/\tau$, which are obviously stable.

3.5. *Entropy Production Variational Method*

The Grad approach as we have presented it still has the problems that the Grad ansatz Eq. (7) does not lead to a non negative one particle distribution function (quite the opposite, since $C_{\mu\nu}p^\mu p^\nu$ must be negative in some direction in momentum space) and that it is unclear how to introduce nonlinear terms. To overcome these problems we need a better motivated closure for Z. We shall try to find it by seeking the value of Z which minimizes entropy production for a given VEMT, and satisfies the constraint Eq. (4) [10]. Adding Lagrange multipliers $\zeta_{\mu\nu}$ we get the equation

$$\left\{I_{\mathrm{col}}\left[\ln\left[1+Z\right]\right]+\frac{I_{\mathrm{col}}\left[Z\right]}{1+Z}\right\}=-\tau\zeta_{\mu\nu}p^\mu p^\nu f_0.$$

The point is that this equation has bounded solutions for any value of the right hand side. We may also regard it as a means to obtain a formal series solution for Z in powers of τ, where, if we keep only the first order term, we recover the Grad ansatz.

It is convenient to introduce a new unknown $\chi=\ln\left[1+Z\right]$, with inverse transformation $Z=e^\chi-1$. In terms of the new unknown, the equation reads

$$I_{\mathrm{col}}\left[\chi\right]=-\frac{\tau}{2}\zeta_{\mu\nu}p^\mu p^\nu f_0-\frac{1}{2}\left[e^{-\chi}I_{\mathrm{col}}\left[e^\chi-1\right]-I_{\mathrm{col}}\left[\chi\right]\right].$$

To lowest order we may neglect the second term in the right hand side. Observe that in any case $\zeta_{\mu\nu}$ is defined up to a multiple of $g_{\mu\nu}$. We use this freedom to require $\zeta_{\mu\nu}u^\mu u^\nu = 0$. Transversality requires

$$0 = \int Dp \,\mathrm{sign}\left(p^0\right) p^\lambda f_0 \zeta_{\mu\nu} p^\mu p^\nu$$
$$= \zeta_{\mu\nu} A_3 \left[u^\lambda u^\mu u^\nu + \frac{1}{3}\left(u^\lambda \Delta^{\mu\nu} + u^\mu \Delta^{\lambda\nu} + u^\nu \Delta^{\lambda\mu}\right)\right].$$

Therefore $\zeta_{\mu\nu}$ must be traceless and transverse. To lowest order,

$$\chi = \frac{-1}{2}\frac{\zeta^{(0)}_{\mu\nu} p^\mu p^\nu}{\left|-u_\rho p^\rho\right|}.$$

Let us further expand the exponential. In the rest frame the energy momentum tensor reads

$$\begin{aligned} T^{00} &= \sigma T^4 \equiv \rho, \\ T^{0i} &= 0, \\ T^{ij} &= \frac{1}{3}\sigma T^4 \delta^{ij} - \frac{A_3}{15}\tau \zeta^{ij}. \end{aligned} \tag{9}$$

The kinetic equation to lowest order in τ reads

$$p^\lambda \partial_\lambda f_0 \left[1 - \tau \frac{1}{2}\frac{\zeta_{\mu\nu} p^\mu p^\nu}{\left|-u_\rho p^\rho\right|}\right] = \frac{1}{2}\left(-u_\mu p^\mu\right) f_0 \frac{\zeta_{\mu\nu} p^\mu p^\nu}{\left|-u_\rho p^\rho\right|}.$$

The meaning of this equation is as a generating equation for its moments. The first order moments give energy momentum conservation. To find an equation for $\zeta_{\mu\nu}$ we go to the rest frame and multiply both sides by $\mathrm{sign}\left(p^0\right) p^i p^j$. Integrating and discarding the trace part, we get

$$\frac{A_4}{15T}\sigma_{ij} - \frac{\tau A_5 \dot{T}}{15T^2}\zeta_{ij} - \frac{\tau A_5}{105T}\left[u_{k,k}\zeta_{ij} + \zeta_{ik}\sigma^k_j + \sigma^k_i \zeta_{kj} - \frac{2}{3}\Delta_{ij}\zeta^{(0)kl}\sigma_{kl}\right]$$
$$= \frac{A_4}{15}\left[\zeta_{ij} + \tau \zeta_{ij,0}\right].$$

3.6. *Non Linear Corrections to Boost Invariant Flow*

We now turn to consider nonlinear corrections to this equations. The strategy we are following consists on improving the stability of the theory by adding new variables obeying dynamical equations of their own. In many approaches, such as the so-called Israel–Stewart theory [33, 34] or Extended Thermodynamics [35] the extra variables are the components of the viscous energy-momentum tensor $\Pi_{\mu\nu}$ itself. The dynamical equations for $\Pi_{\mu\nu}$

have been derived in a number of ways, such as carefully taking moments of the kinetic equation [32], a systematic gradient expansion of the kinetic theory [36], from AdS-CFT correspondence [37] or simply writing down all terms consistent with the symmetries of the theory up to a certain order [38]. We shall call these theories "second order fluid dynamics" (SOFD) for short, and take the presentation in [39] as a suitable representative.

In our approach (EPVM, or entropy production variational method), on the other hand, the new variables are the Lagrange multipliers enforcing the constraints on entropy production; $\Pi_{\mu\nu}$ itself may depend nonlinearly on the Lagrange multipliers. Our approach is therefore closer to the Geroch-Linblom Divergence type theories [40], although we shall not demand that the resulting theory conforms to the dissipative type framework.

To obtain further insight on the meaning of these theories, we shall apply them to the case of boost invariant flow. Adopting Milne coordinates T^{μ}_{ν} is diagonal; we write $T^{\tau}_{\tau} = -\rho$, $T^{\eta}_{\eta} = \rho/3 - \Pi$, $T^{x}_{x} = T^{y}_{y} = \rho/3 + \Pi/2$. Energy-momentum conservation yields

$$\dot{\rho} + \frac{1}{\tau}\left(\frac{4}{3}\rho - \Pi\right) = 0.$$

We also write $\zeta^{(0)\eta}_{\eta} = \zeta$, $\zeta^{(0)x}_{x} = \zeta^{(0)y}_{y} = -\zeta/2$. Recall that $\sigma^{\eta}_{\eta} = 4/3\tau$, $\sigma^{x}_{x} = \sigma^{y}_{y} = -2/3\tau$. We get, discarding a derivative of the temperature,

$$\tau_R\dot{\zeta} + \zeta = \frac{4}{3T\tau} - a_1\tau_R\frac{\zeta}{\tau}, \tag{10}$$

where we have written τ_R for the relaxation time to avoid confusion with Milne time, and

$$a_1 = \frac{A_5}{3A_4T}.$$

The linearized collision term we are using is too simplistic to allow for a reliable derivation of nonlinear terms; however, in this case we may exploit the symmetries of the problem, which indicate that there is a single true degree of freedom ζ underlying the viscous energy momentum tensor. Therefore we generalize Eq. (9) to

$$\Pi = \frac{\tau_R A_3}{15}\zeta + h_1\left(\frac{\tau_R A_3}{15}\zeta\right)^2$$

and Eq. (10) to

$$\tau_R\dot{\zeta} + \zeta = \frac{4}{3T\tau} - a_1\tau_R\frac{\zeta}{\tau} - h_2\frac{\tau_R A_3}{15}\zeta^2.$$

We determine the new transport coefficients $h_{1,2}$ by asking that, in a stationary situation, the relation between Π and the shear $4/3\tau$, to second order in τ_R, agrees with a systematic expansion for a Boltzmann gas [39]. Indeed, to second order we may write

$$\zeta = \frac{15}{\tau_R A_3}\left[\Pi - h_1\Pi^2\right]$$

and so, neglecting time derivatives,

$$\Pi = \frac{\tau_R A_3}{15T}\frac{4}{3\tau} - a_1\tau_R\frac{\Pi}{\tau} - (h_2 - h_1)\,\Pi^2.$$

Matching against SOFD yields [39]

$$\tau_R = \frac{6\eta}{sT}, \qquad \frac{A_3}{15T} = \frac{sT}{6}, \qquad a_1 = \frac{4}{3}, \qquad h_2 - h_1 = \frac{1}{4\pi T\eta},$$

where η is the shear viscosity and s the entropy density, both extracted from a realistic equation of state. An analysis of the possible origins of the $h_{1,2}$ coefficients shows that both are of the same order of magnitude, $\approx \tau_R^{-1}$. When this holds, the behavior of the model is relatively insensitive to the actual values of h_1 and h_2.

To break the degeneracy between h_1 and h_2, let us consider a situation where the shear τ^{-1} terms are negligible and the transport coefficients are constant. The SOFD equation admits a stationary solution with $\Pi = \Pi_{0SOFD} = -1/(h_2 - h_1)$; Π is otherwise unbounded. The EPVM equations on the other hand imply a lower bound $\Pi \geq -1/(4h_1)$, which is realized when $\zeta = -1/(2h_1\eta T)$. There is a steady solution when $\zeta = -1/(h_2\eta T)$, which implies $\Pi = \Pi_{0EPVM} = (-1/h_2)\left[1 - (h_1/h_2)\right]$. In both cases, the steady solutions are unstable. For SOFD, this implies the existence of runaway solutions, namely, solutions which begin below the fixed point (if it is negative) run away to minus infinity. In the EPVM, on the other hand, we may eliminate the runaway solutions by demanding that the steady solution coincides with the lower bound, and adopting, for each allowed value of Π, the value of ζ above the fixed point. This requires $h_1 = h_2/2$, and therefore $h_2 = 2h_1 = 1/(2\pi T\eta)$. Introducing the transport coefficient

$$\lambda = \frac{\eta}{2\pi T},$$

the SOFD equation reads [39]

$$\tau_R\dot{\Pi} + \Pi = \frac{4\eta}{3\tau} - \frac{4}{3}\tau_R\frac{\Pi}{\tau} - \frac{\lambda}{2\eta^2}\Pi^2,$$

while the extended EPVM yields the system [10]

$$\tau_R\dot{\zeta} + \zeta = \frac{4}{3T\tau} - \frac{4}{3}\tau_R\frac{\zeta}{\tau} - \frac{\zeta^2}{2\pi},$$
$$\Pi = \eta T\left[\zeta + \frac{\zeta^2}{4\pi}\right].$$

One way to visualize the difference between these models is to consider the free decay of Π. Suppose Π is observed to have the value Π_0 at some time τ_0 late enough that the τ^{-1} terms in the equations may be neglected, and the transport coefficients regarded as constant. Then, according to SOFD, the further decay of the VEMT is given by

$$\frac{\Pi}{\Pi_0} = \frac{1}{\left(1 + \frac{x_0}{4}\right)e^t - \frac{x_0}{4}},$$

where $t = \tau - \tau_0/\tau_R$ and $x = \Pi/\pi\eta T$, while from EPVM we get

$$\zeta = \frac{\zeta_0}{\left(1 + \frac{\zeta_0}{2\pi}\right)e^t - \frac{\zeta_0}{2\pi}},$$

where

$$\frac{\zeta_0}{2\pi} = \sqrt{1 + x_0} - 1.$$

Besides the absence of runaway solutions already noted, the EPVM provides for a faster decay of large fluctuations (see Fig. 1).

4. Final Remarks

The theory of relativistic real fluids has a curious history because while the stability problems we have discussed have been known for a long time [31], yet they do not seem to have elicited any strong response until fairly recently. There were known ways to improve the theory (foremost the Israel–Stewart [33, 34] and extended thermodynamics theories [35]), and also a family of theories which were known to be free of such problems on a rigorous basis (the Geroch–Lindblom dissipative type theories [40]). However, the former were presented as successive approximations to a yet unknown theory, and the physical foundations of the latter remained elusive [41]. It was only the realization that relativistic real fluids might be produced in RHICs that triggered an all-out attack on the problem, to the extent that it would be impossible to describe all this activity in a short review such as

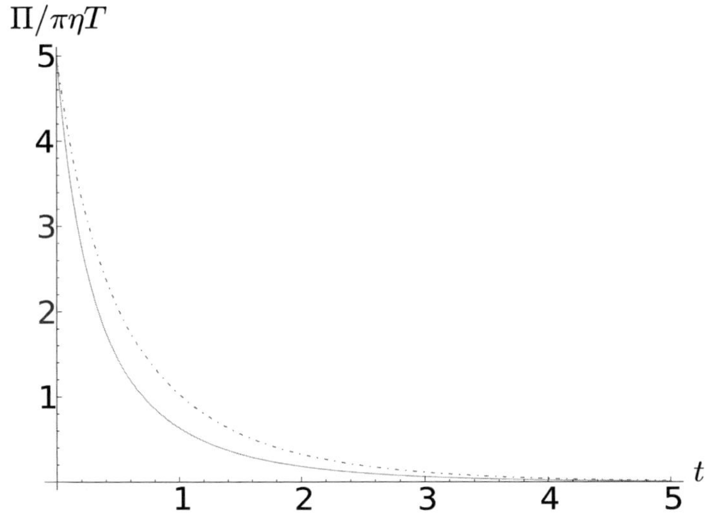

Fig. 1: (Color online) The evolution of the viscous energy momentum tensor, as predicted by SOFD (dashes and dots) and EPVM (full line) starting form $x_0 = 5$. We see that EPVM predicts a faster approach towards $\Pi = 0$.

this. We have therefore aimed to present just the fundamental ideas behind the theory, what the main problems are, and which lines of thought seem to us likely to be fruitful, and this almost entirely from the formal side, leaving phenomenology to more knowledgeable authors.

Concerning the present state of the theory, it seems fair to say that we have a reliable understanding of evolution during the hydrodynamic era, and the beginnings of a theory of the freeze out transition. The early times of the collision are relatively much more poorly understood. In particular, we do not know how hydrodynamic behavior may arise on such short time scales as demanded by theory, though the work on non-abelian instabilities [42] and QGP turbulence [43], on one hand, and on AdS-CFT correspondence on the other [44], makes those scales look not so unrealistic as they used to.

We can only conclude that we are only witnessing the early childhood of the relativistic real fluids - RHICs connection, and this is what makes this such an exciting field to work on.

Acknowledgment

This work has been developed in collaboration with Jerónimo Peralta Ramos. It is supported in part by Universidad de Buenos Aires, CONICET and ANPCYT (Argentina).

Appendix: Kadanoff–Baym Equations and Quantum Kinetic Field Theory

In this appendix we shall discuss the derivation of kinetic theory from quantum field theory. The presentation follows [3].

The CTP generating functional depends on two external sources

$$e^{iW\left[J^1,J^2\right]}=\int D\Phi^1 D\Phi^2\, e^{i\left\{S\left[\Phi^1\right]-S\left[\Phi^2\right]+\int\left(J^1\Phi^1-J^2\Phi^2\right)\right\}}.$$

It defines two background fields through

$$\phi^1\left[x\right]=\frac{\delta W\left[J^1,J^2\right]}{\delta J^1\left[x\right]};\quad \phi^2\left[x\right]=-\frac{\delta W\left[J^1,J^2\right]}{\delta J^2\left[x\right]}.$$

The CTPEA is the full Legendre transform

$$\Gamma\left[\phi^1,\phi^2\right]=W\left[J^1,J^2\right]-\int\left(J^1\phi^1-J^2\phi^2\right).$$

It generates the equations of motion

$$\frac{\delta\Gamma\left[\phi^1,\phi^2\right]}{\delta\phi^1\left[x\right]}=-J^1\left[x\right];\quad \frac{\delta\Gamma\left[\phi^1,\phi^2\right]}{\delta\phi^2\left[x\right]}=J^2\left[x\right]. \tag{11}$$

The equation of motion for the mean field is obtained when $J^1=J^2$ by setting $\phi^1=\phi^2$ in the equations (11) *after* computing the variational derivatives.

To compute the CTPEA, observe that

$$\Gamma\left[\phi^1,\phi^2\right]=S\left[\phi^1\right]-S\left[\phi^2\right]+\text{quantum corrections.}$$

The quantum corrections are the sum of all the one-particle irreducible (1PI) graphs in the theory. The linearized one-particle irreducible $(1PI)$ effective action has the structure

$$\begin{aligned}\Gamma_{1PI}=\int d^dxd^dy\,&\left\{\varphi_-\left(x\right)\left[D\left(x,y\right)+\mathbf{D}\left(x,y\right)\right]\varphi_+\left(y\right)\right.\\&\left.+\frac{i}{2}\varphi_-\left(x\right)\mathbf{N}\left(x,y\right)\varphi_-\left(y\right)\right\},\end{aligned}$$

where $\varphi_- = \left[\varphi^1 - \varphi^2\right]$, $\varphi_+ = \left[\varphi^1 + \varphi^2\right]/2$, and

$$D(x,y) = \left[\partial_x^2 - m_b^2\right]\delta(x-y).$$

$\mathbf{D}$ is causal and $\mathbf{N}$ is even, and both are real. A good deal of our discussion will revolve around the different properties of the propagators of the theory, that is, the expectation values of binary products of field operators with respect to the initial state. Since field operators at different locations do not generally commute, we have several different propagators according to the ordering of the field operators within the expectation value. The equations of motion for the propagators are derived from the identity

$$\frac{\mathcal{D}^2\Gamma_{1PI}}{\mathcal{D}\varphi^a\mathcal{D}\varphi^b}G^{bc} = i\hbar\delta_a^c.$$

The G^{ab} above denote the four basic propagators

Feynman: $G_F \equiv < T\left(\Phi\left(x\right)\Phi\left(x'\right)\right) >= G^{11}$,

Dyson: $G_D \equiv < \tilde{T}\left(\Phi\left(x\right)\Phi\left(x'\right)\right) >= G^{22}$,

Positive frequency: $G^+ \equiv < \Phi\left(x\right)\Phi\left(x'\right) >= G^{21}$,

Negative frequency: $G^- \equiv < \Phi\left(x'\right)\Phi\left(x\right) >= G^{12}$,

where T stands for time ordering and $\tilde{T}$ stands for anti-time ordering. Explicitly,

$$\begin{aligned}
\left[D + \mathbf{D}_{\text{even}} + i\mathbf{N}\right]G^{11} + \left[\mathbf{D}_{\text{odd}} - i\mathbf{N}\right]G^{21} &= i\mathbf{1},\\
\left[\mathbf{D}_{\text{odd}} + i\mathbf{N}\right]G^{11} + \left[D + \mathbf{D}_{\text{even}} - i\mathbf{N}\right]G^{21} &= 0,\\
\left[D + \mathbf{D}_{\text{even}} + i\mathbf{N}\right]G^{12} + \left[\mathbf{D}_{\text{odd}} - i\mathbf{N}\right]G^{22} &= 0,\\
\left[\mathbf{D}_{\text{odd}} + i\mathbf{N}\right]G^{12} + \left[D + \mathbf{D}_{\text{even}} - i\mathbf{N}\right]G^{22} &= -i\mathbf{1},
\end{aligned}$$

and

$$\begin{aligned}
\mathbf{D}_{\text{even}}\left(x,y\right) &= \frac{1}{2}\left[\mathbf{D}\left(x,y\right) + \mathbf{D}\left(y,x\right)\right],\\
\mathbf{D}_{\text{odd}}\left(x,y\right) &= \frac{1}{2}\left[\mathbf{D}\left(x,y\right) - \mathbf{D}\left(y,x\right)\right].
\end{aligned}$$

We obtain a more efficient representation of the dynamics by introducing new propagators: the Hadamard propagator,

$$G_1 = G^{21} + G^{12} \equiv \langle\{\Phi\left(x\right), \Phi\left(x'\right)\}\rangle,$$

which is real and even; the Jordan propagator,

$$G = G^{21} - G^{12} \equiv \langle[\Phi\left(x\right), \Phi\left(x'\right)]\rangle,$$

which is imaginary and odd.

The advanced and retarded propagators are the fundamental solutions to the equations of motion for linear fluctuations in the field.

$$G_{\mathrm{ret}}(x,x') = i\left[G^{11} - G^{12}\right] = iG(x,x')\,\theta(t-t'),$$
$$G_{\mathrm{adv}}(x,x') = -i\left[G^{21} - G^{11}\right] = -iG(x,x')\,\theta(t'-t).$$

All propagators may be expressed in terms of the Jordan and Hadamard ones as

$$G^{\pm}(x,x') = \frac{1}{2}\left[G_1(x,x') \pm G(x,x')\right],$$

$$G_{F,D}(x,x') = \frac{1}{2}\left[G_1(x,x') \pm G(x,x')\,\mathrm{sign}(t-t')\right].$$

In terms of the new propagators the equations of motion are

$$[D + \mathbf{D}_{\mathrm{even}} + \mathbf{D}_{\mathrm{odd}}]\,G_{\mathrm{ret}} = -\mathbf{1},$$
$$[D + \mathbf{D}_{\mathrm{even}} + \mathbf{D}_{\mathrm{odd}}]\,G_1 = \mathbf{N}G_{\mathrm{adv}}.$$

We can now begin the discussion of our subject matter. Our goal is to recast the equations for the propagators in a way suitable to discuss the last stages of the equilibration process. We shall reduce the equations to the form of a kinetic equation, the so-called Kadanoff-Baym equation. Further approximations reduce this to the Boltzmann equation. To succeed, we need a way to identify the important terms and discard the irrelevant ones. This will be provided by the so-called adiabatic expansion, under the assumption that the propagators are almost translation invariant close enough to equilibrium.

We say that a $G^{ab}(x,x')$ is almost translation-invariant if, when partially Fourier transformed with respect to $u = x - x'$, the Fourier transform is weakly dependent on the "centroid" variable $X = (x+x')/2$, *i.e.*,

$$G^{ab}(x,x') = \int \frac{d^dk}{(2\pi)^d}\, e^{iku} G^{ab}(X,k).$$

Observe that

$$\int dy\, A(x,y)\, B(y,x') = \int \frac{d^dk}{(2\pi)^d}\, e^{iku},$$
$$\left\{A(X,k)\,B(X,k) - \frac{i}{2}\{A,B\} + \ldots\right\},$$

where

$$\{A, B\} = \frac{\partial A}{\partial k}\frac{\partial B}{\partial X} - \frac{\partial A}{\partial X}\frac{\partial B}{\partial k}.$$

Expressions involving $G^{ab}(X, k)$ may be classified according to their *adiabatic order*, namely, the number of X derivatives appearing in the expression. We call this the *adiabatic expansion*. When almost translation-invariance is verified, we may further reject all terms above a given adiabatic order. We call such a truncation of an adiabatic expansion an *adiabatic approximation*. In other words, the adiabatic order is used as a tag to bunch together certain terms in the equations of motion in accordance to their derivative orders and the adiabatic approximation determines how many of those terms are kept.
Let

$$i\Gamma(X, k) = i\pi\gamma(X, k)\,\mathrm{sign}(\omega) = \mathbf{D}_{odd}(X, k),$$

where $\omega = k^0$, and

$$R(X, k) = \left(k^2 + m_b^2\right) - \mathbf{D}_{\mathrm{even}}(X, k).$$

Then,

$$G_{\mathrm{ret}}(X, k) = \frac{1}{R - i\Gamma},$$

$$G_{\mathrm{adv}}(X, k) = G_{\mathrm{ret}}(X, -k) = \frac{1}{R + i\Gamma}.$$

The relationship $G_{\mathrm{ret}} = iG\theta(t - t')$ implies that

$$G_{\mathrm{ret}}\left(X, \left(\omega, \vec{k}\right)\right) = -\int \frac{d\omega'}{2\pi}\frac{G\left(X, \left(\omega', \vec{k}\right)\right)}{\omega - \omega' + i\epsilon},$$

or else that

$$\mathrm{Re}\, G_{\mathrm{ret}}\left(X, \left(\omega, \vec{k}\right)\right) = -PV\int \frac{d\omega'}{2\pi}\frac{G\left(X, \left(\omega', \vec{k}\right)\right)}{\omega - \omega'},$$

$$\mathrm{Im}\, G_{\mathrm{ret}}(X, k) = \frac{1}{2}G(X, k).$$

Thus,

$$G(X, k) = \frac{2\Gamma}{R^2 + \Gamma^2}.$$

Under equilibrium conditions the Kubo–Martin–Schwinger theorem implies that

$$G_{1\mathrm{eq}}(X,k) = \mathrm{sign}(\omega)\, G_{\mathrm{eq}}(X,k)\left[1+2f_{BE}\right],$$

where f_{BE} is the Bose–Einstein distribution. We generalize this to a nonequilibrium situation by defining the *density of states* $\mathcal{D}(X,k)$ out of the Fourier transform of the Jordan propagator

$$\mathcal{D}(X,k) \equiv \frac{1}{2\pi} G(X,k)\ \mathrm{sign}(\omega) = \frac{\gamma}{R^2+\Gamma^2}.$$

We now define the distribution function $f(X,k)$ through the partial Fourier transform of the Hadamard propagator

$$G_1(X,k) \equiv 2\pi\mathcal{D}(X,k)\ F_1(X,k),$$

where

$$F_1(X,k) = 1 + 2f(X,k).$$

To obtain the dynamics of the distribution function f, we make use of the equation involving the noise kernel. Let us call $F^{21} = \theta(\omega) + f$, $F^{12} = \theta(-\omega) + f$, $\Sigma_{12} = i(\mathbf{N} - \Gamma)$ and $\Sigma_{21} = i(\mathbf{N} + \Gamma)$. Then to first adiabatic order we get

$$A\{R, F_1\} - B\{\mathbf{\Gamma}, F_1\} = I_{\mathrm{col}}\ \mathrm{sign}\left(k^0\right), \tag{12}$$

where

$$A = \frac{\mathbf{\Gamma}^2}{R^2+\mathbf{\Gamma}^2},$$

$$B = \frac{R\mathbf{\Gamma}}{R^2+\mathbf{\Gamma}^2},$$

and I_{col} is the *collision integral*

$$I_{\mathrm{col}} = -i\left[\Sigma_{12}F^{21} - \Sigma_{21}F^{12}\right].$$

For weakly coupled theories, a series of approximations allow us to reduce the off-shell kinetic equation to the more familiar Boltzmann kinetic equation. We observe that in terms of the coupling constant λ we have, for a generic momentum p, $R \sim O(1)$ while $\mathbf{\Gamma} \sim O\left(\lambda^2\right)$.

A second observation is that in general $\mathbf{\Gamma}$, which involves the coupling constants, will be much smaller than R for a generic choice of p. When the coupling constants go to zero $\mathbf{\Gamma} \to 0$, but the retarded propagator has a well-defined asymptotic value, and the density of states becomes $\mathcal{D} = \delta(R)$.

In this limit the propagators are insensitive to the behavior of the distribution function "off shell" (*i.e.*, when $R \neq 0$), because the distribution function is always multiplied by the density of states, and this is very small there. Therefore, only "on shell" modes (*i.e.*, those for which $R = 0$) really contribute to the field correlation functions. If our only concern is to follow the evolution of the distribution function on shell, we are allowed to replace the A and B coefficients in (12) by their "on shell" values, namely $A = 1$ and $B = 0$. We thus obtain the Kadanoff–Baym equations [45]

$$\{R, F_1\} = -i \operatorname{sign}\left(k^0\right) \left[\Sigma_{12}F^{21} - \Sigma_{21}F^{12}\right]$$

The nontrivial content of the Kadanoff–Baym equations is given by the form of the collision integral, namely, which Feynman graphs contribute to the self energies. We recognize the structure of the collision term as the difference between a gain and a loss term for particles moving in or out of a phase space cell around the point (X, k) per unit time. Taking $\omega > 0$ for simplicity, we see that $\Sigma_{12}F^{21}$ is the gain term, with $F^{21} = 1 + f$ accounting for stimulated emission of particles into the cell, while the other term is the loss term, which is proportional to the number of particles $F^{12} = f$ already there.

If we only keep the first term in the expansion, which for a $\lambda\phi^4$ theory is the setting-sun graph, we recover the Boltzmann's collision term [46].

References

[1] W. Hiscock and L. Lindblom, Ann. Phys. 151, 466 (1983); Phys. Rev. D 31, 725 (1985); Contemporary Mathematics 71, 181 (1988).

[2] D. Rischke, Proceedings of the 11th Chris Engelbrecht Summer School in Theoretical Physics, Cape Town, Feb. 4 - 13, 1998

[3] E. Calzetta and B-L Hu, Nonequilibrium quantum field theory (Cambridge University Press, Cambridge (England), 2008)

[4] P. Romatschke, ArXiv 0912.3663

[5] P. V. Ruuskanen, Acta Phys. Pol. B18, 551 (1986)

[6] P. M. Chaikin and T. C. Lubensky, Principles of condensed matter physics (Cambridge University Press, Cambridge (England), 1995)

[7] W. Israel, in L. O'Raifeartaigh (ed.) General relativity: papers in honour of J. L. Synge (Clarendon Press, Oxford, 1972), p. 201.

[8] W. Israel, in A. Anile and Y. Choquet - Bruhat (eds.), Relativistic fluid dynamics (Springer, New York, 1988).

[9] S. Chapman and T. Cowling, The Mathematical Theory of Non-

uniform Gases (Cambridge University Press, Cambridge, England, 1939) (reissued 1990)).
[10] J. Peralta Ramos and E. Calzetta, Phys. Rev. D 82, 106003 (2010); Phys. Rev. D86, 125024 (2012); Phys. Rev. D 87, 034003 (2013)
[11] S. Ono, in I. Prigogine, ed., Advances in Chemical Physics, Vol III (John Wiley, New York, 1961)
[12] E. T. Jaynes, Ann. Rev. Phys. Chem. 31, 579 (1980).
[13] I. Prigogine, Introduction to thermodynamics of irreversible processes (John Wiley, New York, 1955).
[14] R. Landauer, Phys. Rev. A12, 636 (1975).
[15] S. Bruers, ArXiv:0604482; S. Bruers, C. Maes and K. Netocny, ArXiv:0701035
[16] Yu. L. Klimontovich, Turbulent Motion and the Structure of Chaos (Springer, Berlin, 1991)
[17] Yu. L. Klimontovich, Statistical Theory of Open Systems (Springer, Berlin, 1995)
[18] T. Olson, Ann. Phys. 199, 18 (1990).
[19] T. Olson and W. Hiscock, Ann. Phys. 204, 331 (1990)
[20] G. K. Batchelor, An introduction to fluid dynamics (Cambridge UP, Cambridge (England), 1967)
[21] L. D. Landau and E. M. Lifshitz, Fluid Mechanics (Pergamon Press, Oxford, 1959)
[22] BRAHMS Collaboration, Nucl. Phys. A757, 1 (2005).
[23] PHOBOS Collaboration, Nucl. Phys. A757, 28 (2005).
[24] STAR Collaboration, Nucl. Phys. A757, 102 (2005).
[25] PHENIX Collaboration, Nucl. Phys. A757, 184 (2005).
[26] J.D. Bjorken, Phys. Rev. D **27**, 140 (1983).
[27] T. Sagawa, in M. Nakahara and S. Tanaka (eds.), Lectures on quantum computing, thermodynamics ans statistical physics (World Scientific, Singapore, 2013).
[28] J. L. Anderson and H. R. Witting, Physica 74, 466 (1974); ibid. 74, 489 (1974).
[29] M. Takamoto and S.-I. Inutsuka, Physica A 389, 4580 (2010).
[30] J. C. Maxwell, Philos. Trans. Soc. London 157, 49 (1867).
[31] D. D. Joseph and L. Preziosi, Rev. Mod. Phys. 61, 41 (1989); Addendum, Rev. Mod. Phys. 62, 375 (1990).
[32] G. S. Denicol, T. Koide, and D. H. Rischke, Phys. Rev. Lett. 105, 162501 (2010), G. S. Denicol, E. Molnár, H. Niemi and D. H. Rischke, Eur. Phys. J. A, 48 11 (2012) 170.

[33] W. Israel, Ann. Phys. (NY) **100**, 310 (1976).
[34] W. Israel and J. M. Stewart, Ann. Phys. (NY) **118**, 341 (1979).
[35] D. Jou, J. Casas-Vazquez and G. Lebon, Extended Irreversible Thermodynamics (Springer, Berlin, 2001).
[36] R. Baier, P. Romatschke and U. A. Wiedemann, Phys. Rev. C 73, 064903 (2006).
[37] S. Bhattacharyya, V. E. Hubeny, S. Minwalla, and M. Rangamani, J. High Energy Phys. **02**, 045 (2008); M. Natsuume and T. Okamura, Phys. Rev. D **77**, 066014 (2008); Erratum-ibid. D **78**, 089902(E) (2008).
[38] R. Baier, P. Romatschke, D. T. Son, A. O. Starinets and M. A. Stephanov, JHEP 04, 100 (2008); M. Luzum and P. Romatschke, Phys. Rev. C 78, 034915 (2008) ((E) Phys. Rev. C 79, 039903(E) (2009))
[39] A. El, Z. Xu, and C. Greiner, Phys. Rev. C81, 041901 (2010).
[40] R. Geroch and L. Lindblom, Phys. Rev. D 41, 1855 (1990); R. Geroch and L. Lindblom, Ann. Phys. (NY) 207, 394 (1991); E. Calzetta, Class. Quant. Grav. 15, 653 (1998); J. Peralta-Ramos and E. Calzetta, Phys. Rev. C 82, 054905 (2010).
[41] G. B. Nagy and O. A. Reula, J. Phys. A 28, 6943 (1995); E. Calzetta and M. Thibeault, Phys. Rev. D 63, 103507 (2001); J. Peralta-Ramos and E. Calzetta, Phys. Rev. D 80, 126002 (2009).
[42] S. Mrowczynski, Phys. Lett. B 314, 118 (1993); S. Mrowczynski and M. H. Thoma, Annu. Rev. Nucl. Part. Sci. 57, 61 (2007); B. Schenke, M. Strickland, C. Greiner, and M. H. Thoma, Phys. Rev. D 73, 125004 (2006); C. Manuel and S. Mrowczynski, Phys. Rev. D 74, 105003 (2006); M. Attems, A. Rebhan, and M. Strickland, Phys. Rev. D 87, 025010 (2013); J. Peralta-Ramos and E. Calzetta, ArXiv:1309.5412.
[43] S. Floerchinger and U. A. Wiedemann, JHEP 11, 100 (2011).
[44] R. A. Janik and R. Peschanski, Phys. Rev. D 73, 045013 (2006); Y. V. Kovchegov, Nuclear Physics A 830, 395c (2009); V. Balasubramanian, A. Bernamonti, J. de Boer, N. Copland, B. Craps, E. Keski-Vakkuri, B. Müller, A. Schäfer, M. Shigemori, and W. Staessens Phys. Rev. D 84, 026010 (2011); D. Garfinkle and L. A. Pando Zayas, Phys. Rev. D 84, 066006 (2011).
[45] L. Kadanoff and G. Baym, Quantum Statistical Mechanics (Benjamin, New York, 1962).
[46] E. Calzetta and B-L. Hu, Phys. Rev. D37, 2878 (1988).

FROM SYMPLECTIC GROUPOIDS TO DOUBLE STRUCTURES

Kirill Mackenzie

School of Mathematics and Statistics,
University of Sheffield,
Sheffield S3 7RH, UK
E-mail: K.Mackenzie@sheffield.ac.uk

Foreword

These notes are an introduction to symplectic groupoids and the double structures associated with them. The treatment is intended to lie about midway between the original account of Coste, Dazord and Weinstein [4], which relied on effective use of the symplectic structures, and the account in my book [18], which showed, on the level of Poisson groupoids, that the basic results of the theory follow from 'categorical' compatibility conditions between the associated Lie algebroid and Lie groupoid structures. (See §5 for more details.)

The reader needs to know only the most basic ideas of symplectic geometry, Lie groups, and vector bundles. Conventions are recalled in §1.

In particular, no familiarity with double structures is assumed. Instead the cotangent groupoid — perhaps one of the hardest ideas to assimilate for someone new to this theory — is introduced as a consequence of the isomorphism between the tangent and cotangent bundles. That then leads to the general concept of a $\mathcal{VB}$–groupoid and its duality.

These notes are based on lectures given at the School on *Geometric, Algebraic and Topological Methods for Quantum Field Theory* in Villa de Leyva, Colombia, in July 2013. I am very glad to have had the opportunity to give these lectures, and I thank the organizers most heartily for involving me in a School of such vitality and openness. I want to particularly thank Alexander Cardona for his generous hospitality and for looking after me so well.

These notes aim to introduce the reader to certain important, and relatively new, ideas quickly; accordingly they omit much standard material.

All the main results here are known, but the approach has some new features. Some references which provide alternative treatments and more detail are given at the end.

I am very grateful to Yvette Kosmann–Schwarzbach for her careful reading and valuable comments.

1. Background: Poisson structures, Lie algebroids, Lie groupoids

Definition 1.1. A *Poisson structure* on a manifold P is a bracket of smooth functions

$$\{\ ,\ \}\colon C^\infty(P)\times C^\infty(P)\to C^\infty(P)$$

with respect to which $C^\infty(P)$ is an $\mathbb{R}$–Lie algebra, and such that for all $f_1, f_2, f_3 \in C^\infty(P)$,

$$\{f_1, f_2f_3\} = f_2\{f_1, f_3\} + \{f_1, f_2\}f_3. \tag{1}$$

The bracket $\{f_1, f_2\}$ depends only on df_1 and df_2 and we define a 2-vector field π on P by $\pi(w_1\, df_1, w_2\, df_2) = w_1w_2\{f_1, f_2\}$. This is the *Poisson tensor* and in practice we use π to denote a Poisson structure.

For a symplectic manifold (M,ω), there is an *associated Poisson structure* on M defined by $\{f_1, f_2\} = \omega((df_1)^\sharp, (df_2)^\sharp)$, where $\varphi \mapsto \varphi^\sharp$, $T^*M \to TM$ is the inverse of the map $\omega^\flat\colon TM \to T^*M$, $\langle\omega^\flat(X), Y\rangle = -\omega(X, Y)$.

Returning to a Poisson manifold P, we define $\pi^\#\colon T^*P \to TP$, the *Poisson anchor*, by

$$\langle\varphi, \pi^\#(\psi)\rangle = \pi(\psi, \varphi). \tag{2}$$

If $\pi^\#$ is an isomorphism of vector bundles, then $\omega(X, Y) = \langle(\pi^\#)^{-1}(Y), X\rangle$ is a symplectic structure on P, and the associated Poisson structure is the given one.

Define a bracket of 1-forms by

$$[\varphi, \psi] = \mathfrak{L}_{\varphi^\#}(\psi) - \mathfrak{L}_{\psi^\#}(\varphi) - d(\pi(\varphi, \psi)), \tag{3}$$

where again $\varphi^\# = \pi^\#(\varphi)$. The bracket of 1-forms makes T^*P a Lie algebroid with anchor $\pi^\#$.

Definition 1.2. A *Lie algebroid* is a vector bundle A on base M together with a bracket $[\ ,\ ]\colon \Gamma A\times\Gamma A\to\Gamma A$, with respect to which ΓA is an $\mathbb{R}$-Lie algebra, and a vector bundle map $a\colon A \to TM$, called the *anchor of* A, such that

$$[X, fY] = f[X, Y] + a(X)(f)Y, \qquad a[X, Y] = [aX, aY], \tag{4}$$

for $X, Y \in \Gamma A$ and $f \in C^\infty(M)$.

If A' is also a Lie algebroid on M then a *base-preserving morphism of Lie algebroids* is a morphism of vector bundles $\varphi\colon A \to A'$ such that $a' \circ \varphi = a$ and $\varphi([X, Y]) = [\varphi(X), \varphi(Y)]$ for all $X, Y \in \Gamma A$.

For any vector bundle A on M, sections $X \in \Gamma A$ induce (fibrewise) linear maps $\ell_X\colon A^* \to \mathbb{R}$ by $\ell_X(\varphi) = \langle \varphi, X(q_*(\varphi))\rangle$ where q_* is the bundle projection for A^*. Every fibrewise linear map $A^* \to \mathbb{R}$ is ℓ_X for a unique $X \in \Gamma A$.

When A is a Lie algebroid, there is a unique Poisson structure on the manifold A^*, called the *Lie-Poisson structure corresponding to* A, such that

$$\{\ell_X, \ell_Y\} = \ell_{[X,Y]}, \qquad \{\ell_X, f \circ q_*\} = a(X)(f) \circ q_*, \qquad \{f_1 \circ q_*, f_2 \circ q_*\} = 0, \tag{5}$$

for $X, Y \in \Gamma A$, $f \in C^\infty(M)$. It is not clear that this defines the bracket for all pairs of smooth functions. To bypass this, consider the Poisson tensor for A^*, and note that every 1-form is the sum of $d\ell_X$ for some $X \in \Gamma A$ and $q_*^* df$ for some $f \in C^\infty(M)$. Conversely, if E is a vector bundle on M with a Poisson structure for which: (i) the bracket of fibrewise linear functions is fibrewise linear, (ii) the bracket of a fibrewise linear function with a pullback function is a pullback function, and (iii) the bracket of two pullback functions is zero, then there is a unique Lie algebroid structure on E^* for which this is the Lie-Poisson structure.

For a Lie algebra $\mathfrak{g}$, the second and third definitions in (5) are vacuous. However the first can be extended to any pair of smooth functions. Let $f\colon \mathfrak{g}^* \to \mathbb{R}$ be a smooth function on $\mathfrak{g}^*$. Using an elementary notion of derivative, $D(f)(\varphi)$, for each $\varphi \in \mathfrak{g}^*$, is a linear map $\mathfrak{g}^* \to \mathbb{R}$ (the 'directional derivative') and can thus be identified with an element of $\mathfrak{g}$. Since a Poisson bracket depends only on the derivatives of its arguments, we have

$$\{f_1, f_2\}(\varphi) = \langle \varphi, [D(f_1)(\varphi), D(f_2)(\varphi)]\rangle. \tag{6}$$

A map $\mu\colon P \to Q$ of Poisson manifolds is a *Poisson map* if $\{f_1 \circ \mu, f_2 \circ \mu\}_P = \{f_1, f_2\}_Q \circ \mu$ for all $f_1, f_2 \in C^\infty(Q)$. We will need Poisson maps in §4.

Lie groupoids

Definition 1.3. A *Lie groupoid* consists of a manifold M of *points*, or *objects*, and a manifold $\mathcal{G}$ of *arrows*, together with surjective submersions $\alpha, \beta\colon \mathcal{G} \to M$, called respectively the *source* and *target* maps, and an injective immersion $1\colon M \to \mathcal{G}$, written $m \mapsto 1_m$, together with a smooth *multiplication* or *composition*, $(h, g) \mapsto hg$, $\mathcal{G} \times_M \mathcal{G} \to \mathcal{G}$, defined on

$$\mathcal{G} \times_M \mathcal{G} = \{(h, g) \in \mathcal{G} \times \mathcal{G} \mid \alpha(h) = \beta(g)\}.$$

These maps are subject to modified forms of the group axioms: for all $h, g, f \in \mathcal{G}$ such that $\alpha(h) = \beta(g)$ and $\alpha(g) = \beta(f)$, we have $h(gf) = (hg)f$; for each $g \in \mathcal{G}$ we have $1_n g = g$ and $g1_m = g$ where $n = \beta(g)$, $m = \alpha(g)$; finally, for each $g \in \mathcal{G}$ there is a $g^{-1} \in \mathcal{G}$ such that $g^{-1}g = 1_m$ and $gg^{-1} = 1_n$. We often write $\mathcal{G} \rightrightarrows M$ to indicate a Lie groupoid.

A *morphism of Lie groupoids* consists of two smooth maps $F\colon \mathcal{G} \to \mathcal{G}'$ and $f\colon M \to M'$ which commute with the sources and targets, for which $F(1_m) = 1'_{f(m)}$ for $m \in M$, and for which $F(hg) = F(h)F(g)$ when $\alpha(h) = \beta(g)$. If $M' = M$ and f is the identity map, we say that F is *over* M.

In these lectures we are concerned with symplectic groupoids, rather than the general theory. We include here only four important examples.

Examples 1.4. **(i)** For any manifold M the Cartesian square $M \times M$ is a Lie groupoid with $\alpha(n, m) = m$, $\beta(n, m) = n$, $1_m = (m, m)$ and multiplication $(p, n)(n, m) = (p, m)$. This is the *pair groupoid on* M.

(ii) For a connected manifold M the *fundamental groupoid* $\Pi(M)$ is a Lie groupoid on M. The arrows are homotopy classes, with endpoints fixed, of piecewise smooth paths in M, the source and target maps give the start-point and end-point, and the groupoid multiplication is concatenation. The identity elements are the classes of constant paths. The map $\chi\colon \Pi(M) \to M \times M$ which sends the class of a path $p(t)$ to $(p(1), p(0))$ is a morphism of Lie groupoids over M.

(iii) Let E be a vector bundle on M. Denote by $\Phi(E)$ the set of all linear isomorphisms $\xi\colon E_m \to E_n$ between fibres of E. The local triviality of E can be used to give $\Phi(E)$ a manifold structure. Define source and target $\alpha(\xi) = m$ and $\beta(\xi) = n$, for ξ as above, and write 1_m for the identity map $E_m \to E_m$. Then the standard composition of maps makes $\Phi(E)$ a Lie groupoid on base M, the *frame groupoid of* E.

(iv) Let G be a Lie group and let $G \times M \to M$ be a smooth action of G on a manifold M. Write $\mathcal{G} = G \times M$, the product manifold. Define

$\alpha\colon \mathcal{G} \to M$ by $\alpha(g,m) = m$ and $\beta\colon \mathcal{G} \to M$ by $\beta(g,m) = gm$. For $m \in M$ write $1_m = (1,m)$ and define

$$(h,n)(g,m) = (hg,m)$$

when $n = gm$. Then $\mathcal{G}$ is a Lie groupoid on base M, the *action groupoid* corresponding to the given action. We write $G \lessdot M$ for $\mathcal{G}$; many authors write $G \ltimes M$.

We now describe the Lie algebroid of a Lie groupoid. The construction follows the same general procedure as the construction of the Lie algebra of a Lie group, though we define the bracket using right-invariant vector fields rather than the more usual left-invariant vector fields.

For a Lie group G the Lie algebra $\mathfrak{g}$ may be defined as $T_1(G)$, the tangent space at the identity element, with the bracket on $\mathfrak{g}$ defined by identifying elements of $\mathfrak{g}$ with right-invariant vector fields.

For a Lie groupoid $\mathcal{G} \rightrightarrows M$ there is a regular submanifold of identity elements $\{1_m \mid m \in M\} \subseteq \mathcal{G}$, which we identify with M. Write $T_M\mathcal{G}$ for the restriction of the tangent bundle $T\mathcal{G}$ to M. The fibres of $T_M\mathcal{G}$ are $T_{1_m}\mathcal{G}$ for $m \in M$.

Because the multiplication in $\mathcal{G}$ is only defined on $\mathcal{G} \times_M \mathcal{G}$, right-translations are only defined on the fibres of the source map. For $g \in \mathcal{G}$ we have

$$R_g\colon \alpha^{-1}(\beta g) \to \alpha^{-1}(\alpha g), \quad h \mapsto hg.$$

In order to define right-invariance for vector fields on $\mathcal{G}$, we therefore consider only vector fields which are vertical with respect to the source map. Thus:

Definition 1.5. A vector field ξ on $\mathcal{G}$ is *right-invariant* if $T(\alpha)(\xi(g)) = 0$ for all $g \in \mathcal{G}$, and $T(R_g)(\xi(h)) = \xi(hg)$ for all $(h,g) \in \mathcal{G} \times_M \mathcal{G}$.

Note that for a smooth map $f\colon M \to N$ we usually write $T(f)$ for the tangent map $TM \to TN$, reserving df for when f is real-valued.

The second condition is equivalent to $\xi(g) = T(R_g)(\xi(1_{\beta g}))$ for all $g \in \mathcal{G}$. If the analogy with the Lie algebra of a Lie group is to hold good, it should be sufficient to consider the restrictions of right-invariant vector fields to the manifold of identity elements. We therefore define

$$A\mathcal{G} = \ker(T(\alpha)\colon T\mathcal{G} \to TM) \cap T_M\mathcal{G}.$$

That is, $A\mathcal{G}$ is a vector bundle on M, and the fibre over $m \in M$ is $T_{1_m}(\alpha^{-1}(m))$.

If X is a section of $A\mathcal{G}$ we can define a right-invariant vector field $\overrightarrow{X}$ on $\mathcal{G}$ by $\overrightarrow{X}(g) = T(R_g)(X(\beta g))$. Conversely, every right-invariant vector field is $\overrightarrow{X}$ for some $X \in \Gamma A\mathcal{G}$.

We next need to show that the bracket of right-invariant vector fields is a right-invariant vector field. This follows, as in the case of Lie groups, from the fact that the bracket of projectable vector fields is projectable: note that a vector field ξ on $\mathcal{G}$ is right-invariant if and only if it projects to zero under α and for each $g \in \mathcal{G}$, the restriction of ξ to $\alpha^{-1}(\beta g)$ projects under R_g to the restriction of ξ to $\alpha^{-1}(\alpha g)$. We therefore have a bracket on $\Gamma A\mathcal{G}$ such that, for all $X, Y \in \Gamma A\mathcal{G}$,

$$\overrightarrow{[X,Y]} = [\overrightarrow{X}, \overrightarrow{Y}].$$

Lastly, consider $[X, fY]$ where $f \in C^\infty(M)$. Clearly $\overrightarrow{fY} = (f \circ \beta)\overrightarrow{Y}$. Using the Leibniz identity for vector fields on $\mathcal{G}$, we therefore get that

$$[X, fY] = f[X, Y] + a(X)(f)Y,$$

where $a\colon A\mathcal{G} \to TM$ is the restriction of $T(\beta)\colon T\mathcal{G} \to TM$, called the *anchor of* $A\mathcal{G}$. This completes the construction of the *Lie algebroid of the Lie groupoid* $\mathcal{G}$.

Examples 1.6. **(i)** Consider $\mathcal{G} = M \times M$. The kernel of $T(\alpha)$ is $\beta^!(TM)$, the pullback of TM to $M \times M$ across β. Restricting this to the diagonal in $M \times M$ we get $A\mathcal{G} = TM$ as vector bundles. The right-invariant vector field corresponding to $X \in \Gamma TM$ is $\overrightarrow{X}(n, m) = X(n) \oplus 0_m$ and so the bracket on $A\mathcal{G}$ is the standard bracket of vector fields on M and the anchor is the identity.

(ii) It is easy to see that a base-preserving morphism of Lie groupoids $F\colon \mathcal{G} \to \mathcal{G}'$ induces a map $A(F)\colon A\mathcal{G} \to A\mathcal{G}'$ which is a morphism of Lie algebroids. In the case of $\chi\colon \Pi(M) \to M \times M$ the kernel (that is, the set of elements of $\Pi(M)$ which are mapped to identity elements of $M \times M$), is the union of the fundamental groups of M and is fibre-wise discrete. It follows that $A(\chi)\colon A\Pi(M) \to TM$ is a diffeomorphism, usually identified with the identity.

(iii) The Lie algebroid $A\Phi(E)$, as a vector bundle, is an extension of TM by $\mathrm{End}(E)$. Its sections are those linear differential operators D, of order $\leqslant 1$, for which there exists a vector field X on M such that

$$D(f\mu) = fD(\mu) + X(f)\mu,$$

for all $f \in C^\infty(M)$ and $\mu \in \Gamma E$. These may be called *derivations* on E, and $A\Phi(E)$ denoted by $\mathfrak{D}(E)$. The bracket on $\mathfrak{D}(E)$ is the standard commutator bracket, and the anchor maps each D to its X.

We will not prove this, but it is easy to believe: the sections of $\mathfrak{D}(E)$ are all covariant derivatives ∇_X for all connections ∇ and all vector fields X on M, together with their differences, and $\Phi(E)$ consists of all parallel translations, for all connections in E and all paths in M.

(iv) Let $G \times M \to M$ be a smooth action of a Lie group G on a manifold M and let $X \mapsto X_M$ be the infinitesimal action $\mathfrak{g} \to \mathfrak{X}(M)$.

The Lie algebroid of $G \lessdot M \rightrightarrows M$ is the trivial vector bundle $M \times \mathfrak{g}$ with anchor $M \times \mathfrak{g} \to TM$, $(m, X) \mapsto X_M(m)$ and bracket

$$[V, W] = L_{V_M}(W) - L_{W_M}(V) + [V, W]^{\mathrm{pt}}.$$

Here V and W are $\mathfrak{g}$-valued maps on M; since $\mathfrak{g}$ is a vector space we can take Lie derivatives of $\mathfrak{g}$-valued maps. The final term is the 'pointwise bracket' of V with W as maps into $\mathfrak{g}$.

2. Symplectic groupoids

The following definition is due to Weinstein [28]. Independent definitions were given by Karasëv [11] and Zakrzewski [30, 31].

Definition 2.1. A *symplectic groupoid* is a Lie groupoid $\Sigma \rightrightarrows P$ together with a symplectic structure ω on Σ such that the graph of multiplication

$$Gr = \{(hg, h, g) \mid \alpha h = \beta g\} \tag{7}$$

is Lagrangian in $\overline{\Sigma} \times \Sigma \times \Sigma$. The bar denotes reversal of the symplectic structure.

The following are the fundamental consequences:

Theorem 2.2.

(i) 1_P *is Lagrangian in* Σ.
(ii) Inversion is an antisymplectomorphism.
(iii) There is a unique Poisson structure on P *such that* β *is a Poisson map.*
(iv) There is a canonical isomorphism of Lie algebroids $A\Sigma \cong T^*P$.

The proof occupies the rest of the section. First the fundamental example.

The cotangent groupoid of a Lie group

Consider any Lie group G. The tangent bundle TG has a Lie group structure with multiplication $T(\kappa)\colon TG \times TG \to TG$ where $\kappa\colon G \times G \to G$ is the multiplication in G. Since the domain is a product manifold, we can differentiate in each variable separately and we get $T(\kappa)(Y, X) = T(L_h)(X) + T(R_g)(Y)$ where $Y \in T_hG$ and $X \in T_gG$. In practice we write $Y \bullet X = T(\kappa)(Y, X)$. With this structure, TG is the *tangent group of* G,

It is possible to give the cotangent bundle T^*G a group structure in a similar way. However what we are about to define is distinct from the group structure.

Define a groupoid structure on T^*G with base $\mathfrak{g}^*$ by

$$\begin{gathered}\widetilde{\beta}(\varphi_g) = \varphi \circ T_1(R_g), \qquad \widetilde{\alpha}(\varphi_g) = \varphi \circ T_1(L_g), \\ \psi_h \cdot \varphi_g = \psi \circ T(R_{g^{-1}}) = \varphi \circ T(L_{h^{-1}}).\end{gathered} \tag{8}$$

This structure is isomorphic to an action groupoid. Let G act on $\mathfrak{g}^*$ by the coadjoint action $g\theta := \theta \circ \mathrm{Ad}_{g^{-1}}$ and form $G \lessdot \mathfrak{g}^* \rightrightarrows \mathfrak{g}^*$. Then $G \lessdot \mathfrak{g}^* \to T^*G$, $(g, \theta) \mapsto \theta \circ T(L_{g^{-1}})$ is a groupoid isomorphism.

Whereas the group TG is isomorphic to the semi-direct product group $G \ltimes \mathfrak{g}$ defined by the adjoint representation, T^*G is isomorphic to the action groupoid $G \lessdot \mathfrak{g}^*$. It is important to distinguish the two constructions.

The symplectic structure on T^*G is the standard symplectic structure for a cotangent bundle, $\omega = d\lambda$ where λ is the 1-form on G defined on $\xi \in T(T^*G)$ by

$$\langle \lambda, \xi \rangle = \langle \varphi, T(c)(\xi) \rangle$$

where $\varphi \in T^*G$ is the base-point of ξ and $c\colon T^*G \to G$ is the projection.

Further, ω makes $T^*G \rightrightarrows \mathfrak{g}^*$ a symplectic groupoid and the Poisson structure on $\mathfrak{g}^*$ which arises from Theorem 2.2(iii) is the Lie–Poisson structure dual to $\mathfrak{g}$.

Proof of Theorem 2.2

To see what Definition 2.1 means, we use the tangent groupoid of Σ. This construction applies to any Lie groupoid $\mathcal{G} \rightrightarrows M$. The *tangent groupoid* of $\mathcal{G} \rightrightarrows M$ is the groupoid $T\mathcal{G} \rightrightarrows TM$ obtained by applying the tangent functor to the structure of $\mathcal{G} \rightrightarrows M$. So the source of the tangent groupoid is $T(\alpha)$, the identities are $T(1)(x)$ for $x \in TM$, and the multiplication $\eta \bullet \xi$ of two tangent vectors with $T(\alpha)(\eta) = T(\beta)(\xi)$ is $T(\kappa)(\eta, \xi)$ where

$\kappa\colon \mathcal{G} \times_M \mathcal{G} \to \mathcal{G}$ is the multiplication in $\mathcal{G}$. Because the tangent functor preserves diagrams, the groupoid axioms follow immediately. We write ξ^{-1} for the inverse of $\xi \in T\mathcal{G}$, and $\widetilde{0}_g$ for the zero element of $T_g\mathcal{G}$.

If $\mathcal{G} = G$ is a group, then, as we saw above, TG is a group with $\eta \bullet \xi = T(R_g)(\eta) + T(L_h)(\xi)$ for $\eta \in T_h(G)$ and $\xi \in T_g(G)$. However no Lie group has a symplectic groupoid structure. For general Lie groupoids, the following result is very necessary.

Proposition 2.1. *For elements $\xi_i \in T\mathcal{G}$, $i = 1, \ldots, 4$ with $\xi_1, \xi_2 \in T_g\mathcal{G}$ and $\xi_3, \xi_4 \in T_h\mathcal{G}$, and with $T(\alpha)(\xi_3) = T(\beta)(\xi_1)$ and $T(\alpha)(\xi_4) = T(\beta)(\xi_2)$, we have*

$$(\xi_4 + \xi_3) \bullet (\xi_2 + \xi_1) = \xi_4 \bullet \xi_2 + \xi_3 \bullet \xi_1. \tag{9}$$

PROOF. This is the statement that $T(\kappa)$ is linear as a map of vector bundles $T\mathcal{G} \times_{TM} T\mathcal{G} \to T\mathcal{G}$.

To see this, observe that (ξ_3, ξ_1) and (ξ_4, ξ_2) are elements of $T_h\mathcal{G} \times T_g\mathcal{G}$ and can therefore be added to give $(\xi_4 + \xi_3, \xi_2 + \xi_1) \in T_h\mathcal{G} \times T_g\mathcal{G}$.

The linearity then gives that

$$T(\kappa)(\xi_4 + \xi_3, \xi_2 + \xi_1) = T(\kappa)(\xi_4, \xi_2) + T(\kappa)(\xi_3, \xi_1)$$

which is (9). □

The source and target conditions ensure that each product and sum is defined and is in the domain of $T(\kappa)$.

Consider now a symplectic groupoid $\Sigma \rightrightarrows P$. The tangent to the graph (7) of the multiplication is the graph of the tangent multiplication :

$$T(Gr) = \{(\eta \bullet \xi, \eta, \xi) \mid T(\alpha)(\eta) = T(\beta)(\xi)\}. \tag{10}$$

Recall that a submanifold L of a symplectic manifold (S, σ) is *Lagrangian* if it is isotropic: $\sigma(Y, X) = 0$ for all $X, Y \in TL$, and $\dim L = \frac{1}{2} \dim S$.

Applying the isotropy condition to the graph of the multiplication we have that

$$-\omega(\eta_1 \bullet \xi_1, \eta_2 \bullet \xi_2) + \omega(\eta_1, \eta_2) + \omega(\xi_1, \xi_2) = 0 \tag{11}$$

for all $\xi_1, \xi_2, \eta_1, \eta_2 \in T\Sigma$ for which the multiplications are defined.

Set $\xi_i = \eta_i = T(1)(x_i)$ in (11), where $x_i \in TP$. Then (11) becomes

$$-\omega(T(1)(x_1), T(1)(x_2)) + \omega(T(1)(x_1), T(1)(x_2)) + \omega(T(1)(x_1), T(1)(x_2)) = 0$$

so $\omega(T(1)(x_1), T(1)(x_2)) = 0$ for all $x_1, x_2 \in TP$. This proves that 1_P is isotropic in Σ. Next,

$$\dim(Gr) = \dim(\Sigma \times_P \Sigma) = 2 \dim \Sigma - \dim P$$

and this is to be $\frac{1}{2}\dim(\Sigma\times\Sigma\times\Sigma)$ so $\dim P = \frac{1}{2}\dim\Sigma$. This completes the proof that 1_P is Lagrangian in Σ.

Next consider any ξ_1, ξ_2 and write $x_i = T(\alpha)(\xi_i)$. Then with $\eta_i = \xi_i^{-1}$ in (11) we have

$$-\omega(T(1)(x_1), T(1)(x_2)) + \omega(\xi_1^{-1}, \xi_2^{-1}) + \omega(\xi_1, \xi_2) = 0.$$

So

$$\omega(\xi_1^{-1}, \xi_2^{-1}) = -\omega(\xi_1, \xi_2), \tag{12}$$

which shows that inversion is antisymplectic.

The last two parts of Theorem 2.2 require more work.

Recall the map $\omega^\flat\colon T\Sigma \to T^*\Sigma$ defined by $\omega^\flat(\xi)(\eta) = -\omega(\xi, \eta)$. What is $\omega^\flat(\eta \bullet \xi)$, where $\eta \in T_h\Sigma$, $\xi \in T_g\Sigma$ and $T(\alpha)(\eta) = T(\beta)(\xi)$? It is an element of $T^*_{hg}\Sigma$ so we pair it with an arbitrary element ζ of $T_{hg}\Sigma$.

We can write $\zeta = \zeta_2 \bullet \zeta_1$ where $\zeta_2 \in T_h\Sigma$ and $\zeta_1 \in T_g\Sigma$. To see this, take any $\zeta_1 \in T_g\Sigma$ with $T(\alpha)(\zeta_1) = T(\alpha)(\zeta)$ and define $\zeta_2 = \zeta \bullet \zeta_1^{-1}$. This decomposition is not unique, of course.

Then, by (11),

$$\omega(\eta \bullet \xi, \zeta_2 \bullet \zeta_1) = \omega(\eta, \zeta_2) + \omega(\xi, \zeta_1).$$

Is the RHS well-defined ? Suppose we also have $\zeta = \zeta_4 \bullet \zeta_3$ where $\zeta_4 \in T_h\Sigma$ and $\zeta_3 \in T_g\Sigma$. Then $\zeta_3 \bullet \zeta_1^{-1} = \zeta_4^{-1} \bullet \zeta_2 \in T_{1_m}\Sigma$ where $m = \alpha h = \beta g$. Call this ν. So $\zeta_3 = \nu \bullet \zeta_1$ and $\zeta_4 = \zeta_2 \bullet \nu^{-1}$. Now

$$\omega(\eta \bullet \xi, \zeta_4 \bullet \zeta_3) = \omega(\eta, \zeta_2 \bullet \nu^{-1}) + \omega(\xi, \nu \bullet \zeta_1).$$

Consider the second term on the RHS first. Inserting an identity element and using (11), we have

$$\omega(\xi, \nu \bullet \zeta_1) = \omega(T(1)T(\beta)(\xi) \bullet \xi, \nu \bullet \zeta_1) = \omega(T(1)T(\beta)(\xi), \nu) + \omega(\xi, \zeta_1).$$

Now the first term:

$$\begin{aligned}\omega(\eta, \zeta_2 \bullet \nu^{-1}) &= \omega(\eta \bullet T(1)(T(\alpha)(\eta), \zeta_2 \bullet \nu^{-1})\\ &= \omega(\eta, \zeta_2) + \omega(T(1)(T(\alpha)(\eta)), \nu^{-1}).\end{aligned}$$

Lastly, using (12), $\omega(T(1)(T(\alpha)(\eta)), \nu^{-1}) = -\omega(T(1)(T(\alpha)(\eta)), \nu)$. So

$$\omega(\eta \bullet \xi, \zeta) = \omega(\eta, \zeta_2) + \omega(\xi, \zeta_1) \tag{13}$$

is well-defined.

The cotangent groupoid $T^*\Sigma$

Now we define a groupoid structure on $T^*\Sigma$ so that (13) becomes

$$\omega^\flat(\eta \bullet \xi) = \omega^\flat(\eta) \bullet \omega^\flat(\xi). \tag{14}$$

In fact, we have already done most of the work. The base of $T^*\Sigma$ will be $A^*\Sigma$ and the source and target maps $T^*\Sigma \to A^*\Sigma$ are

$$\langle \widetilde{\beta}(\Phi), Y \rangle = \langle \Phi, T(R_g)(Y) \rangle, \quad \langle \widetilde{\alpha}(\Phi), X \rangle = \langle \Phi, T(L_g)(X - T(1)(aX)) \rangle, \tag{15}$$

where $\Phi \in T^*_g\Sigma$ and $Y \in A_{\beta g}\Sigma$, $X \in A_{\alpha g}\Sigma$.

For $\Phi \in T^*_g\Sigma$ and $\Psi \in T^*_h\Sigma$ define $\Psi \bullet \Phi \in T^*_{hg}\Sigma$ by

$$\langle \Psi \bullet \Phi, \eta \bullet \xi \rangle = \langle \Psi, \eta \rangle + \langle \Phi, \xi \rangle. \tag{16}$$

The proof that the RHS of (16) is well-defined needs precisely the condition that the source and target match, and follows the same pattern as the proof that (13) is well-defined. We leave the reader to check that when $\Sigma = T^*G$ for a Lie group G, (16) reduces to (8).

To define the identity $\widetilde{1}_\varphi \in T^*_{1_m}\Sigma$ corresponding to $\varphi \in A^*_m\Sigma$, observe that every element ξ of $T_{1_m}\Sigma$ can be written uniquely in the form $T(1)(x) + X$ where $x \in T_mM$ and $X \in A_m\Sigma$. We can therefore define

$$\langle \widetilde{1}_\varphi, T(1)(x) + X \rangle = \langle \varphi, X \rangle \tag{17}$$

and it is straightforward to check that $\widetilde{1}_\varphi$ is indeed an identity for the multiplication and that this structure makes $T^*\Sigma \rightrightarrows A^*\Sigma$ a groupoid with inverses

$$\langle \Phi^{-1}, \xi^{-1} \rangle = -\langle \Phi, \xi \rangle. \tag{18}$$

We now prove that $\omega^\flat\colon T\Sigma \to T^*\Sigma$ is a morphism of groupoids,

$$\begin{array}{ccc} T\Sigma & \xrightarrow{\ \omega^\flat\ } & T^*\Sigma \\ \downarrow\downarrow & & \downarrow\downarrow \\ TP & \xrightarrow[\ w\ ]{} & A^*\Sigma \end{array} \tag{19}$$

over some map w. To find w, take the composite of $T(1)$, followed by $\omega^\flat$, followed by $\widetilde{\beta}$ (or $\widetilde{\alpha}$). This gives

$$w(x) = -\,\iota_{T(1)(x)}\omega\big|_{A\Sigma}\,.$$

We must now show that $\widetilde{\beta} \circ \omega^\flat = w \circ T(\beta)$. Take $\xi \in T_g\Sigma$. Then

$$\langle \widetilde{\beta}(\omega^\flat(\xi)), Y \rangle = \langle \omega^\flat(\xi), T(R_g)(Y) \rangle = -\omega(\xi, T(R_g)(Y)).$$

On the other hand,

$$\langle w(T(\beta)(\xi), Y\rangle = -\omega(T(1)(T(\beta)(\xi)), Y).$$

To prove the equality of these, we need to express right-translations in terms of the tangent groupoid multiplication.

Lemma 2.3. *For $X \in A_m\Sigma$ and $g \in \Sigma$ with $\beta(g) = m$,*

$$X \bullet \widetilde{0}_g = T(R_g)(X). \tag{20}$$

PROOF. Write X as the derivative at $t = 0$ of a curve $h_t \in \Sigma_m$ with $h_0 = 1_m$. One can likewise write $\widetilde{0}_g$ as the derivative of the curve constant at g. Then $X \bullet \widetilde{0}_g$ is the derivative of the curve $h_t g$ and this is $T(R_g)(X)$.

We now complete the proof that $\widetilde{\beta} \circ \omega^\flat = w \circ T(\beta)$. We must show that

$$\omega(\xi, T(R_g)(Y)) = \omega(T(1)(T(\beta)(\xi)), Y).$$

Using (11) and (20) the LHS is equal to

$$\omega(T(1)T(\beta)(\xi) \bullet \xi, Y \bullet \widetilde{0}_g) = \omega(T(1)T(\beta)(\xi), Y) + \omega(\xi, \widetilde{0}_g)$$

and $\omega(\xi, \widetilde{0}_g)$ is zero, since $\widetilde{0}_g$ is the zero of $T_g\Sigma$.

The proof that $\widetilde{\alpha} \circ \omega^\flat = w \circ T(\alpha)$ is similar. □

Finally, the morphism property itself is (13). Note that w is an isomorphism of vector bundles since $\omega^\flat$ is.

Poisson structure on P

Consider Figure 1(a). This diagram and (b) are not diagrams of morphisms (as in (19)) but are single mathematical objects, part of the structure of which is indicated by the arrows. (See §5.)

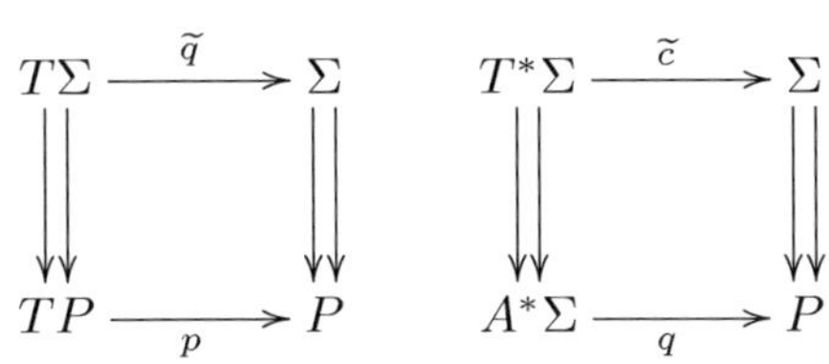

(a) (b)

Fig. 1

Recall that the Lie algebroid of Σ, as a vector bundle, is the intersection of the kernel of the bundle projection $\widetilde{q}$, which is the restriction of $T\Sigma$ to the identity elements, and the kernel of $T(\alpha)$, which is the vector bundle tangent to the source fibres. The bracket on $\Gamma A\Sigma$ is given by associating to a section X of $A\Sigma$ a right-invariant vector field $\overrightarrow{X}$, and showing that the bracket of right-invariant vector fields is right-invariant. We now apply this process to the cotangent structure in Figure 1(b).

First find the kernel of $\widetilde{\alpha}$. To do this we use the following lemma.

Lemma 2.4. *Let $\eta \in T_h\Sigma$ and $\xi \in T_g\Sigma$ have $T(\alpha)(\eta) = 0 = T(\beta)(\xi)$. Then*

$$\eta \bullet \xi = T(L_h)(\xi) + T(R_g)(\eta).$$

Proof. In this case η is tangent to $\alpha^{-1}(m)$ and ξ is tangent to $\beta^{-1}(m)$ where $m = \alpha h = \beta g$, and so $\eta \bullet \xi$ can be calculated from the restriction of multiplication to $\alpha^{-1}(m) \times \beta^{-1}(m) \to \Sigma$, which does not involve a pullback. □

Now take $X \in A\Sigma$. Then, applying the lemma, we have

$$X \bullet (T(1)(aX) - X) = X + (T(1)(aX) - X) = T(1)(aX)$$

and so

$$X^{-1} = T(1)(aX) - X.$$

So we can rewrite the definition of $\widetilde{\alpha}$ as

$$\langle \widetilde{\alpha}(\Phi), X\rangle = -\langle \Phi, T(L_g)(X^{-1})\rangle,$$

and the kernel of $\widetilde{\alpha}$ is therefore the annihilator of $T^\beta\Sigma$.

Suppose $T(\beta)(\xi) = 0$, where $\xi \in T_g\Sigma$, Then $T(\alpha)(\xi^{-1}) = 0$. So $\xi^{-1} = T(R_{g^{-1}})(X)$ for some $X \in A\Sigma$. That is, $\xi^{-1} = X \bullet \widetilde{0}_{g^{-1}}$. Thence $\xi = \widetilde{0}_g \bullet X^{-1}$.

Clearly all pullbacks $\beta^*\mu$, for $\mu \in T^*P$, annihilate $T^\beta\Sigma$. Conversely, suppose that $\Phi \in T^*_{1_m}\Sigma$ annihilates $T^\beta\Sigma$. Then $\langle \Phi, X\rangle = \langle \Phi, T(1)(aX)\rangle$ for all $X \in A_m\Sigma$. Define $\mu \in T^*_mP$ by $\langle \mu, x\rangle = \langle \Phi, T(1)(x)\rangle$ for $x \in T_mP$. Then, for all $T(1)(x) + X \in T_{1_m}\Sigma$,

$$\begin{aligned}\langle \Phi, T(1)(x) + X\rangle &= \langle \Phi, T(1)(x) + T(1)(aX)\rangle = \langle \mu, x + aX\rangle \\ &= \langle \beta^*\mu, T(1)(x) + X\rangle,\end{aligned}$$

so $\Phi = \beta^*\mu$. Thus the bundle for Figure 1(b) which corresponds to $A\Sigma$ in (a), is T^*P.

Now take $\mu \in \Omega^1(P)$. We define $\overrightarrow{\mu} \in \Omega^1(\Sigma)$ by

$$\overrightarrow{\mu}(g) = \mu(\beta g) \bullet \widehat{0}_g,$$

where $\widehat{0}_g$ is the zero element of T_g^*G. In fact $\mu(\beta g)\bullet\widehat{0}_g$ is merely the pullback of $\mu(\beta g)$ to g.

Take $\mu \in T_m^*P$ and g with $\beta g = m$. Then

$$\langle \mu \bullet \widehat{0}_g, \eta \bullet \xi \rangle = \langle \mu, \eta \rangle + 0$$

where $\eta = T(1)(y) + Y$ and $T(\beta)(\xi) = T(\alpha)(\eta) = y$. Now, regarding μ as in $T_{1_m}^*\Sigma$, we have

$$\langle \mu, \eta \rangle = \langle \mu, T(\beta)(\eta) \rangle = \langle \mu, y + aY \rangle.$$

On the other hand,

$$\langle \beta_g^* \mu, \eta \bullet \xi \rangle = \langle \mu, T(\beta)(\eta) \rangle = \langle \mu, y + aY \rangle.$$

We can now define a bracket on $\Omega^1(P)$ by

$$\overrightarrow{[\mu_1, \mu_2]} = [\overrightarrow{\mu_1}, \overrightarrow{\mu_2}], \tag{21}$$

where the bracket on the right is the bracket on $\Omega^1(\Sigma)$ transported via $\omega^\flat$ from $\mathfrak{X}(\Sigma)$. The bracket on $\Omega^1(P)$ is therefore skew-symmetric and satisfies the Jacobi identity. Note that (21) can also be written as

$$\beta^*[\mu_1, \mu_2] = [\beta^*\mu_1, \beta^*\mu_2]. \tag{22}$$

Regarding $A\Sigma \subseteq T\Sigma$ and $T^*P \subseteq T^*\Sigma$ as the intersections of the relevant kernels, it follows from the fact that $\omega^\flat$ is an isomorphism of vector bundles, that the restriction to $A\Sigma \to T^*P$ is also. Denote this restriction temporarily by r. Then,

$$r = -w^*. \tag{23}$$

To see this, take $X \in A\Sigma$ and $y \in TP$. Then

$$\langle w^*(X), y \rangle = \langle w(y), X \rangle = -\omega(T(1)(y), X),$$

whereas

$$\langle r(X), y \rangle = \langle \omega^\flat(X), T(1)(y) \rangle = -\omega(X, T(1)(y)).$$

The brackets in $\Gamma A\Sigma$ and ΓT^*P are defined in terms of those on $\mathfrak{X}(\Sigma)$ and $\Omega^1(\Sigma)$ by analogous processes and it follows that r is an isomorphism of Lie algebroids, to T^*P with the bracket defined by (21) and the anchor $a \circ r^{-1}$, where a is the anchor of $A\Sigma$.

Lemma 2.5.

$$(a \circ r^{-1})^* = -a \circ r^{-1}.$$

Proof. We first prove that $w \circ a = a^* \circ r$. For $X, Y \in A\Sigma$,

$$\langle (w \circ a)(X), Y\rangle = \omega(T(1)(aX), Y)$$

and

$$\begin{aligned}\langle (a^* \circ r)(X), Y\rangle &= \langle r(X), a(Y)\rangle = -\langle w^*(X), a(Y)\rangle \\ &= -\langle w(aY), X\rangle = -\omega(T(1)(aY), X).\end{aligned}$$

We know that $T(1)(aX) = X + X^{-1}$ and $T(1)(aY) = Y + Y^{-1}$. So we have $\langle (w \circ a)(X), Y\rangle = \omega(X, Y) + \omega(X^{-1}, Y)$ and $\langle (a^* \circ r)(X), Y\rangle = \omega(X, Y) + \omega(X, Y^{-1})$. We must show that $\omega(X^{-1}, Y) = -\omega(X, Y^{-1}) = 0$.

Take $X \in A\Sigma$. Since r maps X to $r(X) \in T^*P$, it follows that $\omega^\flat$ maps $\overrightarrow{X}$ to $\overrightarrow{r(X)}$; that is,

$$\omega^\flat(\overrightarrow{X}) = \beta^*(r(X)) = -\beta^*(w^*(X)). \tag{24}$$

So for any $\eta \in T\Sigma$, $\omega(\overrightarrow{X}, \eta) = -\langle \beta^*(w^*(X)), \eta\rangle = -\langle w^*(X), T(\beta)(\eta)\rangle$. In particular if $T(\beta)(\eta) = 0$ then $\omega(\overrightarrow{X}, \eta) = 0$. This completes the proof that $w \circ a = a^* \circ r$.

Rewrite this as $a \circ r^{-1} = w^{-1} \circ a^*$. Now

$$(a \circ r^{-1})^* = (w^{-1} \circ a^*)^* = a \circ (w^*)^{-1} = -a \circ r^{-1},$$

as required. □

Note from the proof that we have also shown that:

Proposition 2.2. *For $\xi, \eta \in T\Sigma$, if $T(\alpha)(\xi) = 0$ and $T(\beta)(\eta) = 0$ then $\omega(\xi, \eta) = 0$.*

Now define a Poisson structure π on P by

$$\{f_1, f_2\} = \langle df_2, (a \circ r^{-1})(df_1)\rangle; \tag{25}$$

that is, $\pi^\# = a \circ r^{-1}$. That $\pi^\#$ is skew-symmetric has just been proved, and the Jacobi identity follows from the isomorphism of T^*P with $A\Sigma$. So π is a Poisson structure on P and, by (22), β is a Poisson map. Since β is a surjective submersion, it is the only such Poisson structure.

This completes the proof of (iii) and (iv) of Theorem 2.2.

Proposition 2.3. *The diffeomorphism $-w\colon TP \to A^*\Sigma$ is a Poisson map from the tangent lift Poisson structure on TP to the Poisson structure dual to the Lie algebroid structure on $A\Sigma$.*

The *tangent lift Poisson structure on* TP can be defined as the Poisson structure which is dual to the Lie algebroid T^*P. It was originally defined by T. Courant [5] directly in terms of functions on TP. For $f \in C^\infty(P)$ denote the function $TP \to \mathbb{R}$ corresponding to $df\colon P \to T^*P$ by ℓ_{df}. Then it is sufficient to define the Poisson brackets for all ℓ_{df} and all p^*f, where $p\colon TP \to P$ is the bundle projection. As in [5], we define:

$$\{\ell_{df_1}, \ell_{df_2}\} = \ell_{d\{f_1,f_2\}}, \ \{\ell_{df_1}, p^*f_2\} = p^*\{f_1, f_2\}, \ \{p^*f_1, p^*f_2\} = 0. \quad (26)$$

Proposition 2.3 now follows from the following general result. The proof is a simple exercise.

Proposition 2.4. *Let A and B be Lie algebroids on the same base M, and let $\varphi\colon A \to B$ be a morphism of vector bundles over M. Then φ is a morphism of Lie algebroids if and only if $\varphi^*\colon B^* \to A^*$ is a Poisson map.*

3. Midword

We have given the proof that the base manifold of a symplectic groupoid has a Poisson structure such that the target is a Poisson map, and that the cotangent Lie algebroid of this Poisson structure is canonically isomorphic to the Lie algebroid of the Lie groupoid. This is a very striking result. It shows, in particular, that the Poisson structure on the base manifold determines the symplectic groupoid up to local isomorphism.

Theorem 2.2 is the apotheosis of a classical question, as to whether a given Poisson manifold can be realized as the quotient of a symplectic manifold. In the example $T^*G \rightrightarrows \mathfrak{g}^*$ for G a Lie group, the Poisson manifold is the quotient of T^*G under an action of G. In the general case of Theorem 2.2, the symplectic manifold Σ is quotiented to its base P by the surjective submersion $\beta\colon \Sigma \to P$. This may be regarded as the quotient over the action of the equivalence relation $\Sigma \times_\beta \Sigma$ on Σ.

The classical question is the 'realizability problem': given a Poisson manifold P, is there a symplectic manifold M and a surjective submersion $M \to P$ which is a Poisson map? In fact, under mild conditions, such a map can be modified to give a symplectic groupoid structure.

The answer to the question — *given a Poisson manifold P is there a symplectic groupoid $\Sigma \rightrightarrows P$?* — was provided by Crainic and Fernandes [6, 7], building on work of Cattaneo and Felder [3].

The modern theory of symplectic realizations was introduced by Weinstein [4, 28] in order to provide a route for the quantization of Poisson manifolds.

In this article, we are concerned with the structures arising from symplectic groupoids. These are, I believe, most easily understood by studying Poisson groupoids, of which symplectic groupoids are a particular case. Poisson groupoids were introduced by Weinstein [29] and gave rise to the theory of Lie bialgebroids. It is to some aspects of this theory that we now turn.

4. Poisson groupoids

Let P be a Poisson manifold. A closed submanifold $C \subseteq P$ is *coisotropic* if the Poisson anchor $\pi^{\#}\colon T^*P \to TP$, when restricted to $(TC)^\circ$, goes into TC. Here $(TC)^\circ$ is the *annihilator of TC in T^*P*; it is isomorphic to the conormal bundle T_CP/TC.

In terms of the bracket of functions, a closed submanifold C is coisotropic in P if, whenever $f, g \in C^\infty(P)$ vanish on C, their bracket $\{f, g\}$ does also.

Coisotropic submanifolds play the role that in symplectic geometry is played by Lagrangian submanifolds. In particular, a map $\varphi\colon P \to Q$ of Poisson manifolds is a Poisson map if and only if the graph of φ is a coisotropic submanifold of $\overline{Q} \times P$.

Definition 4.1 ([29]). *A* Poisson groupoid *is a Lie groupoid $\mathcal{G} \rightrightarrows P$ together with a Poisson structure on $\mathcal{G}$ such that the graph of multiplication*

$$Gr = \{(hg, h, g) \mid \alpha h = \beta g\} \tag{27}$$

is coisotropic in $\overline{\mathcal{G}} \times \mathcal{G} \times \mathcal{G}$.

Using (10), $T(Gr)^\circ$ consists of triples (Θ, Ψ, Φ) of elements of $T^*\mathcal{G}$ such that

$$\Theta(\eta \bullet \xi) + \Psi(\eta) + \Phi(\xi) = 0 \tag{28}$$

for all $\eta, \xi \in T\mathcal{G}$ such that $T(\alpha)(\eta) = T(\beta)(\xi)$. We claim that this implies $\Theta = -\Psi \bullet \Phi$. First we must show that $\widetilde{\alpha}(\Psi) = \widetilde{\beta}(\Phi)$. Let $h, g \in \mathcal{G}$ be the base elements of η, ξ and write $m = \beta(g)$. For $Y \in A_m\mathcal{G}$ we have

$$(\widetilde{\beta}(\Phi))(Y) = \Phi(T(R_g)(Y)) = \Phi(Y \bullet \widetilde{0}_g).$$

From (28) we now get

$$\Phi(Y \bullet \widetilde{0}_g) = \Theta(\widetilde{0}_h \bullet \widetilde{0}_g) - \Psi(\widetilde{0}_h \bullet Y^{-1}) = -\Psi(T(L_h)(Y^{-1}))$$

However $Y^{-1} = T(1)(aY) - Y$, since $Y \in A\mathcal{G}$, and so we have

$$(\widetilde{\beta}(\Phi))(Y) = \Psi(T(L_h)(Y - T(1)(aY))) = (\widetilde{\alpha}(\Psi))(Y).$$

Now (28) can be written

$$T(Gr)^\circ = \{(-\Psi \bullet \Phi, \Psi, \Phi) \mid \widetilde{\alpha}(\Psi) = \widetilde{\beta}(\Phi)\}. \tag{29}$$

The coisotropy condition now is that

$$\pi^\#(\Psi \bullet \Phi) = \pi^\#(\Psi) \bullet \pi^\#(\Phi) \tag{30}$$

for $\Psi, \Phi \in T^*\mathcal{G}$ with $\widetilde{\alpha}(\Psi) = \widetilde{\beta}(\Phi)$. This is evidently a morphism condition on $\pi^\#$. Denote the base map $A^*\mathcal{G} \to TP$ temporarily by b. Then, for $\varphi \in A^*\mathcal{G}$,

$$\pi^\#(\widetilde{1}_\varphi) = T(1)(b(\varphi)). \tag{31}$$

From this it follows that 1_P is coisotropic in $\mathcal{G}$. For if $\Phi \in T^*_{1_m}\mathcal{G}$ annihilates $T_{1_m}(1_P)$ then, by (17), $\Phi = \widetilde{1}_\varphi$ for some $\varphi \in A^*\mathcal{G}$ and (31) is then the coisotropy condition for 1_P.

That 1_P is coisotropic in $\mathcal{G}$ implies, as shown by Weinstein [29], that $A^*\mathcal{G}$ inherits a Lie algebroid structure from $T^*\mathcal{G}$, the anchor of which is the restriction of $\pi^\#$; that is, b. We therefore write a_* for b in what follows.

To define the bracket we must verify two conditions, namely that

- if $\Phi, \Psi \in \Gamma T^*\mathcal{G}$ have $\Phi|_P,\ \Psi|_P \in \Gamma A^*\mathcal{G}$ then $[\Phi, \Psi]|_P \in \Gamma A^*\mathcal{G}$ also;
- if $\Phi, \Psi \in \Gamma T^*\mathcal{G}$ have $\Phi|_P = 0$ and $\Psi|_P \in \Gamma A^*\mathcal{G}$, then $[\Phi, \Psi]|_P = 0$.

For the first, take any $\xi \in \mathfrak{X}(P)$ and extend it to a vector field on $\mathcal{G}$, also denoted ξ. Take $\Phi, \Psi \in \Gamma T^*\mathcal{G}$ such that $\Phi|_P, \Psi|_P$ are sections of $(TP)^\circ$. From (3) we have

$$\langle [\Phi, \Psi], \xi \rangle = \langle \mathfrak{L}_{\Phi^\#}(\Psi), \xi \rangle - \langle \mathfrak{L}_{\Psi^\#}(\Phi), \xi \rangle - \langle d(\pi(\Phi, \Psi)), \xi \rangle. \tag{32}$$

The first term is

$$\langle \mathfrak{L}_{\Phi^\#}(\Psi), \xi \rangle = \mathfrak{L}_{\Phi^\#}(\langle \Psi, \xi \rangle) - \langle \Psi, [\Phi^\#, \xi] \rangle. \tag{33}$$

On the RHS, the first term is zero on P because $\langle \Psi, \xi \rangle$ is constant on P and $\Phi^\#$ is tangent to P. In the second term, both $\Phi^\#$ and ξ are tangent to P, and so their Lie bracket is also; hence the pairing with Ψ is zero on P.

The second term of (32) is likewise zero on P. The third term is $-\xi(\pi(\Phi, \Psi))$ and $\pi(\Phi, \Psi) = \langle \Psi, \Phi^\# \rangle$ vanishes on P since $\Phi^\#$ is tangent to P. Since $\xi \in \mathfrak{X}(P)$ was arbitrary, $[\Phi, \Psi]$ is a section of $(TP)^\circ$.

To verify the second itemized condition, take $\Phi, \Psi \in \Gamma T^*\mathcal{G}$ with $\Phi|_P = 0$ and $\Psi|_P \in \Gamma(TP)^\circ$. Let ξ be any vector field on P.

Now in (33), the first term on the RHS is zero on P because $\Phi^\#$ is zero on P. The second term, $-\langle \Psi, [\Phi^\#, \xi] \rangle$, may not be zero; see below.

In the corresponding equation for $\langle\mathfrak{L}_{\Psi^\#}(\Phi), \xi\rangle$, the first term on the RHS is zero on P because the bracket is zero on P and $\Psi^\#$ is tangent to P, by coisotropy. The second term is zero on P because Φ is zero on P.

The third term is equal to $\mathfrak{L}_\xi(\langle\Psi, \Phi^\#\rangle) = \langle\mathfrak{L}_\xi(\Psi), \Phi^\#\rangle + \langle\Psi, [\xi, \Phi^\#]\rangle$. The first term on the RHS is zero on P because $\Phi^\#$ is so. The second term may be nonzero, and cancels with the term above.

Since ξ was any vector field on $\mathcal{G}$, this proves that $[\Phi, \Psi]$ is zero on P.

This shows that $(TP)^\circ$ is a Lie subalgebroid of $T^*\mathcal{G}$. It is also possible to regard the Lie algebroid structure on $A^*\mathcal{G}$ as a quotient of the Lie algebroid $T^*\mathcal{G}$.

We have now proved part of the following theorem, which should be compared with Theorem 2.2.

Theorem 4.2. *Let $\mathcal{G} \rightrightarrows P$ be a Poisson groupoid. Then:*

(i) The Poisson anchor $\pi^\#\colon T^*\mathcal{G} \to T\mathcal{G}$ is a morphism of Lie groupoids with base map denoted $a_*\colon A^*\mathcal{G} \to TP$.
(ii) 1_P is coisotropic in $\mathcal{G}$.
(iii) The Lie algebroid structure on $T^*\mathcal{G}$ induces a Lie algebroid structure on the dual bundle $A^*\mathcal{G}$ for which the anchor is a_*.
(iv) Inversion is an antiPoisson map.
(v) There is a unique Poisson structure on P such that β is a Poisson map.

The chief difference between Theorem 4.2 and Theorem 2.2 concerns the Lie algebroid structure on $A^*\mathcal{G}$. In 2.2(iv) there is a natural isomorphism of Lie algebroids $A\Sigma \cong T^*P$. The dual of this, $TP \to A^*\Sigma$, may be used to put a Lie algebroid structure on $A^*\Sigma$, though there is little reason to do this if one is working only with symplectic groupoids.

The proof of Theorem 4.2(iv) is similar to that for symplectic groupoids. Note however that whereas inversion $\mathcal{G} \to \mathcal{G}$ is an antiPoisson diffeomorphism, the groupoid inversion $T^*\mathcal{G} \to T^*\mathcal{G}$ is a Lie algebroid isomorphism, without change of sign. This confirms, in a small way, the naturality of working with the cotangent structures.

The Poisson structure on P is defined by $\pi_P^\# = a_* \circ a^* = -a \circ a_*^*$. It follows that $\beta\colon \mathcal{G} \to P$ is a Poisson map.

We need to explicate further the relationship between the Lie algebroid structures on $A\mathcal{G}$ and $A^*\mathcal{G}$; in particular we want to express the relationship between $A\mathcal{G}$ and $A^*\mathcal{G}$ without reference to the underlying Poisson groupoid.

With any Lie algebroid A on base manifold M there is a cochain complex

$$C^\infty(M) \xrightarrow{d} \Gamma(A^*) \xrightarrow{d} \Gamma\Lambda^2(A^*) \xrightarrow{d} \Gamma\Lambda^3(A^*) \xrightarrow{d} \cdots$$

defined by the natural extension of the coboundary operator for de Rham cohomology and for Lie algebra cohomology.

Theorem 4.3. *Let $\mathcal{G} \rightrightarrows P$ be a Poisson groupoid. Then for the coboundary operator d_* of the Lie algebroid $A^*\mathcal{G}$ defined above we have, for $X, Y \in \Gamma A\mathcal{G}$,*

$$d_*[X,Y] = [X, d_*Y] + [d_*X, Y]. \tag{34}$$

We will not give the proof here; see [20] or [18, §12.1].

On the RHS we have the bracket of sections of $A\mathcal{G}$ with sections of $\Gamma\Lambda^2(A\mathcal{G})$. This bracket is defined in terms of decomposable elements by

$$[X, Y \wedge Z] = [X,Y] \wedge Z + Y \wedge [X, Z]$$

and $[X,\eta] = -[\eta, X]$ for $X \in \Gamma A\mathcal{G}$ and $\eta \in \Gamma\Lambda^2(A\mathcal{G})$. The bracket on $\Gamma A\mathcal{G}$ may be extended to the exterior algebra $\Gamma\Lambda(A\mathcal{G})$; this is the *Schouten* or *Gerstenhaber bracket for* $A\mathcal{G}$. Details and references are given in [18, §7.5].

Definition 4.4. Let A be a Lie algebroid on M and suppose that A^* has a Lie algebroid structure. Then (A, A^*) is a *Lie bialgebroid* if (34) holds for all $X, Y \in \Gamma A$.

Lie bialgebroids were defined by Ping Xu and the author in [20], following the construction of the Lie algebroid structure on the dual by Weinstein [29]. Equation (34) was immediately shown by Kosmann-Schwarzbach [12] to extend to elements of arbitrary degree of $\Gamma\Lambda(A)$; therefore d_* is a derivation of the Schouten bracket on $\Gamma\Lambda(A)$. Equation (34) implies its dual form, $d[\varphi,\psi]_* = [\varphi, d\psi]_* + [d\varphi, \psi]_*$, where $\varphi, \psi \in \Gamma A^*$ and $[\ ,\ ]_*$ is the bracket on ΓA^*.

Equation (34) is thus in many respects well-understood, but it remains a very nonlinear equation. We will now show how the relationship between the Lie algebroid structures on $A\mathcal{G}$ and $A^*\mathcal{G}$ may be understood in a 'diagrammatic' way.

5. The diagrammatic approach

We work with a given Poisson groupoid $\mathcal{G} \rightrightarrows P$. The key to the diagrammatic approach is to work with the cotangent bundle $T^*\mathcal{G}$ rather than $\mathcal{G}$ itself. Since $\mathcal{G}$ is a Poisson manifold, $T^*\mathcal{G}$ is a Lie algebroid, and since $\mathcal{G}$ is a Lie groupoid, $T^*\mathcal{G}$ has a Lie groupoid structure. These structures are shown in Figure 2(a).

For any Lie groupoid $\mathcal{G} \rightrightarrows P$, it is straightforward to show that $T^*\mathcal{G}$ is a $\mathcal{VB}$-groupoid in the following sense.

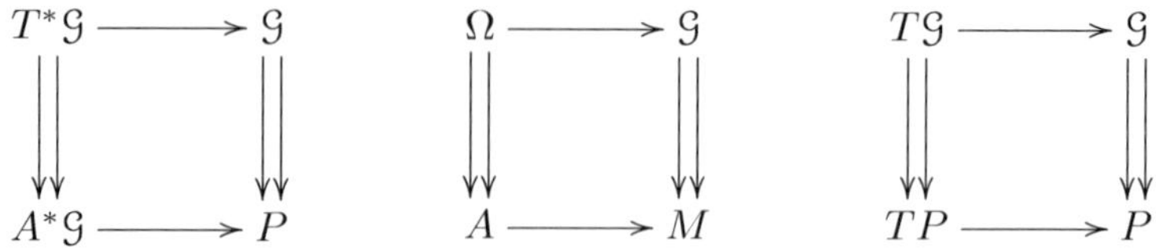

(a) (b) (c)

Fig. 2

Definition 5.1 ([24]). *A $\mathcal{VB}$–groupoid $(\Omega; \mathcal{G}, A; M)$ is a structure as shown in Figure 2(b) in which Ω is a vector bundle over $\mathcal{G}$, which is a Lie groupoid over M, and Ω is also a Lie groupoid over A, which is a vector bundle over M, subject to the condition that the structure maps of the groupoid structure (source, target, identity, multiplication, inversion) are vector bundle morphisms, and the 'double source map' $\Omega \to \mathcal{G} \times_M A$ formed from the bundle projection and the source on Ω, is a surjective submersion.*

According to the duality for $\mathcal{VB}$–groupoids introduced by Pradines [24], the $\mathcal{VB}$–groupoid $T^*\mathcal{G}$ is dual to the $\mathcal{VB}$–groupoid $T\mathcal{G}$ shown in Figure 2(c), the tangent prolongation of $\mathcal{G} \rightrightarrows P$.

Each diagram in Figure 2 denotes the full set of structures just described, and constitutes a single mathematical object; they should not be read, for example, as showing a morphism of groupoids.

For $\mathcal{G} \rightrightarrows P$ a Poisson groupoid, each horizontal structure in Figure 2(a) is a Lie algebroid. These are compatible with the vertical groupoid structures in the following sense.

Definition 5.2 ([16]). *An $\mathcal{LA}$–groupoid $(\Omega; \mathcal{G}, A; M)$ is a structure as shown in Figure 2(b) in which Ω is a Lie algebroid over $\mathcal{G}$, which is a Lie groupoid over M, and Ω is also a Lie groupoid over A, which is a Lie algebroid over M, subject to the condition that the structure maps of the groupoid structure (source, target, identity, multiplication, inversion) are Lie algebroid morphisms, and the 'double source map' $\Omega \to \mathcal{G} \times_M A$ formed from the bundle projection and the source on Ω, is a surjective submersion.*

We proved in §4 that the anchor $\pi^{\#}\colon T^*\mathcal{G} \to T\mathcal{G}$ is a morphism of Lie groupoids; that result may equally be formulated as stating that each of the groupoid structure maps preserves the anchors. It is necessary to also prove that the groupoid structure maps preserve the brackets; for this proof, see [16]. Thus $T^*\mathcal{G}$, for $\mathcal{G} \rightrightarrows P$ a Poisson groupoid, is an $\mathcal{LA}$–groupoid. This has important ramifications: one may differentiate an $\mathcal{LA}$-groupoid, and one may seek to integrate it.

Consider first the result of differentiating the $\mathcal{LA}$-groupoid in Figure 2(b). Applying the Lie functor to the vertical groupoid structures we obtain a structure as shown in Figure 3(a). Here each arrow represents a vector bundle structure; altogether these make $A\Omega$ a double vector bundle, as defined by Pradines [22, 23].

$$
\begin{array}{ccc}
A\Omega & \longrightarrow & A\mathcal{G} \\
\downarrow & & \downarrow \\
A & \longrightarrow & M \\
 & \text{(a)} &
\end{array}
\quad
\begin{array}{ccc}
AT^*\mathcal{G} & \longrightarrow & A\mathcal{G} \\
\downarrow & & \downarrow \\
A^*\mathcal{G} & \longrightarrow & P \\
 & \text{(b)} &
\end{array}
\quad
\begin{array}{ccc}
D & \longrightarrow & B \\
\downarrow & & \downarrow \\
A & \longrightarrow & M \\
 & \text{(c)} &
\end{array}
\quad
\begin{array}{ccc}
T^*A & \longrightarrow & A \\
\downarrow & & \downarrow \\
A^* & \longrightarrow & M \\
 & \text{(d)} &
\end{array}
$$

Fig. 3

The vertical structures are evidently Lie algebroids. It is also true, though not so very quick to prove [17], that the horizontal structures inherit Lie algebroid structures by a process which is an extension of the construction of the tangent prolongation of a Lie algebroid. Thus all four sides of Figure 3(a) have Lie algebroid structures.

Applying this to the cotangent $\mathcal{LA}$-groupoid of a Poisson groupoid, Figure 2(a), we obtain Figure 3(b). Here there is a canonical isomorphism $AT^*\mathcal{G} \cong T^*A\mathcal{G}$ which preserves the Lie algebroid structures [20]. Thus the Lie algebroid structure on $AT^*\mathcal{G} \to A\mathcal{G}$ may be obtained as the cotangent Lie algebroid structure of the Poisson structure on $A\mathcal{G}$ which is dual to the Lie algebroid structure on $A^*\mathcal{G}$.

When $\mathcal{G} \rightrightarrows P$ is a Poisson Lie group G (with $P = \{\cdot\}$), T^*AG reduces as a manifold to $\mathfrak{g} \times \mathfrak{g}^*$. The vertical Lie algebroid structure arises by lifting the Lie algebra structure on $\mathfrak{g}$ to the pullback vector bundle $\mathfrak{g}^* \times \mathfrak{g} \to \mathfrak{g}^*$; similarly with $\mathfrak{g}^*$. In turn the classical Drinfel'd double structure on $\mathfrak{g} \oplus \mathfrak{g}^*$ may be obtained from $T^*A\mathcal{G}$ by a process of diagonalization [19]. Thus Figure 3(b) may be regarded as an extension to Poisson groupoids of the classical Drinfel'd double of a Poisson Lie group.

The relationship between the Lie algebroid structures in diagrams such as those in Figure 3(a)(b) cannot be characterized in the same way as for $\mathcal{VB}$-groupoids and $\mathcal{LA}$-groupoids; it is not possible to say that the Lie algebroid brackets on the (say) vertical structures are morphisms with respect to the horizontal structures. This problem can be solved by using the duality theory of double vector bundles. Very briefly, given a double vector bundle as in Figure 3(c) with Lie algebroid structures on each side, assume that each Lie algebroid structure on D is linear with respect to the other; then dualizing each structure on D leads to a pair of Poisson vector bundles in duality, and D is defined to be a *double Lie algebroid* if the corresponding dual Lie algebroids form a Lie bialgebroid. A full account is in [19]. The case in which one pair of parallel Lie algebroids is zero is variously called an $\mathcal{LA}$–vector–bundle or a $\mathcal{VB}$–algebroid. Thus one may regard a double Lie algebroid as a double vector bundle admitting a horizontal $\mathcal{VB}$–algebroid structure and a vertical $\mathcal{VB}$–algebroid structure which are suitably compatible [10].

Any vector bundle $A \to M$ gives rise to a *cotangent double vector bundle* T^*A as shown in Figure 3(d). The structure $T^*A \to A^*$ arises from using the canonical diffeomorphism $T^*A \cong T^*A^*$ [20]. Now suppose that both A and A^* have Lie algebroid structures, not necessarily related. The Lie algebroid structure on A^* induces a Poisson structure on A and this induces a Lie algebroid structure on $T^*A \to A$. Likewise the Lie algebroid structure on A induces a Lie algebroid structure on $T^*A^* \to A^*$. These structures make $T^*A \cong T^*A^*$ a double Lie algebroid if and only if A and A^* form a Lie bialgebroid [19].

Thus it is reasonable to argue that the cotangent double Figure 3(d) plays the role for Lie bialgebroids which is played for Lie bialgebras by the classical Drinfel'd double. In particular the bialgebroid equation (34) is encapsulated in the double Lie algebroid conditions for the cotangent double T^*A.

The notion of double Lie algebroid provides an alternative to the extension of the classical Drinfel'd double to Lie bialgebroids of Liu, Weinstein and Xu [14]. The theory of Courant algebroids has led to the important concept of Dirac structures, whereas double Lie algebroids arise as second-order invariants of double Lie groupoids. We will describe this briefly below.

Double Lie algebroids have been defined in terms of supergeometry by Th. Voronov [27], and this is an exceptionally clear and elegant formulation. In the late 1990s, Vaĭntrob showed that a homological vector field of degree $+1$ on a supermanifold is equivalent to a Lie algebroid structure on the parity reversed bundle. A double Lie algebroid is then defined by two homological vector fields, of suitable weights, on a double parity reversion, provided that the homological vector fields *commute*. This includes the construction of the super cotangent double by Roytenberg [25].

Very recently a formulation of double Lie algebroids in terms of representations up to homotopy has been given by Gracia-Saz, Jotz Lean, Mehta and the author [9].

It was shown in [21] that given a Lie bialgebroid (A, A^*) in which $A = A\mathcal{G}$ is the Lie algebroid of a source-simply-connected Lie groupoid $\mathcal{G}$, there is a Poisson structure on $\mathcal{G}$ making it a Poisson groupoid, and inducing the given Lie algebroid structure on A^*. In terms of double structures, this shows that the cotangent double Lie algebroid $T^*A\mathcal{G}$ may be integrated to an $\mathcal{LA}$–groupoid $T^*\mathcal{G}$. Very recently Bursztyn, Cabrera and del Hoyo [1] have given a general integrability result for double Lie algebroids in terms of $\mathcal{LA}$–groupoids.

We turn now to the question of integrating an $\mathcal{LA}$–groupoid, such as that in Figure 2(a). It was shown by Lu and Weinstein [15] that underlying any Poisson Lie group is a symplectic double groupoid.

In general a *double Lie groupoid* consists of a manifold S with two Lie groupoid structures, $S \rightrightarrows H$ and $S \rightrightarrows V$, where H and V are themselves Lie groupoids over a manifold M, as shown in Figure 4(a), such that the structure maps of each groupoid structure are morphisms with respect to the other structure, and satisfying a double source condition analogous to those in Definitions 5.1 and 5.2.

In the same way that elements of a groupoid are visualized as arrows, elements of a double groupoid are visualized as squares, the horizontal edges of which come from H, the vertical edges from V, and the four corners from M; see Figure 4(b)(c). If $v_1' = v_2$ one can compose the two elements shown horizontally — the vertical edges of the composite, shown in Figure 4(d), are determined by the groupoid axioms for $S \rightrightarrows V$ and the horizontal edges by the condition that the source and target for $S \rightrightarrows H$ are morphisms.

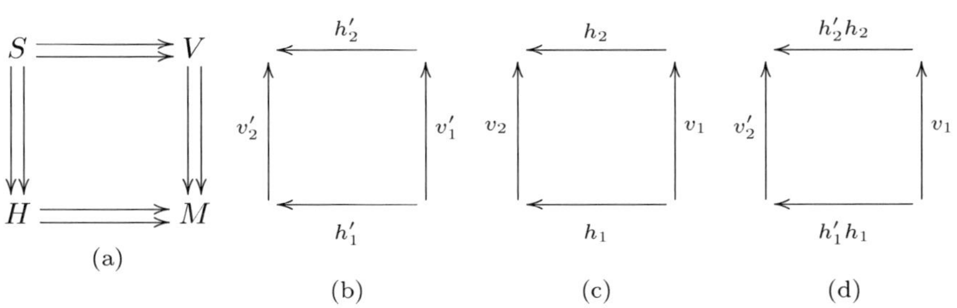

Fig. 4

The condition that each groupoid multiplication is a morphism with respect to the other structure is an interchange law: given four squares which can be arranged to form a larger square with matching inner edges, the result of composing vertically and then horizontally is the same as composing horizontally and then vertically.

Given a double Lie groupoid one may apply the Lie functor to the horizontal structures and obtain an $\mathcal{LA}$-groupoid as in Figure 5(a). As described earlier, the Lie functor may then be applied to the vertical structures to give the double Lie algebroid $A_V A_H S$ of S; see Figure 5(b). Equally, one may apply the Lie functor to the vertical structures and obtain an $\mathcal{LA}$-groupoid as in Figure 5(c), and then to the horizontal structures to give the double Lie algebroid $A_H A_V S$ of S; see Figure 5(d). There is a canonical diffeomorphism from $A_V A_H S$ to $A_H A_V S$.

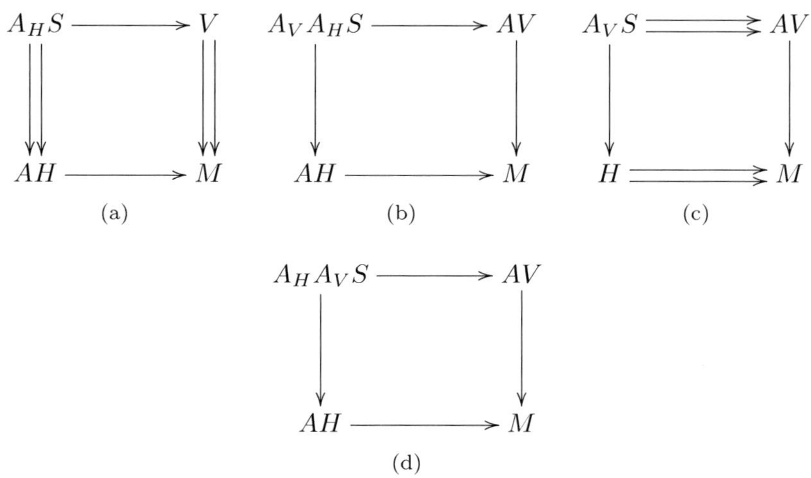

Fig. 5

For example if $S = M^4$, one obtains the $\mathcal{LA}$-groupoid $TM \times TM$ and the double Lie algebroid is the iterated tangent bundle T^2M. Now consider a Poisson Lie group G and let D denote the simply-connected Lie group corresponding to the classical Drinfel'd double $\mathfrak{g} \bowtie \mathfrak{g}^*$. Then the inclusions $\mathfrak{g} \to \mathfrak{g} \bowtie \mathfrak{g}^*$ and $\mathfrak{g}^* \to \mathfrak{g} \bowtie \mathfrak{g}^*$ induce morphisms $G \to D$, $g \mapsto \overline{g}$ and $G^* \to D$, $\varphi \mapsto \overline{\varphi}$, where G^* is the simply-connected Lie group corresponding to $\mathfrak{g}^*$. Lu and Weinstein [15] define a double Lie groupoid S as shown in Figure 6(a), for which the elements are quadruples $(g_2, g_1, \varphi_2, \varphi_1) \in G \times G \times G^* \times G^*$ such that $\overline{g_2}\,\overline{\varphi_1} = \overline{\varphi_2}\,\overline{g_1} \in D$, as shown in Figure 6(b).

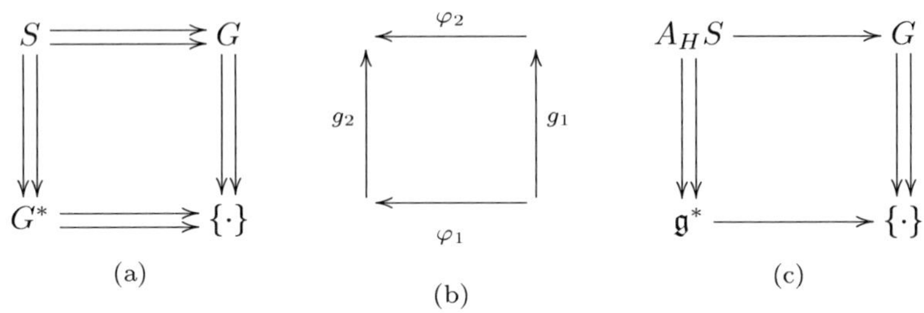

Fig. 6

The Lie algebra structure on $\mathfrak{g} \bowtie \mathfrak{g}^*$ induces a symplectic structure on S which makes S a symplectic groupoid with respect to both groupoid structures. It then follows that $A_H S \cong T^*G$ as Lie algebroids and as Lie groupoids. In effect, [15] has integrated the $\mathcal{LA}$-groupoid in Figure 2(a), in the case of Poisson Lie groups, to the symplectic double groupoid S.

A general integration result, starting with an $\mathcal{LA}$-groupoid Ω as in Figure 2(b), where $\Omega \to \mathcal{G}$ is the Lie algebroid of a suitable Lie groupoid $\Gamma \rightrightarrows \mathcal{G}$, and which constructs a second Lie groupoid structure $\Gamma \rightrightarrows \mathcal{H}$, with $A\mathcal{H} \cong A$, so that Γ becomes a double Lie groupoid over $\mathcal{G}$ and $\mathcal{H}$, seems very far from being accessible.

Endword

In addition to works already cited, other valuable sources on Poisson geometry include the books by Vaisman [26], Cannas da Silva and Weinstein [2], Dufour and Zung [8], and Laurent-Gengoux et al [13].

References

[1] H. Bursztyn, A. Cabrera, and M. del Hoyo. Vector bundles over Lie groupoids and algebroids. arXiv:1410.5135, 2014.

[2] A. Cannas da Silva and A. Weinstein. *Geometric models for noncommutative algebras*, volume 10 of *Berkeley Mathematics Lecture Notes*. American Mathematical Society, Providence, RI; Berkeley Center for Pure and Applied Mathematics, Berkeley, CA, 1999.

[3] A. S. Cattaneo and G. Felder. Poisson sigma models and symplectic groupoids. In *Quantization of singular symplectic quotients*, volume 198 of *Progr. Math.*, pages 61–93. Birkhäuser, Basel, 2001.

[4] A. Coste, P. Dazord, and A. Weinstein. Groupoïdes symplectiques. In *Publications du Département de Mathématiques. Nouvelle Série. A, Vol. 2*, volume 87 of *Publ. Dép. Math. Nouvelle Sér. A*, pages i–ii, 1–62. Univ. Claude-Bernard, Lyon, 1987.

[5] T. Courant. Tangent Lie algebroids. *J. Phys. A*, 27(13):4527–4536, 1994.

[6] M. Crainic and R. L. Fernandes. Integrability of Lie brackets. *Ann. of Math. (2)*, 157(2):575–620, 2003.

[7] M. Crainic and R. L. Fernandes. Integrability of Poisson brackets. *J. Differential Geom.*, 66(1):71–137, 2004.

[8] J.-P. Dufour and N. T. Zung. *Poisson structures and their normal forms*, volume 242 of *Progress in Mathematics*. Birkhäuser Verlag, Basel, 2005.

[9] A. Gracia-Saz, M. Jotz Lean, K. C. H. Mackenzie, and R. A. Mehta. Double Lie algebroids and representations up to homotopy. arXiv:1409.1502, 2014.

[10] A. Gracia-Saz and R. A. Mehta. Lie algebroid structures on double vector bundles and representation theory of Lie algebroids. *Adv. Math.*, 223(4):1236–1275, 2010.

[11] M. V. Karasëv. Analogues of objects of the theory of Lie groups for nonlinear Poisson brackets. *Izv. Akad. Nauk SSSR Ser. Mat.*, 50(3):508–538, 638, 1986.

[12] Y. Kosmann-Schwarzbach. Exact Gerstenhaber algebras and Lie bialgebroids. *Acta Appl. Math.*, 41(1-3):153–165, 1995. Geometric and algebraic structures in differential equations.

[13] C. Laurent-Gengoux, A. Pichereau, and P. Vanhaecke. *Poisson structures*, volume 347 of *Grundlehren der Mathematischen Wissenschaften [Fundamental Principles of Mathematical Sciences]*. Springer, Heidelberg, 2013.

[14] Z.-J. Liu, A. Weinstein, and P. Xu. Manin triples for Lie bialgebroids. *J. Differential Geom.*, 45(3):547–574, 1997.

[15] J.-H. Lu and A. Weinstein. Groupoïdes symplectiques doubles des groupes de Lie-Poisson. *C. R. Acad. Sci. Paris Sér. I Math.*, 309(18):951–954, 1989.

[16] K. C. H. Mackenzie. Double Lie algebroids and second-order geometry. I. *Adv. Math.*, 94(2):180–239, 1992.

[17] K. C. H. Mackenzie. Double Lie algebroids and second-order geometry. II. *Adv. Math.*, 154(1):46–75, 2000.

[18] K. C. H. Mackenzie. *General theory of Lie groupoids and Lie algebroids*, volume 213 of *London Mathematical Society Lecture Note Series*. Cambridge University Press, Cambridge, 2005.

[19] K. C. H. Mackenzie. Ehresmann doubles and Drinfel'd doubles for Lie algebroids and Lie bialgebroids. *J. Reine Angew. Math.*, 658:193–245, 2011.

[20] K. C. H. Mackenzie and P. Xu. Lie bialgebroids and Poisson groupoids. *Duke Math. J.*, 73(2):415–452, 1994.

[21] K. C. H. Mackenzie and P. Xu. Integration of Lie bialgebroids. *Topology*, 39(3):445–467, 2000.

[22] J. Pradines. Fibrés vectoriels doubles et calcul des jets non holonomes. Notes polycopiées, Amiens, 1974.

[23] J. Pradines. Représentation des jets non holonomes par des morphismes vectoriels doubles soudés. *C. R. Acad. Sci. Paris Sér. A*, 278:1523–1526, 1974.

[24] J. Pradines. Remarque sur le groupoïde cotangent de Weinstein-Dazord. *C. R. Acad. Sci. Paris Sér. I Math.*, 306(13):557–560, 1988.

[25] D. Roytenberg. Courant algebroids, derived brackets and even symplectic supermanifolds. Thesis, Univ. California, Berkeley, 1999. arXiv:math.DG/9910078.

[26] I. Vaisman. *Lectures on the geometry of Poisson manifolds*, volume 118 of *Progress in Mathematics*. Birkhäuser Verlag, Basel, 1994.

[27] Th. Th. Voronov. Q-Manifolds and Mackenzie Theory. *Comm. Math. Phys.*, 315(2):279–310, 2012.

[28] A. Weinstein. Symplectic groupoids and Poisson manifolds. *Bull. Amer. Math. Soc. (N.S.)*, 16(1):101–104, 1987.

[29] A. Weinstein. Coisotropic calculus and Poisson groupoids. *J. Math. Soc. Japan*, 40(4):705–727, 1988.

[30] S. Zakrzewski. Quantum and classical pseudogroups. I. Union pseudogroups and their quantization. *Comm. Math. Phys.*, 134(2):347–370, 1990.

[31] S. Zakrzewski. Quantum and classical pseudogroups. II. Differential and symplectic pseudogroups. *Comm. Math. Phys.*, 134(2):371–395, 1990.

LIE ALGEBROIDS AND CHARACTERISTIC CLASSES

Paul Bressler

Mathematics Department,
Universidad de los Andes,
Carrera 1 No. 18A-12,
Bogotá, Colombia
Email: paul.bressler@gmail.com

1. Introduction

The following are expanded notes from a short course on the subject of the title given at "Geometric, Algebraic and Topological Methods for Quantum Field Theory" Summer School in Villa de Leyva. Familiarity with the basics of differential calculus on manifolds and sheaf theory is assumed. Essentially no proofs are given, the main goal being logical exposition of basic constructions and facts. With the explanations given, a dedicated reader should be able to fill in most of the missing details.

The only book on Lie algebroids known to the author is [M]. The reader is advised to consult [BB] and [BK]. Appendix B of the latter reference contains the construction of the Chern-Weil characteristic map sketched in Section 7. We use some basics concerning Picard categories explained in [D]. Chapter 5 of [W] contains sufficient background on sheaves for our purposes. The reference [B] is recommended for the role sheaves, sheaf cohomology and stacks play in geometry.

1.1. *Notation*

Throughout the text X denotes a manifold which can be taken to be C^∞, real- or complex-analytic. The structure sheaf of *complex valued* functions (differentiable of analytic as appropriate) is denoted by $\mathcal{O}_X$. The sheaf of (complex valued) vector fields, *i.e.* of derivations of $\mathcal{O}_X$ is denoted by $\mathcal{T}_X$; the sheaf of (complex valued) differential forms of degree p is denoted Ω^p_X, thus $\mathcal{O}_X = \Omega^0_X$.

For a sheaf of $\mathbb{C}$-algebras $\mathcal{R}$ on X (such as, for example, the constant sheaf $\mathbb{C}$ or $\mathcal{O}_X$) and sheaves $\mathcal{M}$ and $\mathcal{N}$ of $\mathcal{R}$-modules we denote by $\mathrm{Hom}_{\mathcal{R}}(\mathcal{M},\mathcal{N})$ the $\mathbb{C}$-vector space of morphisms from $\mathcal{M}$ to $\mathcal{N}$ in the category of $\mathcal{R}$-modules and set $\mathrm{End}_{\mathcal{R}}(\mathcal{M}) := \mathrm{Hom}_{\mathcal{R}}(\mathcal{M},\mathcal{M})$. The assignment $X \supseteq U \mapsto \mathrm{Hom}_{\mathcal{R}|_U}(\mathcal{M}|_U,\mathcal{N}|_U)$ defines the sheaf denoted by $\underline{\mathrm{Hom}}_{\mathcal{R}}(\mathcal{M},\mathcal{M})$, so that $\mathrm{Hom}_{\mathcal{R}}(\mathcal{M},\mathcal{N}) = \Gamma(X;\underline{\mathrm{Hom}}_{\mathcal{R}}(\mathcal{M},\mathcal{N}))$; we set $\underline{\mathrm{End}}_{\mathcal{R}}(\mathcal{M},\mathcal{M}) := \underline{\mathrm{Hom}}_{\mathcal{R}}(\mathcal{M},\mathcal{M})$.

1.2. *Note to the reader*

The reader interested only in the C^∞ setting and unfamiliar with sheaf-theoretic terminology is advised to replace the word "sheaf" with "space" when the former refers to various differential geometric objects. For example, "the sheaf $\mathcal{O}_X$ of complex valued functions" one can read as "the space of complex valued (C^∞) functions on X". This is possible due to the fact that higher cohomologies vanish for any sheaf of modules over the sheaf C^∞ functions; evidently, this remark does not apply to sheaves of closed forms.

2. Lie algebroids

2.1. *A prototypical example*

Whatever "Lie algebroids" are, the sheaf of vector fields $\mathcal{T}_X$ is supposed to be the prototypical example of such. To this end let us examine the structures present.

(1) $\mathcal{T}_X$ is a $\mathbb{C}$-Lie algebra under the Lie bracket of vector fields;
(2) in this capacity it acts on $\mathcal{O}_X$ by derivations, *i.e.* the Leibniz rule $\xi(f\cdot g) = \xi(f)\cdot g + f\cdot\xi(g)$, $\xi\in\mathcal{T}_X$, $f,g\in\mathcal{O}_X$, holds;
(3) the map of Lie algebras $\mathcal{T}_X \to \underline{\mathrm{End}}_{\mathbb{C}}(\mathcal{O}_X)$ (where the latter is equipped with the commutator bracket) is $\mathcal{O}_X$-linear, *i.e.* $(f\xi)(g) = f\cdot\xi(g)$;[a]
(4) the $\mathcal{O}_X$-module structure on $\mathcal{T}_X$ is compatible with the $\mathbb{C}$-Lie algebra structure in the sense that the Leibniz rule $[\xi, f\eta] = \xi(f)\eta + f[\xi,\eta]$, $\xi,\eta\in\mathcal{T}_X$, $f\in\mathcal{O}_X$, holds.

2.2. *Definitions*

2.2.1.

A *Lie algebroid* on X is a sheaf of vector spaces $\mathcal{A}$ together with

[a]Not all naturally arising actions of the Lie algebra $\mathcal{T}_X$ have this property. For example, the map $\mathcal{T}_X \to \underline{\mathrm{End}}_{\mathbb{C}}(\mathcal{T}_X)$ deduced from the adjoint action is not $\mathcal{O}_X$-linear.

(1) a structure of an $\mathcal{O}_X$-module on $\mathcal{A}$;
(2) a structure of a $\mathbb{C}$-Lie algebra on $\mathcal{A}$ given by a skew-symmetric $\mathbb{C}$-bilinear pairing

$$[\ ,\ ] : \mathcal{A} \times \mathcal{A} \to \mathcal{A}$$

which satisfies the Jacobi identity;
(3) an action of $\mathcal{A}$ on $\mathcal{O}_X$ by derivations given by an $\mathcal{O}_X$-linear map of $\mathbb{C}$-Lie algebras

$$\sigma : \mathcal{A} \to \mathcal{T}_X$$

called *the anchor map*.

These data are subject to the compatibility condition (Leibniz rule)

$$[a, fb] = \sigma(a)(f)b + f[a, b] \ .$$

2.2.2. *First example*

The reader will verify easily that $\mathcal{T}_X$ is a Lie algebroid.

Further examples of Lie algebroids are presented in 3.

2.2.3.

Suppose that $\mathcal{A}$ and $\mathcal{B}$ are Lie algebroids. A *morphism* of Lie algebroids $\varphi : \mathcal{A} \to \mathcal{B}$ is a morphism of $\mathcal{O}_X$-modules $\varphi : \mathcal{A} \to \mathcal{B}$ which is compatible with the respective brackets (*i.e.* is a morphism of Lie algebras) and anchor maps in the sense that the composition $\mathcal{A} \xrightarrow{\varphi} \mathcal{B} \to \mathcal{T}_X$ coincides with the anchor map of $\mathcal{A}$.

2.2.4.

An algebroid is called *transitive* if the anchor map is surjective.

2.2.5.

It is immediate from the definitions that

- $\mathcal{T}_X$ is a (transitive) Lie algebroid, the anchor given by the identity map (see 2.1 for details);
- for any algebroid the anchor map is a morphism of Lie algebroids.

2.3. *Some immediate corollaries*

2.3.1.

With the definition of morphisms given in 2.2.3, Lie algebroids form a category.

Every Lie algebroid admits a unique morphism to $\mathcal{T}_X$, namely the anchor map, *i.e.* the category of Lie algebroids has a terminal object, namely $\mathcal{T}_X$.

2.3.2.

Suppose that $\mathcal{A}$ is a Lie algebroid and $\mathcal{F}$ is an $\mathcal{O}_X$-submodule of $\mathcal{A}$ closed under the Lie bracket, *i.e.* $[\mathcal{F}, \mathcal{F}] \subseteq \mathcal{F}$. Then, equipped with the restrictions of the bracket and the anchor, $\mathcal{F}$ is a Lie algebroid.

2.3.3.

An algebroid with the trivial (zero) anchor map is the same thing as a sheaf of $\mathcal{O}_X$-Lie algebras, *i.e.* the Lie bracket is $\mathcal{O}_X$-bilinear.

In particular, in the case when X is connected and zero-dimensional (*i.e.* consists of just one point), a Lie algebroid on X is the same thing as a Lie algebra.

2.3.4.

Suppose that $\varphi : \mathcal{A} \to \mathcal{B}$ is a morphism of Lie algebroids.

Since φ is $\mathcal{O}_X$-linear, $\ker(\varphi)$ is an $\mathcal{O}_X$-submodule of $\mathcal{A}$. Since φ is a morphism of Lie algebras, it follows that $\ker(\varphi)$ is a Lie ideal in $\mathcal{A}$, *i.e.* $[\mathcal{A}, \ker(\varphi)] \subseteq \ker(\varphi)$. In particular, $\ker(\varphi)$ is closed under bracket (in $\mathcal{A}$). Moreover, the restriction of the anchor map of $\mathcal{A}$ to $\ker(\sigma)$ is trivial. Hence, we can regard $\ker(\varphi)$ as an algebroid with the trivial anchor (see 2.3.3).

2.3.5.

Suppose that $\mathcal{A}$ is a Lie algebroid with the anchor map $\sigma : \mathcal{A} \to \mathcal{T}_X$. Regarding σ as a morphism of Lie algebroids (see 2.2.5), we conclude (as in 2.3.4 and 2.3.3) that $\ker(\sigma)$ is an $\mathcal{O}_X$-Lie algebra.

2.4. *Products*

The category of Lie algebroids has products. We will describe the product of two Lie algebroids leaving to the reader the task of generalization.

Suppose that $\mathcal{A}_i$, $i = 1, 2$, are Lie algebroids with respective anchor maps denoted σ_i. Then, the sheaf $\mathcal{A}_1 \times_{\mathcal{T}_X} \mathcal{A}_2$[b] has a canonical structure of a Lie algebroid. The anchor map is induced by $\sigma_1 \times \sigma_2$. The rest of the structure is given, for $f \in \mathcal{O}_X$, $a_i, b_i \in \mathcal{A}_i$, by the formulas $f(a_1, a_2) = (fa_1, fa_2)$ and $[(a_1, a_2), (b_1, b_2)] = ([a_1, b_1], [a_2, b_2])$.

2.4.1. *Morphisms revisited*

A morphism of Lie algebroids $\varphi\colon \mathcal{A} \to \mathcal{B}$ gives rise to the morphism (the graph of φ) $\mathcal{A} \to \mathcal{A} \times_{\mathcal{T}_X} \mathcal{B}$ defined by $a \mapsto (a, \varphi(a))$, *i.e.* to a splitting of the projection $\mathtt{pr}_\mathcal{A}\colon \mathcal{A} \times_{\mathcal{T}_X} \mathcal{B} \to \mathcal{A}$.

Conversely, a splitting $s\colon \mathcal{A} \to \mathcal{A} \times_{\mathcal{T}_X} \mathcal{B}$ of the projection $\mathtt{pr}_\mathcal{A}$ determines a morphism of Lie algebroid by $s \mapsto \mathtt{pr}_\mathcal{A} \circ s$.

It is clear that the two constructions are mutually inverse and establish a bijection between the morphisms $\mathcal{A} \to \mathcal{B}$ and the splittings of the projection $\mathtt{pr}_\mathcal{A}\colon \mathcal{A} \times_{\mathcal{T}_X} \mathcal{B} \to \mathcal{A}$.

2.5. *The push-out construction*

Applications of the push-out construction outlined below may be found in 3.6 and 3.7.

Suppose that $\mathcal{A}$ is a Lie algebroid with anchor map $\sigma : \mathcal{A} \to \mathcal{T}_X$; let $\mathfrak{g} = \ker(\sigma)$.

Suppose given

(1) an $\mathcal{O}_X$-Lie algebra $\mathfrak{h}$ and a morphism $\varphi : \mathfrak{g} \to \mathfrak{h}$ of such
(2) an action of $\mathcal{A}$ on $\mathfrak{h}$ given by the pairing $\mathcal{A} \times \mathfrak{h} \to \mathfrak{h}$, $(a, s) \mapsto a(s)$ which satisfies
    - the map $\mathcal{A} \to \underline{\mathrm{End}}_\mathbb{C}(\mathfrak{h})$ defined by $a \mapsto (s \mapsto a(s))$ is an $\mathcal{O}_X$-linear map of Lie algebras;
    - for $a \in \mathcal{A}$, $s, t \in \mathfrak{h}$ the identity $a([s, t]) = [a(s), t] + [s, a(t)]$ holds, *i.e.* the action is by derivations;
    - for $a \in \mathcal{A}$, $f \in \mathcal{O}_X$, $s \in \mathfrak{h}$ the identity (Leibniz rule) $a(f \cdot s) = \sigma(a)(f) \cdot s + f \cdot a(s)$ holds;
    - for $a \in \mathfrak{g}$, $s \in \mathfrak{h}$ the identity $b(s) = [\varphi(b), s]$ holds.

Let $\varphi_\bullet\mathcal{A}$ denote the cokernel of the map $\mathfrak{g} \to \mathfrak{h} \oplus \mathcal{A}$ given by $b \mapsto (\varphi(b), -b)$. Thus, a (local) section of $\varphi_\bullet\mathcal{A}$ is an equivalence class of pairs

[b] defined by $X \supseteq U \mapsto \mathcal{A}_1(U) \times_{\mathcal{T}_X(U)} \mathcal{A}_2(U) = \{(a_1, a_2) \mid \sigma_1(a_1) = \sigma_2(a_2)\}$

(s, a), $s \in \mathfrak{h}$, $a \in \mathcal{A}$, with two such pairs considered equivalent if their difference is of the form $(\varphi(b), -b)$, $b \in \mathfrak{g}$.

There are natural maps

$$\widetilde{\varphi} : \mathcal{A} \to \varphi_{\bullet}\mathcal{A}$$

and

$$i : \mathfrak{h} \to \varphi_{\bullet}\mathcal{A}$$

given, respectively, by $a \mapsto (0, a)$ and $s \mapsto (s, 0)$. In addition, the composition

$$\mathfrak{h} \oplus \mathcal{A} \to \mathcal{A} \to \mathcal{T}_X \ ,$$

$(s, a) \mapsto \sigma(a)$ clearly annihilates the image of $\mathfrak{g}$ in $\mathfrak{h} \oplus \mathcal{A}$, hence factors through a uniquely defined map $\varphi_{\bullet}\mathcal{A} \to \mathcal{T}_X$ making the diagram

$$\begin{array}{ccccccc}
0 & \longrightarrow & \mathfrak{g} & \longrightarrow & \mathcal{A} & \xrightarrow{\sigma} & \mathcal{T}_X \\
 & & {\scriptstyle\varphi}\big\downarrow & & {\scriptstyle\widetilde{\varphi}}\big\downarrow & & \big\downarrow{\scriptstyle \mathtt{Id}} \\
0 & \longrightarrow & \mathfrak{h} & \xrightarrow{i} & \varphi_{\bullet}\mathcal{A} & \longrightarrow & \mathcal{T}_X
\end{array}$$

commutative. Moreover, the exactness of the top row implies exactness of the bottom row.

There is a unique structure of a Lie algebroid on $\varphi_{\bullet}\mathcal{A}$ (extending the natural $\mathcal{O}_X$-module structure) such that

- the map $\varphi_{\bullet}\mathcal{A} \to \mathcal{T}_X$ is the anchor map;
- the map $\mathcal{A} \to \varphi_{\bullet}\mathcal{A}$ is a morphism of algebroids
- the composition of $\widetilde{\varphi}$ with the adjoint action of $\varphi_{\bullet}\mathcal{A}$ on $\mathfrak{h}$ coincides with the given action.

It is clear that the above requirements determine the Lie bracket on $\varphi_{\bullet}\mathcal{A}$ uniquely. Namely, we must have

- $[\widetilde{\varphi}(a), \widetilde{\varphi}(b)] = \widetilde{\varphi}([a, b])$;
- $[i(s), i(t)] = i([s, t])$;
- $[\widetilde{\varphi}(a), i(s)] = i(a(s))$.

We leave it to the reader to verify that the above formulas give a well-defined operation on $\varphi_{\bullet}\mathcal{A}$. Since every section of $\varphi_{\bullet}\mathcal{A}$ is of the form $i(s) + \widetilde{\varphi}(a)$ for suitable s and a the above formulas determine the bracket on $\varphi_{\bullet}\mathcal{A}$.

Leaving the remaining details to the reader we have constructed the canonical structure of a Lie algebroid on $\varphi_{\bullet}\mathcal{A}$ which we will refer to as "the push-out of $\mathcal{A}$ by φ".

3. Examples

3.1. *Involutive distributions*

By definition, a *distribution* on X is a sub-bundle of $\mathcal{T}_X$. A distribution $\mathcal{D} \subseteq \mathcal{T}_X$ is called *involutive* if it is closed under the Lie bracket of vector fields, *i.e.* $[\mathcal{D}, \mathcal{D}] \subseteq \mathcal{D}$.

Involutive distributions give rise to examples of Lie algebroids as in 2.3.2.

3.2. *Vector fields tangent to a submanifold*

Suppose that $Y \subseteq X$ is a closed[c] submanifold. Let $\mathcal{I}_Y \subseteq \mathcal{O}_X$ denote the ideal of functions vanishing on Y. Thus, $\mathcal{I}_Y$ is an ideal in $\mathcal{O}_X$.

Let $\mathcal{T}_{Y|X} \subseteq \mathcal{T}_X$ denote the subsheaf of vector fields which preserve $\mathcal{I}_Y$, *i.e.* $\xi \in \mathcal{T}_{Y|X}$ if and only if $\xi(\mathcal{I}_Y) \subseteq \mathcal{I}_Y$. We will refer to $\mathcal{T}_{Y|X}$ as *the sheaf of vector fields tangent to* Y.

The sheaf $\mathcal{T}_{Y|X}$ is an $\mathcal{O}_X$-submodule of $\mathcal{T}_X$. This follows from the fact that $\mathcal{I}_Y$ is an ideal; indeed, for any $f \in \mathcal{O}_X$, $\xi \in \mathcal{T}_{Y|X}$ and $g \in \mathcal{I}_Y$, $(f\xi)(g) = f\xi(g)$ is in $\mathcal{I}_Y$ since $\xi(g)$ is, *i.e.* if ξ is tangent to Y then so is $f\xi$.

Since the Lie bracket of vector fields is the commutator bracket of the corresponding endomorphisms of $\mathcal{O}_X$ it follows that $\mathcal{T}_{Y|X}$ is closed under the Lie bracket. Thus, it is a Lie algebroid as in 2.3.2.

We finish the example with some comments on the structure of $\mathcal{T}_{Y|X}$ in general and, more specifically, in the case when Y is a hypersurface.

Since $\mathcal{I}_Y$ and $\mathcal{O}_X$ coincide on $X \setminus Y$, it follows that so do $\mathcal{T}_{Y|X}$ and $\mathcal{T}_X$. Therefore, the quotient $\mathcal{T}_X/\mathcal{T}_{Y|X}$ is supported on Y.

Vector fields on X which vanish along Y are certainly tangent to Y, *i.e.* $\mathcal{I}_Y \cdot \mathcal{T}_X \subset \mathcal{T}_{Y|X}$. Since $\mathcal{I}_Y$ and $\mathcal{O}_X$ coincide on $X \setminus Y$, it follows that $\mathcal{T}_{Y|X}/\mathcal{I}_Y \cdot \mathcal{T}_X$ is supported on Y.

By the very definition, the restrictions of vector fields tangent to Y give rise to vector fields on Y: $\xi \in \mathcal{T}_{Y|X}$ means that $\xi(\mathcal{I}_Y) \subseteq \mathcal{I}_Y$, therefore induces an endomorphism of $\mathcal{O}_Y = \mathcal{O}_X/\mathcal{I}_Y$. In other words, the "restriction to Y" map $\mathcal{T}_{Y|X} \to \mathcal{T}_X/\mathcal{I}_Y \cdot \mathcal{T}_X$ takes values in $\mathcal{T}_Y$ (embedded in $\mathcal{T}_X/\mathcal{I}_Y \cdot \mathcal{T}_X$ by the differential of the inclusion $Y \hookrightarrow X$). In fact, it is easy to see that the sequence

$$0 \to \mathcal{I}_Y \cdot \mathcal{T}_X \to \mathcal{T}_{Y|X} \to \mathcal{T}_Y \to 0$$

is exact.

[c]in the topology on X

Let us say that $Y \subset X$ is a hypersurface if it is, locally, equal to the set of zeroes of a function $f \in \mathcal{O}_X$ such that df does not vanish. If is real (respectively, complex) valued then Y is a submanifold of real codimension one (respectively, two). In this case $\mathcal{I}_Y = \mathcal{O}_X \cdot f$ (locally), hence locally free of rank one over $\mathcal{O}_X$, *i.e.* is (the sheaf of sections of) a line bundle. Then, $\mathcal{I}_Y \cdot \mathcal{T}_X = \mathcal{I}_Y \otimes_{\mathcal{O}_X} \mathcal{T}_X$ is a vector bundle and so is $\mathcal{T}_{Y|X}$.[d] Note that the inclusion $\mathcal{T}_{Y|X} \to \mathcal{T}_X$ is not of constant rank; in particular, the image is not a subbundle.

3.3. *Poisson structures*

Suppose that π is a bi-vector on X, *i.e.* $\pi \in \Gamma(X; \bigwedge^2 \mathcal{T}_X)$. A bi-vector π defined an alternating bilinear map $\pi\colon \Omega^1_X \otimes \Omega^1_X \to \mathcal{O}_X$ and we denote by $\widetilde{\pi} : \Omega^1_X \to \mathcal{T}_X$ the corresponding adjoint map defined by $\alpha \mapsto \iota_\alpha \pi = (\beta \mapsto \pi(\alpha, \beta))$.

Suppose that π is a *Poisson* bi-vector, *i.e.* the operation

$$\{\ ,\ \} : \mathcal{O}_X \times \mathcal{O}_X \to \mathcal{O}_X, \quad (f, g) \mapsto \iota_{dg}\iota_{df}\pi\ ,$$

the *Poisson bracket*, satisfies the Jacobi identity. The formula

$$\{\alpha, \beta\} = L_{\widetilde{\pi}(\alpha)}\beta - L_{\widetilde{\pi}(\beta)}\alpha - d\iota_\beta\iota_\alpha\pi$$

(where L_ξ denotes the Lie derivative along the vector field ξ) defines an operation

$$\{\ ,\ \} : \Omega^1_X \times \Omega^1_X \to \Omega^1_X$$

which is (obviously) skew-symmetric and satisfies the Jacobi identity. It is easy to check that this operation, together with the anchor map $\widetilde{\pi}$ endows Ω^1_X with the structure of a Lie algebroid.

3.4. *Atiyah algebras of vector bundles*

Suppose that $\mathcal{E}$ is a vector bundle on X. In what follows we will be considering $\underline{\mathrm{End}}_{\mathcal{O}_X}(\mathcal{E})$ and $\underline{\mathrm{End}}_{\mathbb{C}}(\mathcal{E})$ as Lie algebras under the commutator bracket. Note that, in that capacity, the former is a Lie subalgebra of the latter.

Let $\mathcal{D}^{\leqslant 1}_{X,\mathcal{E}} \subset \underline{\mathrm{End}}_{\mathbb{C}}(\mathcal{E})$ denote the subsheaf whose (local) sections are $\mathbb{C}$-linear endomorphisms P of $\mathcal{E}$ such that $[P, f] \in \underline{\mathrm{End}}_{\mathcal{O}_X}(\mathcal{E})$ for all $f \in \mathcal{O}_X$. In other words, $\mathcal{D}^{\leqslant 1}_{X,\mathcal{E}}$ is the sheaf of differential operators of order (at most) one acting on $\mathcal{E}$.

[d]This is not the case if Y is a submanifold of higher codimension in X.

Since the commutator bracket is a derivation with respect to the product in $\underline{\mathrm{End}}_{\mathbb{C}}(\mathcal{E})$, it follows that

- for $f \in \mathcal{O}_X$ the map
$$\mathcal{D}^{\leqslant 1}_{X,\mathcal{E}} \ni P \mapsto [P, f] \in \underline{\mathrm{End}}_{\mathcal{O}_X}(\mathcal{E})$$
is $\mathcal{O}_X$-linear;
- for $P \in \mathcal{D}^{\leqslant 1}_{X,\mathcal{E}}$ the map
$$\mathcal{O}_X \ni f \mapsto [P, f] \in \underline{\mathrm{End}}_{\mathcal{O}_X}(\mathcal{E})$$
is a derivation.

The sheaf of derivations $\mathcal{O}_X \to \underline{\mathrm{End}}_{\mathcal{O}_X}(\mathcal{E})$ is canonically isomorphic to $\mathcal{T}_X \otimes_{\mathcal{O}_X} \underline{\mathrm{End}}_{\mathcal{O}_X}(\mathcal{E})$. Namely, as a consequence of the Leibniz rule any derivation $D\colon \mathcal{O}_X \to \underline{\mathrm{End}}_{\mathcal{O}_X}(\mathcal{E})$ factors uniquely as $\mathcal{O}_X \xrightarrow{d} \Omega^1_X \xrightarrow{\widetilde{D}} \underline{\mathrm{End}}_{\mathcal{O}_X}(\mathcal{E})$ with $\widetilde{D}$ an $\mathcal{O}_X$-linear map. Conversely, any $\mathcal{O}_X$-linear map $\varphi\colon \Omega^1_X \xrightarrow{\widetilde{D}} \underline{\mathrm{End}}_{\mathcal{O}_X}(\mathcal{E})$ give rise to the derivation $\varphi \circ d$. In addition, there is a canonical isomorphism $\underline{\mathrm{Hom}}_{\mathcal{O}_X}(\Omega^1_X, \underline{\mathrm{End}}_{\mathcal{O}_X}(\mathcal{E})) \cong \mathcal{T}_X \otimes_{\mathcal{O}_X} \underline{\mathrm{End}}_{\mathcal{O}_X}(\mathcal{E})$.

Thus, the assignment $P \mapsto (f \mapsto [P, f])$ defines an $\mathcal{O}_X$-linear map
$$\sigma : \mathcal{D}^{\leqslant 1}_{X,\mathcal{E}} \to \mathcal{T}_X \otimes_{\mathcal{O}_X} \underline{\mathrm{End}}_{\mathcal{O}_X}(\mathcal{E})$$
called *the principal symbol map.*

The operators of order zero, *i.e.* those which satisfy the stronger condition $[P, f] = 0$ are precisely the $\mathcal{O}_X$-linear endomorphisms of $\mathcal{E}$ so that we have the inclusion $\underline{\mathrm{End}}_{\mathcal{O}_X}(\mathcal{E}) \to \mathcal{D}^{\leqslant 1}_{X,\mathcal{E}}$ which is an isomorphism onto the kernel of the map σ. In fact, as is easy to see the sequence
$$0 \to \underline{\mathrm{End}}_{\mathcal{O}_X}(\mathcal{E}) \to \mathcal{D}^{\leqslant 1}_{X,\mathcal{E}} \to \mathcal{T}_X \otimes_{\mathcal{O}_X} \underline{\mathrm{End}}_{\mathcal{O}_X}(\mathcal{E}) \to 0 \tag{1}$$
is exact.

There is a natural map $\mathcal{T}_X \to \mathcal{T}_X \otimes_{\mathcal{O}_X} \underline{\mathrm{End}}_{\mathcal{O}_X}(\mathcal{E})$ given by $\xi \mapsto \mathtt{Id} \otimes \xi$. The *Atiyah algebra* of $\mathcal{E}$, denoted $\mathcal{A}_{\mathcal{E}}$, is defined as the the subsheaf of operators with scalar-valued principal symbol, in other words, by the Cartesian square
$$\begin{array}{ccc} \mathcal{A}_{\mathcal{E}} & \longrightarrow & \mathcal{T}_X \\ \big\downarrow & & \big\downarrow {\scriptstyle \xi \mapsto \mathtt{Id} \otimes \xi} \\ \mathcal{D}^{\leqslant 1}_{X,\mathcal{E}} & \xrightarrow{\sigma} & \mathcal{T}_X \otimes_{\mathcal{O}_X} \underline{\mathrm{End}}_{\mathcal{O}_X}(\mathcal{E}) \end{array}$$
It follows from the exactness of (1) that the sequence
$$0 \to \underline{\mathrm{End}}_{\mathcal{O}_X}(\mathcal{E}) \to \mathcal{A}_{\mathcal{E}} \xrightarrow{\sigma} \mathcal{T}_X \to 0$$

is exact.

For $P, Q \in \mathcal{A}_{\mathcal{E}}$, $f \in \mathcal{O}_X$

- $[P, f] = \sigma(P)(f)$ (from the definition of the principal symbol);
- $[[P, Q], f] = [[P, f], Q] + [P, [Q, f]] = [\sigma(P)(f), Q] + [P, \sigma(Q)(f)] = [\sigma(P), \sigma(Q)](f)$;
- $[P, fQ] = [P, f]Q + f[P, Q] = \sigma(P)(f)Q + f[P, Q]$.

The above calculations show that

(1) $\mathcal{A}_{\mathcal{E}}$ is closed under the commutator bracket of operators[e]; therefore, it is a Lie algebra;
(2) the principal symbol map $\sigma : \mathcal{A}_{\mathcal{E}} \to \mathcal{T}_X$ is a morphism of Lie algebras;
(3) the bracket on $\mathcal{A}_{\mathcal{E}}$ is compatible with the $\mathcal{O}_X$-module structure in the sense that the Leibniz rule $[P, fQ] = \sigma(P)(f)Q + f[P, Q]$ holds.

In other words, $\mathcal{A}_{\mathcal{E}}$ is a Lie algebroid with anchor map σ.

3.5. *Atiyah algebras of principal bundles*

Suppose that G is a Lie group with the Lie algebra $\mathfrak{g}$ and $\pi : P \to X$ is a principal G-bundle.

Since π is a submersion, the differential $d\pi : \mathcal{T}_P \to \pi^*\mathcal{T}_X$ is surjective, hence the sequence

$$0 \to \mathcal{T}_{P/X} \to \mathcal{T}_P \to \pi^*\mathcal{T}_X \to 0$$

(where we denote by $\mathcal{T}_{P/X}$ the subsheaf of vector fields tangent to the fibers of π, *i.e.* $\ker(d\pi)$) is exact and so is the sequence

$$0 \to \pi_*\mathcal{T}_{P/X} \to \pi_*\mathcal{T}_P \to \pi_*\pi^*\mathcal{T}_X \to 0.$$

The group G acts naturally on the latter and, passing to the subsheaves of invariants we obtain the exact sequence

$$0 \to (\pi_*\mathcal{T}_{P/X})^G \to (\pi_*\mathcal{T}_P)^G \to (\pi_*\pi^*\mathcal{T}_X)^G \to 0,$$

where, for a sheaf $\mathcal{F}$ with and action for a group G, the notation $(\mathcal{F})^G$ stand for the subsheaf of G-invariant sections.

We denote $(\pi_*\mathcal{T}_P)^G$ by $\mathcal{A}_P$ and call it the *Atiyah algebra* of P. The other two terms of of the sequence are identified as follows.

[e]This is not the case for $\mathcal{D}^{\leqslant 1}_{X,\mathcal{E}}$ unless $\mathcal{E}$ is a line bundle so that $\underline{\mathrm{End}}_{\mathcal{O}_X}(\mathcal{E})$ is commutative.

The "pull-back of sections" map $\pi^* : \mathcal{T}_X \to \pi_*\pi^*\mathcal{T}_X$ is an isomorphism onto the subsheaf of G-invariants due to the fact that π identifies the quotient P/G with X.

The associated infinitesimal action of $\mathfrak{g}$ is given by the map $\mathfrak{g} \to \Gamma(P; \mathcal{T}_{P/X})$ (since the fibers of π are the orbits of the G-action). The corresponding map of sheaves $\mathcal{O}_P \otimes_{\mathbb{C}} \mathfrak{g} \to \mathcal{T}_{P/X}$ is injective because the action is free and surjective because the action on the fibers is transitive.

Thus, the infinitesimal action establishes a G-equivariant isomorphism $\mathcal{O}_P \otimes_{\mathbb{C}} \mathfrak{g} \cong \mathcal{T}_{P/X}$ (with respect to the adjoint action on $\mathfrak{g}$) which induces the isomorphism $(\pi_*\mathcal{O}_P \otimes_{\mathbb{C}} \mathfrak{g})^G \cong (\pi_*\mathcal{T}_{P/X})^G$. Now, the sheaf $\mathfrak{g}^P := (\pi_*\mathcal{O}_P \otimes_{\mathbb{C}} \mathfrak{g})^G$ ("the P-twist of $\mathfrak{g}$") is naturally identified with the sheaf of sections of the vector bundle $P \times_G \mathfrak{g}$ associated to the adjoint representation of G[f].

In terms of these identifications we have the exact sequence

$$0 \to \mathfrak{g}^P \to \mathcal{A}_P \to \mathcal{T}_X \to 0$$

The naturality properties of the Lie bracket of vector fields imply that it is G-equivariant, hence restricts to the respective subsheaves of invariants. In particular, $\mathcal{A}_P = (\pi_*\mathcal{T}_P)^G$ is a Lie algebra under the Lie bracket. The same applies to $(\pi_*\mathcal{T}_{P/X})^G$ (note that $\mathcal{T}_{P/X}$ is a Lie subalgebra of $\mathcal{T}_P$); moreover, the restriction of the Lie bracket to $\pi_*\mathcal{T}_{P/X}$, hence to $(\pi_*\mathcal{T}_{P/X})^G$ is $\mathcal{O}_X$-bilinear. The map $\mathcal{A}_P \to \mathcal{T}_X$ is given by $d\pi$ and is, therefore an $\mathcal{O}_X$-linear map of Lie algebras.

Since G acts on $\mathfrak{g}$ by Lie algebra automorphisms, $\mathfrak{g}^P$ is an $\mathcal{O}_X$-Lie algebra. The sheaf $\pi_*\mathcal{O}_P \otimes \mathfrak{g}$ is a Lie algebra under the $\mathcal{O}_P$-linear extension of the bracket on $\mathfrak{g}$ and G acts on $\pi_*\mathcal{O}_P \otimes \mathfrak{g}$ by Lie algebras automorphisms hence the subsheaf of invariants is closed under the bracket. The identification $(\pi_*\mathcal{O}_P \otimes_{\mathbb{C}} \mathfrak{g})^G \cong \mathfrak{g}^P$ is an isomorphism of Lie algebras.

The infinitesimal action $\mathfrak{g} \to \Gamma(P; \mathcal{T}_{P/X})$ is a map of Lie algebras, therefore so is the isomorphism $(\pi_*\mathcal{O}_P \otimes_{\mathbb{C}} \mathfrak{g})^G \cong (\pi_*\mathcal{T}_{P/X})^G$.

Combining the above remarks we conclude that $\mathcal{A}_P$ is a Lie algebroid.

3.6. *More on Atiyah algebras*

Suppose that G is a Lie group with Lie algebra $\mathfrak{g}$, $\pi : P \to X$ is a principal G-bundle and V is a finite dimensional representation of G given by a

[f]Recall that for a principal G-bundle P and a representation G on a vector space V the associated vector bundle $P \times_G V$ is defined as the quotient of $P \times V$ by the diagonal action of G.

morphism of Lie groups $\rho : G \to GL(V)$. The latter gives rise to the map of Lie algebras $d\rho \colon \mathfrak{g} \to \mathfrak{gl}(V)$.

Let $\mathcal{E}$ denote (the sheaf of sections of) the associated vector bundle $P \times_G V$, *i.e.* $\mathcal{E} = (\pi_* \mathcal{O}_P \otimes V)^G$. There is a canonical isomorphism $\pi^*\mathcal{E} \cong \mathcal{O}_P \otimes_{\mathbb{C}} V$ such that $\mathcal{E} \to \pi_*\pi^*\mathcal{E} \cong \pi_*\mathcal{O}_P \otimes_{\mathbb{C}} V$ coincides with the inclusion of the sub-sheaf of G-invariants.

Therefore, $\mathcal{T}_P$ acts on $\pi^*\mathcal{E}$ and this action is G-equivariant. Thus, $(\mathcal{A}_P = \pi_*\mathcal{T}_P)^G$ acts on $\mathcal{E} = (\pi_*\mathcal{E})^G$. The action is given by a morphism of Lie algebroids $\mathcal{A}_P \to \mathcal{A}_\mathcal{E}$ which restricts to the map $\mathfrak{g}^P \to \underline{\mathrm{End}}_{\mathcal{O}_X}(\mathcal{E}) = \mathfrak{gl}(V)^P$ induced by the map $d\rho \colon \mathfrak{g} \to \mathfrak{gl}(V)$. The commutative diagram

$$\begin{array}{ccc} \mathfrak{g}^P & \longrightarrow & \mathcal{A}_P \\ \downarrow & & \downarrow \\ \underline{\mathrm{End}}_{\mathcal{O}_X}(\mathcal{E}) & \longrightarrow & \mathcal{A}_\mathcal{E} \end{array}$$

is easily seen to be a push-out square exhibiting $\mathcal{A}_\mathcal{E}$ as the push-out (as in 2.5) of $\mathcal{A}_P$ along $(d\rho)^P$.

3.7. *Determinant and trace*

Suppose that $\mathcal{E}$ is a vector bundle. As usual, we denote by $\det\mathcal{E}$ the top non-trivial exterior power of $\mathcal{E}$, a line bundle. The Atiyah algebra of $\det\mathcal{E}$ can be expressed in terms of the Atiyah algebra of $\mathcal{E}$ using the push-out construction of 2.5.

Namely, associating to an endomorphism of $\mathcal{E}$ its trace we obtain the map

$$\mathrm{Tr} \colon \underline{\mathrm{End}}_{\mathcal{O}_X}(\mathcal{E}) \to \mathcal{O}_X$$

of Lie algebras which is compatible with the action of $\mathcal{A}_\mathcal{E}$ on $\underline{\mathrm{End}}_{\mathcal{O}_X}(\mathcal{E})$ by the restriction of the adjoint action and the action of $\mathcal{A}_\mathcal{E}$ on $\mathcal{O}_X$ via the anchor map in the sense of 2.5. The push-out $\mathrm{Tr}_\bullet\mathcal{A}_\mathcal{E}$ is naturally identified with the Atiyah algebra $\mathcal{A}_{\det\mathcal{E}}$.

3.8. *Connections*

Suppose that $\mathcal{A}$ is a Lie algebroid with the anchor map $\sigma \colon \mathcal{A} \to \mathcal{T}_X$. A *connection* on $\mathcal{A}$ is a splitting of the anchor map, *i.e.* an $\mathcal{O}_X$-linear map $\nabla \colon \mathcal{T}_X \to \mathcal{A}$ such that $\sigma \circ \nabla = \mathtt{Id}$. Connections on $\mathcal{A}$ exist only if $\mathcal{A}$ is transitive.

Assume that $\mathcal{A}$ is a transitive Lie algebroid locally free of finite rank over $\mathcal{O}_X$. Then, so is $\mathfrak{g} := \ker(\sigma)$. Let $\mathcal{C}(\mathcal{A})$ denote the sheaf of locally defined connections on $\mathcal{A}$. Note that connections exist locally on X, hence the sheaf of sets $\mathcal{C}(\mathcal{A})$ is locally non-empty. The sheaf of abelian groups $\underline{\mathrm{Hom}}_{\mathcal{O}_X}(\mathcal{T}_X, \mathfrak{g}) \cong \Omega^1_X \otimes \mathfrak{g}$ acts naturally on $\mathcal{C}(\mathcal{A})$. Namely, if ∇ is a connection and A is a section of $\underline{\mathrm{Hom}}_{\mathcal{O}_X}(\mathcal{T}_X, \mathfrak{g})$, then $\nabla + A \colon \mathcal{T}_X \to \mathcal{A}$ satisfies $\sigma \circ (\nabla + A) = \sigma \circ \nabla + \sigma \circ A = \mathtt{Id}$ since $\sigma \circ A = 0$ as A takes values in $\ker(\sigma)$. Note that, if ∇_1 and ∇_2 are two connections, then $\sigma \circ (\nabla_1 - \nabla_2) = 0$, which is to say $\nabla_1 - \nabla_2 \colon \mathcal{T}_X \to \mathfrak{g}$ and $\nabla_1 = \nabla_2 + (\nabla_1 - \nabla_2)$; this shows that the action of $\Omega^1_X \otimes \mathfrak{g}$ on $\mathcal{C}(\mathcal{A})$ is free and transitive. In other words, the above action makes $\mathcal{C}(\mathcal{A})$ a $\Omega^1_X \otimes \mathfrak{g}$-torsor[g]. The curvature $c(\nabla) \in \underline{\mathrm{Hom}}_{\mathcal{O}_X}(\bigwedge^2 \mathcal{T}_X, \mathfrak{g}) \cong \Omega^2_X \otimes \mathfrak{g}$ of a (locally defined) connection $\nabla \in \mathcal{C}(\mathcal{A})$ is defined by the usual formula

$$c(\nabla)(\xi, \eta) = [\nabla(\xi), \nabla(\eta)] - \nabla([\xi, \eta]).$$

A connection is called *flat* if its curvature is equal to zero. In view of 2.4.1 a flat connection on $\mathcal{A}$ is the same thing as a morphism of Lie algebroids $\mathcal{T}_X \to \mathcal{A}$.

Needless to say, a connection on a vector bundle $\mathcal{E}$ (respectively, a principal bundle P) is the same thing as a connection on the Atiyah algebra $\mathcal{A}_{\mathcal{E}}$ (respectively, $\mathcal{A}_P$). The canonical action of $\mathcal{T}_X$ on $\mathcal{O}_X$ defines a connection on $\mathcal{A}_{\mathcal{O}_X}$ which is flat.

3.9. *$\mathcal{O}$-extensions of $\mathcal{T}$*

An $\mathcal{O}_X$-extension of $\mathcal{T}_X$ is a pair $(\mathcal{A}, \mathfrak{c})$, where $\mathcal{A}$ is a Lie algebroid and $\mathfrak{c} \in \Gamma(X; \mathcal{A})$ is a *central* section such that the sequence

$$0 \to \mathcal{O}_X \xrightarrow{\cdot \mathfrak{c}} \mathcal{A} \xrightarrow{\sigma} \mathcal{T}_X \to 0$$

is exact. Morphisms of $\mathcal{O}_X$-extensions of $\mathcal{T}_X$ are morphism of Lie algebroids which preserve the marked central section. It is clear that the resulting category is a groupoid.

Examples of $\mathcal{O}_X$-extension of $\mathcal{T}_X$ are the Atiyah algebras of line bundles (or, equivalently, $\mathbb{C}^\times$-principal bundles). Indeed, for any line bundle $\mathcal{L}$ the canonical map $\mathcal{O}_X \to \underline{\mathrm{End}}_{\mathcal{O}_X}(\mathcal{L})$ is an isomorphism. Hence, $\mathcal{A}_{\mathcal{L}} = \mathcal{D}^{\leqslant 1}_{X,\mathcal{L}}$,

[g]Recall that a torsor $\mathcal{P}$ under a sheaf of groups $\mathcal{G}$ is a sheaf of sets $\mathcal{P}$ equipped with a $\mathcal{G}$-action $\mathcal{G} \times \mathcal{P} \to \mathcal{P}$ which is locally isomorphic, as a sheaf of sets with a $\mathcal{G}$-action, to the sheaf $\mathcal{G}$ with the action by left translations.

where the latter denotes the sheaf of differential operators of order one acting on $\mathcal{L}$.

For $\mathcal{A}$ as above the sheaf of connections $\mathcal{C}(\mathcal{A})$ is a Ω^1_X-torsor. An easy calculation shows that for $\nabla \in \mathcal{C}(\mathcal{A})$ the curvature form $c(\nabla) \in \Omega^2_X$ is closed. Therefore, the assignment $\nabla \mapsto c(\nabla)$ defines a map $c\colon \mathcal{C}(\mathcal{A}) \to \Omega^{2,cl}_X$. It is easily verified that it satisfies $c(\nabla + \alpha) = c(\nabla) + d\alpha$, in other words, is a morphism of torsors compatible with the map $d\colon \Omega^1_X \to \Omega^{2,cl}_X$ of groups.

We briefly recall the requisite facts concerning the classification of torsors. Let $A \xrightarrow{d} B$ be a complex of sheaves of abelian groups on X concentrated in degrees zero and one. Recall that a $(A \xrightarrow{d} B)$-torsor is a pair (T, τ), where T is a A-torsor and $\tau\colon T \to B$ is a map which satisfies $\tau(t + a) = \tau(t) + da$ for $\tau \in T$ and $a \in A$. A morphism of $(A \xrightarrow{d} B)$-torsors $f\colon (T_1, \tau_1) \to (T_2, \tau_2)$ is a morphism of A-torsors $f\colon T_1 \to T_2$ such that $\tau_1 = \tau_2 \circ f$. With these definitions we have the category of $(A \xrightarrow{d} B)$-torsors on X which is a groupoid. The set of isomorphism classes of $(A \xrightarrow{d} B)$-torsors on X is in canonical bijection with the group $H^1(X; A \xrightarrow{d} B)$. A standard proof of this fact relies on explicit calculations with Čech cocycles and will be omitted.

The construction above associates to an $\mathcal{O}_X$-extension $\mathcal{A}$ of $\mathcal{T}_X$ the $\Omega^1_X \to \Omega^{2,cl}_X$-torsor $(\mathcal{C}(\mathcal{A}), c)$. Since it is obviously natural with respect to morphisms of $\mathcal{O}_X$-extensions it extends to a functor from the category of $\mathcal{O}_X$-extensions of $\mathcal{T}_X$ to the category of $(\Omega^1_X \to \Omega^{2,cl}_X)$-torsors. This functor is an equivalence and, hence, set of isomorphism classes of $\mathcal{O}_X$-extensions of $\mathcal{T}_X$ is in canonical bijection with the group $H^1(X; \Omega^1_X \to \Omega^{2,cl}_X)$. The Poincaré Lemma implies that the latter group is isomorphic to $H^1(X; \Omega^{1,cl}_X) \cong H^2(X; \mathbb{C})$.

In the notations introduced above, the assignment $\mathcal{L} \mapsto \mathcal{A}_\mathcal{L}$, where $\mathcal{L}$ is a line bundle (equivalently, a $\mathcal{O}^\times_X$-torsor) and $\mathcal{A}_\mathcal{L}$ its Atiyah algebra, induces, on the sets of isomorphism classes, the "first Chern class" map $H^1(X; \mathcal{O}^\times_X) \to H^1(X; \Omega^1_X \to \Omega^{2,cl}_X)$ which coincides with the map induced by $d\log\colon \mathcal{O}^\times_X \to \Omega^{1,cl}_X$.

4. Modules and universal enveloping algebras

Suppose that $\mathcal{A}$ is a Lie algebroid on X with anchor $\sigma : \mathcal{A} \to \mathcal{T}_X$. Let $\mathfrak{g} = \ker(\sigma)$.

4.1. *The category of $\mathcal{A}$-modules*

4.1.1. *The objects*

Suppose that $\mathcal{M}$ is an $\mathcal{O}_X$-module. A structure of an $\mathcal{A}$-module on $\mathcal{M}$ is given by a $\mathbb{C}$-bilinear pairing $\mathcal{A} \times \mathcal{M} \to \mathcal{M}$, $(a, m) \mapsto am$ such that

- the induced map $\mathcal{A} \to \underline{\mathrm{End}}_{\mathbb{C}}(\mathcal{M})$, $a \mapsto (m \mapsto am)$ is an $\mathcal{O}_X$-linear morphism of Lie algebras, *i.e.* the identity $(fa)m = f(am)$ holds;
- the Leibniz rule $a(fm) = \sigma(a)(f)m + f \cdot am$ holds .

Suppose that the Atiyah algebra $\mathcal{A}_{\mathcal{M}}$ of $\mathcal{M}$ is defined (*e.g.* $\mathcal{M}$ is a vector bundle). Then, clearly, a structure of an $\mathcal{A}$-module on $\mathcal{M}$ is the same thing as a morphism of algebroids $\mathcal{A} \to \mathcal{A}_{\mathcal{M}}$ or, in view of 2.4.1, a splitting of the projection $\mathcal{A} \times_{\mathcal{T}_X} \mathcal{A}_{\mathcal{M}} \to \mathcal{A}$.

4.1.2. *The morphisms*

Suppose that $\mathcal{M}$ and $\mathcal{N}$ are two $\mathcal{A}$-modules. A morphism of $\mathcal{A}$-modules $\varphi : \mathcal{M} \to \mathcal{N}$ is an $\mathcal{O}_X$-linear map $\varphi : \mathcal{M} \to \mathcal{M}$ which satisfies $a\varphi(m) = \varphi(am)$ for $a \in \mathcal{A}$, $m \in \mathcal{M}$.

With the objects and the morphisms defined as above $\mathcal{A}$-modules form a category. The kernel (respectively, cokernel) of a morphism of $\mathcal{A}$ is that of the morphism of the underlying $\mathcal{O}_X$-modules equipped with the obvious structure structure of an $\mathcal{A}$-module. The same remark applies to direct sums and products. From this one deduces easily that the category of $\mathcal{A}$-modules is Abelian.

A perhaps more familiar example of this notion arises from the case when X is equal to a point, in which case the algebroid in question reduces to a Lie algebra and the notion of a module coincides with that of Lie algebra representation.

4.2. *Examples*

4.2.1. *Restriction*

Suppose $\varphi : \mathcal{A} \to \mathcal{B}$ is a morphism of Lie algebroids and $\mathcal{M}$ is a $\mathcal{M}$. The formula $am := \varphi(a)m$, $a \in \mathcal{A}$. $m \in \mathcal{M}$ defines a structure of an $\mathcal{A}$-module on $\mathcal{M}$ which we will denote by $\varphi^\bullet\mathcal{M}$.

It is clear that the assignment $\mathcal{M} \mapsto \varphi^\bullet\mathcal{M}$ extends to a functor from the category of $\mathcal{B}$-modules to that of $\mathcal{A}$-modules. It is clear that the assignment $\varphi \mapsto \varphi^\bullet$ is transitive (*i.e.* natural in compositions): for two morphisms of algebroids

$$\mathcal{A} \xrightarrow{\varphi} \mathcal{B} \xrightarrow{\psi} \mathcal{C}$$

there is a natural isomorphism of functors $\varphi^\bullet \circ \psi^\bullet \cong (\psi \circ \varphi)^\bullet$ which, on a $\mathcal{C}$-module $\mathcal{M}$ is given by the identity map $\mathtt{Id}_{\mathcal{M}}$.

4.2.2.

In particular, any $\mathcal{T}_X$-module $\mathcal{M}$ has a canonical structure of an $\mathcal{A}$-module by restriction via the anchor map, *i.e.* $\sigma^\bullet\mathcal{M}$. Moreover, this canonical structure is preserved under morphisms of algebroids.

4.2.3.

Note that any $\mathcal{A}$-module gives rise to a $\mathfrak{g}$-module in a canonical way (namely, via restriction along the inclusion $\mathfrak{g} \to \mathcal{A}$).

If $\mathfrak{h}$ is a Lie algebroid with trivial anchor (*i.e.* an $\mathcal{O}_X$-Lie algebra) and $\mathcal{M}$ is an $\mathfrak{h}$-module, then $\mathfrak{h}$ acts by $\mathcal{O}_X$-linear endomorphisms of $\mathcal{M}$.

4.2.4. *Lie algebras*

If X consists of just one point, then $\mathfrak{g} = \mathcal{A}$ is a Lie algebra and the notion of a module structure introduced above coincides with the notion of a Lie algebra representation.

4.2.5. $\mathcal{T}_X$*-modules*

Suppose that $\mathcal{E}$ is a vector bundle. A $\mathcal{T}_X$-module structure on $\mathcal{E}$ is the same thing as a flat connection on $\mathcal{E}$. In view of 4.2.2 any flat connection gives rise to an $\mathcal{A}$-module.

4.3. *The universal enveloping algebra*

Suppose that $\mathcal{A}$ is a Lie algebroid and $\mathcal{M}$ is an $\mathcal{A}$-module. The $\mathcal{O}_X$-module structure (respectively, the action of $\mathcal{A}$) gives rise to the map $\mathcal{O}_X \to \underline{\mathrm{End}}_{\mathbb{C}}(\mathcal{M})$ (respectively, $\mathcal{A} \to \underline{\mathrm{End}}_{\mathbb{C}}(\mathcal{M})$). The images of these maps generate an associative subalgebra of $\underline{\mathrm{End}}_{\mathbb{C}}(\mathcal{M})$; these generators are subject to (the universal, *i.e.* independent of the nature of the module $\mathcal{M}$) relations which result from the definition of a module structure given in 4.1.1 as well as additional relations specific to $\mathcal{M}$.

The universal enveloping algebra of a Lie algebroid $\mathcal{A}$ is an associative $\mathbb{C}$-algebra (with unit), denoted by $\mathcal{U}(\mathcal{A})$ which is generated by $\mathcal{O}_X$ and $\mathcal{A}$ subject to the "universal relations". Denoting by i both maps $\mathcal{O}_X \to \mathcal{U}(\mathcal{A})$ and $\mathcal{A} \to \mathcal{U}(\mathcal{A})$ which send a function (respectively, a section of $\mathcal{A}$) to the corresponding generator and using "$\cdot$" to denote the product in $\mathcal{U}(\mathcal{A})$, the relations are

(1) $i(fg) = i(f) \cdot i(g)$
(2) $i([a, b]) = a \cdot b - b \cdot a$
(3) $i(fa) = i(f) \cdot i(a)$
(4) $i(a) \cdot i(f) - i(f) \cdot i(a) = i(\sigma(a)(f))$

For any $\mathcal{A}$-module $\mathcal{M}$, the action map (of Lie algebras) $\mathcal{A} \to \underline{\mathrm{End}}_{\mathbb{C}}(\mathcal{M})$ factors in a unique fashion as

$$\mathcal{A} \xrightarrow{i} \mathcal{U}(\mathcal{A}) \longrightarrow \underline{\mathrm{End}}_{\mathbb{C}}(\mathcal{M})$$

where the second map is defined by $u_1 \cdot u_2 \ldots u_{n-1} \cdot u_n \mapsto i(u_1) \circ i(u_2) \circ \ldots \circ i(u_{n-1}) \circ i(u_n)$, where the u_i are either functions of sections of $\mathcal{A}$ is a morphism of associative algebras. The defining relations in $\mathcal{U}(\mathcal{A})$ imply that the map is well-defined.

Morphisms of Lie algebroids induce morphisms of respective universal enveloping algebras so that the "universal enveloping algebra construction" is a functor from the category of Lie algebroids to the category of (unital) associative $\mathbb{C}$-algebras.

The universal enveloping algebra of the Lie algebroid $\mathcal{T}_X$ is the algebra of differential operators (acting on functions) $\mathcal{D}_X$. Thus, for any Lie algebroid $\mathcal{A}$ there is a canonical map (of algebras) $\mathcal{U}(\mathcal{A}) \to \mathcal{D}_X$.

4.4. *Tensor product of modules*

Suppose that $\mathcal{A}$ is a Lie algebroid. The category of $\mathcal{A}$-modules is endowed with a symmetric monoidal structure: for any two modules $\mathcal{M}$ and $\mathcal{N}$ their tensor product (as $\mathcal{O}_X$-modules) $\mathcal{M} \otimes_{\mathcal{O}_X} \mathcal{N}$ is equipped with the canonical $\mathcal{A}$-module structure given by the Leibniz rule. Namely, for $m \in \mathcal{M}$, $n \in \mathcal{N}$, $a \in \mathcal{A}$, $a.(m \otimes n) = (a.m) \otimes n + m \otimes (a.n)$.

The sheaf $\underline{\mathrm{Hom}}_{\mathcal{O}_X}(\mathcal{M}, \mathcal{N})$ is equipped with the canonical structure of a $\mathcal{A}$-module by $(a.\varphi)(m) = a.(\varphi(m)) - \varphi(a.m)$.

5. Lie algebroid cohomology

5.1. *The de Rham complex*

Suppose that $\mathcal{A}$ is a Lie algebroid *which we will assume to be a vector bundle* (*i.e.* locally free of finite rank over $\mathcal{O}_X$).

For an $\mathcal{A}$-module $\mathcal{M}$, $n = 0, 1, 2, \ldots$ let

$$\mathrm{DR}_{\mathcal{A}}(\mathcal{M})^n = \underline{\mathrm{Hom}}_{\mathcal{O}_X}(\bigwedge^n \mathcal{A}, \mathcal{M}).$$

In the case $\mathcal{M} = \mathcal{O}_X$ with the canonical $\mathcal{A}$-module structure we shall use the suggestive notation $\Omega^n_{\mathcal{A}} := \mathrm{DR}_{\mathcal{A}}(\mathcal{O}_X)^n$. Thus, $\Omega^n_{\mathcal{T}_X} = \Omega^n_X$.

For $\varphi \in \mathrm{DR}_{\mathcal{A}}(\mathcal{M})^n$ we define $d_{\mathcal{A}}\varphi$ by the formula

$$\begin{aligned}(d_{\mathcal{A}}\varphi)(a_0 \ldots a_n) = \sum_{j=0}^{n}(-1)^j a_j \varphi(a_0 \ldots \widehat{a_j} \ldots a_n) + \\ \sum_{0\leqslant k<l\leqslant n} (-1)^{k+l}\varphi([a_k, a_l], a_0 \ldots \widehat{a_k} \ldots \widehat{a_l} \ldots a_n)\end{aligned}$$

Then, $d_{\mathcal{A}}\varphi \in \mathrm{DR}^{n+1}_{\mathcal{A}}(\mathcal{M})$ and $d_{\mathcal{A}} \circ d_{\mathcal{A}} = 0$. Thus, $\mathrm{DR}_{\mathcal{A}}(\mathcal{M})^\bullet$ equipped with the differential $d_{\mathcal{A}}$ is a complex concentrated in non-negative degrees. Note that the map $d_{\mathcal{A}} \colon \mathcal{O}_X \to \Omega^1_{\mathcal{A}}$ coincides with the composition $\mathcal{O}_X \xrightarrow{d} \Omega^1_X \to \Omega^1_{\mathcal{A}}$.

For $\mathcal{A} = \mathcal{T}_X$, $\mathcal{M} = \mathcal{O}_X$ the above construction recovers the de Rham complex $\Omega^\bullet_X$. If X consists of one point the above construction is known as the Chevalley-Eilenberg cochain complex (of a Lie algebra with coefficients in a module).

Clearly, a morphism of Lie algebroids $\mathcal{A} \to \mathcal{B}$ induces a map of de Rham complexes $\mathrm{DR}_{\mathcal{B}}(\mathcal{M}) \to \mathrm{DR}_{\mathcal{A}}(\mathcal{M})$ for any $\mathcal{B}$-module $\mathcal{M}$ (regarded as an $\mathcal{A}$ via the morphism of the algebroids). In particular, for any Lie algebroid $\mathcal{A}$ there is a canonical morphism of complexes $\Omega^\bullet_X = \mathrm{DR}_{\mathcal{T}_X}(\mathcal{O}_X) \to \mathrm{DR}_{\mathcal{A}}(\mathcal{O}_X)$.

5.2. *The Spencer complex*

The above constructions have a natural interpretation in terms of the universal enveloping algebras. Namely, for $n = 0, 1, 2, \ldots$ let $K^{-n} =$

$\mathcal{U}(\mathcal{A}) \otimes_{\mathcal{O}_X} \bigwedge^n \mathcal{A}$; we define the map $\partial : K^{-n} \to K^{-n+1}$ by the formula

$$\partial(P \otimes a_1 \wedge \ldots a_n) = \sum_{j=1}^{n} P \cdot a_j \otimes a_1 \wedge \ldots \widehat{a_j} \ldots \wedge a_n +$$
$$\sum_{1 \leqslant k < l \leqslant n} P \otimes [a_k, a_l] \wedge a_1 \ldots \widehat{a_k} \ldots \widehat{a_l} \ldots \wedge a_n.$$

Then, $\partial \circ \partial = 0$, *i.e.* $K^\bullet$ equipped with the differential ∂ is a complex of $\mathcal{U}(\mathcal{A})$-modules concentrated in non-positive degrees.

The complex $K^\bullet$ satisfies $H^i(K^\bullet) = 0$ for $i \neq 0$. Moreover, the map $K^0 = \mathcal{U}(\mathcal{A}) \to \mathcal{O}_X$, $P \mapsto P1$ extends (by zero in all other degrees) to a map of complexes $K^\bullet \to \mathcal{O}_X$ which induces the canonical isomorphism $H^0(K^\bullet) \cong \mathcal{O}_X$. In other words, $K^\bullet$ is a resolution of $\mathcal{O}_X$ by locally free $\mathcal{U}(\mathcal{A})$-modules sometimes referred to as *the Spencer complex.*

The canonical map (restriction)

$$\underline{\mathrm{Hom}}_{\mathcal{U}(\mathcal{A})}(\mathcal{U}(\mathcal{A}) \otimes_{\mathcal{O}_X} \bigwedge^n \mathcal{A}, \mathcal{M}) \to \underline{\mathrm{Hom}}_{\mathcal{O}_X}(\bigwedge^n \mathcal{A}, \mathcal{M}) = \mathrm{DR}_{\mathcal{A}}(\mathcal{M})^n$$

is an isomorphism. It is easy to see that it is compatible with respective differential, hence there is a natural identification

$$\mathrm{DR}_{\mathcal{A}}(\mathcal{M}) = \underline{\mathrm{Hom}}^\bullet_{\mathcal{U}(\mathcal{A})}(K^\bullet, \mathcal{M}).$$

Since the Spencer complex is a resolution of $\mathcal{O}_X$ by locally free $\mathcal{U}(\mathcal{A})$-modules, it follows that there is a canonical isomorphism

$$H^i(X; \mathrm{DR}_{\mathcal{A}}(\mathcal{M})) \cong \mathrm{Ext}^i_{\mathcal{U}(\mathcal{A})}(\mathcal{O}_X, \mathcal{M}).$$

6. Classification of invertible modules

For the purposes of these notes an invertible $\mathcal{A}$-module is a line bundle (*i.e.* a locally free $\mathcal{O}_X$-module of rank one) equipped with a structure of a $\mathcal{A}$-module. Invertible $\mathcal{A}$-modules form a Picard category (*i.e.* a strict symmetric monoidal groupoid), denoted $\mathbb{P}\mathrm{IC}_{\mathcal{A}}(X)$, with the monoidal structure given by the tensor product.

The following special cases of the above general set-up are well-known. If $\mathcal{A}$ is the trivial Lie algebroid, then $\mathbb{P}\mathrm{IC}_{\mathcal{A}}(X)$ coincides with the category of line bundles on X. The Picard group of X, denoted by $\mathrm{Pic}(X)$, is defined as the group of isomorphism classes of line bundles and is canonically isomorphic to $H^1(X; \mathcal{O}_X^\times)$. If $\mathcal{A} = \mathcal{T}_X$, an $\mathcal{A}$-module structure is a flat connection. The group of isomorphism classes of line bundles equipped with flat connections is canonically isomorphic to $H^1(X; \mathcal{O}_X^\times \to \Omega_X^{1,cl})$. In this section we will outline a general approach to the problem of classification of invertible $\mathcal{A}$-modules based on "general principles", most notably on the passage from local considerations to global results.

Let $\underline{\mathbb{P}\text{IC}}_{\mathcal{A}}$ denote the Picard stack[h] of invertible $\mathcal{A}$-modules, *i.e.* $\underline{\mathbb{P}\text{IC}}_{\mathcal{A}}(U) = \mathbb{P}\text{IC}_{\mathcal{A}}(U)$ for U open in X.

Suppose that $\mathcal{L}$ is a line bundle; we shall denote by $\mathcal{L}^{\times}$ the corresponding $\mathcal{O}_X^{\times}$-torsor. A $\mathcal{A}$-module structure on $\mathcal{L}$ is a splitting $\nabla\colon \mathcal{A} \to \mathcal{A} \times_{\mathcal{T}_X} \mathcal{A}_{\mathcal{L}}$ of the short exact sequence

$$0 \to \mathcal{O}_X \to \mathcal{A} \times_{\mathcal{T}_X} \mathcal{A}_{\mathcal{L}} \to \mathcal{A} \to 0$$

which is a morphism of Lie algebroids. These exist locally, because $\mathcal{L}$ is locally isomorphic to $\mathcal{O}_X$.

For any two such splittings ∇_1 and ∇_2 their difference $\nabla_1 - \nabla_2\colon \mathcal{A} \to \mathcal{A} \times_{\mathcal{T}_X} \mathcal{A}_{\mathcal{L}}$ factors through (the inclusion of) $\mathcal{O}_X$; moreover, the resulting section $\nabla_2 - \nabla_1$ of $\Omega^1_{\mathcal{A}}$ satisfies $d_{\mathcal{A}}(\nabla_2 - \nabla_1) = 0$. The sheaf of such locally defined splittings is a $\Omega^{1,cl}_{\mathcal{A}}$-torsor and will be denoted $\mathcal{C}^{\flat}_{\mathcal{A}}(\mathcal{L})$. Note that a structure of a $\mathcal{A}$-module on $\mathcal{L}$ amounts to a trivialization, equivalently, a global section, of $\mathcal{C}^{\flat}_{\mathcal{A}}(\mathcal{L})$.

There is a canonical morphism of sheaves

$$d_{\mathcal{A}} \log\colon \mathcal{L}^{\times} \to \mathcal{C}^{\flat}_{\mathcal{A}}(\mathcal{L}) \tag{2}$$

defined as follows. A locally defined section of $\mathcal{L}^{\times}$ gives rise to a locally defined isomorphism $\mathcal{O}_{X/\mathcal{P}} \xrightarrow{\cong} \mathcal{L}$. Via this isomorphism the canonical $\mathcal{A}$-module structure on $\mathcal{O}_X$ gives rise to a $\mathcal{A}$-module structure on $\mathcal{L}$ which corresponds to a (locally defined) section of $\mathcal{C}^{\flat}_{\mathcal{A}}(\mathcal{L})$. We leave it to the reader to check that the morphism (2) is compatible with the map of sheaves of groups $d_{\mathcal{A}} \log\colon \mathcal{O}_X^{\times} \to \Omega^{1,cl}_{\mathcal{A}}$ and respective actions, meaning that, for $f \in \mathcal{O}_X^{\times}$ and $s \in \mathcal{L}^{\times}$, $d_{\mathcal{A}} \log(fs) = d_{\mathcal{A}} \log(s) + d_{\mathcal{A}} \log(f)$.

Hence, (2) induces a canonical isomorphism of $\Omega^{1,cl}_{\mathcal{B}}$-torsors

$$\mathcal{C}^{\flat}_{\mathcal{B}}(\mathcal{L}) \cong d_{\mathcal{B}} \log(\mathcal{L}^{\times}) := \mathcal{L}^{\times} \times_{\mathcal{O}^{\times}_{X/\mathcal{P}}} \Omega^{1,cl}_{\mathcal{B}}.$$

In particular, a structure of a $\mathcal{B}$-module on $\mathcal{L}$ amounts to a trivialization of $d_{\mathcal{B}} \log(\mathcal{L}^{\times})$.

The concept of a $(A \xrightarrow{d} B)$-torsor has already appeared in 3.9. Locally defined $(A \xrightarrow{d} B)$-torsors form a Picard stack on X which we will denote by

[h]Just as sheaves are "nice" (from the point of view of the local-to-global transition) examples of presheaves, stacks are "nice" examples of prestacks. A prestack is a (lax) presheaf of categories. A stack is a prestack satisfying certain conditions ("descent"). The reader is referred to [B] and [D] for details.

$(A \xrightarrow{d} B)[1]$. By a theorem of Pierre Deligne all Picard stacks arise in this way. The group (under the operation induced by the monoidal structure) of isomorphism classes of $(A \xrightarrow{d} B)$-torsors on X, *i.e.* $\pi_0(A \xrightarrow{d} B)[1](X)$ is canonically isomorphic to $H^1(X; A \xrightarrow{d} B)$, where the complex $A \xrightarrow{d} B$ is placed in degrees zero and one.

Assigning to an invertible $\mathfrak{D}$-module $\mathcal{L}$ the pair $(\mathcal{L}^\times, \nabla)$, where ∇ is the trivialization of $d_{\mathfrak{D}} \log(\mathcal{L}^\times)$ corresponding to the $\mathcal{A}$-module structure we obtain a morphism of Picard stacks

$$\underline{\mathbb{P}\mathrm{IC}}_{\mathcal{A}} \to (\mathcal{O}_X^\times \xrightarrow{d_{\mathcal{A}} \log} \Omega_{\mathcal{A}}^{1,cl})[1] \tag{3}$$

Lemma 1. *The morphism is an equivalence of Picard stacks.*

Proof. The quasi-inverse equivalence associates to a $\mathcal{O}_X^\times$-torsor T the corresponding line bundle $\mathcal{L} = T \times_{\mathcal{O}_X^\times} \mathcal{O}_X$ and to the trivialization ∇ of $d_{\mathcal{A}} \log(T)$ the $\mathcal{A}$-module structure on $\mathcal{L}$ deduced from the isomorphism $d_{\mathcal{A}} \log(T) \cong \mathcal{C}^\flat_{\mathcal{A}}(\mathcal{L})$. □

Corollary 2. $\pi_0 \, \mathbb{P}\mathrm{IC}_{\mathcal{A}}(X) \cong H^1(X; \mathcal{O}_X^\times \xrightarrow{d_{\mathcal{A}} \log} \Omega_{\mathcal{A}}^{1,cl})$.

This result generalizes the more familiar cases of

(1) the trivial algebroid: $\mathrm{Pic}(X) \cong H^1(X; \mathcal{O}_X^\times)$;
(2) $\mathcal{A} = \mathcal{T}_X$: isomorphism classes of rank one flat connections are in bijective correspondence with $H^1(X; \mathcal{O}_X^\times \xrightarrow{d \log} \Omega_X^{1,cl})$.

7. Characteristic classes

In this section we fix a transitive Lie algebroid $\mathcal{A}$ which is a locally free $\mathcal{O}_X$-module of finite rank (a vector bundle) with anchor map σ. Then $\mathfrak{g} := \ker(\sigma)$ is locally free of finite rank, and there is an exact sequence

$$0 \to \mathfrak{g} \xrightarrow{i} \mathcal{A} \xrightarrow{\sigma} \mathcal{T}_X \to 0.$$

7.1. *The Chern-Weil map*

Let $\mathfrak{A}$ denote the complex concentrated in degrees -1 and 0 with $\mathfrak{A}^{-1} = \mathfrak{g}$, $\mathfrak{A}^0 = \mathcal{A}$ and the only non-trivial differential given by the map i (*i.e.* the cone of the map i).

The complex $\mathfrak{A}$ has a canonical structure of a differential graded Lie algebroid. Moreover, the anchor map $\mathfrak{A} \to \mathcal{T}_X$ (with the only non-trivial component given by σ) is a quasi-isomorphism[i]. This (together with the assumptions on $\mathcal{A}$) implies that

- the induced map of respective differential graded universal enveloping algebras $\mathcal{U}(\mathfrak{A}) \to \mathcal{D}_X$ is a quasi-isomorphism
- the induced map of respective de Rham complexes with coefficients in $\mathcal{O}_X$

$$\Omega_X^\bullet \to \mathrm{DR}_{\mathfrak{A}}(\mathcal{O}_X)$$

  is a quasi-isomorphism.

It is the latter that we are going to be interested in.

The de Rham complex $\mathrm{DR}_{\mathfrak{A}}(\mathcal{O}_X)$ is the total complex of the following double complex:

$$\begin{array}{ccccccccc}
 & & & & & & \cdots & & \\
 & & & & & & \uparrow & & \\
 & & & & S^2\mathfrak{g}^\vee & \longrightarrow & \mathcal{A}^\vee \otimes S^2\mathfrak{g}^\vee & \longrightarrow & \cdots \\
 & & & & \uparrow & & \uparrow & & \\
 & & \mathfrak{g}^\vee & \longrightarrow & \mathcal{A}^\vee \otimes \mathfrak{g}^\vee & \longrightarrow & \bigwedge^2 \mathcal{A}^\vee \otimes \mathfrak{g}^\vee & \longrightarrow & \cdots \\
 & & \uparrow & & \uparrow & & \uparrow & & \\
\mathcal{O}_X & \longrightarrow & \mathcal{A}^\vee & \longrightarrow & \bigwedge^2 \mathcal{A}^\vee & \longrightarrow & \bigwedge^3 \mathcal{A}^\vee & \longrightarrow & \cdots
\end{array}$$

so that

- $\mathrm{DR}_{\mathfrak{A}}(\mathcal{O}_X)^n = \bigoplus_p \bigwedge^p \mathcal{A}^\vee \otimes S^{n-2p}\mathfrak{g}^\vee$;
- the qth row is equal to $\mathrm{DR}_{\mathcal{A}}(S^q\mathfrak{g}^\vee)$, the $\mathcal{A}$-de Rham complex with coefficients in $S^q\mathfrak{g}^\vee$;
- the pth column is the pth exterior power of the complex (concentrated in degrees 0 and 1) $\mathcal{A}^\vee \xrightarrow{i^\vee} \mathfrak{g}^\vee$ (the Koszul complex) with the differential given by the formula

$$a_1 \wedge \cdots \wedge a_k \otimes b_1 \circ \cdots \circ b_l \mapsto \\ \sum_j (-1)^j a_1 \wedge \cdots \wedge \widehat{a_j} \wedge \cdots \wedge a_k \otimes a_j \circ b_1 \circ \cdots \circ b_l.$$

[i] A quasi-isomorphism is a map of complexes which induces an isomorphism on homology.

Let $F_\bullet\Omega^\bullet_X$ (respectively, $F_\bullet \mathrm{DR}_{\mathfrak{A}}(\mathcal{O}_X)$ denote the "stupid" filtration (respectively, the filtration "by columns" of the bi-complex above), *i.e.* $F_k\Omega^\bullet_X = \oplus_{i\geqslant -k}\Omega^i_X$ (respectively, $F_{-q}\mathrm{DR}_{\mathfrak{A}}(\mathcal{O}_X)^n = \bigoplus_{p\geqslant q}\bigwedge^p\mathcal{A}^\vee\otimes S^{n-2p}\mathfrak{g}^\vee$).

The map $\sigma^\vee : \Omega^\bullet_X \to \mathrm{DR}_{\mathfrak{A}}(\mathcal{O}_X)$ is a morphism of filtered complexes. The induced map on the associated graded complexes is a quasi-isomorphism: it is the identity map in degree zero; in degree one this follows from the exactness of the sequence

$$0\to\Omega^1_X\to\mathcal{A}^\vee\to\mathfrak{g}^\vee\to 0.$$

In higher degrees, the map is an exterior power of the map in degree one. Under our assumptions of local freeness the former is a quasi-isomorphism if the latter is. Consequently, the map $\sigma^\vee$ is a quasi-isomorphism and, in particular induces isomorphisms

$$\sigma^\vee : H^n_{dR}(X)\xrightarrow{\cong} H^n(X;\mathrm{DR}_{\mathfrak{A}}(\mathcal{O}_X)).$$

Let $(S^k\mathfrak{g}^\vee)^{\mathcal{A}} := \ker(S^k\mathfrak{g}^\vee\to A^\vee\otimes S^k\mathfrak{g}^\vee)$ denote the subsheaf of $\mathcal{A}$-invariant sections (the 0th $\mathcal{A}$-de Rham cohomology sheaf). The inclusion of the kernel of the de Rham differential give rise to the map of complexes

$$\bigoplus_k (S^k\mathfrak{g}^\vee)^{\mathcal{A}}[-2k]\to\mathrm{DR}_{\mathfrak{A}}(\mathcal{O}_X)$$

which induces the *Chern-Weil* maps

$$\mathtt{CW}:\Gamma(X;(S^k\mathfrak{g}^\vee)^{\mathcal{A}})\to H^{2k}(X;\mathrm{DR}_{\mathfrak{A}}(\mathcal{O}_X))\cong H^{2k}_{dR}(X)$$

which factor through

$$\mathtt{CW}:\Gamma(X;(S^k\mathfrak{g}^\vee)^{\mathcal{A}})\to H^{2k}(X;\tau_{\leqslant 2k}F_{-k}\Omega^\bullet_X).$$

Here, $\tau_{\leqslant p}$ denotes the truncation functor, *i.e.*, for a complex C with differential $d^\bullet$, the complex $\tau_{\leqslant p}C^\bullet$ is defined by $(\tau_{\leqslant p}C^\bullet)^i = C^i$ for $i<p$, $(\tau_{\leqslant p}C^\bullet)^p = \ker d^p$, $(\tau_{\leqslant p}C^\bullet)^i = 0$ for $i>p$ equipped with differential induced by d.

7.2. *Characteristic forms*

For any Lie algebroid (possibly differential graded) $\mathfrak{B}$ the de Rham complex with coefficients in $\mathcal{O}_X$, $\mathrm{DR}_{\mathfrak{B}}(\mathcal{O}_X)$, has a canonical structure of a commutative differential graded algebra (CDGA). Namely, under our assumptions on $\mathfrak{B}$, $\mathrm{DR}_{\mathfrak{B}}(\mathcal{O}_X)$ is identified with the symmetric algebra

$S^{\bullet}_{\mathcal{O}_X}(\Omega^1_{\mathfrak{B}}[-1]) = \bigoplus_k \Omega^k_{\mathfrak{B}}[-k]$ and the latter is a graded commutative algebra. The de Rham differential satisfies the Leibniz rule with respect to the product, hence $\mathrm{DR}_{\mathfrak{B}}(\mathcal{O}_X)$ is a CDGA.

We continue in the notation of 7.1. The commutative graded $\mathcal{O}_X$-algebra underlying $\mathrm{DR}_{\mathfrak{A}}(\mathcal{O}_X)$ freely generated by $\mathcal{A}^\vee$ in degree one and $\mathfrak{g}^\vee$ in degree two.

A connection ∇ on $\mathcal{A}$ determines a morphism of algebras $\mathrm{DR}_{\mathfrak{A}}(\mathcal{O}_X) \to \Omega^{\bullet}_X$ which is defined on generators by $\nabla^\vee : \mathcal{A}^\vee \to \Omega^1_X$ and $c(\nabla)^\vee : \mathfrak{g}^\vee \to \Omega^2_X$. The former is the transpose of the connection $\nabla\colon \mathcal{T}_X \to \mathcal{A}$ whereas the latter is the transpose of the curvature $c(\nabla)\colon \bigwedge^2 \mathcal{T}_X \to \mathfrak{g}$. The resulting map $\mathrm{DR}_{\mathfrak{A}}(\mathcal{O}_X) \to \Omega^{\bullet}_X$ is a morphism of CDGA such that the composition

$$\bigoplus_k (S^k\mathfrak{g}^\vee)^{\mathcal{A}}[-2k] \to \mathrm{DR}_{\mathfrak{A}}(\mathcal{O}_X) \to \Omega^{\bullet}_X$$

sends an invariant polynomial P to the differential form $P(c(\nabla))$.

7.3. *Characteristic classes of principal bundles*

Suppose that G is a Lie group with complexified Lie algebra $\mathfrak{g}$ and $\pi : P \to X$ is a principal G-bundle. Then, the canonical map $\mathfrak{g}^G \to \Gamma(X; \mathfrak{g}^P)$ takes values in the subsheaf of $\mathcal{A}_P$-invariant sections. Thus, applying the constructions above we obtain the maps

$$\mathtt{CW} : (S^k\mathfrak{g}^\vee)^G \to H^{2k}_{dR}(X).$$

These maps assign to a (homogeneous) invariant polynomial P on $\mathfrak{g}$ of degree k a de Rham cohomology class $\mathtt{CW}(P)$ of degree $2k$. All (primary) characteristic classes with values in de Rham cohomology arise in this way.

References

[BB] A. Beilinson, J. Bernstein, The proof of Jantzen conjectures, *Advances in Soviet Mathematics*, Vol. 16, Part 1, AMS, 1993, 1–50

[B] J.-L. Brylinski, Loop Spaces, Characteristic Classes and Geometric Quantization, Progress in Mathematics, Vol. 107, Birkhäuser 1993

[BK] A. Beilinson, D. Kazhdan, Flat projective connections, unpublished manuscript, `http://www.math.sunysb.edu/~kirillov/manuscripts/kazhdan2.pdf`

[D] P. Deligne, La formule de dualité globale, 1973. SGA 4 III, Exposé XVIII

[M] Kirill C. H. Mackenzie, General Theory of Lie Groupoids and Lie Algebroids, Cambridge University Press London Mathematical Society Lecture Note Series 213, 2005

[W] Frank W. Warner, Foundations of differentiable manifolds and Lie groups, Springer, Graduate Texts in Mathematics Volume 94, 1983

INTRODUCTION TO PSEUDODIFFERENTIAL AND K-THEORETIC METHODS IN THE INDEX THEORY FOR BOUNDARY VALUE PROBLEMS

Elmar Schrohe

Institut für Analysis,
Leibniz Universität Hannover,
Welfengarten 1, 30167 Hannover, Germany
Email: schrohe@math.uni-hannover.de

These lecture notes provide basic concepts for the understanding of index theory, in particular on manifolds with boundary. Topics include Fredholm theory, pseudodifferential calculus, Boutet de Monvel's algebra, K-theory for C*-algebras and applications to the index theory for boundary value problems.

Introduction

One of the central goals in analysis is to find solutions to partial differential equations and boundary value problems. In good cases, one can reduce the problem to an equation of the form

$$Au = f,$$

where $A \in \mathcal{L}(E, F)$ is a bounded linear operator between two Banach spaces E and F, and the task is to find, for each $f \in F$, a – preferably unique – solution u of the above equation. Of course, a unique solution for every f exists if and only if A is invertible. When this is not the case, the notion of a Fredholm operator often is a useful alternative.

An operator $A \in \mathcal{L}(E, F)$ is called a Fredholm operator if its kernel is finite dimensional and its image has finite codimension in F. The index then is defined as the difference of the dimension of the kernel and the codimension of the image. While both the kernel dimension and the codimension of the image can vary considerably under small perturbations of the operator, the index is a remarkably stable quantity. It is not only invariant under small perturbations, but also under perturbations by compact operators. It is this stability which in some cases allows us to actually compute it.

The Atiyah-Singer index theorem for example shows how the index of an elliptic differential operator, acting between sections of vector bundles over a smooth compact manifold, can be determined from its principal symbol. In its proof two then newly developed tools were employed: K-theory and pseudodifferential operators. In fact, the central statement of the theorem says that with the principal symbol of the operator, one can associate a K-class. From this an integer can be computed which turns out to be the index of the operator.

The Atiyah-Singer index theorem has found important applications in many fields ranging from number theory to theoretical physics and is considered to be one of the fundamental mathematical achievements of the twentieth century.

If the underlying manifold has a boundary, the situation is more complicated. In 1971, Boutet de Monvel developed a calculus for boundary value problems [10], which by now has become an important tool of modern analysis. One of the results Boutet de Monvel proved was an index theorem for boundary value problems. As a crucial step in the proof, he showed how a K-class can be associated with a boundary value problem. From this class, the index then can be computed as in the case of manifolds without boundary. The construction of this K-class, however, is not easy. Boutet de Monvel used an ingenious combination of analytic deformation and K-theory. It therefore came as a small surprise that this K-class can be also be constructed in a much more elementary way with the help of K-theory for C*-algebras, [28] – a tool which, however, was not yet developed when Boutet de Monvel wrote his groundbreaking paper.

In these notes I will not attempt to reprove the results of Atiyah and Singer or Boutet de Monvel. Instead I will try to sketch the route towards them, to provide the basic prerequisites necessary for the understanding of the original work and to give hints to the literature. The reader is not required to have more than a basic knowledge of analysis and - to some extent - functional analysis.

This article is organized as follows: Section 1 contains the necessary facts on Fredholm operators. Sobolev spaces are the topic of Section 2. Section 3 furnishes a short course on pseudodifferential operators. Boutet de Monvel's calculus is introduced in Section 4. Section 5 provides the basic elements of K-theory for C*-algebras. In Section 6 we will see how K-theory for C*-algebras can be used in the index theory on closed (compact, without boundary) manifolds. In Section 7 we finally treat the index problem for operators in the Boutet de Monvel calculus.

Acknowledgment

These lecture notes grew out of a six hour course at the Summer School 'Geometric, Algebraic and Topological Methods for Quantum Field Theory'. I thank the organizers, Alexander Cardona, Hernán Ocampo, Sylvie Paycha and Andrés Reyes L., for the invitation and their kind hospitality, and the participants of the school for the stimulating atmosphere. Special thanks go to Alejandro Rivera for his useful comments on the manuscript.

1. Fredholm Operators

Definition 1.1. Let E and F be Banach spaces. An operator $A \in \mathcal{L}(E, F)$ is called a Fredholm[a] operator, if

$$\dim \ker A < \infty \text{ and } \operatorname{codim} \operatorname{im} A := \dim(F/\operatorname{im} A) < \infty.$$

In this case, one calls

$$\operatorname{ind} A = \dim \ker A - \operatorname{codim} \operatorname{im} A \in \mathbb{Z}$$

the index of A. The quotient $F/\operatorname{im} A$ is called the cokernel of A.

If A is a Fredholm operator, the equation $Au = f$ has solutions whenever f belongs to a subspace of F which has a finite-dimensional complement, while the space of solutions is at most finite-dimensional. Clearly, an invertible operator has index zero; not every operator of index zero, however, is invertible, as we shall see in a moment.

The notion of a Fredholm operator is only of interest for infinite-dimensional spaces E and F. In fact, the isomorphism theorem tells us that $E/\ker A \cong \operatorname{im} A$. If both E and F have finite dimension, then

$$\dim E - \dim \ker A = \dim \operatorname{im} A = \dim F - \operatorname{codim} \operatorname{im} A$$

and therefore

$$\operatorname{ind} A = \dim E - \dim F.$$

Hence, in this case, *every* operator is a Fredholm operator and the index is actually independent of A.

Example 1.2. Let $E = F = \ell^2(\mathbb{N})$ be the space of square integrable sequences

$$\ell^2(\mathbb{N}) = \{x = (x_1, x_2, \ldots) : x_j \in \mathbb{C}, \sum |x_j|^2 < \infty\}.$$

[a]Named after Erik Ivar Fredholm, 1866–1927.

We define the operators S_L and S_R in $\mathcal{L}(\ell^2(\mathbb{N}))$, which shift the sequences to the left and the right, respectively, by

$$S_L(x_1, x_2, x_3 \dots) = (x_2, x_3, \dots) \text{ and } S_R(x_1, x_2, \dots) = (0, x_1, x_2, \dots).$$

Then it is easy to see that ind $S_L = 1$ and ind $S_R = -1$.

The index behaves well under compositions:

Lemma 1.3. *Let E, F, and G be Banach spaces. If $A \in \mathcal{L}(E, F)$ and $B \in \mathcal{L}(F, G)$ both are Fredholm operators, then so is the composition BA, and*

$$\text{ind }(BA) = \text{ind } B + \text{ind } A.$$

The standard proof is a lengthy but straightforward exercise in linear algebra. A nice alternative proof can be found in Hörmander's book [22, Corollary 19.1.7].

Note that Example 1.2 shows that on $\ell^2(\mathbb{N})$ there exist Fredholm operators of every index.

One remarkable fact about the index is that it is a very stable quantity. It does not change if we subject A to a small or a compact perturbation. We denote by $\mathcal{K}(E, F)$ the compact operators in $\mathcal{L}(E, F)$).

Proposition 1.4. *A Fredholm operator $A \in \mathcal{L}(E, F)$ has closed image. Moreover*

(a) *$A + S$ is a Fredholm operator whenever $S \in \mathcal{L}(E, F)$ has sufficiently small norm. Moreover, we then have*

$$\text{ind }(A + S) = \text{ind } A.$$

(b) *$A + C$ is a Fredholm operator for $C \in \mathcal{K}(E, F)$ and*

$$\text{ind }(A + C) = \text{ind } A.$$

Proof. See Lemma 19.1.1, Theorem 19.1.5 and Corollary 19.1.8 in [22]. □

In particular: If $[0, 1] \ni t \mapsto A(t)$ is a family of Fredholm operators, which is continuous in the topology of $\mathcal{L}(E, F)$, then ind $A(0) =$ ind $A(1)$. Hörmander proves an even stronger stability result in [22, Theorem 19.1.10].

Of course, in practice it is virtually impossible to check whether the dimensions of the kernel and the cokernel are finite. The following theorem simplifies the task considerably. It shows that for proving the Fredholm property it is sufficient to establish invertibility modulo compact operators:

Theorem 1.5. *The following are equivalent:*

(i) $A \in \mathcal{L}(E, F)$ *is a Fredholm operator.*

(ii) *There exists an operator* $B \in \mathcal{L}(F, E)$ *such that*

$$F_1 := AB - I_F \text{ and } F_2 := BA - I_E$$

are operators of finite rank.

(iii) *There exists an operator* $C \in \mathcal{L}(F, E)$ *such that the operators*

$$K_1 := AC - I_F \text{ and } K_2 := CA - I_E$$

are compact.

The proofs of (i) ⇒ (ii) and (ii) ⇒ (iii) are straightforward. The difficult part is (iii) ⇒ (i), known as Atkinson's theorem, see [22, Corollary 19.1.9].

2. Fourier Transform and Sobolev Spaces

2.1. *Rapidly decreasing functions*

The space $\mathcal{S} = \mathcal{S}(\mathbb{R}^n)$ of rapidly decreasing functions on $\mathbb{R}^n$ is defined as the space of all $u \in \mathcal{C}^\infty(\mathbb{R}^n, \mathbb{C})$ such that, for each choice of multi-indices[b] α, β

$$q_{\alpha,\beta}(u) := \sup_{x \in \mathbb{R}^n} |\partial_x^\alpha x^\beta u(x)| < \infty. \tag{1}$$

The seminorms $q_{\alpha,\beta}$, $\alpha, \beta \in \mathbb{N}_0^n$, define a Fréchet (*i.e.* complete metric, translation invariant) topology for $\mathcal{S}$.

Exercise 2.1. Find out what it means for a linear map $T : \mathcal{S} \to \mathbb{C}$ or $S : \mathcal{S} \to \mathcal{S}$ to be continuous.

The dual space of $\mathcal{S}$, *i.e.* the space of all continuous linear maps $\mathcal{S} \to \mathbb{C}$, is denoted by $\mathcal{S}' = \mathcal{S}'(\mathbb{R}^n)$ and called the space of tempered distributions. An example for an element in $\mathcal{S}'$ is the delta distribution δ given by

$$\delta u = u(0), \quad u \in \mathcal{S}.$$

We can also view certain functions as tempered distributions. For example, let f be a function which is measurable and of at most polynomial growth,

[b] We are using multi-index notation: For a multi-index $\alpha = (\alpha_1, \ldots, \alpha_n) \in \mathbb{N}_0^n$ we let $|\alpha| = \alpha_1 + \ldots \alpha_n$, $\alpha! = \alpha_1! \cdots \alpha_n!$, and for $\xi \in \mathbb{R}^n$, $\xi^\alpha = \xi_1^{\alpha_1} \cdots \xi_n^{\alpha_n}$. Moreover, $D_x^\alpha u = D_{x_1}^{\alpha_1} \ldots D_{x_n}^{\alpha_n} u$ and $D_{x_j} = -i\partial_{x_j}$.

i.e. $|f(x)| \le C(1+|x|)^m$ for some $C > 0$ and $m \in \mathbb{N}$ (this condition is stronger than necessary). Define $T_f : \mathcal{S} \to \mathbb{C}$ by

$$T_f u = \int f(x)u(x)\,dx$$

(the integral exists due to the rapid decay of u). Then T_f is a tempered distribution.

Exercise 2.2. Check that indeed $T_f \in \mathcal{S}'$ and that T_f is uniquely determined by the equivalence class of f, *i.e.*, if g is another measurable function with $T_f = T_g$, then $f = g$ a.e.. (Use a book on distribution theory!)

In view of this last fact one often identifies the tempered distribution T_f with the function f. One calls tempered distributions of the form T_f, for a suitable measurable function f, *regular*.

2.2. *The Fourier transform*

The Fourier transform is one of the most important tools in the theory of partial differential equations, due to the fact that it converts differentiation to multiplication and vice versa, see Lemma 2.5, below.

Definition 2.3. The Fourier transform of $u \in \mathcal{S}$ is the function $\mathcal{F}u$ or $\hat{u}$ on $\mathbb{R}^n$ defined by

$$\mathcal{F}u(\xi) = \hat{u}(\xi) = \int e^{ix\xi}u(x)\,\bar{d}x$$

with $\bar{d}x = (2\pi)^{-n/2}dx$.

Remark 2.4. Actually, the integral above makes sense for $u \in L^1(\mathbb{R}^n)$. An application of Lebesgue's theorem on dominated convergence then shows that $\mathcal{F}u$ is a continuous function.

A few easy-to-see facts about the Fourier transform:

Lemma 2.5. *Let* $u \in \mathcal{S}$.

(a) $\mathcal{F}(D^\alpha u)(\xi) = \xi^\alpha(\mathcal{F}u)(\xi)$.

(b) $\mathcal{F}(x^\alpha u)(\xi) = (-D_\xi^\alpha)\mathcal{F}u(\xi)$.

(c) $\mathcal{F} : \mathcal{S} \to \mathcal{S}$ *is continuous.*

As a consequence of 2.1(c), we can extend $\mathcal{F}$ to a continuous map, also denoted $\mathcal{F}$, from $\mathcal{S}'$ to $\mathcal{S}'$: We let

$$(\mathcal{F}U)(u) = U(\mathcal{F}u) \text{ for } U \in \mathcal{S}' \text{ and } u \in \mathcal{S}.$$

A more difficult result is the following theorem. See [34, Section 7] for proofs.

Theorem 2.6. *For $u \in \mathcal{S}$, we have $\mathcal{F}^2(u)(x) = u(-x)$, i.e applying $\mathcal{F}$ twice results in a reflection of the function. Moreover,*

$$\int \mathcal{F}u(\xi)\overline{\mathcal{F}v(\xi)}d\xi = \int u(x)\overline{v(x)}dx, \tag{2}$$

i.e. $\mathcal{F}$ is an isometry with respect to the $L^2(\mathbb{R}^n)$ scalar product.

As a consequence of 2.6, we have $\mathcal{F}^4 = I$ and thus $\mathcal{F}^{-1}u = \mathcal{F}^3 u = \mathcal{F}u(-\cdot)$ on $\mathcal{S}$. Hence

$$(\mathcal{F}^{-1}u)(x) = \int e^{ix\xi}u(\xi)d\xi, \quad u \in \mathcal{S}. \tag{3}$$

Moreover (2) implies

Theorem 2.7. *The Fourier transform extends to an isometric isomorphism of $L^2(\mathbb{R}^n)$.*

This is called Plancherel's theorem. For the proof one uses the fact that each function $u \in L^2(\mathbb{R}^n)$ can be approximated by a sequence (u_k) of functions in $\mathcal{S}$ (even by functions in $\mathcal{C}_c^\infty(\mathbb{R}^n)$). Theorem 2.6 implies that $(\mathcal{F}u_k)$ is a Cauchy sequence in $L^2(\mathbb{R}^n)$ and thus has a limit in $L^2(\mathbb{R}^n)$. One lets $\mathcal{F}u = \lim \mathcal{F}u_k$.

2.3. *Sobolev spaces*

These are the function spaces on which pseudodifferential operators act in a natural way. It will be convenient to work with the function

$$\mathbb{R}^n \ni \xi \mapsto \langle\xi\rangle = (1 + |\xi|^2)^{1/2}. \tag{4}$$

It is smooth, strictly positive and grows like $|\xi|$ as $\xi \to \infty$.

Definition 2.8. For $s \in \mathbb{R}$ the Sobolev space of order s on $\mathbb{R}^n$ is the set $H^s = H^s(\mathbb{R}^n)$ of all tempered distributions U whose Fourier transform $\mathcal{F}U$ is regular and satisfies

$$\|U\|_{H^s}^2 = \int |\mathcal{F}U(\xi)|^2 \langle\xi\rangle^{2s}\, d\xi < \infty.$$

Clearly, $H^0(\mathbb{R}^n) \cong L^2(\mathbb{R}^n)$ through the Fourier transform. For $s > 0$, H^s is a subset of L^2. In fact, we can describe H^s explicitly for nonnegative integer s.

Theorem 2.9. *For $s = k \in \mathbb{N}_0$ we have*

$$H^s = \{u \in L^2 : \partial^\alpha u \in L^2 \ \textit{whenever}\ |\alpha| \le k\}.$$

Recall that the derivative $\partial^\alpha u$ of a tempered distribution such as $u \in L^2$ is defined by $(\partial^\alpha u)(v) = (-1)^{|\alpha|}(\partial^\alpha v)$, $v \in \mathcal{S}$.

Theorem 2.10. ***(Sobolev embedding theorem)*** *For $s > n/2$ the space $H^s(\mathbb{R}^n)$ consists of continuous functions.*[c]

Proof. Let $U \in H^s$. Then $\mathcal{F}U \in L^1(\mathbb{R}^n)$, since

$$\int |\mathcal{F}U(\xi)|\, d\xi = \int |\mathcal{F}U(\xi)| \langle\xi\rangle^s \langle\xi\rangle^{-s}\, d\xi$$
$$\le \left(\int |\mathcal{F}U(\xi)|^2 \langle\xi\rangle^{2s}\, d\xi\right)^{1/2} \left(\int \langle\xi\rangle^{-2s}\, d\xi\right)^{1/2} = C\,\|U\|_{H^s}$$

where the constant C is the value of the square root of the second integral on the right hand side. This integral is finite, since $2s > n$.

By Remark 2.4, $\mathcal{F}^2 U$ is a continuous function. On the other hand, we know from Theorem 2.6 combined with Theorem 2.7 that $\mathcal{F}^2 U = U(-\cdot)$, so that also U is continuous. □

Corollary 2.11. *Iterating the argument in Theorem 2.9 we see that H^{s+k} consists of $\mathcal{C}^k$-functions whenever $k \in \mathbb{N}_0$ and $s > n/2$.*

Remark 2.12. Sobolev spaces can also be defined on a compact manifold M (see also Section 3.4). A distribution on M is a continuous linear map $U : \mathcal{C}^\infty(M) \to \mathbb{C}$. We can multiply U by a function φ by letting $(\varphi U)(u) = U(\varphi u)$ for $\varphi, u \in \mathcal{C}^\infty(M)$.

In order to define Sobolev spaces on M, choose a partition of unity $\{\varphi_j : j = 1, \dots, J\}$, where each member φ_j is supported in a single coordinate neighborhood. Given a distribution U on M, we say that $U \in H^s(M)$, provided $\varphi_j U$, considered in local coordinates, is an element of $H^s(\mathbb{R}^n)$, $n = \dim M$.

Theorem 2.13. *Let M be an n-dimensional compact manifold. Then*

---

[c]As a function in H^s, U is only determined up to a set of measure zero. Hence the precise statement should be that there is a continuous representative for U.

(a) $H^s(M) \hookrightarrow H^t(M)$ *for* $s \ge t$.

(b) $H^{s+k}(M) \hookrightarrow \mathcal{C}^k(M)$ *for* $s > n/2, k \in \mathbb{N}_0$.

(c) *The inclusion* $H^s(M) \hookrightarrow H^t(M)$ *is compact for* $s > t$.

While (a) is trivial and (b) is immediate from Corollary 2.11, (c) is a deeper result known as Rellich's theorem. A proof can be found in [48, Chapter 4, Proposition 3.4].

3. Pseudodifferential Operators

Pseudodifferential operators are an indispensable tool in modern analysis. Understanding the basics of this theory is worthwhile in many respects; I therefore include a short presentation. Pseudodifferential operators originated from the study of singular integral equations. Probably the first paper, where a complete calculus was developed is Kohn and Nirenberg's article [25]. Good sources to read more are the books by Hörmander [22], Kumano-go [26], Shubin [45], and Taylor [47].

3.1. *Symbols and operators*

Let $P = \sum_{|\alpha|\le m} a_\alpha(x) D^\alpha$ be a differential operator on $\mathbb{R}^n$ and $u \in \mathcal{S}$. The properties of the Fourier transform in Lemma 2.5 show that

$$\begin{aligned} Pu(x) &= \sum_{|\alpha|\le m} a_\alpha(x) D^\alpha u(x) = \sum_{|\alpha|\le m} a_\alpha(x) \mathcal{F}^{-1}(\xi^\alpha \hat{u}(\xi))(x) \\ &= \int e^{ix\xi} \sum_{|\alpha|\le m} a_\alpha(x) \xi^\alpha \hat{u}(\xi) đ\xi = \int e^{ix\xi} p(x,\xi) \hat{u}(\xi) đ\xi, \end{aligned}$$

where $p(x,\xi) = \sum_{|\alpha|\le m} a_\alpha(x)\xi^\alpha$. A first idea in the theory of pseudodifferential operators is to replace the polynomial p by more general functions, the so-called *symbols*, in order to obtain much larger classes of operators. There are many different classes of symbols. The most prominent ones are those introduced below.

In a next step one tries to understand operations with these operators like compositions, adjoints, and (approximate) inverses through associated operations on the level of symbols.

Definition 3.1. For $m \in \mathbb{R}$, $S^m = S^m(\mathbb{R}^n \times \mathbb{R}^n)$ is the space of all functions $p \in \mathcal{C}^\infty(\mathbb{R}^n \times \mathbb{R}^n)$ such that for all multi-indices α, β

$$\sup_{x,\xi} |\partial_\xi^\alpha \partial_x^\beta p(x,\xi)| \langle \xi \rangle^{|\alpha|-m} < \infty.$$

These suprema define a set of seminorms, which yield a Fréchet topology for S^m. We call the elements of S^m symbols of order m. We let $S^{-\infty} := \bigcap_m S^m$ and call these symbols smoothing or regularizing.

Of course, if $m \in \mathbb{N}_0$ and $p(x,\xi) = \sum_{|\alpha|\le m} a_\alpha(x)\xi^\alpha$ with $a_\alpha \in \mathcal{C}_b^\infty(\mathbb{R}^n)$[d], then $p \in S^m$.

Exercise 3.2.

(a) Show *Peetre's inequality*: $\langle \xi+\eta\rangle^s \le 2^{|s|/2}\langle\xi\rangle^s\langle\eta\rangle^{|s|}$, $s \in \mathbb{R}$.

(b) Show: $\partial_\xi^\alpha \langle\xi\rangle^s = O(\langle\xi\rangle^{s-|\alpha|})$. So $\langle\xi\rangle^s$ is an x-independent symbol of order s.

(c) Show: If p and q are symbols of orders m_p and m_q, respectively, then pq is a symbol of order $m_p + m_q$, and $\partial_\xi^\alpha\partial_x^\beta p$ is a symbol of order $m_p - |\alpha|$.

We call a function $h : \mathbb{R}^n \setminus \{0\} \to \mathbb{C}$ positively homogeneous of degree m, if

$$h(x,\lambda\xi) = \lambda^m h(x,\xi) \text{ for all } \lambda > 0. \tag{5}$$

Exercise 3.3. Let h be as above. Choose a function $\zeta : \mathbb{R}^n \to \mathbb{R}$ which vanishes near zero and is constant 1 outside a neighborhood of zero.[e] Then $p(\xi) = \zeta(\xi)h(\xi)$ (which is understood to be zero in $\xi = 0$) is a symbol of order m.

More generally, let $p_m(x,\xi) \in \mathcal{C}^\infty(\mathbb{R}^n_\xi \setminus \{0\}, \mathcal{C}_b^\infty(\mathbb{R}^n_x))$ (the subscripts denote with respect to which variable p_m has these properties), be positively homogeneous of degree m in ξ, and let ζ be as above. Then $\zeta p_m \in S^m$.

Definition 3.4. With a symbol p we associate an operator $\mathrm{op}(p)$, defined on $\mathcal{S}$, by

$$(\mathrm{op}(p)u)(x) = \int e^{ix\xi} p(x,\xi)\hat{u}(\xi) đ\xi.$$

We call $\mathrm{op}(p)$ the pseudodifferential operator with symbol p.

Remark 3.5. It is sometimes useful to consider also symbols $q = q(x,y,\xi)$, smooth on $\mathbb{R}^n \times \mathbb{R}^n \times \mathbb{R}^n$ and satisfying

$$\sup_{x,y,\xi} |\partial_\xi^\alpha \partial_x^\beta \partial_y^\gamma q(x,y,\xi)| \langle\xi\rangle^{|\alpha|-m} < \infty.$$

[d]This is the space of all smooth functions which are bounded together with all their derivatives.
[e]Such functions are called zero-excision functions.

They are called double symbols or amplitudes. We associate with q the pseudodifferential operator $\mathrm{op}(q)$ on $\mathcal{S}$ by

$$\mathrm{op}(q)(u)(x) = \iint e^{i(x-y)\xi} q(x,y,\xi)u(y)\,đy\,đ\xi.$$

However, this does not yield more operators. In fact, it is possible to find a symbol $p = p(x,\xi) \in S^m$ with $\mathrm{op}(p) = \mathrm{op}(q)$, see Kumano-go [26, Chapter 2].

3.2. *Continuity*

Exercise 3.6. Show that $\mathrm{op}(p) : \mathcal{S} \to \mathcal{S}$ is continuous.
Hint. Use differentiation under the integral and the identity

$$x^\alpha \,\mathrm{op}(p)u(x) = \int e^{ix\xi} D_\xi^\alpha (p(x,\xi)\hat{u}(\xi))\,đ\xi.$$

Example 3.7. $\mathrm{op}(\langle\xi\rangle^2) = \mathrm{op}(1+|\xi|^2) = I - \Delta$. Moreover, $\mathrm{op}(\langle\xi\rangle^m)\,\mathrm{op}(\langle\xi\rangle^\mu) = \mathrm{op}(\langle\xi\rangle^{m+\mu})$ for arbitrary m, μ.

The operator $\mathrm{op}(\langle\xi\rangle^m)$ defines an isometric isomorphism $H^s \to H^{s-m}$ for each s. Indeed, for $u \in H^s$ we have $\mathrm{op}(\langle\xi\rangle^m)u = \mathcal{F}^{-1}(\langle\xi\rangle^m \hat{u})$ and therefore

$$\begin{aligned}\|\mathrm{op}(\langle\xi\rangle^m)u\|^2_{H^{s-m}} &= \int |\mathcal{F}(\mathrm{op}(\langle\xi\rangle^m)u)|^2 \langle\xi\rangle^{2(s-m)}\,d\xi \\ = \int |\langle\xi\rangle^m \hat{u}(\xi)|^2 \langle\xi\rangle^{2(s-m)}\,d\xi &= \int |u(\xi)|^2 \langle\xi\rangle^{2s}\,d\xi = \|u\|^2_{H^s}.\end{aligned}$$

This shows that $\mathrm{op}(\langle\xi\rangle^m)$ is an isometry. The fact that $\mathrm{op}(\langle\xi\rangle^{-m})\,\mathrm{op}(\langle\xi\rangle^m) = I$ implies the assertion.

Theorem 3.8. *If p is a symbol of order m, then*

$$\mathrm{op}(p) : H^s \to H^{s-m}$$

is continuous for each $s \in \mathbb{R}$. The operator norm can be estimated by finitely many of the symbol seminorms.

Many proofs of this result can be found. Using the isomorphisms in Example 3.7, the task is easily reduced to the case $m = s = 0$. A simple argument is given in [23]. Once the boundedness is established on standard Sobolev spaces, it follows easily on many weighted spaces, too, see *e.g.* [37].

Corollary 3.9. *If p is a regularizing symbol, then $\mathrm{op}(p)$ maps H^s to $\bigcap_t H^t \subseteq C^\infty(\mathbb{R}^n)$ by Corollary 2.11.*

3.3. *Asymptotic expansions*

The calculus of pseudodifferential operators gains much power through another concept, namely that of asymptotic expansions.

Definition 3.10.

(a) Let $p \in S^m$ and $p_{m-j} \in S^{m-j}$. We say that p has the asymptotic expansion

$$p \sim \sum_{j=0}^{\infty} p_{m-j}, \tag{6}$$

provided that, for each N, $p - \sum_{j=0}^{N} p_{m-j} \in S^{m-N-1}$.

(b) A symbol $p \in S^m$ is *classical*, if there exists an expansion (6) with p_{m-j} satisfying $p_{m-j}(x, \lambda\xi) = \lambda^{m-j} p_{m-j}(x, \xi)$ for $|\xi|, \lambda \geq 1$. In this case we call p_m the *principal symbol* of p, provided p_m is non-zero on $\{|\xi| \geq 1\}$.

Remark 3.11. It is easy to see that an asymptotic expansion is unique modulo regularizing symbols. In particular, if all p_{m-j} can be taken to be zero, then p is regularizing.

Remark 3.12.

(a) Instead of asking homogeneity for $|\xi| \geq 1$ in Definition 3.10(b) we could have taken any other positive value.

(b) Sometimes the term 'principal symbol' is not used for p_m but for the homogeneous extension of p_m to $\mathbb{R}^n \times (\mathbb{R}^n \setminus \{0\})$. This avoids ambiguities of the definition of p_m for $|\xi| \leq 1$.

Theorem 3.13. *Let* $(p_{m-j})_{j=0}^{\infty}$ *be a sequence of symbols with* $p_{m-j} \in S^{m-j}$. *Then there exists a symbol* $p \in S^m$ *with* $p \sim \sum p_{m-j}$.

For the proof one chooses a null sequence (ε_j), $\varepsilon_j > 0$, and a zero-excision function ζ and lets $p = \sum_{j=0}^{\infty} \zeta(\varepsilon_j \xi) p_{m-j}(x, \xi)$. If the ε_j tend to zero sufficiently fast, the series converges and furnishes the desired element.

Exercise 3.14. Show that $\langle\xi\rangle^m$ is classical. Its principal symbol is $|\xi|^m$ for $|\xi| \geq 1$.
Hint: For $|\xi| > 1$, $\langle\xi\rangle^m = |\xi|^m (1 + |\xi|^{-2})^{m/2} = |\xi|^m \sum_{j=0}^{\infty} \binom{m/2}{j} |\xi|^{-2j}$.

Example 3.15. Let q and p be as in Remark 3.5. Then p can be shown to have the asymptotic expansion $p(x, \xi) \sim \sum_{\alpha} \frac{1}{\alpha!} \partial_{\xi}^{\alpha} D_y^{\alpha} q(x, y, \xi)|_{y=x}$.

According to 3.6 we can compose pseudodifferential operators as operators on $\mathcal{S}$. The following is a key result in the theory of pseudodifferential operators. A proof can be found in Kumano-go's book [26].

Theorem 3.16. *Let p and q be symbols of orders m_p and m_q, respectively. Then* $\mathrm{op}(p)\,\mathrm{op}(q) = \mathrm{op}(r)$ *for a symbol r of order $m_p + m_q$. It has the asymptotic expansion*

$$r(x,\xi) \sim \sum_\alpha \frac{1}{\alpha!}\partial_\xi^\alpha p(x,\xi) D_x^\alpha q(x,\xi).$$

Moreover, the formal L^2-adjoint[f] $\mathrm{op}(p)^*$ *of* $\mathrm{op}(p)$ *is of the form* $\mathrm{op}(p^*)$ *for a symbol $p^* \in S^m$. It has the expansion*

$$p^*(x,\xi) \sim \sum_\alpha \frac{1}{\alpha!}\partial_\xi^\alpha D_x^\alpha \overline{p(x,\xi)}.$$

If p and q are classical, then so are r and p^*, and their asymptotic expansions can be determined from those of p and q together with the asymptotic expansion formulae above.

Exercise 3.17. Let $\varphi, \psi \in \mathcal{C}_c^\infty(\mathbb{R}^n)$ have disjoint support. Show that $\varphi\,\mathrm{op}(p)\psi$ is a regularizing pseudodifferential operator. *Hint.* $\varphi\,\mathrm{op}(p) = \mathrm{op}(\varphi p)$ together with the expansion formula in Theorem 3.16 and Remark 3.11.

3.4. *Pseudodifferential operators on manifolds*

For the rest of this section let M be a compact manifold of dimension n.

We assume M to be covered by a finite number of coordinate patches (M_j, κ_j), $j = 1, \ldots, J$, where $\kappa_j : T_j \subset \mathbb{R}^n \to M_j$ are homeomorphisms and the T_j are bounded. Moreover, we choose a partition of unity φ_j, $j = 1, \ldots, J$, subordinate to this cover[g], and functions $\psi_j \in \mathcal{C}^\infty(M)$, $j = 1, \ldots, J$, with support in M_j and $\varphi_j\psi_j = \varphi_j$.

Given a function u on M we can write $u = \sum_{j=1}^J \varphi_j u$. Each of the elements $\varphi_j u$ is supported in a single coordinate neighborhood and therefore can be pulled back to a $\mathcal{C}^\infty$-function on $\mathbb{R}^n$. The same is true for $\psi_j u$. An operator $P : \mathcal{C}^\infty(M) \to \mathcal{C}^\infty(M)$ can be written

$$P = \sum_{j=1}^J \varphi_j P \psi_j + \sum_{j=1} \varphi_j P(1-\psi_j). \tag{7}$$

[f]The operator satisfying $\langle \mathrm{op}(p)^* u, v\rangle = \langle u, \mathrm{op}(p)v\rangle$ for $u, v \in \mathcal{S}$.
[g]This means that $\varphi_j \in \mathcal{C}^\infty(M)$ and $\mathrm{supp}\,\varphi_j \subseteq M_j$.

For $j = 1, \ldots, J$, $\varphi_j P \psi_j$ operates in a single coordinate neighborhood. We can therefore transfer these operators to $\mathbb{R}^n$ via the coordinate map κ_j.

Definition 3.18. Let $P : \mathcal{C}^\infty(M) \to \mathcal{C}^\infty(M)$ be a linear operator written as in (7). We say that P is a pseudodifferential operator of order m, provided that

(i) For $j = 1, \ldots, J$, the pullback of the operator $\varphi_j P \psi_j$ is a pseudodifferential operator on $\mathbb{R}^n$ with a symbol in S^m.

(ii) The operators $\varphi_j P(1-\psi_j)$ are given by integral operators with $\mathcal{C}^\infty$-kernels, *i.e.* $\varphi_j P(1-\psi_j)u(x) = \int k(x,y)u(y)dS(y)$ with $k \in \mathcal{C}^\infty(M \times M)$ and a smooth measure dS on M.

We call P classical, if the local symbols are classical.

Remark 3.19. A pseudodifferential operator P on the compact manifold M has order $-\infty$ (that is P has any order $m \in \mathbb{R}$), if and only if P is an integral operator with smooth kernel on M. Indeed, if each of the local symbols p_j is smoothing, then the identity (in local coordinates)

$$\int e^{ix\xi} p(x,\xi)\hat{u}(\xi) đ\xi = \iint e^{i(x-y)\xi} p(x,\xi)u(y) đy đ\xi$$

implies that P can be written with the integral kernel

$$k(x,y) = \int e^{i(x-y)\xi} p(x,\xi) đ\xi.$$

Conversely, an analogous formula shows that an operator with smooth kernel can be written as a pseudodifferential operator with regularizing local symbols.

On $\mathbb{R}^n$ the description of the pseudodifferential operators with regularizing symbols is more difficult as one has to take into account the decay properties at infinity.

The following result is a consequence of the definition of pseudodifferential operators on manifolds together with Theorem 3.8 and the fact that integral operators with smooth kernels map each space $H^s(M)$ to $\mathcal{C}^\infty(M) = \bigcap H^t(M)$.

Theorem 3.20. *Let $P : \mathcal{C}^\infty(M) \to \mathcal{C}^\infty(M)$ be a pseudodifferential operator of order m. Then $P : H^s(M) \to H^{s-m}(M)$ is continuous.*

Definition 3.18 associates with a pseudodifferential operator a collection of symbols, one for each coordinate neighborhood M_j.

Now we make the following observation: Let $\chi : V \to U$ be a diffeomorphism of open sets in $\mathbb{R}^n$. Then χ induces a pullback of operators: It takes an operator $Q : \mathcal{C}_c^\infty(U) \to \mathcal{C}^\infty(U)$ to $\chi^*Q : \mathcal{C}_c^\infty(V) \to \mathcal{C}^\infty(V)$ via

$$(\chi^*Q)u(x) = Q(u \circ \chi^{-1})(\chi(x)), \quad u \in \mathcal{C}_c^\infty(V), x \in V.$$

Suppose additionally that Q is a pseudodifferential operator given by the symbol $q \in S^m$. Then χ^*Q can also be written as a pseudodifferential operator with a symbol $r \in S^m$. If q is classical, so is r, and its principal symbol r_m is given by

$$r_m(x,\xi) = q_m(\chi(x), (\partial\chi(x))^{-t}\xi), \tag{8}$$

where q_m is the principal symbol of q and $(\cdot)^{-t}$ is the inverse of the transpose. In other words:

Corollary 3.21. *The local principal symbols of a classical pseudodifferential operator P of order m on M transform like a function on the cotangent bundle of M (without the zero section) and thus define a map $\sigma(P)$ on $T^*M \setminus 0$, which is homogeneous of degree m, i.e. $\sigma(P)(x,\lambda\xi) = \lambda^m\sigma(P)(x,\xi)$, $x \in M, 0 \neq \xi \in T_x^*M$.*

Theorem 3.22. *Let P_1 and P_2 be two classical pseudodifferential operators of order m on M. Suppose that $\sigma(P_1) = \sigma(P_2)$. Then $P_1 - P_2$ is of order $m-1$ and hence*

$$P_1 - P_2 : H^s(M) \to H^{s-m}(M) \text{ is compact.}$$

Proof. The equality of the principal symbols implies that $P_1 - P_2$ has order $m-1$ in each local coordinate patch and thus globally. So it maps H^s to H^{s-m+1} which is compact in H^{s-m} by Theorem 2.13(c). □

Theorem 3.23. *Let P_1 and P_2 be classical pseudodifferential operators of orders m_1 and m_2 on M. Then the composition P_1P_2 is a pseudodifferential operator of order $m_1 + m_2$, and*

$$\sigma(P_1P_2) = \sigma(P_1)\sigma(P_2).$$

Similarly, the formal L^2-adjoint P_1^* of P_1 is again a classical pseudodifferential operator of order m_1 and

$$\sigma(P_1^*) = \sigma(P_1)^*,$$

the adjoint (i.e. here the complex conjugate) of the symbol of P_1.

Proof. This follows from Theorem 3.16. □

3.5. *Vector bundles*

In general the operators we are considering will not act on scalar functions but on sections of vector bundles over M. Here are a few basic facts, see [3] for more details.

A (complex) vector bundle over M consists of a topological space E, the *total space* of the vector bundle, a continuous map $\pi : E \to M$, the so-called projection, and a finite-dimensional complex vector space structure on each of the sets $E_x = \pi^{-1}\{x\}$, $x \in M$, which is compatible with the topology on E_x induced by E. The space E_x is called the *fiber* of E over x. Moreover one requires E to be *locally trivial*; this means that for every $x \in M$ there exists a neighborhood U of x in M and a homeomorphism

$$\varphi_U : E|_U := \pi^{-1}(U) \to U \times \mathbb{C}^k \tag{9}$$

for suitable k such that φ maps E_x to $\{x\} \times \mathbb{C}^k$ as a linear map. The dimension k then is necessarily locally constant. We shall additionally assume that it is constant on M and call it the dimension of E.

Usually, one does not specify the map π and the vector space structure on the fibers and simply speaks of the vector bundle E over M.

A continuous map $\varphi : E_1 \to E_2$ between two vector bundles E_1, E_2 with projections π_1 and π_2 over M is a (vector bundle) *morphism*, if it is linear on each fiber and satisfies $\pi_2\varphi = \pi_1$. It is called an isomorphism, if it is bijective and the inverse also is a vector bundle morphism. Isomorphic vector bundles will not be distinguished.

It is clear that vector bundles can be defined over arbitrary topological spaces. We shall consider here only smooth vector bundles over M in the sense that the transition maps $\varphi_{U_2}\varphi_{U_1}^{-1}|_{(U_1\cap U_2)\times\mathbb{C}^k}$ for two trivializations of E over subsets U_1 and U_2 with nonempty intersection in the sense of (9) are smooth.

A vector bundle is *trivial*, if it is isomorphic to the bundle $M \times \mathbb{C}^k$ with the product topology and the canonical projection $(x, z) \mapsto x$, $x \in M, z \in \mathbb{C}^k$. The trivial bundle of dimension k is often denoted by $\underline{k}$.

A *section* of E is a map $s : M \to E$ which associates to $x \in M$ an element in the fiber E_x. If E is trivial over $U \subset M$ as above, then there s can be identified with a map $U \to \mathbb{C}^k$. We shall speak of continuous, smooth or H^s sections depending on whether the map s has this property and write $\mathcal{C}(M, E), \mathcal{C}^\infty(M, E)$ or $H^s(M, E)$ for the corresponding spaces of sections.

Example 3.24. (a) Apart from the trivial bundles, well-known vector bundles are the tangent bundle TM and the cotangent bundle T^*M. The associated projection maps associate to a vector or covector, respectively, its base point.

(b) Given a map $f : M_1 \to M_2$ and a vector bundle E over M_2 we obtain a vector bundle f^*E over M_1 (the *pull-back* of E via f) in the following way: The total space of f^*E is the subset of $M_1 \times E$ consisting of all pairs (x, e) such that $f(x) = \pi(e)$ with the projection $(x, e) \mapsto x$. Intuitively, we attach to x in M_1 the fiber $E_{f(x)}$ of E over $f(x) \in M_2$.

Definition 3.25. Let E_1, E_2 be smooth vector bundles over M of dimensions k_1 and k_2, respectively, and let $P : \mathcal{C}^\infty(M, E_1) \to \mathcal{C}^\infty(M, E_2)$ be a linear operator. Similarly as in Section 3.4 cover M by coordinate neighborhoods which are in addition so small that the vector bundles are trivial over them. Then decompose P as in (7). We call P a pseudodifferential operator of order m, if, in local coordinates, the operators $\varphi_j P \psi_j$ are given by $k_2 \times k_1$ matrices of pseudodifferential operators of order m, while the operators $\varphi_j P(1 - \psi_j)$ have smooth integral kernels.

Remark 3.26. Let P be as in Definition 3.25. As in Corollary 3.21 we can associate with P the symbol $\sigma(P)$, which now is a morphism

$$\sigma(P) : \pi_0^* E_1 \to \pi_0^* E_2$$

between the pull-backs of the bundles E_1 and E_2 under the projection map $\pi_0 : T^*M \setminus 0 \to M$ of the cotangent bundle with the zero section removed. (We use the subscript 0 to distinguish π_0 from the projection $\pi : T^*M \to M$.) The results of Theorem 3.23 continue to hold. Note that in order to speak of the adjoint of P, the bundles E_1 and E_2 have to be endowed with a (smooth) hermitian structure. This can always be achieved by patching locally defined hermitian forms with the help of a partition of unity. Locally, the symbol of the adjoint then is given by the adjoint of the symbol matrix.

Theorem 3.27. *We use the notation introduced in Definition 3.25. Given a smooth morphism $p : \pi_0^* E_1 \to \pi_0^* E_2$ which is homogeneous of degree m in the fibers, there exists a classical pseudodifferential operator*

$$P : \mathcal{C}^\infty(M, E_1) \to \mathcal{C}^\infty(M, E_2)$$

with $\sigma(P) = p$. Moreover, we can choose the mapping $p \mapsto P$ in such a way that the operator norm of $P : H^s(M, E_1) \to H^{s-m}(M, E_2)$ depends continuously on p.

Proof. Choose a finite covering of M by coordinate patches M_j over which the bundles E_1 and E_2 are trivial and an associated partition of unity φ_j. Denote by $\tilde{p}_j$, $\tilde{\varphi}_j$ the functions p and φ_j in local coordinates over M_j. Choose $\theta_j \in \mathcal{C}_c^\infty(\mathbb{R}^n)$, supported in the same coordinate patch with $\theta_j(x)\tilde{\varphi}_j(x) = \tilde{\varphi}_j(x)$. Moreover, choose a function η on $\mathbb{R}$ which is constant 1 outside a neighborhood of 0 and vanishes in a smaller neighborhood of 0. Then define a pseudodifferential operator $\tilde{P}_j$ of order m on $\mathbb{R}^n$ by

$$\tilde{P}_j = \operatorname{op}(\theta_j(x)\tilde{p}_j(x,\xi)\eta(|\xi|)\theta_j(y)).$$

It is classical, since p_j is homogeneous of degree m by assumption. This operator can be pulled back to M with the help of the coordinate maps. Call the resulting operator P_j. Finally let $P = \sum \varphi_j P_j$. By construction, $\sigma(P) = p$. □

3.6. *Ellipticity and parametrices*

Definition 3.28. A pseudodifferential operator $P : \mathcal{C}^\infty(M, E_1) \to \mathcal{C}^\infty(M, E_2)$ is elliptic, if the symbol $\sigma(P) : \pi^* E_1 \to \pi^* E_2$ is an isomorphism, *i.e.* if $\sigma(P)(x,\xi)$ is invertible for each (x,ξ) in $T^*M \setminus 0$.

Theorem 3.29. *Let*

$$P : \mathcal{C}^\infty(M; E_1) \to \mathcal{C}^\infty(M, E_2) \tag{10}$$

be an elliptic classical pseudodifferential operator of order m. Then there exists a pseudodifferential operator Q of order $-m$ such that $PQ - I$ and $QP - I$ are pseudodifferential operators of order -1.

Proof. The inverse $q = \sigma(P)^{-1}$ is a smooth morphism $\pi_0^* E_2 \to \pi_0^* E_1$ which is homogeneous of degree $-m$. According to Theorem 3.27 we find a classical pseudodifferential operator Q of order $-m$ with $\sigma(Q) = q$. By Theorem 3.23

$$\sigma(PQ) = \sigma(P)\sigma(Q) = I \text{ and } \sigma(QP) = \sigma(Q)\sigma(P) = I.$$

Hence $PQ - I$ and $QP - I$ have order -1 by Theorem 3.22. □

Remark 3.30. Iterating the above process and using asymptotic summation for symbols, see Theorem 3.13, one can actually find a classical pseudodifferential operator $\tilde{Q}$ of order $-m$ such that $\tilde{Q}P - I$ and $P\tilde{Q} - I$ are regularizing operators.

3.7. *Index theory for pseudodifferential operators*

Theorem 3.31. *Let P be as in Theorem 3.29. Then*

$$P : H^s(M, E_1) \to H^{s-m}(M, E_2)$$

is a Fredholm operator for every $s \in \mathbb{R}$.

Proof. Choose Q as in Theorem 3.29. By Theorem 3.8 it defines a bounded operator $H^{s-m}(M, E_2) \to H^s(M, E_1)$. As $PQ - I$ and $QP - I$ have order -1, they are compact. Hence P is a Fredholm operator and Q is a Fredholm inverse to P. □

Theorem 3.32. ***(Elliptic regularity)**. Given P as in Theorem 3.29 and f in $H^s(M, E_2)$, let u be a solution of the equation $Pu = f$ with u in some Sobolev space $H^{-N}(M, E_1)$, $N \in \mathbb{R}$. Then $u \in H^{s+m}(M, E_1)$.*

Proof. Let Q be as in Theorem 3.29 and $R = QP - I$. Then R is of order -1 and

$$H^{s+m}(M, E_1) \ni Qf = QPu = (I + R)u = u + Ru.$$

In particular, $u \in H^{s+m} + H^{-N+1}$. If $-N + 1 < s + m$ the argument can be iterated until we reach the assertion. □

Corollary 3.33. *If $Pu = 0$, then $u \in \cap H^s$. In other words: The kernel of an elliptic pseudodifferential operator consists of smooth sections.*

Another important fact:

Theorem 3.34. *Let E be a smooth vector bundle over M. For each choice of s and μ there exists a pseudodifferential operator Λ^μ of order μ such that $\Lambda^\mu : H^s(M, E) \to H^{s-\mu}(M, E)$ is an isomorphism.*

Proof. Let $c > 0$. In each local coordinate neighborhood choose the pseudodifferential operator with the symbol $\langle \xi, c \rangle^\mu I_E$, where $\langle \xi, c \rangle = (1 + |\xi|^2 + c^2)^{1/2}$. With the help of a partition of unity patch these operators to a pseudodifferential operator $L_c^{(\mu)}$ of order μ on M. Similarly, we obtain a pseudodifferential operator $L_c^{(-\mu)}$ of order $-\mu$ by starting from the local symbols $\langle \xi, c \rangle^{-\mu} I_E$. Using Theorem 3.16 one can show that $L_c^{(-\mu)} L_c^{(\mu)} = I + R_1(c)$ and $L_c^{(\mu)} L_c^{(-\mu)} = I + R_2(c)$, where the symbol seminorms for $R_1(c)$ and $R_2(c)$ tend to zero as $c \to \infty$. By taking c large, $L_c^{(\mu)}$ will be thus be invertible as an operator $H^s(M, E) \to H^{s-\mu}(M, E)$ and can be taken for Λ^μ. □

Remark 3.35. In fact, if Λ^μ is invertible as an operator $H^{s_0}(M,E) \to H^{s_0-\mu}(M,E)$ for some s_0, then it will be invertible for all $s \in \mathbb{R}$, see Beals [8].

Theorem 3.36. *Let $P : \mathcal{C}^\infty(M,E_1) \to \mathcal{C}^\infty(M,E_2)$ be a classical elliptic pseudodifferential operator of order m. Then the index of $P : H^s(M,E_1) \to H^{s-m}(M,E_2)$ is independent of s and so are the dimensions of the kernel and the complement of the image of P.*

Moreover, the index only depends on the stable homotopy class of the principal symbol within the class of invertible symbols.

Remark 3.37. We call two symbols $\sigma_1, \sigma_2 : \pi_0^* E_1 \to \pi_0^* E_2$ of pseudodifferential operators P_1 and P_2 of order m stably homotopic, if there exists a bundle F over M and a bundle homomorphism λ of $\pi_0^* F$, which is homogeneous of degree m in ξ and a multiple of the identity on S^*M, such that the maps $\sigma_1 \oplus \lambda, \sigma_2 \oplus \lambda : \pi_0^*(E_1 \oplus F) \to \pi_0^*(E_2 \oplus F)$ are homotopic. By the general theory of vector bundles it is no restriction to assume that F is trivial. So we can view $\sigma_1 \oplus \lambda$ as the symbol of $P_1 \oplus \Lambda$, where Λ is an invertible pseudodifferential operator of order m acting on sections of $\underline{k}$ over M with principal symbol λ. The same applies to P_2. Clearly, this change does not affect the index.

Proof of Theorem 3.36. Choose s_1 and s_2 in $\mathbb{R}$, and denote for the moment by P_{s_1} and P_{s_2} the extensions of P to H^{s_1} and H^{s_2} respectively. Let $\mu = s_2 - s_1$ and choose Λ^μ and $\Lambda^{-\mu}$ as in the proof of Theorem 3.34 with mutually inverse principal symbols. As both are isomorphisms, the operator $\Lambda^{-\mu} P_{s_1} \Lambda^\mu$ is an operator on H^{s_2} with the same index as P_{s_1} and the same principal symbol as P_{s_2}, namely $\sigma(P)$. Hence both have the same index by Theorem 3.22.

In view of the fact that the kernel is a subset of $\mathcal{C}^\infty$, its dimension is constant. Since the index is constant, also the dimension of the complement will be constant.

Suppose that $\tilde{P}$ is another classical elliptic pseudodifferential operator and that the symbols $\sigma(P)$ and $\sigma(\tilde{P})$ are homotopic after possibly adding a trivial bundle as explained in Remark 3.37. Let $\sigma_t, 0 \le t \le 1$, be a homotopy through invertible symbols with $\sigma_0 = \sigma(P)$, $\sigma_1 = \sigma(\tilde{P})$. According to Theorem 3.27 we find a continuous family of operators P_t with $\sigma(P_t) = \sigma_t$. In view of the stability of the index in Proposition 1.4, the indices of P_0 and P_1 coincide. Since P_0 and P_1 have the same principal symbols as P and $\tilde{P}$, respectively, we see that ind $P =$ ind $\tilde{P}$. □

Remark 3.38. We have noticed that the kernel of P in Theorem 3.36 consists of smooth functions. In a similar way one can show that there is a space of sections in $\mathcal{C}^\infty(M, E_2)$ which complements the range of P on any space $H^s(M, E_2)$, $s \in \mathbb{R}$. Hence the index of P on any of the Sobolev spaces coincides with that as an operator $\mathcal{C}^\infty(M, E_1) \to \mathcal{C}^\infty(M; E_2)$.

Another observation which is often useful:

Remark 3.39. The existence of order reducing operators allows us to focus on the index theory of elliptic pseudodifferential operators of order zero. This has the additional advantage that the principal symbol is homogenous of degree zero and therefore is in a natural way a function on the cosphere bundle S^*M.

3.8. *The solution of the index problem by Atiyah and Singer*

The results of the previous section leave us with the task of determining the index of an elliptic pseudodifferential operator P as in (10) from the stable homotopy class of

$$\sigma(P) : \pi^* E_1 \to \pi^* E_2.$$

This problem has been solved by Atiyah and Singer in their groundbreaking work. Their crucial idea was to associate to the symbol $\sigma(P)$ a class in $[\sigma(P)] \in K_c(T^*M)$, the K-theory with compact supports on T^*M. Let us first define this:

Definition 3.40. Let Ω be a manifold, not necessarily compact. In order to define $K_c(\Omega)$ we start by considering triples of the form $V = (F_1, F_2, \varphi)$ where F_1 and F_2 are vector bundles over Ω and $\varphi : F_1 \to F_2$ is a vector bundle morphism which is an isomorphism outside a compact subset of Ω.

We actually only have to assume that φ be defined outside a compact set. If Ω is compact, we can take $\varphi = 0$, so that the condition is void.

The triple V is said to be trivial, if φ can be chosen to be an isomorphism everywhere. Two triples $V = (F_1, F_2, \varphi)$ and $W = (G_1, G_2, \psi)$ are *isomorphic*, if there are isomorphisms $f_1 : F_1 \to G_1$ and $f_2 : F_2 \to G_2$ such that $f_2\varphi = \psi f_1$ outside a compact set. The *sum* $U = V \oplus W$ of the triples V and W is the triple $U = (F_1 \oplus G_1, F_2 \oplus G_2, \varphi \oplus \psi)$. The triples V and W are *equivalent*, if there is a third triple Z such that $V \oplus Z \cong W \oplus Z$.

$K_c(\Omega)$, the set of equivalence classes of triples, then is an abelian semigroup. It becomes a group, if we define the inverse of (F_1, F_2, φ) to be the triple (F_2, F_1, φ^{-1}).

In the specific case of an elliptic pseudodifferential operator P as in (10) we can define the class $[\sigma(P)] = [(\pi^* E_1, \pi^* E_2, \sigma(P))] \in K_c(T^* M)$. Now an important observation is:

Theorem 3.41. *There is a group homomorphism* $\chi_{top} : K_c(T^*M) \to \mathbb{Z}$, *called the topological index map.*

This gives us two ways of associating an integer to P: We can associate with P the Fredholm index – in any of the situations studied in Theorem 3.36 or Remark 3.38 – and we can associate with it the topological index of $[\sigma(P)]$.

The Atiyah-Singer index theorem [5] states:

Theorem 3.42. $\operatorname{ind} P = \chi_{top}[\sigma(P)]$, *i.e. the Fredholm index of* P *coincides with the topological index of* $[\sigma(P)]$.

The right hand side can also be expressed in cohomological terms using characteristic classes, see [4, Theorem 1], which we will not treat here.

Theorem 3.43. $\operatorname{ind} P = \int_{T^*M} \operatorname{ch}([\sigma(P)]) \operatorname{Td}(M)$, *where* ch *denotes the Chern character and* Td *the Todd class.*

4. Boundary Value Problems and Boutet de Monvel's Calculus

4.1. *Classical boundary value problems*

Definition 4.1. A classical boundary value problem consists of

(1) a differential operator P of order m, acting between sections of vector bundles E_1 and E_2 over X, where X is a bounded domain in $\mathbb{R}^n$ or an n-dimensional compact manifold with boundary,[h]

$$P : \mathcal{C}^\infty(X, E_1) \to \mathcal{C}^\infty(X, E_2)$$

and

[h]There is a slight inconsistency here in that the domain X does not contain its boundary and thus is not a manifold with boundary. As tradition has the set-ups this way, the reader is kindly requested to think therefore of the closure of the domain while the notation remains unchanged.

(2) a boundary operator, mapping sections of the vector bundle E_1 to those of a vector bundle F_2 over the boundary

$$T : \mathcal{C}^\infty(X, E_1) \to \mathcal{C}^\infty(\partial X, F_2)$$

of the form

$$Tu(x') = \sum_{j=0}^{m-1} B_j(x', D_{x'})\gamma_j u(x') = \gamma_0\Big(\sum_{j=0}^{m-1} B_j(x', D_{x'})\partial_\nu^j u\Big)(x'). \quad (1)$$

Here $x' \in \partial X$ and $\gamma_j u = \partial_\nu^j u|_{\partial X}$ is the evaluation of the j-th derivative in the interior normal direction at the boundary. The B_j are differential operators along the boundary. One generally assumes that the order of B_j is smaller than $m - j$.

We thus obtain an operator

$$\begin{pmatrix} P \\ T \end{pmatrix} : \mathcal{C}^\infty(X, E_1) \to \begin{matrix} \mathcal{C}^\infty(X, E_2) \\ \oplus \\ \mathcal{C}^\infty(\partial X, F_2) \end{matrix} . \qquad (2)$$

Example 4.2. The most prominent examples of boundary value problems are the Dirichlet and the Neumann problem for the Laplacian:

$$\begin{pmatrix} \Delta \\ \gamma_0 \end{pmatrix} : \mathcal{C}^\infty(X) \to \begin{matrix} \mathcal{C}^\infty(X) \\ \oplus \\ \mathcal{C}^\infty(\partial X) \end{matrix} \quad \text{and} \quad \begin{pmatrix} \Delta \\ \gamma_1 \end{pmatrix} : \mathcal{C}^\infty(X) \to \begin{matrix} \mathcal{C}^\infty(X) \\ \oplus \\ \mathcal{C}^\infty(\partial X) \end{matrix}$$

for a bounded domain X. It is well-known that the Dirichlet problem is uniquely solvable, while the Neumann problem on a connected domain with smooth boundary has a one-dimensional kernel, consisting of the constant functions, and one-codimensional range, see *e.g.* [15, Theorem 3.40]. In particular, both are Fredholm operators of order zero.

We next introduce the corresponding scales of Sobolev spaces. Recall that there is a compact n-dimensional manifold $\widetilde{X}$ such that X is embedded as an n-dimensional manifold. By r^+ we denote the operator which restricts distributions on $\widetilde{X}$ to the interior of X. We then let

$$H^s(X) = \{r^+u : u \in H^s(\widetilde{X})\}$$

and write $H^s(X, E)$ for the space of H^s-sections of the vector bundle E.

We recall that restriction to the boundary γ_0 defines a bounded map

$$\gamma_0 : H^s(X) \to H^{s-1/2}(\partial X), \quad s > -1/2; \qquad (3)$$

for a proof see *e.g.* [48, Chapter 4, Proposition 1.6].

4.2. *The condition of Lopatinskij and Shapiro*

When does a boundary value problem define a Fredholm operator between suitable Sobolev spaces? The decisive tool is a condition introduced by Lopatinskij and Shapiro. For the sake of an easier presentation we state it for a simplified situation.

Let X be a smoothly bounded domain in $\mathbb{R}^n$, $n \geq 2$. In a neighborhood of the boundary ∂X introduce local coordinates (x', x_n), where $x' \in \partial X$ and $x_n > 0$ is the distance to the boundary. Moreover let P be a (scalar-valued) differential operator of order 2μ, $\mu \in \mathbb{N}$, with smooth coefficients and let $T_1, \dots T_\mu$ be μ (scalar-valued) boundary operators of orders $m_k < 2\mu$, $k = 1, \dots, \mu$, of the form

$$T_k u(x') = \sum_{j=0}^{m_k} B_{jk}(x', D_{x'})\gamma_j u(x')$$

with differential operators B_{jk} of order $m_k - j$. By (3) we have

Lemma 4.3. *A boundary problem of the described form induces a bounded operator*

$$\begin{pmatrix} P \\ T_1 \\ \vdots \\ T_\mu \end{pmatrix} : H^{s+2\mu}(X) \to \begin{matrix} H^s(X) \\ \oplus \\ H^{s+2\mu-m_1-1/2}(\partial X) \\ \oplus \\ \vdots \\ H^{s+2\mu-m_\mu-1/2}(\partial X) \end{matrix} \tag{4}$$

for every $s > -1/2$.

We need more notation to formulate the ellipticity conditions: By $p^0 = \sigma(P)|_{T^*X\setminus 0}$ denote the principal symbol of P. We associate with it the differential operator

$$p^0(x', 0, \xi', D_n) : H^{2\mu}(\mathbb{R}_+) \to L^2(\mathbb{R}_+).$$

Here, we identify $\mathbb{R}_+$ with the interior normal. In a similar spirit we associate with T_k the operator

$$t_k^0(x', \xi', D_n) = \sum_{j=0}^{\mu-1} \sigma(B_{jk})(x', \xi') i^j \gamma_j : H^{2\mu}(\mathbb{R}_+) \to \mathbb{C}.$$

Definition 4.4. The classical boundary value problem $(P, T_1, \dots, T_\mu)$ is elliptic in the sense of Lopatinskij and Shapiro, if the following two conditions are fulfilled:

(1) $\sigma(P)(x,\xi)$ is invertible for $(x,\xi) \in T^*\overline{X} \setminus 0$.
(2) the so-called boundary symbol operator

$$\begin{pmatrix} p^0(x',0,\xi',D_n) \\ t_1^0(x',\xi',D_n) \\ \vdots \\ t_\mu^0(x',\xi',D_n) \end{pmatrix} : H^{2\mu}(\mathbb{R}_+) \to \begin{matrix} L^2(\mathbb{R}_+) \\ \oplus \\ \mathbb{C}^k \end{matrix} . \tag{5}$$

is invertible for $(x',\xi') \in T^*\partial X \setminus 0$.

Remark 4.5. This is less difficult than it might seem at first glance. What one has to establish is the unique solvability of an initial value problem for system of constant coefficient ordinary differential equations.

Theorem 4.6. *For any $s \geq 0$, the boundary value problem* (4) *is a Fredholm operator if and only if it is elliptic in the sense of Lopatinskij and Shapiro. Both the dimension of the kernel and the cokernel are independent of s.*

Example 4.7. Consider the Dirichlet problem for the Laplacian on a domain X. Then $p^0(x',0,\xi',D_n) = -|\xi'|^2 - D_n^2 = -|\xi'|^2 + \partial_{x_n}^2$ and $t^0(x',\xi',D_n) = \gamma_0$.

A fundamental system for p^0 is given by the functions $e^{\pm|\xi'|x_n}$. Only the solution with the minus-sign actually maps $H^2(\mathbb{R}_+)$ to $L^2(\mathbb{R}_+)$. Injectivity of the operator

$$\begin{pmatrix} \partial_{x_n}^2 - |\xi'|^2 \\ \gamma_0 \end{pmatrix} : H^2(\mathbb{R}_+) \to \begin{matrix} L^2(\mathbb{R}_+) \\ \oplus \\ \mathbb{C} \end{matrix}$$

then is obvious. In order to establish surjectivity use Wronski's theorem.

4.3. *Solving the Dirichlet problem*

As a preparation for the introduction of Boutet de Monvel's calculus for boundary value problems recall how the Dirichlet problem for the Laplacian on a bounded domain X in $\mathbb{R}^n$

$$\Delta u = f \text{ in } X, \qquad \gamma_0 u = g \text{ on } \partial X \tag{6}$$

is solved: It is sufficient to study the two semi-homogeneous problems

$$\begin{aligned} &\Delta u_1 = f \text{ in } X, \ \gamma_0 u_1 = 0 \text{ on } \partial X, \text{ and} \\ &\Delta u_2 = 0 \text{ in } X, \ \gamma_0 u_2 = g \text{ on } \partial X, \end{aligned}$$

for then $u_1 + u_2$ furnishes the solution to (6).

The function u_1 is obtained as $u_1 = \int_X \Gamma(x,y)f(y)\,dy$, where Γ is Green's function. We also recall that $\Gamma(x,y) = N(x-y) + C(x,y)$, where N is the Newton potential and a C a correction term, which ensures that the zero boundary value is attained (more on this in Folland [15]).

The Newton potential is the fundamental solution for the Laplacian, *i.e.* it satisfies $\Delta N = \delta_0$. Taking the Fourier transform, we find that $-|\xi|^2\hat{N}(\xi) = (2\pi)^{-n/2}$, hence $\hat{N}(\xi) = -(2\pi)^{-n/2}/|\xi|^2$. As N acts by $f \mapsto N * f = \mathcal{F}^{-1}\hat{N}\hat{f}$, it can be seen as the pseudodifferential operator with the 'symbol' $\hat{N}$. Of course, $\hat{N}$ does not quite qualify for a symbol, since it is not smooth at $\xi = 0$. However, replacing $\hat{N}$ by $\zeta\hat{N}$, where ζ is a zero excision function, cf. 3.3, we obtain a true pseudodifferential operator, say P, which is a parametrix to the Laplacian on $\mathbb{R}^n$, while the error is an integral operator with a smooth kernel.

There is one more detail to be noted here: P acts on functions on $\mathbb{R}^n$, while f is only defined over X. Moreover, we are only interested in the values of the resulting function over X. Hence we actually consider the operator P_+ given by

$$P_+ = r^+Pe^+,$$

where e^+ denotes extension by zero from X to $\mathbb{R}^n$ and r^+ denotes restriction from $\mathbb{R}^n$ to X.

The correction term appearing in Green's function is a function $C = C(x,y)$ on $X \times X$ such that $C(\cdot,y)$ is harmonic (hence in particular smooth) in X and continuous up to the boundary. Moreover, $N(x-y)+C(x,y) = 0$ for $x \in X$ and $y \in \partial X$. Therefore, the operator induced by Green's function is the sum of two terms:

$$P_+ + G,$$

where P is as above, while G is an integral operator with a kernel function which is smooth over the interior, but not up to the boundary. In fact, it can be shown that G can be viewed as an operator-valued pseudodifferential operator along the boundary, taking values in smoothing operators in the normal direction.

This point of view is based on Schulze's concept of operator-valued pseudodifferential operators, see *e.g.* [43]; more information will be given in the next subsection.

One eventually finds u_2 in (6) as $u_2 = Kg$, where K is the so-called

Poisson operator. It is an integral operator with the kernel function

$$k = k(x', y') = \partial_{\nu_{y'}} \Gamma(x', y'), \; x', y' \in \partial X,$$

i.e. the normal derivative of Green's function with respect to the second variable. The Poisson operator can be seen as an operator-valued pseudodifferential operator along the boundary which acts on functions on the normal to the boundary via multiplication by smooth functions.

Collecting this information we see that the inverse to the operator

$$\begin{pmatrix} \Delta \\ \gamma_0 \end{pmatrix} : \mathcal{C}^\infty(X) \to \begin{matrix} \mathcal{C}^\infty(X) \\ \oplus \\ \mathcal{C}^\infty(\partial X) \end{matrix}$$

can be written in the form

$$\begin{pmatrix} \Delta \\ \gamma_0 \end{pmatrix}^{-1} = (P_+ + G \quad K) : \begin{matrix} \mathcal{C}^\infty(X) \\ \oplus \\ \mathcal{C}^\infty(\partial X) \end{matrix} \to \mathcal{C}^\infty(X).$$

4.4. *Boutet de Monvel's calculus*

In 1971 Boutet de Monvel developed a pseudodifferential calculus on manifolds with boundary which encompasses the classical boundary value problems [10]. It includes a notion of ellipticity extending the condition of Lopatinskij and Shapiro. Ellipticity allows the parametrix construction within the calculus. Moreover, the calculus contains the inverses to invertible elements. Detailed expositions of Boutet de Monvel's calculus have been given in the monographs by Rempel and Schulze [32] or Grubb [18]. We will follow here the short introduction [39].

Let X be a compact, n-dimensional manifold with boundary. We assume that X is embedded in a compact manifold $\widetilde{X}$ of the same dimension.

An operator in Boutet de Monvel's calculus is a matrix

$$A = \begin{pmatrix} P_+ + G & K \\ T & S \end{pmatrix} : \begin{matrix} \mathcal{C}^\infty(X, E_1) \\ \oplus \\ \mathcal{C}^\infty(\partial X, F_1) \end{matrix} \to \begin{matrix} \mathcal{C}^\infty(X, E_2) \\ \oplus \\ \mathcal{C}^\infty(\partial X, F_2) \end{matrix} \tag{7}$$

acting on sections of vector bundles E_1, E_2 over X and F_1, F_2 over ∂X. Let us introduce these operators briefly; details will follow, below.

P is a pseudodifferential operator on $\widetilde{X}$, satisfying the transmission condition at ∂X (see below for details). As before, $P_+ = r^+ P e^+$ with the operators e^+ of extension by zero and r^+ of restriction to X°;

G is a singular Green operator,
T is a trace operator,
K is a potential (or Poisson) operator, and
S is a pseudodifferential operator on ∂X.

All these operators are assumed to be classical; *i.e.* their symbols have polyhomogeneous expansions in the respective classes.

The transmission condition is a condition on the homogenous terms in the local symbols of P. Assume that P is of order m and, in a given coordinate system, has the symbol p with the asymptotic expansion $p \sim \sum_{j=0}^{\infty} p_{m-j}$, where $p_{m-j}(x,\xi)$ is homogeneous of degree $m-j$ for $|\xi| \geq 1$. As in Section 4.2 introduce coordinates (x', x_n) near the boundary with corresponding covariables (ξ', ξ_n). The transmission condition requires that

$$\partial_x^\beta \partial_\xi^\alpha p_{m-j}(x', 0, 0, 1) = (-1)^{m-j-|\alpha|} \partial_x^\beta \partial_\xi^\alpha p_{m-j}(x', 0, 0, -1). \qquad (8)$$

It ensures that the truncated operator P_+ indeed maps functions, which are smooth up to the boundary, to functions which are again smooth up to the boundary.

Operator-valued pseudodifferential operators.

Given the usual Sobolev space $H^s(\mathbb{R})$, $s \in \mathbb{R}$, we denote by $H^{s_1,s_2}(\mathbb{R}^n)$ the weighted space $\langle x\rangle^{-s_2} H^{s_1}(\mathbb{R})$ (for $s_2 = 0$ we recover the standard spaces). We often write $\mathbf{s} = (s_1, s_2)$. Moreover, we let $H^{\mathbf{s}}(\mathbb{R}_+) = \{u|_{\mathbb{R}_+} : u \in H^{\mathbf{s}}(\mathbb{R})\}$ and $H_0^{\mathbf{s}}(\overline{\mathbb{R}}_+) = \{u \in H^{\mathbf{s}}(\mathbb{R}) : \operatorname{supp} u \subseteq \overline{\mathbb{R}}_+\}$. On these spaces we have for $s_1 \geq 0$ the strongly continuous group action

$$\mathbb{R}_+ \ni \lambda \mapsto \kappa_\lambda \text{ given by } (\kappa_\lambda u)(t) = \sqrt{\lambda}\, u(\lambda t), \quad u \in H^{\mathbf{s}}(\mathbb{R});$$

similarly for the other spaces and with the corresponding modifications for non-regular distributions.

On the space $\mathbb{C}$ we use the trivial group action $\kappa_\lambda c = c$, $c \in \mathbb{C}$.

Definition 4.8. Let E and F be Banach spaces (*e.g.* any of the spaces $H^{\mathbf{s}}(\mathbb{R})$ or $H^{\mathbf{s}}(\mathbb{R}_+)$ or $H_0^{\mathbf{s}}(\overline{\mathbb{R}}_+)$ or $\mathbb{C}$) with strongly continuous group actions κ_λ^E and κ_λ^F, respectively.

A smooth function $a = a(y,\eta)$, $y, \eta \in \mathbb{R}^q$, with values in $\mathcal{L}(E,F)$ is called an operator-valued pseudodifferential operator of order m, provided that, for all multi-indices α, β,

$$\|\kappa^F_{\langle\eta\rangle^{-1}} D_\eta^\alpha D_y^\beta a(y,\eta) \kappa^E_{\langle\eta\rangle}\|_{\mathcal{L}(E,F)} = \mathcal{O}(\langle\eta\rangle^{m-|\alpha|}).$$

We write $a \in S^m(\mathbb{R}^q \times \mathbb{R}^q; E, F)$ and let $S^{-\infty}(\ldots) = \bigcap_{m\in\mathbb{R}} S^m(\ldots)$.

To a we associate the operator $\operatorname{op}(a) : \mathcal{S}(\mathbb{R}^q, E) \to \mathcal{S}(\mathbb{R}^q, F)$ by

$$\operatorname{op}(a)u(y) = \int e^{iy\eta} a(y,\eta)\hat{u}(\eta) đ\eta, \quad u \in \mathcal{S}(\mathbb{R}^q, E), y \in \mathbb{R}^q.$$

This yields the usual results on compositions, adjoints and asymptotic expansions, including the formulae for the asymptotic expansion of the symbols.

We will use this concept in order to express the operators in Boutet de Monvel's calculus as operator-valued pseudodifferential operators along the boundary, acting on functions or distributions in the normal direction. Thus q will in general be $n-1$. An important point here is that the usual pseudodifferential operators also fit into this concept. This is the contents of the following exercise.

Exercise 4.9. Let p be a pseudodifferential symbol of order m on $\mathbb{R}^n$. Fix $x', \xi' \in \mathbb{R}^{n-1}$ and let $p_{x',\xi'}(x_n, \xi_n) = p(x', x_n, \xi', \xi_n)$. Define the operator

$$\operatorname{op}_n(p)(x', \xi') : \mathcal{S}(\mathbb{R}) \to \mathcal{S}(\mathbb{R})$$

by

$$(\operatorname{op}_n(p)(x', \xi'))\, u(x_n) = \int e^{ix_n\xi_n} p_{x',\xi'}(x_n, \xi_n)\hat{u}(\xi_n) đ\xi_n.$$

Show: $(x', \xi') \mapsto \operatorname{op}_n(p)(x', \xi')$ is an element of $S^m(\mathbb{R}^{n-1} \times \mathbb{R}^{n-1}; H^s(\mathbb{R}), H^{s-m}(\mathbb{R}))$ for every $s \in \mathbb{R}$.

Remark 4.10. Let $\mathcal{S}(\mathbb{R}_+) = \{r^+u : u \in \mathcal{S}(\mathbb{R}_+)\}$. It follows more or less from the definition that

$$\mathcal{S}(\mathbb{R}_+) = \operatorname*{proj\,lim}_{\mathbf{s}\in\mathbb{R}^2} H^{\mathbf{s}}(\mathbb{R}_+).$$

The dual space $\mathcal{S}'(\mathbb{R}_+)$ then can be shown to be the inductive limit of the spaces $H_0^{\mathbf{s}}(\overline{\mathbb{R}}_+)$. Hence it makes sense to let, with $\mathbf{s}, \mathbf{t}$ ranging over $\mathbb{R}^2$,

$$\begin{aligned} S^m(\mathbb{R}^q \times \mathbb{R}^q, \mathbb{C}, \mathcal{S}(\mathbb{R}_+)) &= \operatorname*{proj\,lim}_{\mathbf{s}} S^m(\mathbb{R}^q \times \mathbb{R}^q, \mathbb{C}, H^{\mathbf{s}}(\mathbb{R}_+)) \\ S^m(\mathbb{R}^q \times \mathbb{R}^q, \mathcal{S}'(\mathbb{R}_+), \mathbb{C}) &= \operatorname*{proj\,lim}_{\mathbf{s}} S^m(\mathbb{R}^q \times \mathbb{R}^q, H_0^{\mathbf{s}}(\overline{\mathbb{R}}_+), \mathbb{C}) \text{ and} \\ S^m(\mathbb{R}^q \times \mathbb{R}^q, \mathcal{S}'(\mathbb{R}_+), \mathcal{S}(\mathbb{R}_+)) &= \operatorname*{proj\,lim}_{\mathbf{s},\mathbf{t}} S^m(\mathbb{R}^q \times \mathbb{R}^q, H_0^{\mathbf{s}}(\overline{\mathbb{R}}_+), H^{\mathbf{t}}(\mathbb{R}_+)). \end{aligned}$$

The operators G, T and K in (7) can now be defined within this framework. Note that G and T not only have an order, but also a class $d \in \mathbb{N}_0$.

Sometimes, *e.g.* in the work of Rempel and Schulze, the notion 'class' is replaced by 'type'.

Definition 4.11. Let $m \in \mathbb{R}$ and $d \in \mathbb{N}_0$.

(a) A singular Green operator (sGo) of order m and class 0 on $\mathbb{R}^n_+$ is an operator $G : \mathcal{S}(\mathbb{R}^n_+) \to \mathcal{S}(\mathbb{R}^n_+)$ of the form

$$G = \operatorname{op} g \text{ with } g \in S^m(\mathbb{R}^q \times \mathbb{R}^q, \mathcal{S}'(\mathbb{R}_+), \mathcal{S}(\mathbb{R}_+)).$$

An sGo of order m and class d is of the form $G = \sum G_j \partial_+^j$, where $\partial_+ : \mathcal{S}(\mathbb{R}^n_+) \to \mathcal{S}(\mathbb{R}^n_+)$ is the truncated normal derivative $r^+\partial_{x_n} e^+$ on $\mathbb{R}^n_+$ and the G_j are sGo's of order $m - j$ and class 0.

(b) A trace operator of order m and class 0 is an operator $T : \mathcal{S}(\mathbb{R}^n_+) \to \mathcal{S}(\mathbb{R}^{n-1})$ given by

$$T = \operatorname{op} t \text{ with } t \in S^m(\mathbb{R}^q \times \mathbb{R}^q, \mathcal{S}'(\mathbb{R}_+), \mathbb{C}).$$

A trace operator of order m and class d is of the form $T = \sum T_j \partial_+^j$ with trace operators T_j of order $m - j$ and class 0.

(c) A potential operator of order m is an operator $K : \mathcal{S}(\mathbb{R}^{n-1}) \to \mathcal{S}(\mathbb{R}^n_+)$ with

$$K = \operatorname{op} k \text{ with } k \in S^m(\mathbb{R}^q \times \mathbb{R}^q, \mathbb{C}, \mathcal{S}(\mathbb{R}_+)).$$

Exercise 4.12. For classical boundary value problems the operator γ_j, *i.e.* the restriction of the j-th normal derivative to the boundary plays an important role. Show that γ_j is a trace operator of order $j + 1/2$ and class $j + 1$ in the above sense.

Remark 4.13. Fix $x', \xi' \in \mathbb{R}^{n-1}$. A singular Green symbol $g = g(x', \xi')$ of class zero can be shown to act on, say, $L^2(\mathbb{R}_+)$ as an integral operator with a kernel $\tilde{g}(x', \xi'; \cdot, \cdot)$ in $\mathcal{S}(\mathbb{R}_+ \times \mathbb{R}_+)$:

$$g(x', \xi')u(x_n) = \int \tilde{g}(x', \xi'; x_n, y_n)u(y_n)\, dy_n, \quad u \in L^2(\mathbb{R}_+).$$

A trace symbol $t = t(x', \xi')$ of class zero acts by integration against a kernel $\tilde{t}(x', \xi'; \cdot) \in \mathcal{S}(\mathbb{R}_+)$:

$$t(x', \xi')u = \int \tilde{t}(x', \xi'; y_n)u(y_n)\, dy_n, \quad u \in L^2(\mathbb{R}_+).$$

A potential symbol $k = k(x', \xi')$ acts via multiplication by a function $\tilde{k}(x', \xi'; \cdot) \in \mathcal{S}(\mathbb{R}_+)$:

$$(k(x', \xi')c)(x_n) = \tilde{k}(x', \xi'; x_n)c, \quad c \in \mathbb{C}.$$

Definition 4.14. Let

$$A = \begin{pmatrix} P_+ + G & K \\ T & S \end{pmatrix} : \begin{matrix} \mathcal{C}^\infty(\overline{\mathbb{R}}^n_+) \\ \oplus \\ \mathcal{C}^\infty(\mathbb{R}^{n-1}) \end{matrix} \to \begin{matrix} \mathcal{C}^\infty(\overline{\mathbb{R}}^n_+) \\ \oplus \\ \mathcal{C}^\infty(\mathbb{R}^{n-1}) \end{matrix}$$

be a matrix of operators. We say that A is an operator in Boutet de Monvel's calculus on $\overline{\mathbb{R}}^n_+$ of order m and class d, provided that all entries are as in Definition 4.11, of order m and class d.

Remark 4.15. The regularizing elements in Boutet de Monvel's calculus are the operators of order $-\infty$ and class d. Here all the symbols involved are of order $-\infty$.

Theorem 4.16. *Let A be an operator of order m and class d as in 4.14 and let*

$$A' = \begin{pmatrix} P'_+ + G' & K' \\ T' & S' \end{pmatrix}$$

be a corresponding operator of order m' and class d'. Then the composition AA' is an operator of order $m+m'$ and class $\max\{m'+d, d'\}$. *In particular, the operators of order and class zero form an algebra.*

The proof of this theorem is not obvious. The most subtle point is the composition of the truncated pseudodifferential operators P_+ and P'_+. In fact, writing

$$P_+P'_+ = (PP')_+ + L(P, P'),$$

the so-called leftover term $L(P, P')$ turns out to be a singular Green operator.

Remark 4.17. The operators of order and class zero form a Fréchet algebra which is submultiplicative, spectrally invariant and closed under holomorphic functional calculus, see [17] and [38] for more details.

Theorem 4.18. *Let A be an operator of order m and class d as in* 9. *Then*

$$A : \begin{matrix} H^s(\mathbb{R}^n_+) \\ \oplus \\ H^s(\mathbb{R}^{n-1}) \end{matrix} \to \begin{matrix} H^{s-m}(\mathbb{R}^n_+) \\ \oplus \\ H^{s-m}(\mathbb{R}^{n-1}) \end{matrix}$$

is continuous for $s > d - 1/2$.

Remark 4.19.

(a) The reason for restricting s to values $> d - 1/2$ is that extension by zero (as required by the pseudodifferential part) only makes sense for distributions in $H^s(\mathbb{R}^n_+)$, if $s > -1/2$; the same is true for the action of the singular Green and trace symbols of class zero via integration against kernel functions, which are smooth up to $x_n = 0$ as in Remark 4.13.

(b) There also is a theory of operators of negative class leading to better mapping properties. It was developed independently by Franke and Grubb. See [18] for more details.

Definition 4.20. A symbol $a \in S^m(\mathbb{R}^q \times \mathbb{R}^q; E, F)$ is said to be homogeneous of degree m, if

$$\kappa^F_{\lambda^{-1}} a(y, \lambda\eta)\kappa^E_\lambda = \lambda^m a(y, \eta).$$

It is classical, if it has an asymptotic expansion $a \sim \sum a_{m-j}$ with a_{m-j} homogeneous of degree $m - j$ in η for $|\eta| \geq 1$. The top order term a_m is called the principal symbol.

Remark 4.21. In the sequel, we shall only consider *classical* elements in Boutet de Monvel's calculus, *i.e.* all symbols involved are assumed to be classical.

4.5. *Operators acting on vector bundles over compact manifolds*

It is well-known that there exists a *collar neighborhood* $\mathcal{U}$ of ∂X in X which is diffeomorphic to $\partial X \times [0, 1)$. As before we choose local coordinates on $\mathcal{U}$ of the form (x', x_n) with $x' \in \partial X$ and $x_n \in [0, 1)$. Moreover we assume that any chart for X will either not intersect ∂X (a so-called *interior chart*) or else be of the form $V \times [0, 1)$ with V a chart for ∂X and the globally defined function x_n (a *boundary chart*). In particular, we assume that changes of coordinates between boundary charts are of the form

$$(x', x_n) \mapsto (\chi(x'), x_n) \tag{9}$$

for a diffeomorphism χ of charts on ∂X. For simplicity, we also assume that the coordinates for the interior charts are taken in $\mathbb{R}^n_+$.

As in Section 3.4 we fix a partition of unity φ_j, $j = 1, \ldots, J$, subordinate to a finite cover of X by coordinate charts over which the bundles are trivial, and fix cut-off functions ψ_j, $j = 1, \ldots, J$, supported in the same chart such that $\varphi_j\psi_j = \varphi_j$. By φ'_j and ψ'_j denote the restrictions of φ_j and ψ_j, respectively, to the boundary.

Let E_1 and E_2 be vector bundles over X and F_1, F_2 vector bundles over ∂X.

Definition 4.22. Consider an operator matrix A as in (7). With the help of the above partition of unity and cut-off functions we can write

$$A = \sum_{j=1}^{J} \begin{pmatrix} \varphi_j & 0 \\ 0 & \varphi_j' \end{pmatrix} A \begin{pmatrix} \psi_j & 0 \\ 0 & \psi_j' \end{pmatrix} + \sum_{j=1}^{J} \begin{pmatrix} \varphi_j & 0 \\ 0 & \varphi_j' \end{pmatrix} A \begin{pmatrix} 1-\psi_j & 0 \\ 0 & 1-\psi_j' \end{pmatrix}.$$

We say that A is an operator in Boutet de Monvel's calculus on X of order m and class d, provided that

- The terms $\begin{pmatrix} \varphi_j & 0 \\ 0 & \varphi_j' \end{pmatrix} A \begin{pmatrix} \psi_j & 0 \\ 0 & \psi_j' \end{pmatrix}$ are given in local coordinates by operators of order m and class d in Boutet de Monvel's calculus on $\overline{\mathbb{R}}^n_+$.
- The terms $\begin{pmatrix} \varphi_j & 0 \\ 0 & \varphi_j' \end{pmatrix} A \begin{pmatrix} 1-\psi_j & 0 \\ 0 & 1-\psi_j' \end{pmatrix}$ are of the form $\begin{pmatrix} P_+' + G' & K' \\ T' & S' \end{pmatrix}$, where P', G', K' have smooth integral kernels, and $G' = \sum_{k=0}^{d} G_k \partial_+^k$, $T' = \sum_{k=0}^{d} T_k \partial_+^k$, where the G_k and the T_k have smooth integral kernels.

Remark 4.23. The classical boundary value problems are operators of order and class m in Boutet de Monvel's calculus: Take $F_1 = 0$, P and T as in Definition 4.1 and $G = 0$. Note that taking $F_1 = 0$ makes the matrix shrink to a column vector; the entries K and S do not appear.

From Theorem 4.16 we obtain the following result:

Theorem 4.24. *Let A be as in (7) and let*

$$A' = \begin{pmatrix} P_+' + G' & K' \\ T' & S' \end{pmatrix} : \begin{matrix} \mathcal{C}^\infty(X, E_0) \\ \oplus \\ \mathcal{C}^\infty(\partial X, F_0) \end{matrix} \to \begin{matrix} \mathcal{C}^\infty(X, E_1) \\ \oplus \\ \mathcal{C}^\infty(\partial X, F_1) \end{matrix} \tag{10}$$

be an operator of order m' and class d' acting on sections of vector bundles E_0, E_1 over X and F_0, F_1 over ∂X. Then the composition AA' is an operator of order $m + m'$ and class $\max\{m' + d, d'\}$.

From Theorem 4.18 we also obtain:

Theorem 4.25. *Let A be an operator of order m and class d acting as in*

(7). *Then A extends to a bounded operator*

$$A : \begin{matrix} H^s(X, E_1) \\ \oplus \\ H^s(\partial X, F_1) \end{matrix} \rightarrow \begin{matrix} H^{s-m}(X, E_2) \\ \oplus \\ H^{s-m}(\partial X, F_2) \end{matrix} \tag{11}$$

for every $s > d - 1/2$.

Remark 4.26.

(a) For $E_1 = E_2$ and $F_1 = F_2$ the operators of order and class zero form an algebra by Theorem 4.24. The classical operators, on which we will focus here, form a subalgebra with a natural Fréchet topology; it is a subalgebra of the C^*-algebra $\mathcal{L}(L^2(X, E_1) \oplus L^2(\partial X, F_1))$.

(b) It follows from Remark 4.13 and Definition 4.22 that the regularizing elements, *i.e.* the operators of order $-\infty$ and class 0 are those operators, whose entries are all given by integral operators with smooth kernel functions.

We will next clarify when an operator A acting as in Theorem 4.25 is a Fredholm operator. Denote by p^0, g^0, k^0, t^0 and s^0 the local principal symbols of the entries of A. To make things slightly easier, we consider them as homogeneous functions on $T^*\partial X \setminus 0$, not just for $|\xi'| \geq 1$.

Since we assumed that x_n is globally defined and changes of coordinates for boundary charts are of the form (9), $g^0(x', \xi')$, $k^0(x', \xi')$, and $t^0(x', \xi')$ are well-defined as operators acting on the normal to the boundary.

More precisely, let F_j^*, $j = 1, 2$, be the pullbacks of F_j via $\pi_\partial : T^*\partial X \setminus 0 \rightarrow \partial X$ and let E_j^* be the pullbacks of E_j restricted to ∂X via $\pi_0 : T^*_{|\partial X}X \rightarrow \partial X$. Then $g^0(x', \xi')$ defines an operator

$$g^0(x', \xi') : \mathcal{S}(\mathbb{R}_+, E^*_{1,(x',\xi')}) \rightarrow \mathcal{S}(\mathbb{R}_+, E^*_{2,(x',\xi')})$$

where the interior normal to ∂X is identified with $\mathbb{R}_+$. Similarly, we obtain operators

$$\begin{aligned} t^0(x', \xi') &: \mathcal{S}(\mathbb{R}_+, E^*_{1,(x',\xi')}) \rightarrow F^*_{2,(x',\xi')} \text{ and} \\ k^0(x', \xi') &: \quad F^*_{1,(x',\xi')} \rightarrow \mathcal{S}(\mathbb{R}_+, E^*_{2,x',\xi'}). \end{aligned}$$

The principal symbol $s^0 = s^0(x', \xi)$ of S furnishes a morphism

$$s^0(x', \xi') : F^*_{1,(x',\xi')} \rightarrow F^*_{2,(x',\xi')}.$$

All these operators are homogenous of degree m in the sense of Definition 4.20.

Moreover, we evaluate the principal symbol p^0 at the boundary. For fixed (x',ξ') we obtain the operator $p^0(x',0,\xi',D_n)$ from the action of $p^0(x',0,\xi',\xi_n)$ in the normal direction, and hence

$$p^0(x',0,\xi',D_n)_+ : \mathcal{S}(\mathbb{R}_+, E^*_{1,(x',\xi')}) \to \mathcal{S}(\mathbb{R}_+, E^*_{2,(x',\xi')}),$$

which is homogeneous of degree m in ξ' in the sense of 4.20.

Definition 4.27. Let A be an operator of order m and class d in Boutet de Monvel's calculus as in (7). We associate to A two symbols:

(a) The *principal pseudodifferential symbol* of A is $\sigma(A) := \sigma(P)|_{T^*X\setminus 0}$.

(b) The *principal boundary symbol* of A is the operator matrix

$$\gamma(A)(x',\xi') = \begin{pmatrix} p^0(x',0,\xi',D_n)_+ + g^0(x',\xi') & k^0(x',\xi') \\ t^0(x',\xi') & s^0(x',\xi') \end{pmatrix} : \tag{12}$$

$$\begin{matrix} \mathcal{S}(\mathbb{R}_+, E^*_{1,(x',\xi')}) & & \mathcal{S}(\mathbb{R}_+, E^*_{2,(x',\xi')}) \\ \oplus & \to & \oplus \\ F^*_{1,(x',\xi')} & & F^*_{2,(x',\xi')} \end{matrix}$$

defined for $(x',\xi') \in T^*\partial X \setminus 0$.

(c) The operator A is called elliptic, provided

  (i) $\sigma(A)(x,\xi)$ is invertible for all $(x,\xi) \in T^*X \setminus 0$ and

  (ii) $\gamma(A)(x',\xi')$ is invertible for all $(x',\xi') \in T^*\partial X \setminus 0$.

The following statement is obvious:

Theorem 4.28. *Let A be an operator of order m and class d. If both principal symbols vanish, then A is in fact of order $m-1$ and class d.*

Theorem 4.29. *Let A be an operator of order m and class $d \le m_+ = \max\{m,0\}$ in Boutet de Monvel's calculus. If A is elliptic, then there exists a parametrix*

$$A' = \begin{pmatrix} P'_+ + G' & K' \\ T' & S' \end{pmatrix} : \begin{matrix} C^\infty(X,E_2) & & C^\infty(X,E_1) \\ \oplus & \to & \oplus \\ C^\infty(\partial X,F_2) & & C^\infty(\partial X,F_1) \end{matrix}$$

to A, i.e. an operator of order $-m$ and class $(-m)_+$ in Boutet de Monvel's calculus such that

$$AA' = I + R_1 \text{ and } A'A = I + R_2,$$

where R_1 and R_2 are of order $-\infty$ and class $(-m)_+$ and m_+, respectively.

Corollary 4.30. *In particular, A acting as in* (11) *is a Fredholm operator. According to Theorem* 4.28 *the index only depends on the principal symbols.*

Theorem 4.31. *If the operator A of order m and class $d \leq m_+$ in* (11) *is invertible, then the inverse is an element of order $-m$ and class $(-m)_+$ in Boutet de Monvel's calculus.*

A proof for classical operators can be found in [42]; non-classical elements are covered by [38].

The theorem, below, shows that there are order reducing operators also in Boutet de Monvel's calculus. The construction of the operators Λ can be found in [18].

Theorem 4.32. *There is a first order pseudodifferential operator Λ on $\widetilde{X}$ satisfying the transmission condition with respect to ∂X such that*

$$\Lambda_+ : H^s(X, E_1) \to H^{s-1}(X, E_1)$$

is an isomorphism for each $s \in \mathbb{R}$. In fact, we can replace here the operator e^+ of extension by zero by any other extension operator $\tilde{e}$; the result of $r^+\Lambda\tilde{e}$ is independent of the choice. The restriction $s > -1/2$ therefore is not necessary.

Note that Λ_+ is an operator of order 1 *and class zero in Boutet de Monvel's calculus. Moreover, $(\Lambda_+)^k = (\Lambda^k)_+$ is of order k and class zero for each $k \in \mathbb{Z}$.*

Corollary 4.33. *For many purposes, it is sufficient to consider operators of order and class zero. In fact, let A be of order $m \in \mathbb{Z}$ and class $d \leq m_+$, let Λ be as in Theorem 4.32 and let $\tilde{\Lambda}_\partial^m$ be a corresponding operator on ∂X as in Theorem 3.34. Consider the composition*

$$A \begin{pmatrix} \Lambda_+^m & 0 \\ 0 & \tilde{\Lambda}_\partial^m \end{pmatrix}.$$

It is of order and class zero. Moreover, it is elliptic, a Fredholm operator and invertible, respectively, if and only if A is.

Remark 4.34. While this text focuses primarily on the Fredholm property and index theory, it should be pointed out that Boutet de Monvel's calculus is an extremely versatile tool also for many other tasks in modern analysis. Most prominently features Grubb's work, of which many references can be found in her book [18]. Moreover, there are applications in noncommutative geometry, *e.g.* on the existence of noncommutative residues [13, 14, 20],

Dixmier's trace [31], and Kontsevich-Vishik type traces [21]. For a version of Boutet de Monvel's calculus on noncompact manifolds with an SG-structure in the sense of [36], see [38], for a groupoid picture [1, 2]. Finally, the pseudodifferential character of Boutet de Monvel's calculus makes it very suitable for extensions to other settings, say the case of manifolds with conical singularities [40, 41, 44] or Fourier integral operators [7].

5. Elements of the K-theory for C*-algebras

We shall introduce the basic notions needed to understand the K-theory of C^*-algebras. A good source for this part of the text are the books by Wegge-Olsen [49], Rørdam, Larsen and Laustsen [33] and Blackadar [9].

5.1. *C*-algebras*

Definition 5.1. A C^*-algebra is a Banach algebra $\mathcal{A}$ over[i] $\mathbb{C}$ together with an involution $* : \mathcal{A} \to \mathcal{A}$, such that $(a+b)^* = a^* + b^*$, $(\lambda a)^* = \overline{\lambda} a^*$ and $(ab)^* = b^* a^*$ for all $a, b \in \mathcal{A}$ and $\lambda \in \mathbb{C}$. In addition, one requires that

$$\|a^* a\| = \|a\|^2, \quad a \in \mathcal{A}.$$

A C^*-algebra is *unital*, if it contains a multiplicative unit e with $\|e\| = 1$.

Exercise 5.2.

(a) Let H be a Hilbert space and $* : a \mapsto a^*$ be the operation which associates to an operator $a \in \mathcal{L}(H)$ its adjoint. Show that this makes $\mathcal{L}(H)$ a unital C^*-algebra. In particular $\mathbb{C} = \mathcal{L}(\mathbb{C})$ is a C^*-algebra.

(b) More generally, let $\mathcal{A}$ be a closed subalgebra of $\mathcal{L}(H)$, which is invariant under taking adjoints. Then $\mathcal{A}$ is a C^*-algebra.

In particular, the algebra $\mathcal{K}(H)$ of compact operators on H is a C^*-algebra.

(c) Let X be a locally compact Hausdorff space and let $C_0(X)$ denote the space of all continuous functions which *vanish at infinity.* The latter means that for every function $f \in C_0(X)$ and every $\varepsilon > 0$ there is a compact set K such that $|f(x)| < \varepsilon$ for all $x \notin K$. Of course, this last condition is void, if X is compact.

Show that $C_0(X)$ becomes a C^*-algebra, if we choose the involution $*$ to be given by $f^*(x) = \overline{f(x)}$, the complex conjugate of f, and the

[i] One can also consider C^*-algebras over $\mathbb{R}$, but these will not play a role here.

norm to be the supremum norm. Clearly, $C_0(X)$ is unital if and only if X is compact.

(d) Show that $\|a^*\| = \|a\|$ for each element of a C^*-algebra.

Quite surprisingly, there are no other C^*-algebras. This is

Theorem 5.3. *Let $\mathcal{A}$ be a C^*-algebra.*

(a) Then $\mathcal{A}$ is isomorphic to a closed $*$-invariant subalgebra of $\mathcal{L}(H)$ for some Hilbert space H.

(b) If $\mathcal{A}$ is commutative, then $\mathcal{A}$ is isomorphic to $C_0(X)$ for a suitable locally compact Hausdorff space X.

Part (a) is the Theorem of Gelfand-Naimark-Segal, part (b) the Theorem of Gelfand-Naimark.

Lemma 5.4. *Let $\mathcal{A}$ be a C^*-algebra and denote by $M_n(\mathcal{A})$ the algebra of $(n \times n)$-matrices with entries in $\mathcal{A}$. Then $M_n(\mathcal{A})$ also is a C^*-algebra.*

If the algebra A does not have a unit, then it is always possible to adjoin one. In fact, there are many ways to do this. We will need two of these constructions.

Theorem 5.5. *Let $\mathcal{A}$ be a nonunital C^*-algebra. Then there is a smallest unital C^*-algebra $\mathcal{A}^\sim$ containing $\mathcal{A}$.*

Proof. We may consider $\mathcal{A}$ as a subalgebra of $\mathcal{L}(\mathcal{A})$, the Banach algebra of all bounded operators on $\mathcal{A}$ by associating with $a \in \mathcal{A}$ the operator $\mathcal{M}_a$ of multiplication by a: $\mathcal{M}_a(b) = ab$ for $b \in \mathcal{L}(\mathcal{A})$. Then $\mathcal{M}_a$ is an element of $\mathcal{L}(\mathcal{A})$ and the map $a \mapsto \mathcal{M}_a$ is an isometry. In fact, $\|\mathcal{M}_a b\| = \|ab\| \le \|a\|\|b\|$, and $\|\mathcal{M}_a a^*\| = \|aa^*\| = \|a\|^2 = \|a\|\|a^*\|$, so that $\|\mathcal{M}_a\| = \|a\|$. Hence $\mathcal{A}$ is isomorphic to a subalgebra $\iota(\mathcal{A})$ of $\mathcal{L}(\mathcal{A})$. We can then let $\mathcal{A}^\sim = \iota(\mathcal{A}) + \mathbb{C}I$, where I is the identity in $\mathcal{L}(\mathcal{A})$ and endow $\mathcal{A}^\sim$ with the multiplication given by the composition in $\mathcal{L}(\mathcal{A})$. We finally define an adjoint on $\mathcal{A}^\sim$ by letting

$$(\mathcal{M}_a + \lambda I)^* = \mathcal{M}_{a^*} + \overline{\lambda} I \tag{13}$$

(see the exercise, below).

Note that we could have also started with a unital C^*-algebra $\mathcal{A}$. In that case, the construction would have given us $\mathcal{A}^\sim = \iota(\mathcal{A})$, as the identity is already contained in $\iota(\mathcal{A})$.

Why is this the smallest unitization? Let $\overline{\mathcal{A}}$ be another unitization. Since $\mathcal{A}$ is an ideal in $\overline{\mathcal{A}}$, we can embed $\overline{\mathcal{A}}$ in $\mathcal{L}(\mathcal{A})$ by the above procedure. Then the identity of $\overline{\mathcal{A}}$ is mapped to the identity of $\mathcal{L}(\mathcal{A})$, and hence $\overline{\mathcal{A}} \cong \iota(\overline{\mathcal{A}}) \subseteq \mathcal{A}^{\sim}$. □

Exercise 5.6. Show that (13) indeed defines an adjoint, in particular that it satisfies the C^*-property.

Hint. Given $\mathcal{M}_a + \lambda I \in \mathcal{A}^{\sim}$ and $\varepsilon > 0$, there exists an element $b \in \mathcal{L}(\mathcal{A})$ with $\|b\| \leq 1$ such that

$$\|\mathcal{M}_a + \lambda I\|^2 \leq \|(\mathcal{M}_a + \lambda I)b\|^2 + \varepsilon = \|(ab + \lambda b)^*(ab + \lambda b)\| + \varepsilon.$$

Exercise 5.7. Let $\mathcal{A}$ and $\mathcal{B}$ be C^*-algebras. Define the *orthogonal direct sum* $\mathcal{A} \oplus \mathcal{B}$ of $\mathcal{A}$ and $\mathcal{B}$ as the cartesian product $\mathcal{A} \times \mathcal{B}$ with component-wise addition, multiplication and adjoint. Endow it with the norm

$$\|(a, b)\| = \max\{\|a\|, \|b\|\}.$$

Show that this makes $\mathcal{A} \oplus \mathcal{B}$ a C^*-algebra. It is unital whenever both $\mathcal{A}$ and $\mathcal{B}$ are unital. In that case, the unit in $\mathcal{A} \oplus \mathcal{B}$ is $(1_{\mathcal{A}}, 1_{\mathcal{B}})$.

Definition 5.8. Let $\mathcal{A}$ be a C^*-algebra and $\mathcal{A}^+ = \mathcal{A} \times \mathbb{C}$ with pointwise addition and adjoint. Define a multiplication by

$$(a_1, \lambda_1)(a_2, \lambda_2) = (a_1 a_2 + \lambda_1 a_2 + \lambda_2 a_1, \lambda_1 \lambda_2).$$

Exercise 5.9. Show:

(a) $\mathcal{A}^+$ a unital $*$-algebra (what is the unit?)

(b) If $\mathcal{A}$ is non-unital, then the map $\mathcal{A}^+ \to \mathcal{A}^{\sim}$, $(a, \lambda) \mapsto \mathcal{M}_a + \lambda I$ is a $*$-preserving algebra isomorphism. How does this make $\mathcal{A}^+$ a C^*-algebra?

(c) If $\mathcal{A}$ is unital with unit $1_{\mathcal{A}}$, then $(a, \lambda) \mapsto (a + \lambda 1_{\mathcal{A}}, \lambda)$ provides a $*$-preserving algebra isomorphism $\mathcal{A}^+ \to \mathcal{A} \oplus \mathbb{C}$.

5.2. *Projections and* $\mathbf{K_0}$

Definition 5.10.

(a) A *projection* in a C^*-algebra $\mathcal{A}$ is an element $p \in \mathcal{A}$ such that $p = p^* = p^2$.

(b) A *unitary* in a unital C^*-algebra is an element u with $u^*u = uu^* = 1$.

(c) A *partial isometry* is an element v such that v^*v is a projection.

(d) Let p and q be two projections. They are

(i) *orthogonal*, if $pq = 0$

(ii) *equivalent*, denoted $p \sim q$, if there exists a partial isometry v such that $vv^* = p$ and $v^*v = q$

(iii) *unitarily equivalent*, denoted $p \overset{u}{\sim} q$, if there exists a unitary u in $\mathcal{A}^\sim$ such that $p = u^*qu$

(iv) *homotopic*, denoted $p \overset{h}{\sim} q$, provided there is a norm continuous path of projections $\pi(t)$, $0 \le t \le 1$, such that $\pi(0) = p$ and $\pi(1) = q$.

Given a C^*-algebra $\mathcal{A}$, the C^*-algebra $M_n(\mathcal{A})$ of $(n \times n)$-matrices with entries in $\mathcal{A}$ can be considered as a subset of $M_{n+1}(\mathcal{A})$ by

$$\begin{pmatrix} a_{11} & \dots & a_{1n} \\ \vdots & & \vdots \\ a_{n1} & \dots & a_{nn} \end{pmatrix} \mapsto \begin{pmatrix} a_{11} & \dots & a_{1n} & 0 \\ \vdots & & \vdots & \vdots \\ a_{n1} & \dots & a_{nn} & 0 \\ 0 & \dots & 0 & 0 \end{pmatrix}.$$

By $M_\infty(\mathcal{A})$ we denote the inductive limit with respect to this identification.

Lemma 5.11. *In $M_\infty(\mathcal{A})$, the three notions in 5.7(d) are equivalent.*

In fact, one can prove that

$$p \overset{h}{\sim} q \Rightarrow p \overset{u}{\sim} p \Rightarrow p \sim q,$$

and if $p \sim q$, then $\begin{pmatrix} p & 0 \\ 0 & 0 \end{pmatrix} \overset{h}{\sim} \begin{pmatrix} q & 0 \\ 0 & 0 \end{pmatrix}$ in $M_4(\mathcal{A})$. See [49, Proposition 5.2.10 and Proposition 5.2.12] for details.

For a projection p in $M_n(\mathcal{A})$ denote by $[p]$ the equivalence class of p in $M_\infty(\mathcal{A})$. We define the addition of equivalence classes of projections in $M_\infty(\mathcal{A})$ by

$$[p] \oplus [q] = \left[\begin{pmatrix} p & 0 \\ 0 & q \end{pmatrix} \right].$$

This gives us an abelian semigroup $V(\mathcal{A})$ with identity $[0]$.

Exercise 5.12. Show that $V(\mathcal{A}) = \mathbb{N}_0$ for $\mathcal{A} = \mathbb{C}$ and $\mathcal{A} = \mathcal{K}(H)$.

By Grothendieck's group construction, see Appendix G in [49], we obtain an abelian group $K_{00}(\mathcal{A})$. The elements of $K_{00}(\mathcal{A})$ may be visualized as formal differences

$$[p] - [q], \quad [p], [q] \in V(\mathcal{A}),$$

where $[p]-[q]=[p']-[q']$ for $[p'],[q']$ in $V(\mathcal{A})$, if there exists $[r]$ in $V(\mathcal{A})$ such that $[p]+[q']+[r]=[p']+[q]+[r]$.

Clearly, $K_{00}(\mathbb{C})=K_{00}(\mathcal{K}(H))=\mathbb{Z}$.

In order to introduce K-theory classes we need the following observation: If

$$\alpha:\mathcal{A}\to\mathcal{B}$$

is a C^*-algebra homomorphism, then it induces a map $\alpha_*: M_n(\mathcal{A})\to M_n(\mathcal{B})$ by

$$\begin{pmatrix} a_{11} & \dots & a_{1n} \\ \vdots & & \vdots \\ a_{n1} & \dots & a_{nn} \end{pmatrix} \mapsto \begin{pmatrix} \alpha(a_{11}) & \dots & \alpha(a_{1n}) \\ \vdots & & \vdots \\ \alpha(a_{n1}) & \dots & \alpha(a_{nn}) \end{pmatrix}.$$

Definition 5.13. Let $\mathcal{A}$ be a C^*-algebra.

(a) If $\mathcal{A}$ is unital, we let $K_0(\mathcal{A})=K_{00}(\mathcal{A})$.

(b) If $\mathcal{A}$ is not unital, we let $K_0(\mathcal{A})=\ker(\pi_*:K_{00}(\mathcal{A}^+)\to\mathbb{Z})\subseteq K_{00}(\mathcal{A})$

Here $\pi:\mathcal{A}^+\to\mathbb{C}$ is the map $(a,\lambda)\mapsto\lambda$. It induces maps $M_n(\mathcal{A}^+)\to M_n(\mathbb{C})$, $V(\mathcal{A}^+)\to V(\mathbb{C})$ and $K_{00}(\mathcal{A}^+)\to K_{00}(\mathbb{C})$ all denoted by π_*.

Remark 5.14. By construction, $K_0(\mathcal{A})$ is an abelian group.

An element in $K_0(\mathcal{A})$ is a formal difference $[p]-[q]$, where p and q are projections in $M_n(\mathcal{A}^+)$ for some $n\in\mathbb{N}$ and $p-q\in M_n(\mathcal{A})$; by $[\cdot]$ we denote the equivalence classes in $M_\infty(\mathcal{A})$ modulo one of the relations in Definition 5.10. If $\mathcal{A}$ is unital, p and q may be chosen in $M_n(\mathcal{A})$.

The formal difference $[p]-[q]$ represents the zero element of $K_0(\mathcal{A})$ if and only if there exist $m,k\in\mathbb{N}$, $m\le k$ such hat

$$\begin{pmatrix} p & 0 & 0 \\ 0 & 1_m & 0 \\ 0 & 0 & 0_{k-m} \end{pmatrix} \overset{h}{\sim} \begin{pmatrix} q & 0 & 0 \\ 0 & 1_m & 0 \\ 0 & 0 & 0_{k-m} \end{pmatrix} \quad \text{in } M_{n+k}(\mathcal{A}^+).$$

Here 1_m and 0_{k-m} denote the $m\times m$ unit matrix and the $(k-m)\times(k-m)$ zero-matrix, respectively.

Corollary 5.15. *From the results of Exercise 5.12 we obtain immediately that* $K_0(\mathbb{C})\cong\mathbb{Z}$.

K_0 is stable under tensoring with the C^*-algebra $\mathcal{K}$ of the compact operators on a separable Hilbert space. The theorem, below, is [49, Corollary 6.2.11].

Theorem 5.16. *Let e be a projection of rank one in $\mathcal{K}$. Then*

$$\mathcal{A} \ni a \mapsto a \otimes e \in \mathcal{A} \otimes \mathcal{K}$$

induces an isomorphism $K_0(\mathcal{A}) \to K_0(\mathcal{A} \otimes \mathcal{K})$. In particular, $K_0(\mathcal{K}) \cong K_0(\mathbb{C}) \cong \mathbb{Z}$.

The following result, known as Swan's Theorem, shows the connection between the classical K-theory for vector bundles and K-theory for C^*-algebras. A proof can be found in [49, Section 13].

Theorem 5.17. *For a locally compact Hausdorff space Y we have*

$$K_0(C_0(Y)) \cong K_c^0(Y).$$

5.3. $\mathbf{K_1}$ *and its relation to* $\mathbf{K_0}$

Let x be an element of $\mathrm{Gl}_n(\mathcal{A}^+)$, the invertible $(n \times n)$-matrices with entries in $\mathcal{A}$. We may consider $\mathrm{Gl}_n(\mathcal{A})$ as a subset of $\mathrm{Gl}_{n+m}(\mathcal{A})$, $m \in \mathbb{N}$, by identifying x with the element $\begin{pmatrix} x & 0 \\ 0 & 1_m \end{pmatrix}$. By $\mathrm{Gl}_\infty(\mathcal{A})$ we denote the inductive limit with respect to these embeddings.

Definition 5.18. For any C^*-algebra $\mathcal{A}$ we define

$$K_1(\mathcal{A}) = \mathrm{Gl}_\infty(\mathcal{A}^+)/\,\mathrm{Gl}_\infty(\mathcal{A}^+)_0,$$

where $\mathrm{Gl}_\infty(\mathcal{A}^+)_0$ denotes the connected component of the identity in $\mathrm{Gl}_\infty(\mathcal{A}^+)$.

$K_1(\mathcal{A})$ becomes an abelian group with the multiplication

$$[x][y] = [xy] = \left[\begin{pmatrix} x & 0 \\ 0 & y \end{pmatrix}\right].$$

Remark 5.19.

(a) In other words: An element of $K_1(\mathcal{A})$ is an equivalence class for an element x in $\mathrm{Gl}_n(\mathcal{A}^+)$ for some n modulo homotopy in $\mathrm{Gl}_\infty(\mathcal{A}^+)$.

(b) Instead of $\mathrm{Gl}_\infty(\mathcal{A}^+)/\,\mathrm{Gl}_\infty(\mathcal{A}^+)_0$ we could have taken $\mathrm{Gl}_\infty^+(\mathcal{A})/\,\mathrm{Gl}_\infty^+(\mathcal{A})_0$ or $U_\infty(\mathcal{A}^+)/U_\infty(\mathcal{A}^+)_0$ or $U_\infty^+(\mathcal{A})/U_\infty^+(\mathcal{A})_0$, where U_∞ stands for the unitary elements in Gl_∞, see [49, Proposition 7.1.5].

(c) $K_1(\mathcal{A}) = \mathrm{Gl}_\infty(\mathcal{A})/\,\mathrm{Gl}_\infty(\mathcal{A})_0$ for unital $\mathcal{A}$, see [49, Remark 7.1.3].

Example 5.20. $K_1(\mathbb{C}) = 0$. Indeed, any two invertible matrices in $\mathrm{Gl}_n(\mathbb{C})$ are connected by a homotopy in $\mathrm{Gl}_n(\mathbb{C})$.

Theorem 5.21.

(a) $K_1(\mathcal{A}\otimes\mathcal{K}) \cong K_1(\mathcal{A})$ *for every* C^**-algebra* $\mathcal{A}$. *In particular,* $K_1(\mathcal{K}) \cong K_1(\mathbb{C}) = 0$.

(b) $K_1(\mathcal{A}\oplus\mathcal{B}) = K_1(\mathcal{A}) \oplus K_1(\mathcal{B})$.

A proof is given in [49, Theorem 7.1.9 and Example 7.1.11].

Definition 5.22.

(a) Two morphisms $\alpha_0, \alpha_1 : \mathcal{A} \to \mathcal{B}$ between C^*-algebras are *homotopic*, provided they can be joined by a norm continuous path $t \mapsto \alpha_t$, $0 \le t \le 1$, of C^*-algebra morphisms.

(b) A C^*-algebra $\mathcal{A}$ is *contractible*, if the identity map on $\mathcal{A}$ is homotopic to the zero map $\mathcal{A} \to \mathcal{A}$.

Lemma 5.23. *If* α_0 *and* α_1 *are homotopic morphisms between* C^**-algebras* $\mathcal{A}$ *and* $\mathcal{B}$*, then the induced maps*

$$\alpha_{0*}, \alpha_{1*} : K_j(\mathcal{A}) \to K_j(\mathcal{B}), \quad j = 0, 1,$$

coincide. In particular, $K_j(\mathcal{A}) = 0$, $j = 0, 1$, *whenever* $\mathcal{A}$ *is contractible.*

Remark 5.24. K_0 and K_1 are covariant, homotopy invariant functors from the category of C^*-algebras to the category of abelian groups. They are continuous with respect to directed limits, see Proposition 6.2.4, Theorem 6.4.3 and Proposition 7.1.6 in [49].

This functor is not exact, as we shall next see.

Definition 5.25. A sequence of C^*-algebras and C^*-algebra morphisms

$$\ldots \to \mathcal{A}_{-1} \to \mathcal{A}_0 \to \mathcal{A}_1 \to \ldots$$

is called *exact*, if the image of one of the maps equals the kernel of the next one. Of interest are the short exact sequences of C^*-algebras $\mathcal{A}, \mathcal{B}, \mathcal{C}$:

$$0 \to \mathcal{A} \overset{\alpha}{\to} \mathcal{B} \overset{\beta}{\to} \mathcal{C} \to 0 \tag{14}$$

with α injective, β surjective and $\operatorname{im}\alpha = \ker\beta$. In particular, $\alpha(\mathcal{A})$, being the kernel of a C^*-algebra morphism, then is an ideal in $\mathcal{B}$.

The fundamental theorem of K-theory is the following:

Theorem 5.26. *Given a short exact sequence* (14) *of C^*-algebras there exist maps* ind *and* exp *such that*

$$\begin{array}{ccccc} K_0(\mathcal{A}) & \stackrel{\alpha_*}{\to} & K_0(\mathcal{B}) & \stackrel{\beta_*}{\to} & K_0(\mathcal{C}) \\ \uparrow \text{ind} & & & & \downarrow \text{exp} \\ K_1(\mathcal{C}) & \stackrel{\beta_*}{\leftarrow} & K_1(\mathcal{B}) & \stackrel{\alpha_*}{\leftarrow} & K_1(\mathcal{A}) \end{array} \tag{15}$$

is an exact sequence of abelian groups.

This diagram is called the six-term exact sequence associated with the short exact sequence (14). In general the maps 'exp' and 'ind' are hard to determine, and we make here no effort to even define them. There are two cases, when we easily can say more. The first is the following:

Theorem 5.27. *Let H be a separable Hilbert space and $\mathcal{A}$ be a C^*-subalgebra of $\mathcal{L}(H)$ containing $\mathcal{K} = \mathcal{K}(H)$. Consider the short exact sequence*

$$0 \to \mathcal{K} \to \mathcal{A} \to \mathcal{A}/\mathcal{K} \to 0.$$

$K_1(\mathcal{A}/\mathcal{K})$ consists of equivalence classes of matrices of operators which are invertible modulo $\mathcal{K}$, hence Fredholm operators, see Theorem 1.5. The map $ind : K_1(\mathcal{A}/\mathcal{K}) \to K_0(\mathcal{K}) = \mathbb{Z}$ then is the Fredholm index.

For the second case we need some preparations.

5.4. *Mapping cones and suspensions*

Example 5.28. Let $\mathcal{A}$ be a C^*-algebra. The *cone* over $\mathcal{A}$ is the C^*-algebra

$$C\mathcal{A} = \{f \in C([0,1], \mathcal{A}) : f(1) = 0\}.$$

$C\mathcal{A}$ is contractible: Define $\alpha_t : C\mathcal{A} \to C\mathcal{A}$ by $(\alpha_t f)(s) = f(t + (1-t)s)$. This furnishes a homotopy between the identity and the zero map.

Definition 5.29.

(a) The *suspension* of $\mathcal{A}$ is the C^*-algebra

$$S\mathcal{A} = \{f \in C([0,1], \mathcal{A}) : f(0) = f(1) = 0\} = \{f \in C\mathcal{A} : f(0) = 0\}.$$

(b) Let $\alpha : \mathcal{A} \to \mathcal{B}$ be a C^*-algebra morphism. The *mapping cone* is the set

$$C\alpha = \{(a, \varphi) \in \mathcal{A} \oplus C\mathcal{A} : \alpha(a) = \varphi(0)\}.$$

(c) With α we can further associate the map $S_\alpha : S\mathcal{A} \to S\mathcal{B}$ by

$$(S_\alpha f)(t) = \alpha(f(t)).$$

Theorem 5.30. *S is an exact functor: If*

$$0 \to \mathcal{A} \overset{\alpha}{\to} \mathcal{B} \overset{\beta}{\to} \mathcal{C} \to 0$$

is a short exact sequence of C^*-algebras, then so is

$$0 \longrightarrow S\mathcal{A} \overset{S_\alpha}{\longrightarrow} S\mathcal{B} \overset{S_\beta}{\longrightarrow} S\mathcal{C} \longrightarrow 0.$$

For a proof see *e.g.* [33, Proposition 10.1.2].

Theorem 5.31. *For every C^*-algebra $\mathcal{A}$, there exist isomorphisms $\theta_j^{\mathcal{A}}$: $K_j(\mathcal{A}) \to K_{1-j}(S\mathcal{A})$, $j = 0, 1$. Given a morphism $\alpha : \mathcal{A} \to \mathcal{B}$, the diagram, below, commutes:*

$$\begin{array}{ccc} K_j(\mathcal{A}) & \overset{\alpha_*}{\longrightarrow} & K_j(\mathcal{B}) \\ \downarrow \theta_j^{\mathcal{A}} & & \downarrow \theta_j^{\mathcal{B}} \\ K_{1-j}(S\mathcal{A}) & \overset{S_{\alpha *}}{\longrightarrow} & K_{1-j}(S\mathcal{B}) \end{array} \tag{16}$$

In particular, $K_j(\mathcal{A}) \cong K_j(S^2\mathcal{A})$, where $S^2\mathcal{A} = S(S\mathcal{A})$. This is Bott periodicity *in C^*-algebra K-theory.*

For a proof see [49, Theorem 7.2.5 and Proposition 9.1.3].

Example 5.32. We have

$$S\mathbb{C} = \{f \in \mathcal{C}([0,1], \mathbb{C}) : f(0) = f(1) = 0\} \cong \mathcal{C}_0(\mathbb{R})$$

and

$$K_0(S\mathbb{C}) \cong K_1(\mathbb{C}) \cong 0$$
$$K_1(S\mathbb{C}) \cong K_0(\mathbb{C}) \cong \mathbb{Z}.$$

We shall use this to compute $K_j(\mathcal{C}(S^1))$, $j = 0, 1$, for the unit circle S^1 in $\mathbb{C}$. To this end consider the short exact sequence

$$0 \to S\mathbb{C} \to \mathcal{C}(S^1) \overset{\pi}{\to} \mathbb{C} \to 0,$$

where we identified $\mathcal{C}(S^1)$ with $\{f \in \mathcal{C}([0,1], \mathbb{C}) : f(0) = f(1)\}$. Here ι is the inclusion and $\pi : \mathcal{C}(S^1) \to \mathbb{C}$ is given by $f \mapsto f(0)$. It has a right inverse s, which associates to $c \in \mathbb{C}$ the constant function c. As $\pi s = id$, the induced maps π_*, s_* in K-theory satisfy also $\pi_* s_* = id$, hence π_* is surjective.

We next study the associated six-term exact sequence

$$\begin{array}{ccccc} 0 = K_0(S\mathbb{C}) & \stackrel{\iota_*}{\to} & K_0(\mathcal{C}(S^1)) & \stackrel{\pi_*}{\twoheadrightarrow} & K_0(\mathbb{C}) = \mathbb{Z} \\ \uparrow 0 & & & & \downarrow 0 \\ 0 = K_1(\mathbb{C}) & \stackrel{\pi_*}{\twoheadleftarrow} & K_1(\mathcal{C}(S^1)) & \stackrel{\iota_*}{\leftarrow} & K_1(S\mathbb{C}), \end{array}$$

where we have inserted the information we have on $K_j(S\mathbb{C})$ and π_*. The arrow $\twoheadrightarrow$ indicates that a map is surjective. We conclude that

$$\begin{aligned} K_0(\mathcal{C}(S^1)) &\stackrel{\pi_*}{\cong} K_0(\mathbb{C}) \cong \mathbb{Z} \text{ and} \\ K_1(\mathcal{C}(S^1)) &\stackrel{\iota_*}{\cong} K_1(S\mathbb{C}) \cong \mathbb{Z}. \end{aligned}$$

We will next come to the second case, in which one can determine the maps in the six-term exact sequence. Given a C^*-algebra morphism $\alpha : \mathcal{A} \longrightarrow \mathcal{B}$ we obtain a natural short exact sequence

$$0 \longrightarrow S\mathcal{B} \stackrel{\iota}{\longrightarrow} C\alpha \stackrel{\pi}{\longrightarrow} \mathcal{A} \longrightarrow 0, \tag{17}$$

where $\iota(f) = (0, f)$ and $\pi(a, f) = a$. The associated six-term exact sequence is

$$\begin{array}{ccccc} K_0(S\mathcal{B}) & \stackrel{\iota_*}{\to} & K_0(C\alpha) & \stackrel{\pi_*}{\to} & K_0(\mathcal{A}) \\ \uparrow \text{ind} & & & & \downarrow \exp \\ K_1(\mathcal{A}) & \stackrel{\pi_*}{\leftarrow} & K_1(C\alpha) & \stackrel{\iota_*}{\leftarrow} & K_1(S\mathcal{B}). \end{array}$$

Theorem 5.33. *In the above situation, the map ind: $K_1(\mathcal{A}) \to K_0(S\mathcal{B})$ is given as the composition of $\alpha_* : K_1(\mathcal{A}) \to K_1(\mathcal{B})$ and $\theta_1^{\mathcal{B}} : K_1(\mathcal{B}) \to K_0(S\mathcal{B})$; the map exp: $K_0(\mathcal{A}) \to K_1(S\mathcal{B})$ is the composition of $\alpha_* : K_0(\mathcal{A}) \to K_0(\mathcal{B})$ and $\theta_0^{\mathcal{B}} : K_0(\mathcal{B}) \to K_1(S\mathcal{B})$.*

This is essentially a consequence of the naturality of all the maps involved. For a proof see [28, Lemma 10].

The lemma, below, will be important later on.

Lemma 5.34. *Assume that the following is a commutative diagram of short exact sequences of C^*-algebras:*

$$\begin{array}{ccccccccc} 0 & \longrightarrow & \mathcal{A} & \longrightarrow & \mathcal{B} & \longrightarrow & \mathcal{C} & \longrightarrow & 0 \\ & & \uparrow \alpha & & \uparrow \beta & & \uparrow \gamma & & \\ 0 & \longrightarrow & \mathcal{A}' & \longrightarrow & \mathcal{B}' & \longrightarrow & \mathcal{C}' & \longrightarrow & 0 \end{array}$$

Then we get an induced commutative grid of short exact sequences of C^*-algebras

$$\begin{array}{ccccccccc}
 & & 0 & & 0 & & 0 & & \\
 & & \uparrow & & \uparrow & & \uparrow & & \\
0 & \longrightarrow & \mathcal{A}' & \longrightarrow & \mathcal{B}' & \longrightarrow & \mathcal{C}' & \longrightarrow & 0\\
 & & \uparrow & & \uparrow & & \uparrow & & \\
0 & \longrightarrow & C\alpha & \longrightarrow & C\beta & \longrightarrow & C\gamma & \longrightarrow & 0\\
 & & \uparrow & & \uparrow & & \uparrow & & \\
0 & \longrightarrow & S\mathcal{A} & \longrightarrow & S\mathcal{B} & \longrightarrow & S\mathcal{C} & \longrightarrow & 0\\
 & & \uparrow & & \uparrow & & \uparrow & & \\
 & & 0 & & 0 & & 0. & &
\end{array}$$

The proof is a diagram chase, using that the maps $C\mathcal{B} \to C\mathcal{C}$ and $S\mathcal{B} \to S\mathcal{C}$ are surjective. Recall the exactness of S in Theorem 5.30!

5.5. *Toeplitz operators*

We know that every function $u \in L^2(S^1)$ has a Fourier series

$$u = \sum_{k=-\infty}^{\infty} u_k z^k, \quad z \in S^1,$$

with a sequence $(u_k) \in \ell^2(\mathbb{Z})$. Often one writes e^{ikt} instead of z^k and considers u a 2π-periodic function of t on $\mathbb{R}$.

Definition 5.35. The *Hardy space* $\mathcal{H}$ on S^1 is the closed subspace consisting of all $u \in L^2(S^1)$ such that $u_k = 0$ for $k < 0$. We denote by $\pi : L^2(S^1) \to \mathcal{H}$ the orthogonal projection onto $\mathcal{H}$.

Remark 5.36. The functions in $\mathcal{H}$ extend to holomorphic functions on the unit disk $B(0,1)$ in $\mathbb{C}$.

Definition 5.37. For $f \in \mathcal{C}(S^1)$ define the Toeplitz operator T_f with symbol f in $\mathcal{L}(\mathcal{H})$ by

$$T_f u = \pi(fu).$$

The Toeplitz algebra $\mathcal{T}$ is the C^*-algebra generated by the operators T_f, $f \in \mathcal{C}(S^1)$.

The essential properties of the Toeplitz operators are the following:

Proposition 5.38. *Let $f, g \in \mathcal{C}(S^1)$.*

(a) *(i)* $T_f + T_g = T_{f+g}$

(ii) $T_f^* = T_{\overline{f}}$

(iii) $T_f T_g - T_{fg} \in \mathcal{K}(\mathcal{H})$

(iv) $T_f \in \mathcal{K}(\mathcal{H})$, *if and only if* $f \equiv 0$

(b) $\mathcal{K}$ *is the commutator ideal*[j] *of* $\mathcal{T}$, *and each Toeplitz operator can be uniquely written in the form* $T_f + K$, *with* $f \in \mathcal{C}(S^1)$ *and* $K \in \mathcal{K}$.

(c) $\mathcal{T}/\mathcal{K} \cong \mathcal{C}(S^1)$, *where the isomorphism is given by the map* $[T_f] \mapsto f$, *and*

$$\|[T_f]\|_{\mathcal{T}/\mathcal{K}} = \inf_{K \in \mathcal{K}} \|T_f + K\| = \|f\|_{\sup}.$$

(d) T_f *is a Fredholm operator if and only if* $f(z) \neq 0$ *for all* $z \in S^1$. *In this case*

$$\text{ind } T_f = -\text{wind}(f),$$

the winding number of f *around* $z = 0$.

Proofs for all these statements can be found in Chapter 7 of [11].

We can now compute the K-theory of $\mathcal{T}$ from the short exact sequence

$$0 \to \mathcal{K} \to \mathcal{T} \to C(S^1) \to 0.$$

We know already that $K_0(\mathcal{K}) = \mathbb{Z}$, $K_1(\mathcal{K}) = 0$, $K_0(\mathcal{C}(S^1)) \cong \mathbb{Z} \cong K_1(\mathcal{C}(S^1))$. Hence we obtain the six-term exact sequence

$$\begin{array}{ccc} \mathbb{Z} = K_0(\mathcal{K}) & \overset{\iota_*}{\to} K_0(\mathcal{T}) \to & K_0(\mathcal{C}(S^1)) \cong \mathbb{Z} \\ \uparrow \text{ind} & & \downarrow \\ \mathbb{Z} \cong K_1(\mathcal{C}(S^1)) & \overset{\pi_*}{\leftarrow} K_1(\mathcal{T}) \leftarrow & K_1(\mathcal{K}) = 0 \end{array}$$

By Theorem 5.27, the index map is given by the Fredholm index, which is surjective according to Proposition 5.38(d). Hence ι_* is the zero map and $K_0(\mathcal{T}) \cong \mathbb{Z}$. Being surjective from $\mathbb{Z}$ to $\mathbb{Z}$, the index map is also injective. Thus π_* is the zero map, and the exactness implies that $K_1(\mathcal{T}) = 0$. We obtain:

Proposition 5.39. $K_0(\mathcal{T}) \cong \mathbb{Z}$ *and* $K_1(\mathcal{T}) = 0$.

[j] *i.e.* the closed ideal generated by the commutators

Below, we will encounter an ideal in the Toeplitz algebra, namely the algebra $\mathcal{T}_0$ of all operators whose symbol vanishes in $z = -1$. In order to compute its K-theory we consider the short exact sequence

$$0 \to \mathcal{T}_0 \overset{\iota}{\to} \mathcal{T} \overset{\pi}{\to} \mathbb{C} \to 0 \tag{18}$$

with $\iota : \mathcal{T}_0 \hookrightarrow \mathcal{T}$ and $\pi : T_f + K \mapsto f(-1)$. We obtain a right inverse to π by associating with $c \in \mathbb{C}$ the operator $T_c = cI$. The corresponding six-term exact sequence is

$$\begin{array}{ccccc} K_0(\mathcal{T}_0) & \overset{\iota_*}{\to} & K_0(\mathcal{T}) & \overset{\pi_*}{\twoheadrightarrow} & K_0(\mathbb{C}) \cong \mathbb{Z} \\ \uparrow \text{ind} & & & & \downarrow \exp \\ 0 \cong K_1(\mathbb{C}) & \overset{\pi_*}{\twoheadleftarrow} & K_1(\mathcal{T}) & \overset{\iota_*}{\leftarrow} & K_1(\mathcal{T}_0), \end{array}$$

where π_* is surjective. So both the exponential map and the index map are zero. Moreover, π_* is an isomorphism by Proposition 5.39. Hence $\iota_* : K_0(\mathcal{T}_0) \to K_0(\mathcal{T})$ is injective with range 0, so that $K_0(\mathcal{T}_0) = 0$. Also $\iota_* : K_1(\mathcal{T}_0) \to K_1(\mathcal{T}) = 0$ is injective, thus $K_1(\mathcal{T}_0) = 0$. We collect the results in the following proposition.

Proposition 5.40. $K_0(\mathcal{T}_0) = K_1(\mathcal{T}_0) = 0$.

6. Index Theory for Closed Manifolds via K-theory

6.1. *The set-up*

Let M be a closed manifold and E a hermitian vector bundle over M. We shall study the algebra $\Psi^0(M, E)$ of classical zero order pseudodifferential operators acting on sections of E. In view of the existence of order reducing operators shown in Theorem 3.34 this is no restriction. From Section 3 we know:

Theorem 6.1. *$\Psi^0(M, E)$ is an adjoint invariant subalgebra of $\mathcal{L}(L^2(M, E))$. Its closure $\mathcal{A}$ in $\mathcal{L}(L^2(M, E))$ therefore is a C^*-algebra.*

Concerning the subalgebra $\Psi^{-1}(M, E)$ of classical pseudodifferential operators of order -1, which also is adjoint invariant, we find:

Lemma 6.2. *The closure of $\Psi^{-1}(M, E)$ in $\mathcal{L}(L^2(M, E))$ is the ideal $\mathcal{K}$ of compact operators in $\mathcal{L}(L^2(M, E))$.*

Proof. Every operator of order -1 is compact, hence the closure is a subset of the compact operators. On the other hand, we know that each Hilbert-Schmidt operator can be approximated in $\mathcal{L}(L^2(M, E))$ by an integral operator with a C^∞-kernel. Since Ψ^{-1} contains all these and since the closure of the Hilbert-Schmidt operators gives the compact operators, we obtain the assertion. □

A crucial result is the following theorem, which goes back to Gohberg [16] and Seeley [46]. We first recall that the principal symbol of an operator in $\Psi^0(M,E)$ is a zero-homogeneous function on $T^*X\setminus 0$ taking values in endomorphisms of E, which we can also consider as a smooth function on S^*X.

Theorem 6.3. *For $P\in\Psi^0(M,E)$ we have*

$$\inf\{\|P+K\| : K\in\mathcal{K}\} = \|\sigma(P)\|_{\sup},$$

where the right hand side is the sup-norm of the principal symbol in $C(S^*M,\mathrm{End}(E))$.

6.2. *K-theory*

Corollary 6.4. *The principal symbol map σ furnishes a short exact sequence*

$$0\to\Psi^{-1}(M,E)\to\Psi^0(M,E)\xrightarrow{\sigma}C^\infty(S^*M,\mathrm{End}(E))\to 0.$$

Going over to the closure, we therefore obtain the short exact sequence

$$0\to\mathcal{K}\to\mathcal{A}\to C(S^*M,\mathrm{End}(E))\to 0.$$

Hence $\mathcal{A}/\mathcal{K}\cong C(S^*M,\mathrm{End}(E))$ and $K_j(\mathcal{A}/\mathcal{K})\cong K_j(C(S^*M,\mathrm{End}(E))$, $j=0,1$.

In view of the result of Atiyah and Singer, Theorem 3.42, and Swan's Theorem, Theorem 5.17, we still need a map

$$K_j(C(S^*M,\mathrm{End}(E))\to K_j(C_0(T^*M)).$$

To this end, we first note that it can be shown that $K_j(C(S^*M,\mathrm{End}(E))$ is canonically isomorphic to $K_j(C(S^*M))$, so that we are actually looking for a map

$$K_j(C(S^*M))\to K_j(C_0(T^*M)).$$

To this end consider the cosphere bundle as the boundary of the ball compactification B^*M of the cotangent bundle. This gives rise to a short exact sequence of C^*-algebras:

$$0\to C_0(T^*M)\to C(B^*M)\to C(S^*M)\to 0$$

which in turn induces a six-term exact sequence

$$\begin{array}{ccc} K_0(C_0(T^*M))) \to K_0(C(B^*M)) \to K_0(C(S^*M)) \\ \uparrow \text{ind} \qquad\qquad\qquad\qquad \downarrow \exp \\ K_1(C(S^*M)) \leftarrow K_1(C(B^*M)) \leftarrow K_1(\mathcal{C}_0(T^*M)) \end{array} \tag{19}$$

We then conclude with:

Theorem 6.5. *Under the identification* $K_c^0(T^*M) \cong K_0(\mathcal{C}_0(T^*M))$ *the index map in* (19) *is the map* $\sigma(P) \mapsto [\sigma(P)]$.

See the corresponding proof of Lemma 16 in [27].

7. K-theory and Index Theory for Boutet de Monvel's Algebra

7.1. *The set-up*

As before, let X be a compact manifold with boundary. We shall assume X to be connected with $\partial X \neq \emptyset$. In order to apply operator algebraic techniques, we will consider operator matrices in Boutet de Monvel's calculus of the form

$$A = \begin{pmatrix} P_+ + G & K \\ T & S \end{pmatrix} : \begin{array}{c} \mathcal{C}^\infty(X,E) \\ \oplus \\ \mathcal{C}^\infty(\partial X, F) \end{array} \to \begin{array}{c} \mathcal{C}^\infty(X,E) \\ \oplus \\ \mathcal{C}^\infty(\partial X, F) \end{array} \tag{20}$$

with vector bundles E over X and F over ∂X and all entries classical of order and class zero.

We know from Section 4 that these operators form an adjoint-invariant subalgebra of $\mathcal{L}(L^2(X,E) \oplus L^2(\partial X, F))$.

We shall now sketch how the K-theory of this algebra can be computed and an index theorem in the spirit of Atiyah and Singer can be derived, following [27], [28] and [30].

Remark 7.1. It is a legitimate question whether really all index problems can be covered by this set-up, where one considers only operators of order zero acting between the same vector bundles on both sides and X is assumed to be connected with $\partial X \neq \emptyset$.

The last assumption presents no loss of generality: If X had several components, we could consider them separately, since operators acting between different components are smoothing by Definition 4.22, and if X had no boundary, we would be in the closed manifold case studied by Atiyah and Singer.

The reduction to zero order operators is clearly possible in view of Corollary 4.33. As for having the same vector bundles on both sides, a more subtle argument has to be applied, see Section 1.1 of [28] for details.

In order to keep the exposition simple, we shall restrict ourselves to the case $F = 0$, so that we are dealing with operators of the form

$$A = P_+ + G : \mathcal{C}^\infty(X, E) \to \mathcal{C}^\infty(X, E) \tag{21}$$

with a classical pseudodifferential operator P of order zero satisfying the transmission condition and a classical singular Green operator G of order and class zero. Again this is not a loss of generality; the argument can be found in Section 1.5 of [27].

We can even go one step further: There is a bundle $E^\perp$ such that the sum $E \oplus E^\perp$ over X is a trivial bundle of rank, say, N, see e.g [3, Corollary 1.4.14]. Then the operators in (21) embed into the algebra of corresponding operators acting on sections of $\mathbb{C}^N$, which are just $N \times N$-matrices of scalar $(N = 1)$ operators. As the transition from scalar operators to matrices by definition does not change the K-theory, we may as well study the algebra $\mathcal{A}^\circ$ of scalar operators of order and class zero in Boutet de Monvel's calculus

$$A = P_+ + G : \mathcal{C}^\infty(X) \to \mathcal{C}^\infty(X). \tag{22}$$

Definition 7.2. We denote by $\mathcal{A}$ the C^*-closure (*i.e.* in our case just the closure) of the algebra $\mathcal{A}^\circ$ in the topology of $\mathcal{L}(L^2(X))$.

By [27, Sections 1.2, 1.3] we have:

Theorem 7.3. *The principal pseudodifferential symbol and and the principal boundary symbol extend to continuous maps on $\mathcal{A}$ with values in zero-homogeneous maps*

$$\sigma : \mathcal{A} \to \mathcal{C}(T^*X \setminus 0) \text{ and}$$
$$\gamma : \mathcal{A} \to \mathcal{C}(T^*\partial X \setminus 0, \mathcal{W}),$$

where $\mathcal{W}$ is an algebra of Wiener-Hopf type operators in $\mathcal{L}(L^2(\mathbb{R}_+))$ described below.

In view of the zero-homogeneity, we will also consider σ and γ to be functions on S^*X and $S^*\partial X$, respectively.

7.2. *The algebra of Wiener-Hopf operators*

We recall that the boundary symbol associated with the operator $P_+ + G$ is the operator

$$p^0(x', 0, \xi', D_n)_+ + g^0(x', \xi')$$

defined from the principal symbols p^0 and g^0 of p and g, respectively, for $(x', \xi') \in T^*\partial X \setminus 0$.

Fixing (x', ξ') we thus obtain operators of the form

$$q(D_n)_+ + K : L^2(\mathbb{R}_+) \to L^2(\mathbb{R}_+) \tag{23}$$

with a classical symbol $q = q(\xi_n) \in S^0_{tr}(\mathbb{R} \times \mathbb{R})$ independent of x_n, satisfying the transmission property, and a compact operator K on $L^2(\mathbb{R}_+)$.

Definition 7.4. By $\mathcal{W}$ we denote the C^*-algebra generated in $\mathcal{L}(L^2(\mathbb{R}_+))$ by the operators (23).

Wiener-Hopf operators are closely related to Toeplitz operators. In order to see this, we define

$$\begin{aligned} \mathcal{U} : L^2(S^1) &\to L^2(\mathbb{R}) \\ (\mathcal{U}g)(t) &= \frac{\sqrt{2}}{1+it}\, g\left(\frac{1-it}{1+it}\right). \end{aligned}$$

Then $\mathcal{U}$ links the Hardy space $\mathcal{H}$ and $L^2(\mathbb{R}_+)$:

$$\mathcal{U}(\mathcal{H}) = \mathcal{F}(e^+ L^2(\mathbb{R}_+)). \tag{24}$$

Exercise 7.5. Show (24). Help can be found in Section 2.1 of [32].

Moreover, we have

$$q(D_n)_+ = \mathcal{F}^{-1}\mathcal{U}\, T_{\tilde{q}}\, \mathcal{U}^{-1}\mathcal{F}, \tag{25}$$

where $\tilde{q} \in \mathcal{C}(S^1)$ is given by $\tilde{q}(z) = q\big(i\frac{z-1}{z+1}\big)$. Note that the transmission condition implies that

$$\lim_{\xi_n \to -\infty} q(\xi_n) = \lim_{\xi_n \to +\infty} q(\xi_n),$$

so that indeed $\tilde{q}$ in $\mathcal{C}(S^1)$.

Corollary 7.6. *Conjugation by $\mathcal{F}^{-1}\mathcal{U}$ transforms the Wiener-Hopf algebra $\mathcal{W}$ into the Toeplitz algebra $\mathcal{T}$.*

Corollary 7.7. *Let p be as above. Then, by Proposition 5.38(c),*

$$\inf\{\|p^0(x',0,\xi',D_n)_+ + K : K \in \mathcal{K}(L^2(\mathbb{R}_+))\} = \sup_{\xi_n} |p^0(x',0,\xi',\xi_n)|.$$

The next result is an analog of the Gohberg-Seeley estimate in Theorem 6.3. It is shown in [32, 2.3.4.4, Theorem 1] based on work by Grubb and Geymonat [19].

Theorem 7.8. *Let $A = P_+ + G$ be an operator of order and class zero in Boutet de Monvel's calculus. Then*

$$\inf\{\|A + K\| : K \in \mathcal{K}\} = \max\{\|\sigma(A)\|_{\sup}, \|\gamma(A)\|_{\sup}\}.$$

In particular, A is compact, if and only if $\sigma(A)$ and $\gamma(A)$ both vanish.

Corollary 7.9. *The maps σ and γ induce maps $\bar\sigma$ and $\bar\gamma$ on $\mathcal{A}/\mathcal{K}$.*

7.3. *Kernel and image of the boundary symbol map γ*

We will next study the short exact exact sequence

$$0 \longrightarrow \ker\bar\gamma \stackrel{\iota}{\longrightarrow} \mathcal{A}/\mathcal{K} \stackrel{\bar\gamma}{\longrightarrow} \operatorname{im}\bar\gamma \longrightarrow 0. \tag{26}$$

There exists an adjoint-invariant ideal $\mathcal{I}^\circ$ in the algebra of all operators $A = P_+ + G$ in (22) consisting of those operators, where

$$\sigma(P)(x,\xi) = 0 \quad \text{on } S^*_{\partial X}X \text{ and} \tag{27}$$

$$\gamma(G) = g^0 = 0. \tag{28}$$

We denote by $\mathcal{I}$ the closure of $\mathcal{I}^\circ$ in $\mathcal{A}$. Clearly, $\mathcal{I}$ contains the closure of all operators of order -1 and hence the compact operators.

Theorem 7.10. $\ker\bar\gamma = \mathcal{I}/\mathcal{K} \cong \mathcal{C}_0(S^*X^\circ)$, *where the last isomorphism is given by the map $\bar\sigma$.*

Proof. We first show that $\ker\bar\gamma = \mathcal{I}/\mathcal{K}$.

'$\supseteq$'. By definition, the boundary symbol map γ vanishes on all operators (22) satisfying (27) and (28). Hence its extension to $\mathcal{A}$ vanishes on $\mathcal{I}$, and $\bar\gamma$ vanishes on $\mathcal{I}/\mathcal{K}$.

'$\subseteq$' is more delicate. A full proof is given in [27]. Here is the main idea: Let $p^0(x',0,\xi',D_n)_+ + g^0(x',\xi') = 0$ for $A = P_+ + G \in \mathcal{A}^\circ$. As $g(x',\xi)$ is compact, we infer from Corollary 7.7 that p^0 vanishes at ∂X. Hence $p^0(x',0,\xi',D_n) = 0$ and therefore also $g^0(x',\xi') = 0$. Thus $A \in \mathcal{I}^\circ$.

To conclude the proof, we use the estimate

$$\inf\{\|A + A^\circ\| : A^\circ \in \mathcal{I}^\circ\} = \|\gamma(A)\|_{\sup}, \quad A \in \mathcal{A}.$$

It implies that every $A \in \ker \gamma$ is in the closure of $\mathcal{I}^\circ$ and thus in $\mathcal{I}$.

Next it is clear that $\sigma(A)(x, \xi) = 0$ for $A \in \mathcal{I}$ and $x \in \partial X$. Hence $\bar{\sigma}$ maps $\mathcal{I}/\mathcal{K}$ to $\mathcal{C}_0(S^*X^\circ)$. The map is actually surjective: Given $q \in \mathcal{C}_0(S^*X^\circ)$ there is a sequence $(P_k)_k$ of zero order pseudodifferential operators with $\sigma(P_k)$ vanishing near ∂X and $\sigma(P_k) \to q$ in sup-norm. As the P_k belong to $\mathcal{I}^\circ$, we obtain the assertion. □

It is a natural conjecture that the image of $\bar{\gamma}$ is $\mathcal{C}(S^*\partial X, \mathcal{W})$. This, however, is not true. We need notation to describe the range of $\bar{\gamma}$.

Definition 7.11. Let $\mathcal{W}_0$ be the C^*-closure of the set of Wiener-Hopf operators of the form $q(D_n)_+ + K$ with $q(\xi_n) \to 0$ as $\xi_n \to \pm\infty$, K compact.

Clearly, under the isomorphism $\mathcal{W} \cong \mathcal{T}$ above, $\mathcal{W}_0$ is isomorphic to the ideal $\mathcal{T}_0$ in $\mathcal{T}$ introduced after Proposition 5.39.

Theorem 7.12. *The image of $\bar{\gamma}$ fits into a short exact sequence*

$$0 \longrightarrow \mathcal{C}(S^*\partial X, \mathcal{W}_0) \xrightarrow{\iota} \operatorname{im}\bar{\gamma} \xrightarrow{\pi} \mathcal{C}(\partial X) \longrightarrow 0. \tag{29}$$

Here ι is the inclusion map and π is the map which associates with $q(x', \xi', D_n)_+ + K$ the value of q at $\xi_n = \pm\infty$ (both values coincide due to the transmission condition).

The map π has a right inverse b. It associates to $f \in \mathcal{C}(\partial X)$ the image under $\bar{\gamma}$ of the class $[\mathcal{M}_{\tilde{f}}]$, where $\mathcal{M}_{\tilde{f}}$ is the operator of multiplication by an arbitrary function $\tilde{f}$ in $\mathcal{C}(\partial X)$, which coincides with f on ∂X.

A priori it is neither clear that ι maps into $\operatorname{im}\bar{\gamma}$ nor that π maps into $\mathcal{C}(\partial X)$; in fact, $q(x', \xi', \pm\infty)$ seems to depend also on ξ'. To see the former, we note that it is true for functions of the form $r(x', \xi')\varphi(D_n/\langle\xi'\rangle)\eta(\xi')$, where $r \in \mathcal{C}^\infty(S^*\partial X)$ is extended to $T^*\partial X \setminus 0$ by zero-homogeneity, η is a zero excision function, and $\varphi \in \mathcal{S}(\mathbb{R})$; for details see [27, Lemma 5]. These elements generate $\mathcal{C}(S^*X, \mathcal{W}_0)$.

In order to see the latter, let $f \in \mathcal{C}(S^*X, \mathcal{W})$ be an element in $\operatorname{im}\bar{\gamma}$. As $\mathcal{W} = \mathcal{W}_0 \oplus \mathbb{C}$, and since $\mathcal{C}(S^*\partial X, \mathcal{W}_0) \subset \operatorname{im}\bar{\gamma}$, we may subtract an element in $\mathcal{C}(S^*\partial X, \mathcal{W}_0)$ and assume that $f \in \mathcal{C}(S^*\partial X, \mathbb{C})$. As $\mathcal{A}^\circ$ is dense in $\mathcal{A}$, we find for arbitrary $\varepsilon > 0$ an element $A = P_+ + G$ such that

$$\|\gamma(P_+)(x', \xi') + \gamma(G)(x', \xi') - f(x', \xi')\| < \varepsilon.$$

Since $\gamma(G)$ is compact, we deduce that

$$\inf_K \|\gamma(P_+)(x',\xi') - f(x',\xi') + K\| < \varepsilon,$$

where the infimum is over all $K \in \mathcal{K}(L^2(\mathbb{R}_+))$. Expressing the left hand side via Corollary 7.7, and using the zero-homogeneity, we can estimate it from below:

$$\begin{aligned}&\sup_{\xi_n} |p^0(x',0,\xi',\xi_n) - f(x',\xi')|\\ &\quad = \sup_{\xi_n\neq 0} |p^0(x',0,\xi'/|\xi_n|,\pm 1) - f(x',\xi')| \geq |p^0(x',0,0,\pm 1) - f(x',\xi')|.\end{aligned}$$

Hence $f(x',\xi')$ is independent of ξ' and an element of $\mathcal{C}(\partial X)$. This argument also shows the exactness of the sequence.

Lemma 7.13. $K_j(S^*\partial X, \mathcal{W}_0) = 0$, $j = 0, 1$.

Proof. This follows from two facts. First we note that $K_j(\mathcal{W}_0) = 0$ by Proposition 5.40, since $\mathcal{W}_0$ is isomorphic to $\mathcal{T}_0$. In particular, $K_j(\mathcal{W}_0)$ is torsion free.

The second fact is that $\mathcal{C}(S^*\partial X)$ is a nuclear[k] C^*-algebra and so $\mathcal{C}(S^*\partial X, \mathcal{W}_0) \cong \mathcal{C}(S^*\partial X) \otimes \mathcal{W}_0$. This allows us to apply Künneth's theorem, see [9, 23.1.3]. We obtain that

$$\begin{aligned}K_0(\mathcal{C}(S^*\partial X)\otimes\mathcal{W}_0) &= K_0(\mathcal{C}(S^*\partial X))\otimes K_0(\mathcal{W}_0) + K_1(\mathcal{C}^(S^*\partial X))\otimes K_1(\mathcal{W}_0),\\ K_1(\mathcal{C}(S^*\partial X)\otimes\mathcal{W}_0) &= K_0(\mathcal{C}(S^*\partial X))\otimes K_1(\mathcal{W}_0) + K_1(\mathcal{C}(S^*\partial X))\otimes K_0(\mathcal{W}_0).\end{aligned}$$

In both cases, the result is zero. □

As an important consequence we find by considering the six-term exact sequence associated with (29):

Proposition 7.14. *The map* $b : \mathcal{C}(\partial X) \to \operatorname{im}\bar{\gamma}$ *defined in Theorem 7.12 induces an isomorphism in* K*-theory.*

7.4. *K-theory of* $\mathcal{A}/\mathcal{K}$

The key to determining the K-theory of $\mathcal{A}/\mathcal{K}$ now lies in combining the short exact sequence (26) with another short exact sequence to a commutative diagram. In order to make things fit better, we insert the fact that

---

[k]see Definition T.6.18 and Theorem T.6.20 in [49]. The latter states that every commutative C^*-algebra is nuclear.

$\ker \bar{\gamma} \cong \mathcal{C}_0(S^*X^\circ)$.

$$\begin{array}{ccccccccc} 0 & \longrightarrow & C_0(S^*X^\circ) & \longrightarrow & \mathcal{A}/\mathcal{K} & \xrightarrow{\bar{\gamma}} & \operatorname{im}\bar{\gamma} & \longrightarrow & 0 \\ & & \uparrow{\scriptstyle m^\circ} & & \uparrow{\scriptstyle m} & & \uparrow{\scriptstyle b} & & \\ 0 & \longrightarrow & C_0(X^\circ) & \longrightarrow & C(X) & \xrightarrow{r} & C(\partial X) & \longrightarrow & 0. \end{array} \tag{30}$$

Here m associates to a function f the operator of multiplication by f; m° is induced by the corresponding map from $\mathcal{C}_0(X^\circ)$ to $\ker\bar{\gamma}$ and thus associates to f its composition with the bundle projection $S^*X^\circ \to X^\circ$. Finally r denotes the restriction to the boundary.

According to Lemma 5.34, we obtain from the commutative diagram (30) a commutative grid

$$\begin{array}{ccccccccc} & & 0 & & 0 & & 0 & & \\ & & \uparrow & & \uparrow & & \uparrow & & \\ 0 & \longrightarrow & \mathcal{C}_0(X^\circ) & \longrightarrow & \mathcal{C}(X) & \xrightarrow{r} & \mathcal{C}(\partial X) & \longrightarrow & 0 \\ & & \uparrow & & \uparrow & & \uparrow & & \\ 0 & \longrightarrow & Cm^\circ & \longrightarrow & Cm & \longrightarrow & Cb & \longrightarrow & 0\,. \\ & & \uparrow & & \uparrow & & \uparrow & & \\ 0 & \longrightarrow & S(\mathcal{C}_0(S^*X^\circ)) & \longrightarrow & S(\mathcal{A}/\mathcal{K}) & \xrightarrow{S\pi} & S(\operatorname{im}\bar{\gamma}) & \longrightarrow & 0 \\ & & \uparrow & & \uparrow & & \uparrow & & \\ & & 0 & & 0 & & 0 & & \end{array} \tag{31}$$

Let us first consider the column on the right. The associated six-term exact sequence in K-theory is

$$\begin{array}{ccccc} K_0(S(\operatorname{im}\bar{\gamma})) & \to & K_0(Cb) & \to & K_0(\mathcal{C}(\partial X)) \\ \uparrow{\scriptstyle \mathrm{ind}} & & & & \downarrow{\scriptstyle \exp} \\ K_1(\mathcal{C}(\partial X)) & \leftarrow & K_1(Cb) & \leftarrow & K_1(S(\operatorname{im}\bar{\gamma})) \end{array} \tag{32}$$

We know from Theorem 5.33 that the connecting maps 'ind' and 'exp' are the compositions of b_* with the isomorphisms $\theta_j : K_j(\operatorname{im}\bar{\gamma})) \to K_j(S(\operatorname{im}\bar{\gamma}))$. As b_* is an isomorphism by Proposition 7.14, so are 'ind' and 'exp', and we conclude that

$$K_j(Cb) = 0, \quad j = 0, 1.$$

Next we study the six-term exact sequence associated with the middle row. Since the K-theory of Cb vanishes, we find that there exists an isomorphism

$$K_j(Cm^\circ) \overset{\varphi}{\cong} K_j(Cm).$$

Now we write the six-term exact sequences for the two columns on the left. For better legibility, we write them in one line each (not in a rectangle) and we take into account the isomorphisms between the K-theory of a C^*-algebra and that of its suspension given in Theorem 5.33. This yields the commutative diagram, below.

$$\begin{array}{ccccccc} \ldots \to K_j(\mathcal{C}(X)) & \overset{m_*}{\to} & K_j(\mathcal{A}/\mathcal{K}) & \overset{\beta}{\to} & K_{1-j}(Cm) & \to \ldots \\ \uparrow & & \uparrow{\scriptstyle \iota_*} & & \uparrow{\scriptstyle \varphi} & \\ \ldots \to K_j(\mathcal{C}(X^\circ)) & \overset{m^\circ_*}{\to} & K_j(\mathcal{C}_0(S^*X^\circ)) & \overset{\alpha}{\to} & K_{1-j}(Cm^\circ) & \to \ldots \end{array} \tag{33}$$

A crucial point now is the following well-known result; for a proof see *e.g.* [27, Proposition 9]:

Proposition 7.15. *Let X be connected with $\partial X \neq \emptyset$. Then S^*X has a continuous section.*

Calling this section s, we obtain a left inverse $s^* : \mathcal{C}_0(S^*(X^\circ)) \to \mathcal{C}_0(X^\circ)$ to m° by $(s^*f)(x) = f(s(x))$ and hence a left inverse to m°_*. Consequently, there is a right inverse to α and, via inversion of φ, a right inverse r to β. The upper exact sequence therefore yields short exact sequences

$$0 \to K_j(\mathcal{C}(X)) \xrightarrow{m_*} K_j(\mathcal{A}/\mathcal{K}) \xrightarrow{\beta} K_{1-j}(Cm) \to 0, \quad j = 0,1. \tag{34}$$

This has an immediate consequence:

Lemma 7.16. *Every element in $K_j(\mathcal{A}/\mathcal{K})$ can be written as the sum of an element in the image of m_* and one in the image of the map i_* in* (33).

Proof. Every element in $K_j(\mathcal{A}/\mathcal{K})$ can be written as the sum of an element in the image of m_* and one in the image of the right inverse r to β: Write $x = (x - r(\beta x)) + r(\beta x)$. Then the first summand is in the kernel of β, hence the image of m_*. By construction, the image of r is a subset of the image of i_*. □

In [28, Section 3] the proposition, below, is established

Proposition 7.17. *There is a natural way to identify $K_j(Cm)$ and $K_j(C_0(T^*X^\circ))$, $j = 0, 1$.*

This identification, however, uses the inverse of the isomorphism $K_j(Cm^\circ) \to K_j(Cm)$, therefore it is not explicitly given.

Putting these results together, we see:

Theorem 7.18. *$K_j(\mathcal{A}/\mathcal{K})$ fits into a short exact sequence*

$$0 \to K_j(\mathcal{C}(X)) \xrightarrow{m_*} K_j(\mathcal{A}/\mathcal{K}) \xrightarrow{p} K_{1-j}(\mathcal{C}_0(T^*X^\circ)) \to 0, \quad j = 0, 1.$$

The sequence splits, i.e. there is a (non-canonical) right inverse for p. In particular, we can write (in a non-canonical way)

$$K_j(\mathcal{A}/\mathcal{K}) \cong K_j(\mathcal{C}(X)) \oplus K_{1-j}(\mathcal{C}_0(T^*X^\circ)).$$

Considering the six-term exact sequence associated with the short exact sequence

$$0 \to \mathcal{K} \to \mathcal{A} \to \mathcal{A}/\mathcal{K} \to 0,$$

we obtain the following result:

Corollary 7.19. *$K_0(\mathcal{A})$ is canonically isomorphic with $K_0(\mathcal{A}/\mathcal{K})$, while $K_1(\mathcal{A})$ is isomorphic to the kernel of the surjective index map* ind : $K_1(\mathcal{A}/\mathcal{K}) \to K_0(\mathcal{K}) \cong \mathbb{Z}$.

There is a more intuitive picture of the K-theory of $\mathcal{A}/\mathcal{K}$. It goes back to [10, Theorem 5.14] and is explained in [29, Theorem 2.11] and [30, Section 4].

Definition 7.20. Denote by $\mathcal{B}$ the subalgebra of $\mathcal{C}(S^*X)$ consisting of those functions $f = f(x, \xi)$ which are independent of ξ for $x \in \partial X$ and by $\mathcal{A}^\dagger$ the subalgebra of $\mathcal{A}$ of all operators A of the form

$$A = \begin{pmatrix} P_+ + K_1 & K_2 \\ K_3 & K_4 \end{pmatrix},$$

where the principal symbol of P is in $\mathcal{B}$ and K_1, K_2, K_3, K_4 are compact.

Then the following holds:

Theorem 7.21. $K_j(\mathcal{B}) \cong K_j(\mathcal{A}^\dagger/\mathcal{K}) \cong K_j(\mathcal{A}/\mathcal{K}), \quad j = 0, 1.$

7.5. *Index theory for Boutet de Monvel's algebra*

Theorem 7.18 allows to express the index of an elliptic operator $A \in \mathcal{A}$ in purely K-theoretic terms. To this end we recall the topological index map

$$\text{ind}_t : K_0(\mathcal{C}_0(T^*X^\circ)) \cong K^0_c(T^*X^\circ) \to \mathbb{Z}.$$

Theorem 7.22. *The index of a Fredholm operator $A \in \mathcal{A}$ is given by*

$$\text{ind } A = \text{ind}_t(p[A]),$$

where $[A]$ is the class of A in $K_1(\mathcal{A}/\mathcal{K})$ and p is defined in Theorem 7.18.

A full proof of this theorem can be found in [28]. It relies on Lemma 7.16: Every element in $K_1(\mathcal{A}/\mathcal{K})$ can be written as the sum of an element in the image of m_* and one in the image of i_*. Hence it suffices to show that the composition $\text{ind}_t \circ p$ furnishes the Fredholm index on both, the image of m_* and the image of i_*.

The first is easy to see: An element in the image of m_* is the K_1-class associated with the image in $\mathcal{A}/\mathcal{K}$ of a multiplication operator. As a multiplication operator is elliptic if and only if it is invertible, its Fredholm index is zero. On the other hand, $\text{im}\, m_* = \ker p$, hence also $\text{ind}_t \circ p$ vanishes.

Now suppose the K_1-class $[A]$ of A lies in the image of $K_1(\mathcal{C}_0(S^*X^\circ))$ under i_*. We recall that X is embedded in a closed manifold $\widetilde{X}$ of the same dimension. The inclusion $\iota : \mathcal{C}_0(S^*X^\circ) \hookrightarrow \mathcal{C}(S^*\widetilde{X})$ induces a map

$$\iota_* : K_1(\mathcal{C}_0(S^*X^\circ)) \to K_1(\mathcal{C}(S^*\widetilde{X})).$$

The elements of the right hand side are K_1-classes associated with elliptic operators on $\widetilde{X}$. With the help of the Atiyah-Singer index theorem we can then show the assertion.

Remark 7.23. In [12] Fedosov obtained a cohomological formula for the index of an operator in $\mathcal{A}$:

$$\text{ind } A = \int_{T^*X} \text{ch}(\sigma(A))\text{Td}(X) + \int_{T^*\partial X} \text{ch}'(\gamma(A))\text{Td}(X). \qquad (35)$$

Here $\text{Td}(X)$ is the differential form given by the Todd class of the tangent bundle (or its restriction to the boundary), ch is the Chern character, and $\text{ch}'\,\gamma(A)$ is a differential form constructed from the K-class of the boundary symbol; it is given by a formula analogous to that for the usual Chern character.

References

[1] J. Aastrup, R. Nest and E. Schrohe. *A continuous field of C^*-algebras and the tangent groupoid for manifolds with boundary.* J. Funct. Anal. **237** (2006), no. 2, 482–506.

[2] J. Aastrup, R. Nest and E. Schrohe. *Index theory for boundary value problems via continuous fields of C^*-algebras.* J. Funct. Anal. **257** (2009), no. 8, 2645–2692.

[3] M. F. ATIYAH. *K-Theory.* Lecture notes by D. W. Anderson. W. A. Benjamin, Inc., New York-Amsterdam, 1967.

[4] M. F. ATIYAH AND I. M. SINGER. *The index of elliptic operators on compact manifolds.* Bull. Amer. Math. Soc. **69** (1963), 422–433.

[5] M. F. ATIYAH AND I. M. SINGER. *The index of elliptic operators.* I. Ann. of Math. (2), **87** (1968), 484–530.

[6] M. F. ATIYAH & I. M. SINGER. *The index of elliptic operators IV.* Ann. of Math. (2) **93** (1971), 119–138.

[7] U. BATTISTI, S. CORIASCO, E. SCHROHE. *Fourier integral operators and the index of symplectomorphisms on manifolds with boundary.* J. Funct. Anal. 269:3528?3574 (2015).

[8] BEALS, R.: Characterization of pseudodifferential operators and applications, *Duke Math. Journal* **44**, 45–57 (1977), *ibid.* **46**, p. 215 (1979).

[9] B. BLACKADAR. *K-Theory for Operator Algebras.* Cambridge University Press, Cambridge, 1998.

[10] L. BOUTET DE MONVEL. *Boundary problems for pseudo-differential operators.* Acta Math. **126** (1971), no. 1-2, 11–51.

[11] R. DOUGLAS. *Banach algebra techniques in operator theory.* Second edition. Graduate Texts in Mathematics, 179. Springer-Verlag, New York, 1998

[12] B. V. FEDOSOV. *Index theorems.* In *Partial differential equations.* VIII, 155–251, Encyclopaedia Math. Sci., 65, Springer, Berlin, 1996.

[13] B. V. FEDOSOV, F. GOLSE, E. LEICHTNAM, E. SCHROHE. *Le résidu non commutatif pour les variétés à bord.* C.R. Acad. Sc. Paris **320** Série I, 669 - 674, 1995

[14] B.V. FEDOSOV, F. GOLSE, E. LEICHTNAM, AND E. SCHROHE. *The noncommutative residue for manifolds with boundary.* J. Funct. Anal. **142** (1996), no. 1, 1–31.

[15] G. FOLLAND. *Introduction to Partial Differential Equations.* Second edition. Princeton University Press, Princeton, NJ, 1995.

[16] I. C. GOHBERG. *On the theory of multidimensional singular integral operators.* Dokl. Akad. Nauk SSSR **133** (1960), 1279–1282.

[17] B. GRAMSCH, E. SCHROHE. *Submultiplicativity of Boutet de Monvel's algebra for boundary value problems.* Pseudo-differential calculus and mathematical physics, 235–258, Math. Top., 5, Akademie Verlag, Berlin, 1994.

[18] G. GRUBB. *Functional Calculus of Pseudodifferential Boundary Problems.* Second Edition. Birkhäuser, Boston, 1996.

[19] G. Grubb and G. Geymonat. *The essential spectrum of elliptic systems of mixed order.* Math. Ann. **227** (1977), 247–276.

[20] G. Grubb and E. Schrohe. *Trace expansions and the noncommutative residue for manifolds with boundary.* J. Reine Angew. Math. **536** (2001), 167–207.

[21] G. Grubb and E. Schrohe. *Traces and quasi-traces on the Boutet de Monvel algeb ra.* Ann. Inst. Fourier (Grenoble) **54** (2004), no. 5, 1641–1696

[22] L. Hörmander. *The analysis of linear partial differential operators.* III. *Pseudo-differential operators.* Corrected reprint of the 1985 original. Grundlehren der Mathematischen Wissenschaften 274. Springer-Verlag, Berlin, 1994.

[23] I. Hwang. *The L^2-boundedness of pseudodifferential operators.* Trans. Amer. Math. Soc., **302** (1987), 55–76.

[24] M. Karoubi. K-theory. An introduction. Grundlehren der Mathematischen Wissenschaften, Band 226. Springer-Verlag, Berlin-New York, 1978.

[25] J. J. Kohn and L. Nirenberg. *An algebra of pseudo-differential operators.* Comm. Pure Appl. Math. **18** (1965), 269–305.

[26] H. Kumano-go. *Pseudo-differential operators.* The MIT-Press, Cambridge, Mass. 1982.

[27] S. T. Melo, R. Nest, E. Schrohe. *C^*-structure and K-theory of Boutet de Monvel's algebra.* J. reine angew. Math. **561** (2003), 145–175.

[28] S. T. Melo, T. Schick, E. Schrohe. *A K-theoretic proof of Boutet de Monvel's index theorem for boundary value problems.* J. reine angew. Math. **599** (2006), 217–233.

[29] S. T. Melo, T. Schick, E. Schrohe. *Families index for Boutet de Monvel operators.* Münster J. Math. **6**, (2013), 343–364.

[30] S. T. Melo, T. Schick, E. Schrohe. *C^*-algebra approach to the index theory of boundary value problems.* In 'Analysis, Geometry, and Quantum Field Theory', Contemp. Math. **584**, 2012, 129–146,.

[31] R. Nest and E. Schrohe. *Dixmier's trace for boundary value problems.* Manuscripta Math. **96** (1998), 203–218.

[32] S. Rempel & B.-W. Schulze. *Index Theory of Elliptic Boundary Problems.* Akademie Verlag, Berlin, 1982.

[33] M. Rørdam, F. Larsen, and N. J. Laustsen. *An Introduction to K-Theory for C^*-Algebras.* London Mathematical Society Student Texts **49**, Cambridge University Press, Cambridge 2000.

[34] W. RUDIN. *Functional Analysis.* Second Edition. McGraw-Hill, Inc., New York, 1991.

[35] C. SCHOCHET. *Topological methods for C^*-algebras II: geometric resolutions and the Künneth formula.* Pacific J. Math. **98** (1982), no. 2, 443–458.

[36] E. SCHROHE. Spaces of weighted symbols and weighted Sobolev spaces on manifolds, pp. 360 - 377 in *"Pseudo-Differential Operators"*, Springer LN Math. 1256, Berlin, Heidelberg, New York 1987

[37] E. SCHROHE. *Boundedness and spectral invariance for standard pseudodifferential operators on anisotropically weighted L^p-Sobolev spaces,* Integral Equations OperatorTheory **13** (1990), 271 – 284.

[38] E. SCHROHE *Fréchet algebra techniques for boundary value problems on noncompact manifolds: Fredholm criteria and functional calculus via spectral invariance.* Math. Nachr. **199** (1999), 145–185.

[39] E. SCHROHE. *A short introduction to Boutet de Monvel's calculus*; Approaches to Singular Analysis (Berlin, 1999). Oper. Theory Adv. Appl. **125**, 85-116, Birkhäuser, Basel, 2001.

[40] E. SCHROHE AND B.-W. SCHULZE. *Boundary value problems in Boutet de Monvel's algebra for manifolds with conical singularities.* I. Pseudo-differential calculus and mathematical physics, 97–209, Math. Top., 5, Akademie Verlag, Berlin, 1994.

[41] E. SCHROHE AND B.-W. SCHULZE. *Boundary value problems in Boutet de Monvel's algebra for manifolds with conical singularities.* II. Boundary value problems, Schrödinger operators, deformation quantization, 70–205, Math. Top., 8, Akademie Verlag, Berlin, 1995.

[42] B.-W. SCHULZE. *Topologies and invertibility in operator spaces with symbolic structures.* Proceedings of the 9th Conference on Problems and Methods in Mathematical Physics (9.TMP) (Karl-Marx-Stadt, 1988), 257–270, Teubner-Texte Math., 111, Teubner, Leipzig, 1989.

[43] B.-W. SCHULZE. *Pseudo-Differential Operators on Manifolds with Singularities.* North Holland, Amsterdam, 1991.

[44] B.-W. SCHULZE. *Pseudo-differential boundary value problems, conical singularities, and asymptotics.* Mathematical Topics, 4. Akademie Verlag, Berlin, 1994.

[45] M.A. SHUBIN *Pseudodifferential operators and spectral theory.* Translated from the 1978 Russian original by Stig I. Andersson. Second edition. Springer-Verlag, Berlin, 2001

[46] R. T. SEELEY. *Integro-differential operators on vector bundles.* Trans. Amer. Math. Soc. **117** (1965), 167–204.

[47] M. TAYLOR. *Pseudodifferential operators.* Princeton Mathematical Series, 34. Princeton University Press, Princeton, N.J., 1981.

[48] M. TAYLOR. *Partial Differential Equations* I. Basic theory. Applied Mathematical Sciences, 115. Springer-Verlag, New York, 1996.

[49] N. E. WEGGE-OLSEN. *K-theory and C^*-algebras.* Oxford Science Publications. The Clarendon Press, Oxford University Press, New York, 1993.

COUNTING AN INFINITE NUMBER OF POINTS: A TESTING GROUND FOR RENORMALIZATION METHODS

Li Guo

Department of Mathematics and Computer Science,
Rutgers University,
Newark, NJ 07102, USA
Email: liguo@rutgers.edu

Sylvie Paycha

Institute of Mathematics,
University of Potsdam,
Am Neuen Palais 10,
D-14469 Potsdam, Germany
Email: paycha@math.uni-potsdam.de

Bin Zhang

Yangtze Center of Mathematics,
Department of Mathematics,
Sichuan University,
Chengdu, 610064, P. R. China
Email: zhangbin@scu.edu.cn

This is a leisurely introductory account addressed to non-experts and based on previous work by the authors, on how methods borrowed from physics can be used to "count" an infinite number of points. We begin with the classical case of counting integer points on the non-negative real axis and the classical Euler-Maclaurin formula. As an intermediate stage, we count integer points on product cones where the roles played by the coalgebra and the algebraic Birkhoff factorization can be appreciated in a relatively simple setting. We then consider the general case of (lattice) cones for which we introduce a connected cograded coalgebra of cones, with applications to renormalization of conical zeta values. When evaluated at zero arguments conical zeta functions indeed "count" integer points on cones.

Keywords: Cones, coalgebra, renormalization, Birkhoff decomposition, Euler-Maclaurin formula, meromorphic functions

Introduction

"Counting" an infinite number of points might seem pointless and a lost cause; it has nevertheless been the concern of many a mathematician as far back as Leonhardt Euler and Bernhardt Riemann and relates to renormalization issues in quantum field theory.

We want to "count" lattice points on rational polyhedral convex cones, starting from the one dimensional cone $\mathbb{R}_+$ with lattice points given by the positive integers studied in the first section. Evaluating the Riemann zeta function at zero provides one way of "counting" the positive integers. It indeed assigns a finite value $\zeta(0) = -\frac{1}{2}$ to the ill-defined sum "$\sum_{n=1}^{\infty} n^0$" by means of an analytic continuation $\zeta(z)$ of the regularized sum $\sum_{n=1}^{\infty} n^{-z}$. Alternatively the "number" $\frac{1}{2} = 1 - \frac{1}{2}$ of non-negative integers can be derived using an alternative approximation $S(\varepsilon) = \sum_{n=0}^{\infty} e^{-\varepsilon n}$ by an exponential sum. Its analytic extension (denoted by the same symbol S) presents a simple pole at $\varepsilon = 0$ with residue 1 so that $S(\varepsilon) = \frac{1}{\varepsilon} + S_+(\varepsilon)$ where S_+ is holomorphic at zero. Coincidentally, the "polar part" $\frac{1}{\varepsilon}$ equals the integral $I(\varepsilon) = \int_0^{\infty} e^{-\varepsilon x} dx$ leading to the Euler-Maclaurin formula $S = I + \mu$ which relates the sum and the integral of the map $x \mapsto e^{-\varepsilon x}$ by means of the interpolator $\mu = S_+$. Using the terminology borrowed from physicists, we refer to the decomposition $S(\varepsilon) = \frac{1}{\varepsilon} + S_+(\varepsilon)$ into a "polar part" $\frac{1}{\varepsilon}$ and a holomorphic part $S_+(\varepsilon)$ as the minimal subtraction scheme applied to S. For this particular function, it coincides with the Euler-Maclaurin formula and we have $S_+(0) = \mu(0) = \zeta(0) + 1 = \frac{1}{2}$.

The coincidence in the case of the case of the exponential sum, between the minimal subtraction scheme and the Euler-Maclaurin formula, carries out to higher dimensions. The second section is dedicated to "counting" the lattice points $\mathbb{Z}_{\geqslant 0}^k$ of a (closed) product cone $\mathbb{R}_{\geqslant 0}^k$ of dimension $k \in \mathbb{N}$. One expects the "number" of points of $\mathbb{Z}_{\geqslant 0}^k$ to be the k-th power of the "number" of points of $\mathbb{Z}_{\geqslant 0}$ and this is indeed the case provided one "counts carefully". By this we mean that one should not naively evaluate the "holomorphic part" of the k-th power $S^k(\varepsilon)$ at zero of the exponential sum but instead take the k-th power $S_+^k(0)$ of the holomorphic part S_+ evaluated at zero, which is a straightforward procedure in the rather trivial case of product cones. However there is a general algebraic construction which derives S_+^k from S^k, known as the algebraic Birkhoff factorization that can be viewed as a generalization to higher dimensions of the minimal subtraction scheme mentioned above. It relies on a coproduct on (product) cones built from a complement map on the poset of convex cones as described in Section 2 (Compare with [GP]), which uses a projection onto an orthogonal complement set (see Lemma 3.14), separating a face of the cone from the remaining faces. When applied to the multivariable exponential sum $\widetilde{S}_k : (\varepsilon_1, \cdots, \varepsilon_k) \mapsto \prod_{i=1}^k S(\varepsilon_i)$ on the product cone $\mathbb{R}^k_{\geqslant 0}$, the general algebraic Birkhoff factorization on coalgebras described in Section 4 gives $(\varepsilon_1, \cdots, \varepsilon_k) \mapsto \prod_{i=1}^k S_+(\varepsilon_i)$ as the "renormalized holomorphic" part of the map $\widetilde{S}_k$. This algebraic Birkhoff factorization can also be interpreted as an Euler-Maclaurin formula for it factorizes the sum as a (convolution) product of integrals and interpolators on product cones.

We close this presentation by briefly mentioning the corresponding result on rational polyhedral (lattice) cones, namely that the Euler-Maclaurin formula (first derived in [BV], see also [B]) for the exponential sum is given by its algebraic Birkhoff factorization, leaving out the precise statement for which we refer to reader to [GPZ3]. Renormalized conical zeta values associated to a cone C correspond to the Taylor coefficients of the "holomorphic part" $S_+(C)$ of the multivariable exponential sum $S(C)$ on the cone. For Chen cones $x_k \leqslant \cdots \leqslant x_1$ they yield renormalized multiple zeta values [GPZ5]; in Section 5 we illustrate our approach with the computation of renormalized multiple zeta values with 2 and 3 arguments.

The algebraic Birkhoff factorization on cones, seen as a general device which renormalizes any conical zeta value at non-positive integers, therefore yields a geometric approach to renormalize multiple zeta values at non-positive integers. This geometric approach contrasts with other approaches such as [MP] and [GZ] to the renormalization of multiple zeta values at non-positive arguments, where the algebraic Birkhoff factorization is carried out on the summands (functions $(x_1, \cdots, x_k) \mapsto x_1^{-s_1} \cdots x_1^{-s_k}$) rather than on the domain (the cones) of summation as in our present construction or [Sa] where a purely analytic renormalization method is implemented, which does not use algebraic Birkhoff factorization.

To conclude, "counting" lattice points on cones which might a priori seem like a very specific issue, actually brings together i) renormalization methods à la Connes and Kreimer [CK] borrowed from quantum field theory in the form of algebraic Birkhoff factorization, ii) the Euler-Maclaurin formula on cones and hence on polytopes used to study the geometry of toric varieties, iii) number theory with the conical zeta values (introduced in [GPZ2]) that generalize multiple zeta values [Ho, Za], and which arise in our context as the Taylor coefficients of the interpolator in the Euler-Maclaurin formula. We hope that this presentation which does not claim to be neither exhaustive nor new since it relies on previous work by the authors, will act as an incentive for the lay reader to get further acquainted with renormalization methods.

1. Counting integers

We want to count the non-negative integer points *i.e.* to evaluate the ill-defined sum "$1 + 1 + \cdots + 1 + \cdots = \sum_{n=0}^{\infty} n^0$" and more generally the no better defined sum $\sum_{n=0}^{\infty} n^k$ for any non-negative integer k.

1.1. *Approximated sums over integers*

We first approximate these ill-defined sums; there are at least three ways to do so, we refer the reader to for a more detailed description of these various regularization methods:

(a) The **cut-off regularization** only considers a finite number of terms of the sum. For $N \in \mathbb{N}$ we set $S_k(N) := \sum_{n=0}^{N} n^k$;

(b) The **heat-kernel type regularization** approximates the summand by an exponential expression. For positive ε we set

$$S(\varepsilon) := \sum_{n=0}^{\infty} e^{-\varepsilon n} \tag{1}$$

and $S_k(\varepsilon) := \sum_{n=0}^{\infty} n^k e^{-\varepsilon n} = (-1)^k \partial_k S(\varepsilon)$;

(c) The **zeta-function type regularization** approximates the summand by a complex power. For a complex number z whose real part is larger than 1, the expression

$$\tilde{S}(z) := \sum_{n=1}^{\infty} n^{-z} =: \zeta(z)$$

called the ζ-function converges and $\tilde{S}_k(z) := \sum_{n=1}^{\infty} n^{k-z} = \zeta(z-k)$ converges for any complex number z whose real part is larger than $k+1$.

The sums S and $\tilde{S}$ relate via the Mellin transform; for any positive number λ the map $f_\lambda : \varepsilon \mapsto e^{-\lambda\varepsilon}$ defines a Schwartz function whose **Mellin transform** reads

$$\mathcal{M}\left(f_\lambda\right)(z) := \frac{1}{\Gamma(z)} \int_0^{\infty} \varepsilon^{z-1} f_\lambda(\varepsilon)\, d\varepsilon = \lambda^{-z}.$$

$\tilde{S}(z) = \sum_{n=1}^{\infty} \mathcal{M}\left(f_n\right)(z) = \mathcal{M}\left(S-1\right)(z)$. This extends to an identity of meromorphic functions

$$\tilde{S} = \mathcal{M}(S-1)$$

with simple poles at integers smaller or equal 1. It turns out that the residue at $z = 1$ is one and zero elsewhere.

The sum $S(\varepsilon) = \frac{1}{1-e^{-\varepsilon}}$ can be expressed in terms of the **Todd function**[a].

$$\mathrm{Td}(\varepsilon) := \frac{\varepsilon}{e^{\varepsilon} - 1} \tag{2}$$

as

$$S(\varepsilon) = \frac{\mathrm{Td}(-\varepsilon)}{\varepsilon}.$$

The **Todd function** is the exponential generating function for the **Bernoulli numbers**[b] that correspond to the Taylor coefficients[c]

$$\mathrm{Td}(\varepsilon) = \sum_{n=0}^{\infty} B_n \frac{\varepsilon^n}{n!}. \tag{3}$$

We have

$$\mathrm{Td}(\varepsilon) = \frac{\varepsilon}{e^{\varepsilon} - 1} = \frac{\varepsilon}{\varepsilon + \frac{\varepsilon^2}{2} + o(\varepsilon^2)} = \frac{1}{1 + \frac{\varepsilon}{2} + o(\varepsilon)} = 1 - \frac{\varepsilon}{2} + o(\varepsilon)$$

so $B_0 = 1; B_1 = -\frac{1}{2}$. Since $\frac{\varepsilon}{e^{\varepsilon}-1} + \frac{\varepsilon}{2} = \frac{\varepsilon}{2} \frac{e^{\frac{\varepsilon}{2}} + e^{-\frac{\varepsilon}{2}}}{e^{\frac{\varepsilon}{2}} - e^{-\frac{\varepsilon}{2}}}$ is an even function, $B_{2k+1} = 0$ for any positive integer k.

Consequently, for any positive integer K we have

$$\mathrm{Td}(\varepsilon) = 1 - \frac{\varepsilon}{2} + \sum_{k=1}^{K} \frac{B_{2k}}{(2k)!} \varepsilon^{2k} + o(\varepsilon^{2K}) \tag{4}$$

and

$$S(\varepsilon) = \frac{\mathrm{Td}(-\varepsilon)}{\varepsilon} = \frac{1}{\varepsilon} + \frac{1}{2} + \sum_{k=1}^{K} \frac{B_{2k}}{(2k)!} \varepsilon^{2k-1} + o(\varepsilon^{2K}). \tag{5}$$

---

[a]There are two variants of the Todd function; in topology it is defined as the map $\tau : \varepsilon \mapsto \frac{\varepsilon}{1-e^{-\varepsilon}}$, an alternative definition one finds in the literature is $\varepsilon \mapsto \tau(-\varepsilon)$, which we opt for in these notes.

[b]They were discovered by Jakob Bernoulli and independently by a Japanese mathematician Seki Kōwa, both of whose discoveries were posthumously published (in 1712 for Seki Kōwa, in his work Katsuyo Sampo, in 1713 for Bernoulli, in his Ars Conjectandi).

[c]They also arise in the Taylor series expansions of the tangent and hyperbolic tangent functions.

1.2. *The one-dimensional Euler-Maclaurin formula*

As a consequence of formula (5), the discrete sum $S(\varepsilon) := \sum_{k=0}^{\infty} e^{-\varepsilon k} = \frac{1}{1-e^{-\varepsilon}}$ for positive ε relates to the integral

$$I(\varepsilon) := \int_0^\infty e^{-\varepsilon x}\, dx = \frac{1}{\varepsilon} \tag{6}$$

by means of the interpolator

$$\mu(\varepsilon) := S(\varepsilon) - I(\varepsilon) = S(\varepsilon) - \frac{1}{\varepsilon} = \frac{1}{2} + \sum_{k=1}^{K} \frac{B_{2k}}{(2k)!}\varepsilon^{2k-1} + o(\varepsilon^{2K}) \quad \text{for all } K \in \mathbb{N},$$

which is holomorphic at $\varepsilon = 0$. This interpolation formula between the sum and the integral

$$S(\varepsilon) = I(\varepsilon) + \mu(\varepsilon) \tag{7}$$

generalizes to other L^1 functions by means of the Euler-Maclaurin formula.

As a motivation for the Euler-Maclaurin formula with remainder, let us first derive a formal Euler-Maclaurin formula using the Todd function $\mathrm{Td}(D)$ obtained by inserting the derivation map $D : f \mapsto f'$ on $C^\infty(\mathbb{R})$ in formula (2).

Let $\nabla f(x) = f(x) - f(x-1)$ denote the discrete derivation. Using a formal Taylor expansion, we have

$$\nabla f(x) = f(x) - f(x-1) = \sum_{k=1}^{\infty} \frac{(-1)^{k+1} D^k f(x)}{k!} = \left(1 - e^{-D}\right)(f)(x)$$

and hence at any non-negative integer n

$$\begin{aligned}\left(\nabla^{-1} f\right)(n) &= \left(1 - e^{-D}\right)^{-1}(f)(n)\\ &= \left(D^{-1} f\right)(n) + \frac{1}{2} + \sum_{k=1}^{K} \frac{B_{2k}}{(2k)!}\left(D^{2k-1} f\right)(n) + o(\varepsilon^{2K}).\end{aligned}$$

Here $\left(\nabla^{-1} f\right)(n) = \sum_{k=0}^{n} f(k) + C$ stands for the discrete primitive of f defined modulo a constant; it satisfies $\left(\nabla \circ \nabla^{-1}\right) f(n) = f(n)$ for any $n \in \mathbb{Z}_{\geqslant 0}$. Similarly, $\left(D^{-1} f\right)(x) = \int_0^x f(y)\, dy + C$ stands for the continuous integration map defined modulo a constant; it satisfies $\left(D \circ D^{-1}\right) f(x) = f(x)$ for any $x \in \mathbb{R}$. This gives a first formal expansion

$$\begin{aligned}\sum_{n=a}^{b} f(n) &= \left(\nabla^{-1} f\right)(b) - \left(\nabla^{-1} f\right)(a)\\ &= \left(D^{-1} f\right)(b) - \left(D^{-1} f\right)(a) + \frac{f(a) + f(b)}{2}\end{aligned}$$

$$
\begin{aligned}
&+\sum_{j=1}^{J} \frac{B_{2j}}{(2j)!}\left(D^{2j-1}f(b)-D^{(2j-1)}f(a)\right)\\
&=\int_a^b f(x)\,dx+\frac{f(a)+f(b)}{2}\\
&+\sum_{j=1}^{J} \frac{B_{2j}}{(2j)!}\left(f^{(2j-1)}(b)-f^{(2j-1)}(a)\right),
\end{aligned}
$$

for any two non-negative integers a and b.

We are now ready to state the **Euler-Maclaurin formula** with remainder [Ha]

Proposition 1.1. *For any function f in $C^\infty(\mathbb{R})$ and any two integers $a < b$,*

$$
\begin{aligned}
\sum_{n=a}^{b} f(n) = \frac{f(a)+f(b)}{2} &+ \int_a^b f(x)\,dx\\
&+\sum_{j=1}^{J} \frac{B_{2j}}{(2j)!}\left(f^{(2j-1)}(b)-f^{(2j-1)}(a)\right)\\
&-\frac{1}{(2J)!}\int_a^b \overline{B_{2J}}(x)\,f^{(2J)}(x)\,dx
\end{aligned}
\tag{8}
$$

where J is any positive integer and $\overline{B}_n(x) := B_n\left(x-\lfloor x\rfloor\right)$ built from the Bernoulli polynomials (see e.g. [A]) $B_n(x) := \sum_{k=0}^{n}\binom{n}{k} B_{n-k}\, x^k$ and $\lfloor x\rfloor$ the integral part of x.

In particular, for $f(x) = x^k$ and $a = 0, b = N$ we have $f^{(2j-1)}(x) = \frac{k!}{(k-2j+1)!}x^{k-2j+1}$ and hence

$$
\begin{aligned}
S_k(N) :&= \sum_{n=0}^{N} n^k = \frac{\delta_k + N^k}{2}\\
&+\int_0^N x^k\,dx + \sum_{j=1}^{\left[\frac{k+1}{2}\right]} \frac{B_{2j}}{2j!}\left(\frac{k!}{(k-2j+1)!}\left(N^{k-2j+1}-\delta_{k-2j+1}\right)\right)\\
&=\frac{N^{k+1}}{k+1}+\frac{N^k}{2}+\sum_{j=1}^{\left[\frac{k+1}{2}\right]}\binom{k}{2j-1}\frac{B_{2j}}{2j}\left(N^{k-2j+1}-\delta_{k-2j+1}\right)+\frac{\delta_k}{2},
\end{aligned}
\tag{9}
$$

and we recover this way the well-known formulae

$$\sum_{n=0}^{N} n^0 = N+1; \quad \sum_{n=0}^{N} n = \frac{N(N+1)}{2}; \quad \sum_{n=0}^{N} n^2 = \frac{N(N+1)(2N+1)}{6}$$

using the fact that $B_2 = \frac{1}{6}$ for the last one. More generally, it follows from Eq. (9) that

Corollary 1.2. *The cut-off discrete sum $S_k(N)$ is a polynomial of order $k+1$ in N which vanishes at zero for any positive integer k.*

1.3. *Evaluating meromorphic functions at poles*

Let $\mathrm{Mer}_0^k(\mathbb{C})$ be the set of germs of meromorphic functions at zero [d] with poles at zero of order no larger than k, and let

$$\mathrm{Mer}_0(\mathbb{C}) = \cup_{k=0}^{\infty} \mathrm{Mer}_0^k(\mathbb{C}).$$

Let $\mathrm{Hol}_0(\mathbb{C})$ (also denoted by $\mathrm{Mer}_0^0(\mathbb{C})$) be the set of germs of holomorphic functions at zero.

If f in $\mathrm{Mer}_0^k(\mathbb{C})$ reads $f(z) = \sum_{i=-k}^{\infty} a_i z^i$, then for any $j \in \{1, \cdots, k\}$ we set $\mathrm{Res}_0^j(f) := a_{-j}$, called the **$j$-th residue** of f at zero.

The projection map

$$\pi_+ : \mathrm{Mer}_0(\mathbb{C}) \to \mathrm{Hol}_0(\mathbb{C})$$
$$f \mapsto \left(z \mapsto f(z) - \sum_{j=1}^{k} \frac{\mathrm{Res}_0^j(f)}{z^j} \right) \quad \text{for} \quad f \in \mathrm{Mer}_0^k(\mathbb{C})$$

corresponds to what physicists call a **minimal subtraction scheme**. Whereas $\pi_+(f)$ corresponds to the holomorphic part of f, $\pi_-(f) := (1-\pi_+)(f)$ corresponds to the "polar part" of f.

Example 1.3. With the notation of the previous paragraphs, we have

$$S_+ := \pi_+ \circ S(\varepsilon) = \mu(\varepsilon); \quad S_- := \pi_- \circ S(\varepsilon) = I(\varepsilon). \tag{10}$$

Thus the Euler-Maclaurin formula (7) amounts to the minimal subtraction scheme applied to S:

$$S = S_+ + S_- = \mu + I. \tag{11}$$

---

[d] *i.e.* equivalence classes of meromorphic functions defined on a neighborhood of zero for the equivalence relation $f \sim g$ if f and g coincide on some open neighborhood of zero.

An easy computation further shows that

$$\pi_+ \circ S_k(\varepsilon) = (-1)^k \, \mu^{(k)}(\varepsilon); \quad \pi_- \circ S_k(\varepsilon) = (-1)^k \, I^{(k)}(\varepsilon). \tag{12}$$

The holomorphic part $\pi_+(f\, g)$ of the product of two meromorphic functions f and g differs from the product $\pi_+(f)\,\pi_+(g)$ of the holomorphic parts of f and g by contributions of the poles through $\pi_-(f)$ and $\pi_-(g)$ and we have

$$\pi_+(f\, g) = \pi_+(f)\,\pi_+(g) + \pi_+(f\,\pi_-(g)) + \pi_+(g\,\pi_-(f)),$$

which implies that the maps π_+ and π_- are both **Rota-Baxter operators** of weight -1 on $\mathrm{Mer}_0(\mathbb{C})$, *i.e.*

$$\pi_\pm(f)\,\pi_\pm(g) = \pi_\pm(\pi_\pm(f)\,g) + \pi_\pm(f\,\pi_\pm(g)) - \pi_\pm(fg). \tag{13}$$

We refer the reader to [G] for a survey on Rota-Baxter operators.

Combining the evaluation map at zero $\mathrm{ev}_0 : f \mapsto f(0)$ on holomorphic germs at zero with the map π_+ provides a first regularized evaluator at zero on $\mathrm{Mer}_0^k(\mathbb{C})$. The map

$$\begin{aligned} \mathrm{ev}_0^{\mathrm{reg}} : \mathrm{Mer}_0^k(\mathbb{C}) &\to \mathbb{C} \\ f &\mapsto \mathrm{ev}_0 \circ \pi_+(f), \end{aligned} \tag{14}$$

is a linear form that extends the ordinary evaluation map ev_0 defined on the space $\mathrm{Hol}_0(\mathbb{C})$.

Definition 1.4. We call a **regularized evaluator** any linear extension of the evaluation map ev_0 to the space $\mathrm{Mer}_0(\mathbb{C})$.

The following result provides a classification of regularized evaluators.

Proposition 1.5. *Regularized evaluators at zero on* $\mathrm{Mer}_0^k(\mathbb{C})$ *are of the form:*

$$\lambda_0 = \mathrm{ev}_0^{\mathrm{reg}} + \sum_{j=1}^{k} \mu_j \, \mathrm{Res}_0^j \tag{15}$$

for some constants $\mu_1, \cdots, \mu_k$. *In particular, regularized evaluators at zero on* $\mathrm{Mer}_0^1(\mathbb{C})$ *are of the form*[e]

$$\lambda_0 = \mathrm{ev}_0^{\mathrm{reg}} + \mu \, \mathrm{Res}_0. \tag{16}$$

for some constant μ.

[e]The parameter μ that arises here is related to the renormalization group parameter in quantum field theory.

Proof. A linear form λ_0 which extends ev_0 coincides with ev_0 on the range of π_+ and therefore fulfills the following identity:

$$\lambda_0 \circ \pi_+ = \mathrm{ev}_0 \circ \pi_+ = \mathrm{ev}_0^{\mathrm{reg}}.$$

Thus, for any $f \in \mathrm{Mer}_0^k(\mathbb{C})$, using the linearity of λ_0 we get

$$\lambda_0(f) = \lambda_0\left(\pi_+(f)\right) + \lambda_0(\pi_-(f)) = \mathrm{ev}_0^{\mathrm{reg}} + \sum_{j=1}^{k} \mu_j \,\mathrm{Res}_0^j(f)$$

where we have set $\mu_j := \lambda_0(z^{-j})$. □

Example 1.6. We have

$$\mathrm{ev}_0^{\mathrm{reg}}(S) = \mu(0) = \frac{1}{2} = 1 + B_1.$$

Similarly, the higher Taylor coefficients of the holomorphic function μ at zero relate to the value of the zeta function at negative integers

$$\mathrm{ev}_0^{\mathrm{reg}}(S_k) = (-1)^k \,\mu^{(k)}(0) = -\frac{B_{k+1}}{k+1}$$

and yield the renormalized polynomial sums "$\sum_{k=0}^{\infty} k^n$" on integer points of the one dimensional closed cone $[0, +\infty)$.

1.4. *The zeta function at non-positive integers*

Let us start with some notation. Given $\alpha \in \mathbb{C}$ we consider smooth functions f on $\mathbb{R}_+$ with the following asymptotic behavior at infinity

$$f(R) \sim_{R\to\infty} \sum_{j=0}^{\infty} a_j R^{\alpha-j} + b \log R \tag{17}$$

by which we mean

$$f(R) - \sum_{j=0}^{N-1} a_j R^{\alpha-j} - b\log R = o\left(R^{\mathrm{Re}(\alpha)-N+\varepsilon}\right)$$

for any positive ε and any positive integer N. We call such a function **asymptotically log-polyhomogeneous** at infinity of logarithmic type 1. If $b = 0$ we call it asymptotically polyhomogeneous at infinity; let us consider the class $\mathcal{S}_\infty^\alpha(\mathbb{R}_+)$ of **asymptotically polyhomogeneous functions** at infinity of logarithmic type 1.

Example 1.7. The logarithmic function $f : R \mapsto \log R$ is asymptotically log-polyhomogeneous at infinity, of logarithmic type 1. Physicists say that the integral $\int_1^R \frac{1}{x}\, dx = \log R$ has a logarithmic divergence as $R \to \infty$.

The **Hadamard finite part of f in $\mathcal{S}_\infty^\alpha(\mathbb{R}_+)$ at infinity**

$$\mathrm{fp}_{R\to\infty} f(R) := \begin{cases} a_\alpha, & \text{if } \alpha \in \mathbb{Z}_{\geqslant 0}, \\ 0, & \text{otherwise.} \end{cases}$$

defines a linear map

$$\begin{aligned} \mathrm{ev}_\infty^{\mathrm{reg}} : \mathcal{S}_\infty^\alpha(\mathbb{R}_+) &\longrightarrow \mathbb{C} \\ f &\longmapsto \mathrm{fp}_{R\to\infty} f(R) \end{aligned}$$

which extends the ordinary limit at infinity whenever it exists. We call such a linear extension of the ordinary limit a **regularized evaluator at infinity.**

Setting $R = \frac{1}{r}$ in Eq. (17) with $r > 0$, and choosing $\beta = -\alpha$, $b_j = a_j$, $c = -b$ leads to smooth functions f on $(0, +\infty)$ with the following log-polyhomogeneous asymptotic behavior at zero:

$$f(r) \sim_{r\to 0} \sum_{j=0}^{\infty} b_j r^{\beta+j} + c \log r \tag{18}$$

by which we mean

$$f(r) - \sum_{j=0}^{N-1} b_j r^{\beta+j} - c \log r = o\left(r^{\mathrm{Re}(\beta)+N+\varepsilon}\right)$$

for any positive ε and any positive integer N. We call such a function **asymptotically log-polyhomogeneous** at zero of logarithmic type 1. If $c = 0$ we call it asymptotically polyhomogeneous at zero; let us consider the class $\mathcal{S}_0^\beta(\mathbb{R}_+)$ of **asymptotically polyhomogeneous** functions at zero of logarithmic type 1.

The **Hadamard finite part of f $\mathcal{S}_0^\beta(\mathbb{R}_+)$ at zero**

$$\mathrm{fp}_{r\to 0} f(r) := \begin{cases} b_{-\beta}, & \text{if } \beta \in \mathbb{Z}_{\geqslant 0}, \\ 0, & \text{otherwise.} \end{cases}$$

defines a linear map

$$\begin{aligned} \mathrm{ev}_0^{\mathrm{reg}} : \mathcal{S}_0^\beta(\mathbb{R}_+) &\longrightarrow \mathbb{C} \\ f &\longmapsto \mathrm{fp}_{r\to 0} f(r) \end{aligned}$$

which extends the ordinary limit at zero whenever it exists. We call such a linear extension of the ordinary limit a **regularized evaluator at zero.**

We recall here well known results on the Mellin transform [f], see *e.g.* [Je].

Proposition 1.8. *Let f be a Schwartz function in $\mathcal{S}_0^\beta(\mathbb{R}_+)$ for some $\beta \in \mathbb{C}$. Its Mellin transform defines a holomorphic function on the half plane* $\mathrm{Re}(z) + \beta > 0$ *which extends to a meromorphic function on the whole complex plane which is holomorphic at zero. We have* $\mathcal{M}(f^{(k)})(z) = (-1)^k \mathcal{M}(f)(z-k)$ *for any $k \in \mathbb{Z}_{\geqslant 0}$ and the value at zero is given by*

$$\mathrm{ev}_0^{\mathrm{reg}} \circ \mathcal{M}(f) = \mathcal{M}(f)(0) = \mathrm{ev}_0^{\mathrm{reg}}(f). \tag{19}$$

Proof. We split the Mellin transform

$$\mathcal{M}(f)(z) = \frac{1}{\Gamma(z)} \left(\int_0^A \varepsilon^{z-1} f(\varepsilon)\, d\varepsilon + \int_A^\infty \varepsilon^{z-1} f(\varepsilon)\, d\varepsilon \right)$$

for some positive real number A. The function $\frac{1}{\Gamma}$ is holomorphic at zero and we have $\frac{1}{\Gamma(z)} \sim_0 z$.

Since f is a Schwartz function, the second term in the bracket yields a holomorphic function $I_2 : z \mapsto \frac{1}{\Gamma(z)} \int_A^\infty \varepsilon^{z-1} f(\varepsilon)\, d\varepsilon$ which vanishes at zero. For $f(\varepsilon) = \sum_{j=0}^J b_j \varepsilon^{\beta+j}$ and $\mathrm{Re}(z)+\beta+j > 0$, the first term in the bracket gives rise to

$$I_1^J(z) := \frac{1}{\Gamma(z)} \sum_{j=0}^J b_j \int_0^A \varepsilon^{z+\beta+j-1}\, d\varepsilon = \frac{1}{\Gamma(z)} \sum_{j=0}^J b_j \frac{A^{z+\beta+j}}{z+\beta+j}$$

which extends to a meromorphic function denoted by the same symbol. Hence

$$\mathcal{M}(f)(z) := I_1(z) + I_2^J(z) + o\left(\varepsilon^{z+\beta+J}\right) \text{ for } J \in \mathbb{N}$$

defines a meromorphic function on the whole plane. Integrating by parts k times and implementing the property $\Gamma(z) = (z-1)\Gamma(z-1) = (z-1)\cdots(z-k)\Gamma(z-k)$ gives $\mathcal{M}(f^{(k)})(z) = (-1)^k \mathcal{M}(f)(z-k)$.

Since $\Gamma(z) \sim \frac{1}{z}$, the value of $I_2^J(z)$ at $z = 0$ is $b_{-\beta}$ if $\beta \in \mathbb{Z}_{\leqslant 0} \cap [0, J]$ and zero otherwise so the same holds for $\mathcal{M}(f)(z)$. Since $\mathrm{fp}_{\varepsilon=0} f(\varepsilon) = b_{-\beta}$ if $\beta \in \mathbb{Z}_{\leqslant 0}$ and zero otherwise, this yields Eq. (19). $\square$

The Mellin transform of the Schwartz function $f_n : \varepsilon \mapsto e^{-\varepsilon n}$ on $\mathbb{R}_+$ reads $n^{-z} = \mathcal{M}(f_n)(z)$ for any $n \in \mathbb{N}$.

---

[f]Note that definitions of the Mellin transform differ according to the reference by a multiplicative factor $\Gamma(z)$.

Corollary 1.9. *The function* $z \mapsto \sum_{n=1}^{\infty} n^{-z}$ *defined on the half-plane* $\mathrm{Re}(z) > 1$ *extends meromorphically on the whole plane to the* ***zeta function*** ζ*, which has only one simple pole at* -1 *and its value at non-positive integers is expressed in terms of Bernoulli numbers*

$$\zeta(0) = \mathrm{ev}_0^{\mathrm{reg}}(S) - 1 = -\frac{1}{2};$$
$$\zeta(-k) = (-1)^k\, \mathrm{ev}_0^{\mathrm{reg}}(\partial^k S) = -\frac{B_{k+1}}{k+1} \text{ for all } k \in \mathbb{N}. \tag{20}$$

Proof. This follows from applying Proposition 1.8 to the function $\varepsilon \mapsto S(\varepsilon) - 1 = \sum_{n=1}^{\infty} e^{-\varepsilon n}$ and its derivatives $(-1)^k S_k, k \in \mathbb{N}$. □

1.5. *Conclusion*

By means of the heat-kernel regularization method we evaluated

$$\text{“}\left(\sum_{n=1}^{\infty} n^0\right)\text{”} = \text{“}\left(\sum_{n=0}^{\infty} n^0\right)\text{”} - 1 = \mathrm{ev}_0^{\mathrm{reg}} \circ S - 1 = \mu(0) - 1 = B_1 - 1 = -\frac{1}{2}.$$

In this paper, $B_1 = \frac{1}{2}$. By means of the zeta-function regularization method we evaluated

$$\text{“}\left(\sum_{n=1}^{\infty} n^0\right)\text{”} = \mathrm{ev}_0 \circ S = \zeta(0) = -\frac{1}{2},$$

so these two methods agrees in the case $k = 0$. The two methods actually coincide for any $k \in \mathbb{Z}_{\geqslant 0}$.

$$\text{“}\left(\sum_{n=0}^{\infty} n^k\right)\text{”} = \text{“}\left(\sum_{n=1}^{\infty} n^k\right)\text{”} = \mathrm{ev}_0^{\mathrm{reg}} \circ S_k = \zeta(-k) = -\frac{B_{k+1}}{k+1}.$$

Moreover, combining Eqs. (20) and (5) yields the Laurent expansion of the exponential sum in terms of ζ-values at non-positive arguments

$$S(\varepsilon) = \frac{1}{\varepsilon} - \zeta(0) - \sum_{k=1}^{K} \zeta(-(2k-1))\, \varepsilon^{2k-1} + o(\varepsilon^{2K}) \quad \forall K \in \mathbb{N}. \tag{21}$$

In contrast, the cut-off method gives

$$\mathrm{fp}_{N\to\infty} S_k(N) = P_k(0) = \delta_k,$$

where $\delta_k = 1$ if $k = 0$ and zero otherwise.

2. Counting lattice points on product cones

Given a positive integer k, we now want to "count" the number "$\left(\sum_{\vec{n}\in\mathbb{Z}^k_{\geqslant 0}} \vec{n}^{\vec{0}}\right)$" of lattice points $\vec{n}\in\mathbb{Z}^k_{\geqslant 0}$ in the product cone $\mathbb{R}^k_{\geqslant 0}$, where for $\vec{n}=(n_1,\cdots,n_k)\in\mathbb{Z}^k_{\geqslant 0}$ and $\vec{r}=(r_1,\cdots,r_k)\in\mathbb{Z}^k_{\geqslant 0}$ we have set $\vec{n}^{\vec{r}}=n_1^{r_1}\cdots n_k^{r_k}$. We first describe the algebra of product cones.

2.1. *The exponential summation and integration map on product cones*

Given a basis $\mathcal{B}_n:=(e_1,\cdots,e_n)$ of $\mathbb{R}^n$, let $\mathcal{P}_{\mathcal{B}_n}(\mathbb{R}^n)$ be the set of **product cones**

$$\langle e_I\rangle := \sum_{i\in I}\mathbb{R}_{\geq 0}e_i,\quad I\subseteq[n]:=\{1,\cdots,n\},$$

viewed as subsets of $\mathbb{R}^n$. Extending this basis to a basis $\mathcal{B}_{n+1}=(e_1,\cdots,e_{n+1})$ of $\mathbb{R}^{n+1}$, a product cone in $\mathbb{R}^n$ can be viewed as a product cone in $\mathbb{R}^{n+1}$. Setting $\mathcal{P}_{\mathcal{B}_0}(\mathbb{R}^0)=\{0\}$, we define the set

$$\mathcal{P}_{\mathcal{B}}(\mathbb{R}^\infty) := \cup_{n=0}^\infty \mathcal{P}_{\mathcal{B}_n}(\mathbb{R}^n)$$

of product cones in $\mathbb{R}^\infty$ equipped with a basis $\mathcal{B}=\{e_n\,|\,n\in\mathbb{N}\}$. Equivalently,

$$\mathcal{P}_{\mathcal{B}}(\mathbb{R}^\infty)=\{\langle e_I\rangle\,|\,I\subset\mathbb{N}\text{ finite }\}\ \text{ with } \langle e_\emptyset\rangle:=\{0\}.$$

It is $\mathbb{Z}_{\geqslant 0}$-filtered by the dimension $\mathrm{card}(I)$ (here card stands for cardinal) of the cone $\langle e_I\rangle$ and it is equipped with a partial product

$$\langle e_I\rangle\bullet\langle e_J\rangle := \langle e_{I\cup J}\rangle$$

for two disjoint subsets I,J of $\mathbb{N}$. This product is compatible with the filtration since the dimension of the product of two cones is the sum of their dimensions.

Unless otherwise specified, we take $\mathcal{B}$ to be the canonical basis of $\mathbb{R}^\infty$, in which case we drop the subscript $\mathcal{B}$ in the notation.

The linear map $\mathrm{ev}_0^{\mathrm{reg}}:\mathrm{Mer}_0^1(\mathbb{C})\to\mathbb{C}$ extends multiplicatively to the subspace $\mathrm{Mer}_{\mathrm{sep}}(\mathbb{C}^\infty)$ of the space of meromorphic germs at zero with linear poles, spanned by separable functions [g]:

$$\left\{ f = \prod_{i \in I} f_i \;\middle|\; I \subseteq \mathbb{N}\,\text{finite},\ f_i \in \mathrm{Mer}_0(\mathbb{C}e_i) \right\}$$

by

$$\mathrm{ev}_{\vec{0}}^{\mathrm{ren}} \left(\prod_{i \in I} f_i \right) := \prod_{i \in I} \mathrm{ev}_0^{\mathrm{reg}}(f_i). \tag{22}$$

We refer the reader to [GPZ1] for a more detailed study of renormalized (or generalized) evaluators. Note that even though the subspace $\mathrm{Mer}_{\mathrm{sep}}(\mathbb{C}^\infty)$ is closed under the multiplication of functions, the map $\mathrm{ev}_{\vec{0}}^{\mathrm{reg}}$ on $\mathrm{Mer}_{\mathrm{sep}}(\mathbb{C}^\infty)$ resulting from Eq. (22) is multiplicative only for a product with disjoint variables. More precisely, for a separable function f (resp. g) with variables in a finite subset I (resp. J) of $\mathbb{N}$, with I and J disjoint, we have

$$\mathrm{ev}_{\vec{0}}^{\mathrm{reg}}(fg) = \mathrm{ev}_{\vec{0}}^{\mathrm{reg}}(f)\mathrm{ev}_{\vec{0}}^{\mathrm{reg}}(g).$$

However, $\mathrm{ev}_0^{\mathrm{reg}}(\frac{1}{\varepsilon_1}) \neq \mathrm{ev}_0^{\mathrm{reg}}(\frac{1+\varepsilon_1}{\varepsilon_1})\mathrm{ev}_0^{\mathrm{reg}}(\frac{1}{1+\varepsilon_1})$ even though $\frac{1}{\varepsilon_1} = \frac{1+\varepsilon_1}{\varepsilon_1}\frac{1}{1+\varepsilon_1}$.

The summation map (1) and the integration map (6), which lie in the linear space $\mathrm{Mer}_0^1(\mathbb{C})$ of meromorphic germs in one complex variable with a simple pole at zero, induce linear maps on the linear space $\mathbb{R}\mathcal{P}_{\mathcal{B}}(\mathbb{R}^\infty)$ spanned by $\mathcal{P}_{\mathcal{B}}(\mathbb{R}^\infty)$ as follows

$$\mathcal{S} : \mathbb{R}\mathcal{P}_{\mathcal{B}}(\mathbb{R}^\infty) \longrightarrow \mathrm{Mer}_{\mathrm{sep}}(\mathbb{C}^\infty), \quad \langle e_I \rangle \mapsto \prod_{i \in I} S(\varepsilon_i)$$

and

$$\mathcal{I} : \mathbb{R}\mathcal{P}_{\mathcal{B}}(\mathbb{R}^\infty) \longrightarrow \mathrm{Mer}_{\mathrm{sep}}(\mathbb{C}^\infty), \quad \langle e_I \rangle \mapsto \prod_{i \in I} I(\varepsilon_i).$$

For simplicity and emphasizing the dependence on the variables, we also use the notations

$$\mathcal{S}_i : \mathbb{R}\langle e_i \rangle \to \mathrm{Mer}_0(\mathbb{C}\varepsilon_i), \quad \langle e_i \rangle \mapsto S(\varepsilon_i) \tag{23}$$

and

$$\mathcal{I}_i : \mathbb{R}\langle e_i \rangle \to \mathrm{Mer}_0(\mathbb{C}\varepsilon_i), \quad \langle e_i \rangle \mapsto I(\varepsilon_i). \tag{24}$$

[g] $\mathrm{Mer}_{\mathrm{sep}}(\mathbb{C}^\infty)$ is isomorphic to the filtered vector space $\mathcal{F} := \varinjlim (\mathrm{Mer}_0^1(\mathbb{C}))^n$ by assigning $f_1 \otimes \cdots \otimes f_n$ to $f_1(\varepsilon_1) \cdots f_n(e_n)$. But the map does not respect the tensor product. For example, $(f \otimes f)(\varepsilon_1, \varepsilon_2) = f(\varepsilon_1)f(\varepsilon_2) \neq f(\varepsilon_1)^2$.

The maps $\mathcal{S}$ and $\mathcal{I}$ are compatible with the partial product on cones. Indeed, for two disjoint finite sets I and J of $\mathbb{N}$ we have

$$\mathcal{S}\left(\langle e_{I\cup J}\rangle\right)=\prod_{i\in I\cup J}S(\varepsilon_i)=\left(\prod_{i\in I}S(\varepsilon_i)\right)\left(\prod_{j\in J}S(\varepsilon_j)\right)$$

and similarly for the integration map. We further set $\mathcal{S}(\{0\}):=\mathcal{I}(\{0\}):=1$. Set $I:=\{i_1,\cdots,i_J\}$ and $\varepsilon_I:=(\varepsilon_{i_1},\cdots,\varepsilon_{i_J})$. As a consequence of Eq. (21) we have the following iterated Laurent expansion

$$\begin{aligned}&\mathcal{S}\left(\langle e_I\rangle\right)(\varepsilon_I)=\\&=\prod_{j=1}^{J}\left(\frac{1}{\varepsilon_{i_j}}-\zeta(0)-\sum_{k_j=1}^{K_j}\zeta(-(2k_j-1))\,\varepsilon_{i_j}^{2k_j-1}+o(\varepsilon_{i_j}^{2K_j})\right).\end{aligned}\qquad(25)$$

In order to "count" the number of lattice points

$$\text{“}\left(\sum_{\vec{n}\in\sum_{i\in I}\mathbb{Z}_{\geqslant 0}e_i}\vec{n}^{\vec{0}}\right)\text{”}$$

in the product cone $\langle e_I\rangle$ we want to evaluate $\mathcal{S}\left(\langle e_I\rangle\right)$ at $(\varepsilon_{i_1},\cdots,\varepsilon_{i_k})=\vec{0}$. Since $\mathcal{S}\left(\langle e_I\rangle\right)\in\mathrm{Mer}_{\mathrm{sep}}(\mathbb{C}^\infty)$ a first guess is to assign the value

$$\mathcal{S}_{\vec{0}}^{\mathrm{ren}}:=\mathrm{ev}_{\vec{0}}^{\mathrm{ren}}\circ\mathcal{S},\qquad(26)$$

where $\mathrm{ev}_{\vec{0}}^{\mathrm{ren}}$ is defined in (22). This "renormalized value" at zero is multiplicative as a result of the multiplicativity of $\mathrm{ev}_{\vec{0}}^{\mathrm{reg}}$. Indeed, given two disjoint index sets I and J, we have

$$\begin{aligned}\mathcal{S}_{\vec{0}}^{\mathrm{ren}}\left(\langle e_I\rangle\bullet\langle e_J\rangle\right)&=\mathrm{ev}_{\vec{0}}^{\mathrm{reg}}\circ\left(\mathcal{S}\left(\langle e_I\rangle\right)\mathcal{S}\left(\langle e_J\rangle\right)\right)\\&=\left(\mathrm{ev}_{\vec{0}}^{\mathrm{reg}}\circ\left(\mathcal{S}\left(\langle e_I\rangle\right)\right)\right)\cdot\left(\mathrm{ev}_{\vec{0}}^{\mathrm{reg}}\circ\left(\mathcal{S}\left(\langle e_J\rangle\right)\right)\right)\\&=\mathcal{S}_{\vec{0}}^{\mathrm{ren}}\left(\langle e_I\rangle\right)\cdot\mathcal{S}_{\vec{0}}^{\mathrm{ren}}\left(\langle e_J\rangle\right).\end{aligned}$$

We shall now describe the underlying algebraic framework, which might seem somewhat artificial in the rather trivial product cone situation. However, on the one hand even in this simple situation it is useful to control the "polar part" which one needs to extract in order to define the finite part, on the other hand it offers a good toy model to motivate otherwise relatively sophisticated techniques which can be generalized beyond product cones, namely to general convex cones [GPZ3].

2.2. *A complement map on product cones*

Let us first recall the properties of the set complement map.

Let $\mathcal{P}_f(E)$ be the set of finite subsets of a given set E equipped with the inclusion $\subseteq$ which defines a partial order compatible with the filtration of $\mathcal{P}_f(E)$ by the cardinal in the sense that $J \subseteq I$ implies $|J| \leqslant |I|$. For $I \in \mathcal{P}_f(E)$ let

$$\mathfrak{s}(I) := \{J \in \mathcal{P}_f(E) \,|\, J \subseteq I\}$$

be the set of subsets of I.

The **set complement map** assigns to any $I \subseteq E$ a map

$$\complement_I : \mathfrak{s}(I) \longrightarrow \mathfrak{s}(I)$$

$$J \longmapsto I \setminus J := I \cap \overline{J}.$$

The complement $I \setminus J$ satisfies the following properties:

(a) **Compatibility with the partial order:** Let $I, J \in \mathcal{P}_f(E)$ be such that $J \subseteq I$. Then for any $H \in \mathcal{P}_f(E)$ with $H \subseteq I \setminus J$ there exists unique $K \in \mathcal{P}_f(E)$; $J \subseteq K \subseteq I$ such that $H = I \setminus K$.

(b) **Transitivity:** Let $I, J, K \in \mathcal{P}_f(E)$ be such that $K \subseteq J \subseteq I$. Then

$$(I \setminus K) \setminus (J \setminus K) = I \setminus J$$

(c) **Compatibility with the filtration:** Let $I, J \in \mathcal{P}_f(E)$ be such that $J \subseteq I$. Then

$$\mathrm{card}(J) + \mathrm{card}(I \setminus J) = \mathrm{card}(I),$$

where card stands for the cardinality.

The set complement map on $\mathbb{Z}_{\geq 0}$ induces a complement map on the product cones. Let us first introduce some notations. Faces of the product cone $C := \langle e_I \rangle$ are of the form

$$F_J := \langle e_J \rangle$$

with $J \subset I$, each of them defining a cone with faces $F_{J'}$ where $J' \subset J$. The cone C therefore has $2^{|I|}$ faces, as many as subsets of I. The set $\mathcal{F}(C)$ of faces of the cone C is equipped with a partial order

$$F' \subset F \text{ if and only if } F' \text{ is a face of } F'$$

or equivalently, $F_{J'} \subset F_J$ if and only if $J' \subset J$. For $F' = F_{J'} \subset F = F_J$ we consider the complement set $\overline{F'}^F := F_{J \setminus J'}$, which again defines an element of $\mathcal{F}(C)$ and hence a cone. We define the complement map

$$\begin{aligned} \mathcal{F}(C) &\longrightarrow \mathcal{F}(C) \\ F_J &\longmapsto \overline{F}_J^{\,C} = F_{I \setminus J}, \end{aligned} \tag{27}$$

which is an involution. As a consequence of the properties of the set complement map, it enjoys the following properties. Let $F \in \mathcal{F}(C)$.

(a) **Compatibility with the partial order:** There is a one-to-one correspondence between the set of faces of C containing a given face F and the set of faces of the cone $\overline{F}^C$; for any face H of $\overline{F}^C$, there is a unique face G of C containing F such that $H = \overline{F}^G$.

(b) **Transitivity:** $\overline{F}^C = \left(\overline{F'}^F\right)^{\overline{F'}^C}$ if $F' \subset F$.

(c) **Compatibility with the filtration by the dimension:** For any face F of C we have $\dim(F) + \dim\left(\overline{F}^C\right) = \dim(C)$.

There is an alternative description of this complement map which is generalizable to general convex cones, those not necessarily obtained as product cones. For this we observe that for a face $F = F_J = \langle e_J \rangle$ of a product cone $C = \langle e_I \rangle$, we have

$$\overline{F}^C = F_{I \setminus J} = \pi_{F^\perp}(C), \tag{28}$$

where $F^\perp$ denotes the orthogonal space of the linear space spanned by the cone F in the linear space $\langle C \rangle$ spanned by C, and $\pi_{F^\perp}$ is the orthogonal projection from $\langle C \rangle$ onto $F^\perp$. Here the orthogonal projection is taken with respect to the canonical Euclidean product on $\mathbb{R}^\infty$. Eq. (28) follows from the fact that $\pi_{F^\perp}(e_i)$ is 0 for $i \in J$ and e_i for $i \notin J$.

2.3. *Algebraic Birkhoff factorization on product cones*

For each $i \geqslant 1$, the algebra $\mathcal{A}_i := \mathrm{Mer}_0(\mathbb{C}\varepsilon_i)$ is naturally isomorphic to $\mathcal{A} := \mathrm{Mer}_0(\mathbb{C})$ as the algebra of Laurent series. Following the minimal subtraction scheme we have a direct sum $\mathcal{A}_i = \mathcal{A}_{i,+} \oplus \mathcal{A}_{i,-}$ of two subalgebras $\mathcal{A}_{i,\pm} := \pi_\pm(\mathcal{A}_i)$. The maps $\mathcal{S}_i : \mathbb{R}\langle e_i \rangle \longrightarrow \mathrm{Mer}_0(\mathbb{C}\varepsilon_i)$ defined in Eq. (23) split accordingly $\mathcal{S}_i = \mathcal{S}_{i,+} + \mathcal{S}_{i,-}$ into a sum of maps $\mathcal{S}_{i,\pm} : \mathbb{R}\langle e_i \rangle \longrightarrow \mathcal{A}_{i,\pm}$.

We next consider separable functions in several variables. For disjoint subsets $I, J \subseteq \mathbb{N}$, define

$$\mathcal{A}_{I,+,J,-} := \left(\prod_{i\in I} \mathcal{A}_{i,+}\right)\left(\prod_{j\in J} \mathcal{A}_{j,-}\right).$$

Also denote $\mathcal{A}_{I,+,J,-} = \mathcal{A}_{I,+}$ if $J = \emptyset$. Then we have

$$\mathcal{A}_I := \prod_{i\in I} \mathcal{A}_i = \oplus_{I_1 \sqcup I_2 = I} \mathcal{A}_{I_1,+,I_2,-}.$$

Further denote

$$\mathcal{A}_{I,+} := \prod_{i\in I} \mathcal{A}_{i,+}, \quad \mathcal{A}_{I,-} := \prod_{J\subsetneq I} \mathcal{A}_{J,+,I\setminus J,-}$$

and

$$\mathcal{A}_\infty := \varinjlim \mathcal{A}_I, \quad \mathcal{A}_{\infty,\pm} := \varinjlim \mathcal{A}_{I,\pm}.$$

Then we have

$$\mathcal{A}_\infty = \mathcal{A}_{\infty,+} \oplus \mathcal{A}_{\infty,-}.$$

$\mathcal{A}_{\infty,+}$ is a subalgebra but not $\mathcal{A}_{\infty,-}$. For example, $\mathcal{A}_{1,+}\mathcal{A}_{2,-}$ and $\mathcal{A}_{1,-}$ are in $\mathcal{A}_{\infty,-}$, but their product is not. This decomposition corresponds to the restriction of the decomposition on meromorphic functions with linear poles studied in [GPZ4].

As we saw in the previous section, since $S_- = I$, such a splitting $S = S_+ + S_-$ of the exponential sum corresponds to the Euler-Maclaurin formula $S = \mu + I$ with $\mu = S_+$.

We are now ready to generalize the minimal subtraction scheme and the Euler-Maclaurin formula to product cones. In the product cone framework, the minimal subtraction scheme generalizes to an elementary form of the more general algebraic Birkhoff factorization on coalgebras which we shall describe in the next section.

Proposition 2.1. *Given a product cone $C = \langle e_I \rangle$ in $\mathcal{P}(\mathbb{R}^\infty)$ the map $\mathcal{S}(C)$ extends to a meromorphic map in $\mathrm{Mer}_{\mathrm{sep}}(\mathbb{C}^\infty)$ with simple poles on the intersections of hyperplanes $\cap_{j\in J}\{\varepsilon_j = 0\}$ corresponding to faces $F_J = \langle e_I \rangle, J \subseteq I$ of the cone C. It decomposes as*

$$\mathcal{S}(C) = \sum_{F\in\mathcal{F}(C)} \mathcal{S}_+(\overline{F}^C)\mathcal{S}_-(F) \quad \textit{(algebraic Birkhoff factorization)} \tag{29}$$

$$= \sum_{F\in\mathcal{F}(C)} \mu(\overline{F}^C)\mathcal{I}(F) \quad \textit{(Euler-Maclaurin formula)}, \tag{30}$$

where for a face $F = \langle e_K \rangle$ of the cone C, $\overline{F}^C = F_{I \setminus K}$ is the "complement face" defined in the previous paragraph and where we have set

$$\mathcal{S}(F) = \mathcal{S}\left(\langle e_K \rangle\right) := \prod_{i \in K} S(\langle e_i \rangle), \quad \mathcal{S}_\pm(F) = \mathcal{S}_\pm\left(\langle e_K \rangle\right) := \prod_{i \in K} \mathcal{S}_{i,\pm}.$$

Remark 2.2. Eq. (29) which arises from the one-dimensional minimal substraction scheme can be viewed as a higher dimensional minimal subtraction scheme and Eq. (30) as a higher dimensional Euler-Maclaurin formula. When $k = 1$ they yield back the one dimensional minimal subtraction scheme and the Euler-Maclaurin formula applied to $S(\varepsilon)$.

Proof. Let $C = \langle e_I \rangle$ for some finite subset I in $\mathbb{N}$. We have

$$\begin{aligned}
\mathcal{S}(C) &= \prod_{i \in I} \mathcal{S}_i\left(\langle e_i \rangle\right) \quad \text{(a product of sums)} \\
&= \prod_{i \in I} \left(\mathcal{S}_{i,+} + \mathcal{S}_{i,-}\right)\left(\langle e_i \rangle\right) \quad \text{(a sum of products)} \\
&= \sum_{J \subset I} \left(\prod_{j \in I \setminus J} \mathcal{S}_{j,+}\left(\langle e_j \rangle\right)\right) \left(\prod_{j \in J} \mathcal{S}_{j,-}\left(\langle e_j \rangle\right)\right) \\
&= \sum_{F \in \mathcal{F}(C)} \mathcal{S}_+(\overline{F}^C)\, \mathcal{S}_-(F) \\
&= \sum_{F \in \mathcal{F}(C)} \mu(\overline{F}^C)\, \mathcal{I}(F).
\end{aligned}$$

□

The fact that the algebraic Birkhoff factorization (29) and the Euler-Maclaurin formula (30) coincide for product cones is a consequence of Eq. (11) which shows how, in the one dimensional case, the minimal subtraction scheme and the Euler-Maclaurin formula coincide for the exponential sum. From formula (25) we derive a formal Taylor expansion at zero of $\mathcal{S}_+\left(\langle e_I \rangle\right)$

$$\mathcal{S}_+\left(\langle e_I \rangle\right)(\varepsilon_I) = \sum_{J \subset I} \left(\zeta(-k_J)\, \varepsilon_J^{k_J}\right) \tag{31}$$

where for $J = \{i_1, \cdots, i_j\} \subset I$ and any multiindex $k_J = (k_{i_1}, \cdots, k_{i_j}) \in \mathbb{Z}_{\geqslant 0}^j$ we have set $\varepsilon_J^{k_J} = \prod_{l=1}^j \varepsilon_{i_l}$, whose coefficients

$$\prod_{j \in J} \zeta(-k_{i_j}) \tag{32}$$

are the so called **(renormalized) product zeta values** at non-positive integers.

The renormalized discrete sum in Eq. (26), which can be rewritten as

$$\mathcal{S}_{\vec{0}}^{\mathrm{reg}} = \mathrm{ev}_{\vec{0}} \circ \mathcal{S}_+ = \mathrm{ev}_{\vec{0}} \circ \mu,$$

is obtained from evaluating at zero the renormalized "holomorphic part" $\mathcal{S}_+$ of the exponential sum derived from the algebraic Birkhoff factorization (see (29)) or equivalently from evaluating at zero the renormalized interpolator μ derived from the Euler-Maclaurin formula (see (30)).

We have gone a long way around to recover our first guess (26). This approach using Birkhoff-Hopf factorization, even if somewhat artificial in the case of product cones, is nevertheless useful for it can be generalized to all rational polyhedral convex (lattice) cones [GPZ3] a case which will be briefly discussed at the end of the paper.

3. From complement maps to coproducts

We now set up an algebraic framework to derive an algebraic Birkhoff factorization from a complement map in a more general set up than the specific example of product cones which served as a toy model in the previous section.

3.1. *Posets*

Let $(\mathcal{P}, \leqslant)$ be a poset, *i.e.* a set $\mathcal{P}$ together with a partial order $\leqslant$. We do not assume that the poset is finite.

The poset is **filtered** if $\mathcal{P} = \bigcup_{n=0}^{\infty} \mathcal{P}_n$ with $\mathcal{P}_n \subset \mathcal{P}_{n+1}$. The degree of $A \in \mathcal{P}$ denoted by $|A|$ is the smallest integer n such that $A \in \mathcal{P}_n$. The partial order $\leqslant$ is compatible with the filtration if $A \leqslant B$ implies $|A| \leqslant |B|$.

We call a filtered poset $\mathcal{P}$ **connected** if $\mathcal{P}$ has a least element 1, called the **bottom** of $\mathcal{P}$, and we have $\mathcal{P}_0 = \{1\}$.

Example 3.1. For a given set X (finite or infinite), the set $\mathcal{P}_f(X)$ of finite subsets of X equipped with the inclusion relation is a poset $(\mathcal{P}_f(X), \subseteq)$ filtered by the cardinal. It is connected since $\emptyset$ is the only subset of cardinal 0 and $\emptyset \subseteq A$ for any $A \in \mathcal{P}_f(X)$.

Example 3.2. This example can be regarded as a special case of the previous example but its pertinence for convex cones justifies that we treat it separately. The set $\mathcal{P}(\mathbb{R}^\infty) = \cup_{n=0}^{\infty} \mathcal{P}(\mathbb{R}^n)$ of closed product cones described in the previous section is filtered by the dimension and partially ordered by the partial order on the index sets. Equivalently, $F \leqslant C$ if the product cone F is a face of the product cone C. $\mathcal{P}(\mathbb{R}^\infty)$ is connected since the zero cone $\{0\}$ is the only cone of dimension 0 and $\{0\} \leqslant C$ for any $C \in \mathcal{P}(\mathbb{R}^\infty)$ as 0 is a vertex of any product cone.

Example 3.3. A closed (polyhedral) convex cone in $\mathbb{R}^n$ is the convex set

$$\langle v_1, \cdots, v_k \rangle := \mathbb{R}_{\geqslant 0} v_1 + \cdots + \mathbb{R}_{\geqslant 0} v_k, \tag{33}$$

where $v_i \in \mathbb{R}^n$, $i = 1, \cdots, k$.

Let $\mathcal{C}(\mathbb{R}^\infty) = \cup_{n=0}^{\infty} \mathcal{C}(\mathbb{R}^n)$ be the set of closed polyhedral convex cones in $\mathbb{R}^\infty$ see [GPZ2]. We have $\mathcal{P}(\mathbb{R}^\infty) \subset \mathcal{C}(\mathbb{R}^\infty)$. It is filtered by the dimension $|C|$ of the cone C defined as the dimension of the linear subspace spanned by C. A face of a cone $C = \langle v_1, \cdots, v_k \rangle$ is a subset of the form $\langle v_1, \cdots, v_k \rangle \cap \{u = 0\}$, where $u : \mathbb{R}^n \to \mathbb{R}$ is a linear form which is non-negative on $\langle v_1, \cdots, v_k \rangle$. A face F of a cone is itself a cone and we equip $\mathcal{C}(\mathbb{R}^\infty)$ with the following partial order which extends the partial order on product cones:

$$F \leqslant C \text{ if and only if } F \text{ is a face of C},$$

which is compatible with the filtration since $F \leqslant C$ implies $\mathrm{lin}(F) \subset \mathrm{lin}(C)$ which implies $|F| \leqslant |C|$. The filtered poset $(\mathcal{C}(\mathbb{R}^\infty), \leqslant)$ is connected since the zero cone $\{0\}$ is the only cone of dimension 0 and $\{0\} \leqslant C$ for any $C \in \mathcal{C}(\mathbb{R}^\infty)$ since 0 is a vertex of any cone pointed at zero.

Example 3.4. A **planar rooted tree** (see *e.g.* [CK, F, M1]) is a finite connected directed graph, without cycles, together with an embedding of it into the plane, such that only one vertex (the root) has outgoing edges only. We consider the set $\mathcal{T}$ of planar rooted trees filtered by the number of vertices. Concatenations of trees give rise to forests.

An elementary cut on a tree is a cut on some edge of the tree and an **admissible cut** on a tree consists of elementary cuts on some edges of the tree such that any path starting from the root contains at most one of them. For such a cut c, the tree $R^c(\mathrm{t})$ which contains the root of t is called the trunk of the tree, and the product $P^c(\mathrm{t})$ of the remaining trees, which is a forest, is called the crown. We define a partial order on trees by $\mathrm{t}' \leqslant \mathrm{t}$ (we say t' is a subtree of t) if there is an admissible cut c such that $\mathrm{t}' = P^c(t)$. It is compatible with the filtration since the subtree has fewer vertices than the original tree. The filtered poset $(\mathcal{T}, \leqslant)$ is connected since the empty tree is the only tree without vertices and it is clearly a subtree of any tree.

Example 3.5. A **Feynman graph** (see *e.g.* [CK, M1]) is a (non-oriented, non-planar) graph with a finite number of vertices and edges. We shall assume that the edges (internal or external) are of some given type which depends on the quantum field model we are considering, see *e.g.* [M1]. A **one-particle irreducible graph** (1PI graph) is a connected graph which remains connected when we cut an internal edge. The residue of a connected graph is the graph left over after shrinking all internal edges to a point.

The set $\mathcal{F}$ of Feynman graphs is filtered by the loop number $L := I - V + 1$ where I is the number of internal edges and V is the number of vertices of a given graph.

For a connected graph F in $\mathcal{F}$, we write $G \leqslant F$ if G is a subgraph of F, which should be 1 PI if F is 1 PI. This partial order is compatible with the filtration. However, the poset $(\mathcal{F}, \leqslant)$ is not connected since there are many graphs with zero loop number.

3.2. *Complement maps on posets*

Definition 3.6. Let $(\mathcal{P}, \leqslant)$ be a poset such that for any $E \in \mathcal{P}$

$$\mathfrak{s}(E) := \{A \in \mathcal{P} \,|\, A \leqslant E\} \tag{34}$$

is a finite set. A **complement** map on $\mathcal{P}$ assigns to any element $E \in \mathcal{P}$ a map

$$\begin{aligned} \complement_E : \mathfrak{s}(E) &\longrightarrow \mathcal{P} \\ A &\longmapsto E\backslash A \end{aligned}$$

satisfying the following properties

(a) **Compatibility with the partial order:** Let A, C in $\mathcal{P}$ be such that $A \leqslant C$. Then

$$\mathfrak{s}(C\backslash A) = \{B\backslash A \,|\, A \leqslant B \leqslant C\}.$$

(b) **Transitivity:** Let A, B, C in $\mathcal{P}$ be such that $A \leqslant B \leqslant C$. Then

$$(C\backslash A) \setminus (B\backslash A) = C\backslash B.$$

(c) **Compatibility with the filtration:** Assume that the poset is filtered: $\mathcal{P} = \cup_{n\in\mathbb{N}} \mathcal{P}_n$. Then the complement map is compatible with the filtration in the sense that

$$A \leqslant C \Longrightarrow |C\backslash A| = |C| - |A|\,.$$

(d) **Compatibility with the bottom:** Assume that the poset is connected and let $\mathcal{P}_0 = \{1\}$. Then

$$C\backslash 1 = C \quad \text{for all } C \in \mathcal{P}.$$

Condition (d) is obviously satisfied by previous examples of complement maps.

Remark 3.7. If the poset is connected, it follows from Condition (c) that $C\backslash C = \{1\}$ for any $C \in \mathcal{P}$ since

$$|C\backslash C| = |C| - |C| = 0 \Longrightarrow C\backslash C = \{1\}.$$

Note that by (a), from $|\mathfrak{s}(C)| < \infty$ we have $|\mathfrak{s}(C\backslash A)| < \infty$ and $B\backslash A \in \mathfrak{s}\,(C\backslash A)$. Thus the expressions in (b) are well-defined.

Example 3.8. Let E be a set. For $X \in \mathcal{P}_f(E)$, the complement set map:

$$\begin{aligned} \mathcal{P}_f(X) &\longrightarrow \mathcal{P}_f(X) \\ Y &\longmapsto X \setminus Y := X \cap \overline{Y} \end{aligned}$$

defines a complement map compatible with the filtration by the dimension.

Example 3.9. As we saw in the previous section, the set complement map on $\mathbb{Z}_{\geq 0}$ induces a complement map on product cones which we recall here for convenience. Given a product cone $\langle e_I \rangle$ and a subset $J \subseteq I$, the map $\langle e_J \rangle \longmapsto \langle e_{I\backslash J} \rangle$ defines a complement map on $\mathcal{P}(\mathbb{R}^\infty)$ compatible with the filtration by the dimension of the cone.

Example 3.10. Let E be a separable Hilbert vector space equipped with a countable orthonormal basis $(e_1, \cdots, e_n, \cdots)$. For any $I \subseteq \mathbb{N}$ we define the set

$$\mathcal{V}(I) := \{\text{lin}(e_J) \,|\, J \in \mathcal{P}_f(I)\}$$

of finite dimensional vector subspaces of E spanned by basis vectors indexed by finite subsets J of I. The set $\mathcal{V}(\mathbb{N})$ is equipped with a partial order given by the inclusion on the index sets which is compatible with the filtration given by the cardinal of the index set. The map

$$\begin{aligned} \mathcal{V}(I) &\longrightarrow \mathcal{V}(I) \\ \text{lin}(e_J) &\longmapsto \text{lin}(e_I) \setminus \text{lin}(e_J) := \text{lin}(\pi_{\text{lin}(e_J)^\perp}(e_I)\rangle = \text{lin}(e_{I\backslash J}) \end{aligned}$$

defines a complement map compatible with the filtration by the dimension. Here $\pi_{\mathrm{lin}(e_J)^\perp}$ stands for the orthogonal projection onto the orthogonal complement to the linear space $\mathrm{lin}(e_J)$. The notation $\pi_{\mathrm{lin}(e_J)^\perp}(e_I)$ means that the projection is applied to each basis vector indexed by an element of I.

Example 3.11. This orthogonal complement map on linear spaces also induces the complement map on product cones as can be seen from Eq. (28).

3.3. *A complement map on convex cones*

We now generalize the complement map built on product cones to general convex cones by means of an orthogonal projection.

Let $\mathcal{F}(C)$ be the set of all faces of a convex cone $C \subseteq \mathbb{R}^k$. We borrow the following concept from [BV] (see also [GPZ3]) which we refer the reader to for further details. The **transverse cone** to $F \in \mathcal{F}(C)$ is

$$t(C, F) := (C + \mathrm{lin}(F))/\,\mathrm{lin}(F), \tag{35}$$

(where lin stands for the linear span) which we identify to the cone in $\mathcal{C}\left(\mathbb{R}^\infty\right)$ defined by the projection $\pi_{F^\perp}(C)$ of C onto the orthogonal complement[h] $\mathrm{lin}(F)^\perp$ in $\mathrm{lin}(C)$ for the canonical scalar product on $\mathbb{R}^\infty$.

Example 3.12. The transverse cone to a face $F = \langle e_J \rangle$ of a product cone $\langle e_I \rangle$ is the cone $\langle e_{I \setminus J} \rangle$, which corresponds to the transverse cone $t\left(\langle e_I \rangle, \langle e_J \rangle\right)$.

Example 3.13. The transverse cone to the face $F = \langle e_1 + e_2 \rangle$ in the cone $C = \langle e_1, e_1 + e_2 \rangle$ is the cone $t(C, F) = \langle e_1 - e_2 \rangle$. Note that $t(C, F)$ is not a face of C.

Lemma 3.14. *The map*

$$\begin{aligned} \mathcal{F}(C) &\longrightarrow \mathcal{C}(\mathbb{R}^\infty) \\ F &\longmapsto t(C, F) \end{aligned}$$

which to a face F of a cone C assigns the transverse cone $t(C, F)$, is a complement map. More precisely, it enjoys the following properties.

---

[h]Our approach, like the one of Berline and Vergne in [BV], actually requires a choice a rational lattice which consists of a pair built from a cone and a rational lattice in the linear space spanned by the cone. We refer the reader to [GPZ3] for a detailed description.

(a) ***Compatibility with the partial order:*** *The set of faces of the cone* $t(C, F)$ *equals* $\{t(G, F) \mid G$ *a face of* C *containing* $F\}$.
(b) ***Transitivity:*** $t(C, F) = t\,(t(C, F'), t(F, F'))$ *if* F' *is a face of* F.
(c) ***Compatibility with the dimension filtration:*** $\dim(C) = \dim(F) + \dim\,(t(C, F))$ *for any face* F *of* C.
(d) ***Compatibility with the bottom:*** $\mathcal{C}$ *is connected since* $\mathcal{C}_0$ *is reduced to* $1 := \{0\}$ *and for any cone* C *we have* $t(C, \{0\}) = C$.

Proof. (a) Assume that F is defined by the linear form u_F on $\mathbb{R}^\infty$, *i.e.*,

$$F = \{v \in C \mid \langle u_F, v\rangle = 0\}.$$

Let G be any face of C containing F that is defined by a linear form u_G on $\mathbb{R}^\infty$, then $u_G|_F = 0$. Since a linear form u on $\mathbb{R}^\infty$ with $u|_F = 0$ induces a linear form u on $\mathrm{lin}(F)^\perp$, we can view u_G as a linear form on $\mathrm{lin}(F)^\perp$. It therefore defines a face $t(G, F)$ of $t(C, F)$. We can therefore define a map

$$t(\bullet, F) : \{\text{faces of } C \text{ containing } F\} \to \{\text{faces of } t(C, F)\}$$

$$G \mapsto t(G, F) = t(C, F) \cap \{v \in \mathbb{R}^\infty \mid \langle u_G, v\rangle = 0\}.$$

To check the bijectivity of $t(\bullet, F)$, we first note that any face of $t(C, F)$ is defined by some linear form u on $\mathrm{lin}(F)^\perp$ which can be viewed as a linear form on $\mathbb{R}^\infty$ that vanishes on lin(F). Hence u defines a face G of C containing F. Thus $t(\bullet, F)$ is surjective.

For two different faces G_1, G_2 containing F defined by linear forms u_1, u_2 on $\mathbb{R}^\infty$, there are vectors v_1 in G_1 and v_2 in G_2 such that $\langle u_1, v_2\rangle > 0$ and $\langle u_2, v_1\rangle > 0$. Thus $t(G_1, F)$ and $t(G_2, F)$ are different since the image of v_1 is not in $t(G_2, F)$ and the image of v_2 is not in $t(G_1, F)$. Hence the map $t(\bullet, F)$ is one-to-one.

(b) The linear space $\mathrm{lin}(t(C, F))$ spanned by the transverse cone is the orthogonal space $\mathrm{lin}(F)^{\perp_{\mathrm{lin}(C)}}$ in $\mathrm{lin}(C)$ to $\mathrm{lin}(F)$. The transitivity then follows from the "transitivity" of the orthogonal complement map on linear spaces:

$$\begin{aligned}\mathrm{lin}\,(t(t(C, F'), t(F, F'))) &= \mathrm{lin}(t(F, F'))^{\perp_{\mathrm{lin}(t(C,F'))}}\\ &= \left((\mathrm{lin}\,F')^{\perp_{\mathrm{lin}\,F}}\right)^{\perp_{\left(\mathrm{lin}(F')^{\perp}_{\mathrm{lin}\,C}\right)}} = \mathrm{lin}(F)^{\perp_{\mathrm{lin}\,C}} = \mathrm{lin}\,(t(F, C)).\end{aligned}$$

(c) follows the fact that $\mathrm{lin}(t(C, F))$ and $\mathrm{lin}(F)$ are orthogonal complements in $\mathrm{lin}(C)$. □

Example 3.15. We use the notations of Example 3.4 above. See [F] for further details. In $\mathcal{T}$, the map

$$\mathfrak{t}' = P^c(\mathfrak{t}) \leqslant \mathfrak{t} \longmapsto R^c(\mathfrak{t}) = \mathfrak{t} \setminus \mathfrak{t}'$$

defines a complement map. Let us first check the transitivity; let $\mathfrak{t}_3 \preceq \mathfrak{t}_2 \preceq \mathfrak{t}_1$, then cutting the smaller trunk $\mathfrak{t}_3$ off both $\mathfrak{t}_2$ and $\mathfrak{t}_1$, before cutting off the remaining crown $\mathfrak{t}_2 \setminus \mathfrak{t}_3$ off $\mathfrak{t}_1$ amounts to cutting off the whole trunk $\mathfrak{t}_2$ from $\mathfrak{t}_1$. We now check the compatibility with the partial order; if $\mathfrak{f}$ is a forest made of trunks cut off from the crown $R^c(\mathfrak{t})$ of a tree $\mathfrak{t}$ –*i.e.*, if $\mathfrak{f} \leqslant \mathfrak{f}' = R^c(\mathfrak{t}) = \mathfrak{t} \setminus P^c(\mathfrak{t})$– then there is a unique tree $\mathfrak{t}'$ larger than $\mathfrak{t}$ *i.e.*, $\mathfrak{t} \preceq \mathfrak{t}'$, such that $\mathfrak{f} = \mathfrak{t} \setminus \mathfrak{t}'$; $\mathfrak{t}'$ is built from gluing $\mathfrak{f}'$ as a crown onto $P^c(\mathfrak{t})$.

Example 3.16. We use the notation of Example 3.5. See [M1] for further details. In $\mathcal{F}$, the complement $\Gamma \setminus \gamma$ of $\gamma \leqslant \Gamma$ in Γ is the diagram obtained after "shrinking" the subdiagram γ to a point. There is a bijection $\gamma \mapsto \tilde{\gamma} = \gamma \setminus \delta$ from subgraphs of a graph Γ containing δ onto subgraphs of $\Gamma \setminus \delta$ which shows the compatibility of the complement map with the partial order. The shrinking procedure is also clearly transitive $\Gamma \setminus \gamma = (\Gamma \setminus \delta) \setminus (\gamma \setminus \delta)$.

3.4. *Coproducts derived from complement maps*

Loosely speaking, coalgebras are objects dual to algebras. More precisely, algebras are dual to coalgebras but the converse only holds in finite dimensions (see *e.g.* [Ca]).

Definition 3.17. A **(counital) coalgebra** is a linear space $\mathcal{C}$ (here over $\mathbb{R}$) equipped with two linear maps:

(a) a **comultiplication** $\Delta : \mathcal{C} \to \mathcal{C} \otimes \mathcal{C}$ written in Sweedler's notation [Sw]

$$\Delta c = \sum_{(c)} c_{(1)} \otimes c_{(2)},$$

which is **coassociative**

$$(I \otimes \Delta) \otimes \Delta = (\Delta \otimes I) \otimes \Delta.$$

The coassociativity of Δ translates to the following commutative diagram

$$\begin{array}{ccc} \mathcal{C} & \xrightarrow{\quad \Delta \quad} & \mathcal{C} \otimes \mathcal{C} \\ {\scriptstyle \Delta}\downarrow & & \downarrow{\scriptstyle I \otimes \Delta} \\ \mathcal{C} \otimes \mathcal{C} & \xrightarrow{\quad \Delta \otimes I \quad} & \mathcal{C} \otimes \mathcal{C} \otimes \mathcal{C} \end{array}$$

and can be expressed in the following compact notation:

$$\sum_{(c)} c_{(1)} \otimes \left(\sum_{(c_{(2)})} (c_{(21)}) \otimes (c_{(22)}) \right) = \sum_{(c)} \left(\sum_{(c_{(1)})} (c_{(11)}) \otimes (c_{(12)}) \right) \otimes c_{(2)}.$$

With Sweedler's notation [Sw], both these expressions read

$$\sum_{(c)} c_{(1)} \otimes c_{(2)} \otimes c_{(3)}.$$

(b) a **counit** $\varepsilon : \mathcal{C} \to \mathbb{R}$ satisfying the **counitarity** property

$$(I_{\mathcal{C}} \otimes \varepsilon) \circ \Delta = (\varepsilon \otimes I_{\mathcal{C}}) \circ \Delta = I_{\mathcal{C}}, \tag{36}$$

with the identification $\mathcal{C} \otimes \mathbb{R} \simeq \mathcal{C} \simeq \mathbb{R} \otimes \mathcal{C}$. This translates to the following commutative diagram:

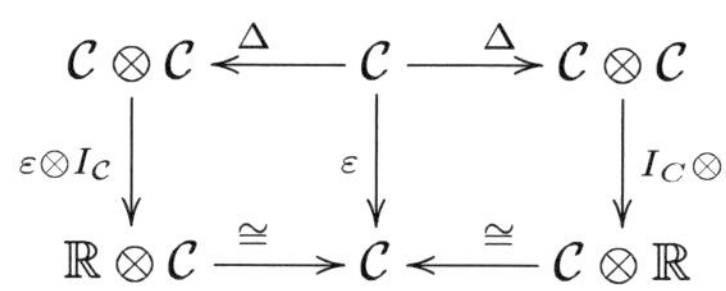

The fact that ε is a counit can be expressed by means of the following formula

$$c = \sum_{(c)} \varepsilon(c_{(1)}) c_{(2)} = \sum_{(c)} c_{(1)} \varepsilon(c_{(2)}).$$

The coalgebra is **cograded** if there is a grading $\mathcal{C} = \bigoplus_{n \geqslant 0} \mathcal{C}^{(n)}$ such that

$$\Delta(\mathcal{C}^{(n)}) \subseteq \bigoplus_{p+q=n} \mathcal{C}^{(p)} \otimes \mathcal{C}^{(q)}, \quad n \geqslant 0.$$

Elements in $\mathcal{C}^{(n)}$ are said to have **degree** n.

The coalgebra is **cocommutative** if $\tau \circ \Delta = \Delta$ where $\tau : \mathcal{C} \otimes \mathcal{C} \to \mathcal{C} \otimes \mathcal{C}$ is the flip $c_1 \otimes c_2 \longmapsto c_2 \otimes c_1$. This translates to the following commutative diagram:

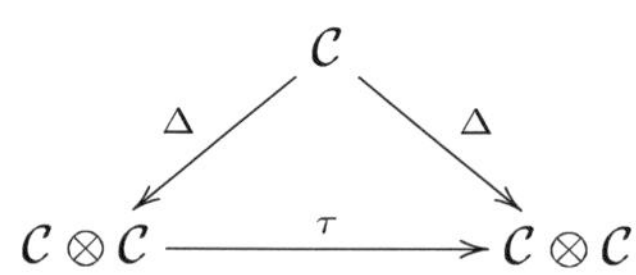

and the equation

$$\sum_{(c)} c_{(1)} \otimes c_{(2)} = \sum_{(c)} c_{(2)} \otimes c_{(1)}.$$

The coalgebra $\mathcal{C}$ is **coaugmented** if there is a morphism of coalgebras $u : \mathbb{R} \to \mathcal{C}$ in which case we have $\varepsilon \circ u = I_{\mathbb{R}}$ and we set $1_C := u(1_{\mathbb{R}})$ where $1_{\mathbb{R}}$ is the unit in $\mathbb{R}$. If $\mathcal{C}$ is coaugmented, then $\mathcal{C}$ is canonically isomorphic to $\mathrm{Ker}\ \varepsilon \oplus \mathbb{R}1_{\mathcal{C}}$. The kernel $\mathrm{Ker}\ \varepsilon$ is often denoted by $\overline{\mathcal{C}}$ so $\mathcal{C} = \overline{\mathcal{C}} \oplus \mathbb{R}1_{\mathcal{C}}$.

The coalgebra $\mathcal{C}$ is **connected** if it is coaugmented and $\mathcal{C}^{(0)} = \mathbb{R}1_{\mathcal{C}}$.

We are ready to build a coproduct from a complement map.

Proposition 3.18. *Let a poset $(\mathcal{P}, \leqslant)$ be such that for any E in $\mathcal{P}$ the set $\mathfrak{s}(E)$ defined as in Eq. (34) is finite and let it be equipped with a complement map, which assigns to any element $E \in \mathcal{P}$ a map*

$$\complement_E : \mathfrak{s}(E) \longrightarrow \mathcal{P}$$

$$A \longmapsto E\backslash A.$$

Then the map

$$\begin{aligned} \Delta : \mathcal{P} &\longrightarrow \mathcal{P} \otimes \mathcal{P} \\ E &\longmapsto \sum_{A \in \mathfrak{s}(E)} E\backslash A \otimes A, \end{aligned}$$

extends linearly to a coassociative coproduct on the space $\mathbb{K}\mathcal{P}$ freely generated over a field $\mathbb{K}$ by $\mathcal{P}$.

If the poset is filtered $\mathcal{P} = \cup_{n \in \mathbb{N}} \mathcal{P}_n$ and the complement map is compatible with the filtration then so is the coproduct, that is, if C is in $\mathcal{P}_n$, then ΔC is in $\sum_{p+q=n} \mathcal{P}_p \otimes \mathcal{P}_q$.

Let $\varepsilon : \mathcal{P} \to \mathbb{K}$ be zero outside $\mathcal{P}_0$ where it takes the value one and let us denote its linear extension to $\mathbb{K}\mathcal{P}$ by the same symbol. If moreover the poset $\mathcal{P}$ is connected, then the linear space $(\mathbb{K}\mathcal{P}, \Delta, \varepsilon)$ is a counital cograded connected coalgebra.

Proof. We first check the coassociativity.

Let $C \in \mathcal{P}$. On the one hand we have

$$(I \otimes \Delta)\Delta(C) = \sum_{B \leqslant C} (I \otimes \Delta)(C \setminus B \otimes B) = \sum_{D \leqslant B \leqslant C} C \setminus B \otimes B \setminus D \otimes D.$$

On the other hand,

$$\begin{aligned}(\Delta\otimes I)\Delta(C) &= \textstyle\sum_{D\leqslant C}(\Delta\otimes I)(C\setminus D\otimes D)\\ &= \textstyle\sum_{D\leqslant C}\sum_{H\leqslant C\setminus D}((C\setminus D)\setminus H)\otimes H\otimes D\\ &= \textstyle\sum_{D\leqslant B\leqslant C}(C\setminus D)\setminus(B\setminus D)\otimes B\setminus D\otimes D\\ &\qquad\qquad\text{(compatibility with the partial order)}\\ &= \textstyle\sum_{D\leqslant B\leqslant C}C\setminus B\otimes B\setminus D\otimes D. \quad \text{(transitivity)}\end{aligned}$$

Let us check the counitarity. For any $C\in\mathcal{P}$, setting $\mathcal{P}_0=\{1\}$ and using the fact that $C\backslash 1=C$ (see item (d) in Definition 3.6), for any $C\in\mathcal{P}$ we have

$$\sum_{B\leqslant C}\varepsilon(B)\ (C\backslash B)=\sum_{|B|=0}\varepsilon(B)\ (C\backslash B)=\varepsilon(1)\ (C\backslash 1)=1_{\mathbb{K}}\cdot C=C.$$

Furthermore, since $|C\backslash B|=0\Longrightarrow C\backslash B=1\Longrightarrow B=C$, using the fact that $C\backslash C=1$ (see Remark 3.7) we have

$$\sum_{B\leqslant C}B\,\varepsilon\,(C\backslash B)=\sum_{|C\backslash B|=0}B\,\varepsilon\,(C\backslash B)=C\,\varepsilon(C\backslash C)=C\cdot 1_{\mathbb{K}}=C.$$

□

Example 3.19. The vector space $\mathbb{R}\mathcal{P}_f(E)$ spanned by finite subsets of a finite set E defines a cograded connected coalgebra.

Example 3.20. The free algebra $\mathbb{R}\mathcal{C}(\mathbb{R}^\infty)$ spanned by closed convex cones pointed at zero in $\mathbb{R}^\infty$ defines a cograded connected coalgebra.

Example 3.21. The free algebra $\mathbb{R}\mathcal{T}$ generated by planar rooted trees defines a cograded connected coalgebra.

4. Algebraic Birkhoff factorization on a graded connected coalgebra

We give a generalization ([GPZ3]) of the algebraic Birkhoff factorization used for renormalization purposes in quantum field theory (see [CK, M1]) in so far as we weaken the assumptions on the source space which is not anymore assumed to be a Hopf algebra but only a coalgebra, as well as on the target algebra which is not anymore required to decompose into two subalgebras. We first define the convolution product and give its main properties.

For a given algebra $\mathcal{A}$ and coalgebra $\mathcal{C}$, let $\mathcal{L}(\mathcal{C}, \mathcal{A})$ denote the algebra of $\mathcal{A}$- valued linear maps on $\mathcal{C}$.

4.1. *The convolution product*

Let $(\mathcal{A}, m_{\mathcal{A}}, 1_{\mathcal{A}})$ be an (unital) commutative algebra over $\mathbb{R}$.

Proposition 4.1. *(see e.g. [M1, Proposition II.3.1]) Let $(\mathcal{C}, \Delta_{\mathcal{C}}, \varepsilon_{\mathcal{C}})$ be a (counital) coalgebra over $\mathbb{R}$.*

(a) The **convolution** *product on $\mathcal{L}(\mathcal{C}, \mathcal{A})$ defined as*

$$\varphi * \psi = m_{\mathcal{A}} \circ (\varphi \otimes \psi) \circ \Delta_{\mathcal{C}}$$

is associative. In Sweedler's notation it reads:

$$\varphi * \psi(x) = \sum_{(x)} \varphi(x_{(1)})\psi(x_{(2)}).$$

(b) $e := u_{\mathcal{A}} \circ \varepsilon_{\mathcal{C}}$ is a unit for the convolution product on $\mathcal{L}(\mathcal{C}, \mathcal{A})$.

Proof. (a) Using the coassociativity of $\Delta_{\mathcal{C}}$ and the associativity of $m_{\mathcal{A}}$ (we omit the explicit mention of the product in the computation below), for three $\varphi, \psi, \chi \in \mathcal{L}(\mathcal{C}, \mathcal{A})$ and using Sweedler's notations $\Delta x = \sum_{(x)} x_{(1)} \otimes x_{(2)}$; $\Delta x_{(i)} = \sum_{(x_{(i)})} x_{(i1)} \otimes x_{(i2)}$ for each $i \in \{1, 2\}$ we have

$$\begin{aligned}
((\varphi * \psi) * \chi)(x) &= \sum_{(x)} \left(\varphi(x_{(11)})\psi(x_{(12)})\right) \chi(x_{(2)}) = \\
&= \sum_{(x)} \varphi(x_{(11)}) \left(\psi(x_{(12)})\chi(x_{(2)})\right) \quad \text{(since } m_{\mathcal{A}} \text{ is associative)} \\
&= \sum_{(x)} \varphi(x_{(1)}) \left(\psi(x_{(21)})\chi(x_{(22)})\right) \quad \text{(since } \Delta_{\mathcal{C}} \text{ is coassociative)} \\
&= \sum_{(x)} \varphi(x_{(1)})\, \psi(x_{(21)})\chi(x_{(22)}) \quad \text{(since } m_{\mathcal{A}} \text{ is associative)} \\
&= (\varphi * (\psi * \chi))(x).
\end{aligned}$$

(b) Let $\varphi \in \mathcal{L}(\mathcal{C}, \mathcal{A})$. Since $(\varepsilon_{\mathcal{C}} \otimes I) \circ \Delta_{\mathcal{C}} = I = (I \otimes \varepsilon_{\mathcal{C}}) \circ \Delta_{\mathcal{C}}$ we have

$$e * \varphi(x) = \sum_{(x)} u_{\mathcal{A}} \circ \varepsilon_{\mathcal{C}}(x_{(1)})\, \varphi(x_{(2)}) = u_{\mathcal{A}}(1_{\mathcal{C}})\, \varphi(x) = \varphi(x),$$

and similarly, we show that $\varphi * e(x) = \varphi(x)$ for any $x \in \mathcal{C}$. $\square$

Example 4.2. The convolution product of two maps φ and ψ in $\mathcal{L}\left(\mathcal{P}(\mathbb{R}^\infty),\mathcal{A}\right)$ on a product cone $C=\langle e_I\rangle$ derived from the complement map described in Example 3.9 reads

$$\varphi*\psi\left(\langle e_I\rangle\right)=\sum_{J\subset I}\varphi\left(\langle e_{I\setminus J}\rangle\right)\psi\left(\langle e_J\rangle\right)=\sum_{F\in\mathcal{F}(C)}\varphi(\overline{F}^C)\,\psi(F)$$

with the notation of Eq. (28).

Setting $\mathcal{A}=\mathrm{Mer}_{\mathrm{sep}}(\mathbb{C}^\infty)$, then Eqs. (29) and (30) seen as identities of maps on product cones read

$$\mathcal{S}=\mathcal{S}_+*\mathcal{S}_-=\mu*\mathcal{I}. \tag{37}$$

Proposition 4.3. *([M1, Proposition II.3.1.]) Let $\mathcal{C}$ be a cograded connected coalgebra and $\mathcal{A}$ an algebra. The set*

$$\mathcal{G}(\mathcal{C},\mathcal{A}):=\{\varphi\in\mathcal{L}(\mathcal{C},\mathcal{A}),\quad \varphi(1_\mathcal{C})=1_\mathcal{A}\}$$

endowed with the convolution product is a group with unit $e:=\varepsilon_\mathcal{C}\circ u_\mathcal{A}$ and inverse

$$\varphi^{*(-1)}(x):=\sum_{k=0}^\infty(e-\varphi)^{*k}(x) \tag{38}$$

is well defined as a finite sum.

Proof. We saw that e is a unit for the convolution product. Let us now build an inverse $\varphi^{*(-1)}$ for any φ in $\mathcal{G}\left(\mathcal{C},\mathcal{A}\right)$:

$$\varphi^{*(-1)}(x)=\left(e-(e-\varphi)\right)^{*(-1)}(x)=\sum_{k=0}^\infty(e-\varphi)^{*k}(x).$$

Note that we indeed have

$$\begin{aligned}\varphi*\sum_{k=0}^\infty(e-\varphi)^{*k}&=(\varphi-e)*\sum_{k=0}^\infty(e-\varphi)^{*k}+\sum_{k=0}^\infty(e-\varphi)^{*k}\\&=-\sum_{k=1}^\infty(e-\varphi)^{*k}+\sum_{k=0}^\infty(e-\varphi)^{*k}=1.\end{aligned}$$

On the one hand, since $\varepsilon(1_\mathcal{C})=u_\mathcal{A}\circ\varepsilon(1_\mathcal{C})=1u_\mathcal{A}(1)=1_\mathcal{A}$ we have $(e-\varphi)(1_\mathcal{C})=0$ by assumption on φ. Hence $(e-\varphi)^{*k}(1_\mathcal{C})=0$ for $k>1$ so that the above formula yields $\varphi^{*(-1)}(1_\mathcal{C})=1_\mathcal{A}$ and $\varphi^{*(-1)}\in\mathcal{G}\left(\mathcal{C},\mathcal{A}\right)$. On the other hand, for any x in $\mathrm{Ker}(\varepsilon)$ we have

$$(e-\varphi)^{*k}(x)=m_{\mathcal{A},k-1}(\varphi\otimes\cdots\otimes\varphi)\circ\overline{\Delta}^{k-1}(x).$$

Since element x lies in some $\overline{\mathcal{C}}_n$, this expression vanishes for $k \geqslant n+1$. Hence the above power series is finite and therefore defines an inverse of φ which lies in $\mathcal{G}$ since $\varphi^{*(-1)}(1_{\mathcal{C}}) = (e-\varphi)^{*0}(x) = 1_{\mathcal{A}}$ □

Example 4.4. Back to Example 4.2, we can rewrite the renormalized holomorphic part (37) of S as

$$\mathcal{S}_+ = \mathcal{S} * \mathcal{S}_-^{*(-1)} = \mathcal{S} * \mathcal{I}^{*(-1)}.$$

4.2. *Algebraic Birkhoff factorization*

We quote the following result from [GPZ3, Theorem 3.2].

Theorem 4.5. *Let $\mathcal{C} = \bigoplus_{n\geqslant 0} \mathcal{C}^{(n)}$ be a cograded connected coalgebra and let $\mathcal{A}$ be a unitary algebra. Let $\mathcal{A} = \mathcal{A}_1 \oplus \mathcal{A}_2$ be a linear decomposition such that $1_{\mathcal{A}} \in \mathcal{A}_1$ and let P be the induced projection onto $\mathcal{A}_1$ parallel to $\mathcal{A}_2$. Given $\varphi \in \mathcal{G}(\mathcal{C},\mathcal{A})$, we define two maps $\varphi_i \in \mathcal{G}(\mathcal{C},\mathcal{A}), i=1,2$ defined by the following recursive formulae on* $\ker\varepsilon$*:*

$$\varphi_1(x) = -P\left(\varphi(x) + \sum_{(x)} \varphi_1(x')\varphi(x'')\right), \tag{39}$$

$$\varphi_2(x) = (\mathrm{id}_A - P)\left(\varphi(x) + \sum_{(x)} \varphi_1(x')\varphi(x'')\right), \tag{40}$$

where, following Sweedler's notation, we have set $\overline{\Delta} x = \sum x' \otimes x''$.

(a) We have $\varphi_i(\ker\varepsilon) \subseteq \mathcal{A}_i$ and hence $\varphi_i : \mathcal{C} \to \mathbb{K}1_{\mathcal{A}} + \mathcal{A}_i$. Moreover, the following factorization holds

$$\varphi = \varphi_1^{*(-1)} * \varphi_2. \tag{41}$$

(b) $\varphi_i, i = 1,2$, are the unique maps in $\mathcal{G}(\mathcal{C},\mathcal{A})$ such that $\varphi_i(\ker\varepsilon) \subseteq \mathcal{A}_i$ for $i = 1,2$, and satisfying Eqn. (41).
(c) If moreover $\mathcal{A}_1$ is a subalgebra of $\mathcal{A}$ then $\varphi_1^{*(-1)}$ lies in $\mathcal{G}(\mathcal{C},\mathcal{A}_1)$.

Proof. (a) The proof is the same as in [GPZ3, Theorem 3.2] ignoring the differential structure discussed there. We reproduce the proof for the sake of completeness.

The inclusion $\varphi_i(\ker\varepsilon) \subseteq \mathcal{A}_i, i = 1, 2$, follows from the definitions. Further

$$\begin{aligned}\varphi_2(x) &= (\mathrm{id}_{\mathcal{A}} - P)\left(\varphi(x) + \sum_{(x)} \varphi_1(x')\varphi(x'')\right)\\ &= \varphi(x) + \varphi_1(x) + \sum_{(x)} \varphi_1(x')\varphi(x'')\\ &= (\varphi_1 * \varphi)(x).\end{aligned}$$

Since $\varphi_1(J) = 1_{\mathcal{A}}$, φ_1 is invertible for the convolution product in $\mathcal{A}$ as a result of Proposition 4.3 applied to φ_1, from which Eq. (41) then follows.
(b) Suppose there are $\psi_i \in \mathcal{G}(\mathcal{C}, \mathcal{A}), i = 1, 2$, with $\psi_i(\ker\varepsilon) \subseteq \mathcal{A}_i$ such that $\varphi = \psi_1^{*(-1)} * \psi_2$ and $\psi_i(J) = 1_{\mathcal{A}}$. We prove $\varphi_i(x) = \psi_i(x)$ for $i = 1, 2, x \in \mathcal{C}^{(k)}$ by induction on $k \geqslant 0$. These equations hold for $k = 0$. Assume that the equations hold for $\mathcal{C}^{(k)}$. For $x \in \mathcal{C}^{(k+1)} \subseteq \ker(\varepsilon)$, by $\varphi_2 = \varphi_1 * \varphi$ and $\psi_2 = \psi_1 * \varphi$, we have

$$\varphi_2(x) = \varphi_1(x) + \varphi(x) + \sum_{(x)} \varphi_1(x')\varphi(x'')$$

and

$$\psi_2(x) = \psi_1(x) + \varphi(x) + \sum_{(x)} \psi_1(x')\varphi(x''),$$

where we have used $\varphi_1(J) = \psi_1(J) = \varphi(J) = 1_A$. Hence by the induction hypothesis, we have

$$\begin{aligned}\varphi_2(x) - \psi_2(x) &= \varphi_1(x) - \psi_1(x) + \sum_{(x)}(\varphi_1(x') - \psi_1(x'))\varphi(x'')\\ &= \varphi_1(x) - \psi_1(x) \in A_1 \cap A_2 = \{0\}.\end{aligned}$$

Thus

$$\varphi_i(x) = \psi_i(x) \text{ for all } x \in \ker(\varepsilon), i = 1, 2.$$

(c) If $\mathcal{A}_1$ is a subalgebra, then it follows from Proposition 4.1 applied to $\mathcal{A}_1$ instead of $\mathcal{A}$, that φ_1 is invertible in $\mathcal{A}_1$. □

5. Application to renormalized conical zeta values

5.1. *Algebraic Birkhoff factorization on cones*

The algebraic Birkhoff factorization can be carried out from exponential sums on product cones to exponential sums on general convex polyhedral cones using the complement map described in Lemma 3.14 built from the transverse cone to a face [GPZ3], which generalizes the orthogonal complement used in the case of product cones. Here we consider both closed convex (polyhedral) cones in $\mathbb{R}^k$

$$\langle v_1, \cdots, v_n\rangle := \mathbb{R}\{v_1, \cdots, v_n\} = \mathbb{R}_{\geqslant 0} v_1 + \cdots + \mathbb{R}_{\geqslant 0} v_n,$$

where $v_i \in \mathbb{R}^k$, $i = 1, \cdots, n$ defined previously and open cones defined in a similar manner replacing $\mathbb{R}_{\geqslant 0}$ by $\mathbb{R}_+$. Product cones $\langle e_i, i \in I\rangle$ with $I \subset \{1, \cdots, k\}$ and $\{e_i \,|\, i \in \{1, \cdots, k\}\}$ the canonical basis of $\mathbb{R}^k$ are convex cones. We shall focus here on Chen cones $\langle e_{i_1}, e_{i_1} + e_{i_2}, \cdots, e_{i_1} + \cdots + e_{i_n}\rangle$ with $\{i_1, \cdots, i_n\} \subset \{1, \ldots, k\}$, which are closed convex cones as well as their open counterparts.

Both the complement map defined by means of the transverse map in Lemma 3.14 and the corresponding coproduct defined in Proposition 3.18 (see also Example 3.20) are compatible with subdivisions in a suitable sense. Recall that a **subdivision** of a cone C is a set $\{C_1, \cdots, C_r\}$ of cones such that

(i) $C = \cup_{i=1}^r C_i$,
(ii) $C_1, \cdots, C_r$ have the same dimension as C and
(iii) $C_1, \cdots, C_r$ intersect along their faces, *i.e.*, $C_i \cap C_j$ is a face of both C_i and C_j.

Example 5.1. The product cone $\langle e_1, e_2\rangle$ can be subdivided into two Chen cones $\langle e_1, e_1 + e_2\rangle$ and $\langle e_1 + e_2, e_2\rangle$.

To a simplicial convex (closed) cone $C \subset \mathbb{Z}^k$, namely one whose generators are linearly independent, one can assign an exponential sum and an exponential integral which can informally be described as follows

$$S^c(C)(\vec{\varepsilon}) := \sum_{\vec{n}\in C\cap Z^k} e^{\langle \vec{n}, \vec{\varepsilon}\rangle};\ S^o(C)(\vec{\varepsilon}) := \sum_{\vec{n}\in C^o\cap \mathbb{Z}^k} e^{\langle \vec{n}, \vec{\varepsilon}\rangle};\ I(C)(\vec{\varepsilon}) = \int_C e^{\langle \vec{x}, \vec{\varepsilon}\rangle}\, d\vec{x}.$$

Here C^o is the open cone given by the interior of C and ε is taken in

$$\check{C}_k^- := \left\{ \vec{\varepsilon} := \sum_{i=1}^k \varepsilon_i e_i^* \,\middle|\, \langle \vec{x}, \vec{\varepsilon}\rangle < 0 \text{ for all } \vec{x} \in C \right\},$$

where $\{e_i^* \,|\, i \in \{1, \ldots, k\}\}$ is the dual canonical basis and $\langle \vec{x}, \vec{\varepsilon}\rangle$ the natural pairing $R^k \otimes \left(\mathbb{R}^k\right)^* \to \mathbb{R}$. Keep in mind that a precise formulation requires introducing a lattice attached to the cone, so considering lattice cones instead of mere cones (see [GPZ3]). This then extends to any convex cones by additivity on subdivisions.

The exponential sums on general convex cones take their values in the space of meromorphic germs with simple linear poles supported by the faces of the cone. It turns out that a meromorphic germs with linear poles also decomposes as a sum of a holomorphic part and a polar part. A decomposition of the algebra of meromorphic germs with linear poles into the holomorphic part and a linear complement was shown in [GPZ4] by means of an inner product using our results on cones and pure fractions in an essential way. We shall denote by π_+ the corresponding projection onto the holomorphic part.

Consequently, one can implement an algebraic Birkhoff factorization [GPZ3] on the coalgebra of convex polyhedral cones. [i] Just as the algebraic Birkhoff factorization gave rise to an Euler-Maclaurin formula on product cones, when the inner product used to defined the coproduct on cones coincides with the inner product used to decompose the space of meromorphic germs, the algebraic Birkhoff factorization of the exponential sum on a convex (lattice) cone yields back Berline and Vergne's local Euler-Maclaurin formula [BV]. To prove this identification which is easy to see on smooth cones, we subdivide a general convex cone into simplicial ones and use the compatibility of S_- in the factorization procedure with subdivisions. This compatibility is shown by means of a rather involved combinatorial proof.

Recall from Eq. (32) that the "holomorphic part" of the exponential sum on product cones generates products of renormalized zeta values at non-positive integers as coefficients of its Taylor expansion at zero. Similarly [GPZ5], the "holomorphic part" of the exponential sums on general convex polyhedral cones obtained from an algebraic Birkhoff factorization, generates what we call renormalized *conical* zeta values at non-positive integers which arise as coefficients of its Taylor expansion at zero. It turns out that the "holomorphic part" of the exponential sums $S^c(C)$ and $S^o(C)$ on a cone C derived from the algebraic Birkhoff factorization actually coincides with the projection $\pi_+(S^c(C))$ and $\pi_+(S^o(C))$, when the inner product used to defined the coproduct on cones coincides with the inner product used to decompose the space of meromorphic germs respectively, onto their holomorphic part when seen as meromorphic functions with linear poles.

[i] We actually carry out the algebraic Birkhoff factorization on lattice cones.

5.2. *Meromorphic functions with linear poles*

For later use, we define the projection π_+ somewhat informally; a precise definition can be found in [GPZ4]. One shows that a meromorphic function $f = \frac{h}{L_1 \cdots L_n}$ on $\mathbb{C}^k$ with linear poles $L_i, i = 1, \cdots, n$ given by linear forms and h a holomorphic function at zero, uniquely decomposes as

$$f = \sum_{i=1}^{n} \left(\frac{h_i(\vec{\ell}_i)}{\vec{L}_i^{\vec{s}_i}} + \varphi_i(\vec{\ell}_i, \vec{L}_i) \right), \tag{42}$$

with $|\vec{s}_i| > 0$ and where $\vec{L}_i = (L_{i1}, \cdots, L_{im_i})$, $\{L_{i1}, \cdots, L_{im_i}\}$ is a linear independent subset of $\{L_1, \cdots, L_n\}$, extended to a basis $\{\vec{L}_i, \vec{\ell}_i\}$ of $\mathbb{C}^k$, with $\vec{\ell}_i = (\ell_{i(m_i+1)}, \cdots \ell_{ik})$, L_{ij}, ℓ_{im} orthogonal for the canonical inner product on $\mathbb{C}^k$ and $h_i(\vec{\ell}_i)$ holomorphic (reduced to a constant when $k = 1$). Then we call $f_+ := \pi_+(f) = \sum_{i=1}^{n} \varphi_i$, which is a germ of holomorphic function in the independent variables $\vec{\ell}_i$ and $\vec{L}_i$, the **holomorphic part** of f and $f_- := (1 - \pi_+)(f) = \sum_{i=1}^{n} \frac{h_i(\vec{\ell}_i)}{\vec{L}_i^{\vec{s}_i}}$ the **polar part** of f.

In order to discuss examples, it is convenient to set the following notation. Given k linear forms $L_1, \cdots, L_k$, we set

$$[L_1, \cdots, L_k] := \frac{e^{L_1}}{1 - e^{L_1}} \frac{e^{L_1+L_2}}{1 - e^{L_1+L_2}} \cdots \frac{e^{L_1+L_2+\cdots+L_k}}{1 - e^{L_1+L_2+\cdots+L_k}}. \tag{43}$$

So, for any (closed) Chen cone $C_k = \langle e_1, e_1 + e_2, \cdots, e_1 + \cdots + e_k \rangle$ (here $e_1, \ldots, e_k$ is the canonical basis of $\mathbb{R}^k$), we have

$$S^o(C_k)(\varepsilon_1, \varepsilon_2, \cdots, \varepsilon_k) = [\varepsilon_1, \varepsilon_2, \cdots, \varepsilon_k].$$

Example 5.2.

(a) Take $k = 1$. Let $f(\varepsilon) = \frac{e^{\varepsilon}}{1-e^{\varepsilon}} = \frac{1}{e^{-\varepsilon}-1} = -\frac{\mathrm{Td}(-\varepsilon)}{-\varepsilon}$ on $\mathbb{C}$. Then by Eq. (5) we have

$$f(\varepsilon) = -\frac{1}{\varepsilon} - \frac{1}{2} - \sum_{k=1}^{K} \frac{B_{2k}}{(2k)!}\varepsilon^{2k-1} + o(\varepsilon^{2K}) = -\frac{1}{\varepsilon} + \varphi(\varepsilon), \tag{44}$$

with

$$\varphi(\varepsilon) := -\frac{1}{2} - \sum_{k=1}^{K} \frac{B_{2k}}{(2k)!}\varepsilon^{2k-1} + o(\varepsilon^{2K}) = -\frac{1}{2} - \frac{1}{12}\varepsilon + \frac{1}{720}\varepsilon^3 + \cdots \tag{45}$$

holomorphic at zero so $\pi_+(f) = \varphi(\varepsilon)$.

(b) Let $k = 2$ and let $f(\varepsilon) = [\varepsilon_1, \varepsilon_2]$. Applying Eq. (44) we write

$$
\begin{aligned}
\pi_+\left([\varepsilon_1,\varepsilon_2]\right) &= \pi_+\left(\Big(-\frac{1}{\varepsilon_1}+\varphi(\varepsilon_1)\Big)\Big(-\frac{1}{\varepsilon_1+\varepsilon_2}+\varphi(\varepsilon_1+\varepsilon_2)\Big)\right)\\
&= \pi_+\left(-\frac{\varphi(\varepsilon_1+\varepsilon_2)}{\varepsilon_1}-\frac{\varphi(\varepsilon_1)}{\varepsilon_1+\varepsilon_2}+\varphi(\varepsilon_1)\varphi(\varepsilon_1+\varepsilon_2)\right)\\
&= -\frac{\varphi(\varepsilon_1+\varepsilon_2)-\varphi(\varepsilon_2)}{\varepsilon_1}-\frac{\varphi(\varepsilon_1)-\varphi\left(\frac{\varepsilon_1-\varepsilon_2}{2}\right)}{\varepsilon_1+\varepsilon_2}\\
&\quad+\varphi(\varepsilon_1)\varphi(\varepsilon_1+\varepsilon_2),
\end{aligned}
$$

(c) Let $k = 2$ and let $f(\varepsilon) = [\varepsilon_1 + \varepsilon_3, \varepsilon_2]$. Using Eq. (44) we write

$$
\begin{aligned}
&\pi_+\left([\varepsilon_1+\varepsilon_3,\varepsilon_2]\right)\\
&= \pi_+\left(\Big(-\frac{1}{\varepsilon_1+\varepsilon_3}+\varphi(\varepsilon_1+\varepsilon_3)\Big)\Big(-\frac{1}{\varepsilon_1+\varepsilon_2+\varepsilon_3}+\varphi(\varepsilon_1+\varepsilon_2+\varepsilon_3)\Big)\right)\\
&= \pi_+\left(-\frac{\varphi(\varepsilon_1+\varepsilon_2+\varepsilon_3)}{\varepsilon_1+\varepsilon_3}-\frac{\varphi(\varepsilon_1+\varepsilon_3)}{\varepsilon_1+\varepsilon_2+\varepsilon_3}+h(\varepsilon_1+\varepsilon_3)\,\varphi(\varepsilon_1+\varepsilon_2+\varepsilon_3)\right)\\
&= -\frac{\varphi(\varepsilon_1+\varepsilon_2+\varepsilon_3)-\varphi(\varepsilon_2)}{\varepsilon_1+\varepsilon_3}-\frac{\varphi(\varepsilon_1+\varepsilon_3)-\varphi\left(\frac{\varepsilon_1+\varepsilon_3-2\varepsilon_2}{3}\right)}{\varepsilon_1+\varepsilon_2+\varepsilon_3}\\
&\quad+\varphi(\varepsilon_1+\varepsilon_3)\varphi(\varepsilon_1+\varepsilon_2+\varepsilon_3).
\end{aligned}
\tag{46}
$$

Similarly, for $f(\varepsilon) = [\varepsilon_1, \varepsilon_2 + \varepsilon_3]$, we have

$$
\begin{aligned}
&\pi_+\left([\varepsilon_1,\varepsilon_2+\varepsilon_3]\right)\\
&= \pi_+\left(\Big(-\frac{1}{\varepsilon_1}+\varphi(\varepsilon_1)\Big)\Big(-\frac{1}{\varepsilon_1+\varepsilon_2+\varepsilon_3}+\varphi(\varepsilon_1+\varepsilon_2+\varepsilon_3)\Big)\right)\\
&= \pi_+\left(-\frac{\varphi(\varepsilon_1+\varepsilon_2+\varepsilon_3)}{\varepsilon_1}-\frac{\varphi(\varepsilon_1)}{\varepsilon_1+\varepsilon_2+\varepsilon_3}+\varphi(\varepsilon_1)\varphi(\varepsilon_1+\varepsilon_2+\varepsilon_3)\right)\\
&= -\frac{\varphi(\varepsilon_1+\varepsilon_2+\varepsilon_3)-\varphi(\varepsilon_2+\varepsilon_3)}{\varepsilon_1}-\frac{\varphi(\varepsilon_1)-\varphi\left(\frac{2\varepsilon_1-\varepsilon_2-\varepsilon_3}{3}\right)}{\varepsilon_1+\varepsilon_2+\varepsilon_3}\\
&\quad+\varphi(\varepsilon_1)\varphi(\varepsilon_1+\varepsilon_2+\varepsilon_3).
\end{aligned}
\tag{47}
$$

5.3. *Renormalized conical zeta values: the case of Chen cones*

To a (closed) convex polyhedral (lattice) cone C, one can assign closed (resp. open) renormalized conical zeta values $\zeta^c(C; -a_1, \cdots, -a_k)$ (resp. $\zeta^o(C; -a_1, \cdots, -a_k)$) with $a_i \in \mathbb{Z}_{\geqslant 0}$, $i \in \{1, \cdots, k\}$ corresponding to the coefficient in $\varepsilon^{a_1} \cdots \varepsilon^{a_k}$ of the Taylor expansion at zero of $\pi_+ \left(S^c(C)\right)$ (resp. $\pi_+ \left(S^o(C)\right)$).

This applied to an open (resp. closed) Chen cone $C_k = \langle e_1, e_1 + e_2, \cdots, e_1 + \cdots + e_k\rangle$ (seen as a lattice cone when equipped with the canonical lattice $\mathbb{Z}^k$) gives rise to multiple zeta values

$$\zeta(-a_1, \cdots, -a_k) := \zeta^o(C_k; -a_1, \cdots, -a_k)$$

given by the Taylor coefficient in $\varepsilon^{a_1} \cdots \varepsilon^{a_k}$ of $\pi_+ (S^o(C_k))$ and multiple zeta-star values

$$\zeta^\star(-a_1, \cdots, -a_k) := \zeta^c(C_k; -a_1, \cdots, -a_k)$$

given by the Taylor coefficient in $\varepsilon^{a_1} \cdots \varepsilon^{a_k}$ of $\pi_+ (S^c(C_k))$. The latter are algebraic expressions in the former.

With the notations of Eq. (43) we have

$$\zeta(-a_1, -a_2, \ldots, -a_k) = \pi_+ \left(\partial_1^{a_1} \partial_2^{a_2} \cdots \partial_k^{a_k} [\varepsilon_1, \varepsilon_2, \cdots, \varepsilon_k]\right)_{|_{\vec{\varepsilon}=\vec{0}}}. \tag{48}$$

Example 5.3.

(a)

$$\zeta(-a) = \varphi^{(a)}(0) = -\frac{B_{a+1}}{a+1}. \tag{49}$$

(b)

$$\zeta(-a_1, -a_2) = \tag{50}$$

$$\left(\partial_1^{a_1} \partial_2^{a_2} \left(-\frac{\varphi(\varepsilon_1+\varepsilon_2)-\varphi(\varepsilon_2)}{\varepsilon_1} - \frac{\varphi(\varepsilon_1)-\varphi\left(\frac{\varepsilon_1-\varepsilon_2}{2}\right)}{\varepsilon_1+\varepsilon_2} + \varphi(\varepsilon_1)\varphi(\varepsilon_1+\varepsilon_2)\right)\right)_{|_{\varepsilon_1=\varepsilon_2=0}}.$$

As we pointed out in the introduction, our geometric approach contrasts with other approaches such as [GZ, MP] to the renormalization of multiple zeta values at non-integer arguments, where the algebraic Birkhoff factorization is carried out on the summands (functions $(x_1, \cdots, x_k) \mapsto x_1^{-s_1} \cdots x_1^{-s_k}$) rather than on the domain (the cones) of summation as in our present construction or [Sa] where a purely analytic renormalization method is implemented, which does not use algebraic Birkhoff factorization. The generality and the geometric nature of our approach nevertheless have a cost; whereas in [GZ, MP] the renormalized multiple zeta values obey the stuffle relations, these are not preserved in our approach. They nevertheless do hold for two arguments; as one could expect from the relation

$$[\varepsilon_1][\varepsilon_2] = [\varepsilon_1, \varepsilon_2] + [\varepsilon_2, \varepsilon_1] + [\varepsilon_1 + \varepsilon_2], \tag{51}$$

for any non-positive integers s_1, s_2 we have

$$\zeta(s_1)\,\zeta(s_2) = \zeta(s_1, s_2) + \zeta(s_2, s_1) + \zeta(s_1 + s_2).$$

Similarly,

$$\zeta^\star(s_1)\,\zeta^\star(s_2) = \zeta^\star(s_1, s_2) + \zeta^\star(s_2, s_1) - \zeta^\star(s_1 + s_2).$$

However, as we shall see below, they do not necessarily hold for three arguments.

Table 1 displays values for the double zeta values $\zeta(-a, -b)$ with $a, b \in \mathbb{Z}_{\neq 0}$ derived using our conical approach from the *open Chen cone* $\langle e_1, e_1+e_2\rangle$ in $\mathbb{R}^2$, where $\{e_1, e_2\}$ is the canonical basis of $\mathbb{R}^2$. It is then followed by Table 2 of values for the conical double zeta $\star$-values $\zeta^*(-a, -b)$ with $a, b \in \mathbb{Z}_{\neq 0}$ corresponding to the *closed Chen cone* $\langle e_1, e_1 + e_2\rangle$ in $\mathbb{R}^2$. The values in boldface of the first table coincide with the ones obtained in [GZ, MP].

Table 1: Values of renormalized conical double zeta values

$\zeta(-a_1, -a_2)$	$a_1 = 1$	$a_1 = 2$	$a_1 = 3$	$a_1 = 4$	$a_1 = 5$	$a_1 = 6$
$a_2 = 1$	$\mathbf{\frac{1}{288}}$	$-\mathbf{\frac{1}{240}}$	$\frac{101}{80640}$	$\mathbf{\frac{1}{504}}$	$-\frac{169}{96768}$	$-\mathbf{\frac{1}{480}}$
$a_2 = 2$	$-\mathbf{\frac{1}{240}}$	$\mathbf{0}$	$\mathbf{\frac{1}{504}}$	$-\frac{7127}{9676800}$	$-\mathbf{\frac{1}{480}}$	$\frac{7097}{3870720}$
$a_2 = 3$	$-\frac{157}{80640}$	$\mathbf{\frac{1}{504}}$	$\mathbf{\frac{1}{28800}}$	$-\mathbf{\frac{1}{480}}$	$\frac{1543}{1892352}$	$\mathbf{\frac{1}{264}}$
$a_2 = 4$	$\mathbf{\frac{1}{504}}$	$\frac{7127}{9676800}$	$-\mathbf{\frac{1}{480}}$	$\mathbf{0}$	$\mathbf{\frac{1}{264}}$	$-\frac{9280679}{5960908800}$
$a_2 = 5$	$\frac{67}{32256}$	$-\mathbf{\frac{1}{480}}$	$-\frac{72251}{85155840}$	$\mathbf{\frac{1}{264}}$	$\mathbf{\frac{1}{127008}}$	$-\mathbf{\frac{691}{65520}}$
$a_2 = 6$	$-\mathbf{\frac{1}{480}}$	$-\frac{7097}{3870720}$	$\mathbf{\frac{1}{264}}$	$\frac{9280679}{5960908800}$	$-\mathbf{\frac{691}{65520}}$	$\mathbf{0}$

Table 2: Values of conical double zeta star values

$\zeta^\star(-a_1, -a_2)$	$a_1 = 1$	$a_1 = 2$	$a_1 = 3$	$a_1 = 4$	$a_1 = 5$	$a_1 = 6$
$a_2 = 1$	$\frac{1}{288}$	$\frac{1}{240}$	$\frac{101}{80640}$	$-\frac{1}{504}$	$-\frac{169}{96768}$	$\frac{1}{480}$
$a_2 = 2$	$\frac{1}{240}$	0	$-\frac{1}{504}$	$-\frac{7127}{9676800}$	$\frac{1}{480}$	$\frac{7097}{3870720}$
$a_2 = 3$	$-\frac{157}{80640}$	$-\frac{1}{504}$	$\frac{1}{28800}$	$\frac{1}{480}$	$\frac{1543}{1892352}$	$-\frac{1}{264}$
$a_2 = 4$	$-\frac{1}{504}$	$\frac{7127}{9676800}$	$\frac{1}{480}$	0	$-\frac{1}{264}$	$-\frac{9280679}{5960908800}$
$a_2 = 5$	$\frac{67}{32256}$	$\frac{1}{480}$	$-\frac{72251}{85155840}$	$-\frac{1}{264}$	$\frac{1}{127008}$	$\frac{691}{65520}$
$a_2 = 6$	$\frac{1}{480}$	$-\frac{7097}{3870720}$	$-\frac{1}{264}$	$\frac{9280679}{5960908800}$	$\frac{691}{65520}$	0

Stuffle relations hold for two arguments but fail to hold for three arguments. For example, from the following relation

$$[\varepsilon_1, \varepsilon_2][\varepsilon_3] = [\varepsilon_3, \varepsilon_1, \varepsilon_2] + [\varepsilon_1, \varepsilon_3, \varepsilon_2] + [\varepsilon_1, \varepsilon_2, \varepsilon_3] + [\varepsilon_1 + \varepsilon_3, \varepsilon_2] + [\varepsilon_1, \varepsilon_2 + \varepsilon_3], \tag{52}$$

one might expect the following stuffle relation

$$\begin{aligned}\zeta(-a_1,-a_2)\zeta(-a_3) &= \zeta(-a_3,-a_1,-a_2)+\zeta(-a_1,-a_3,-a_2)\\ &\quad+\zeta(-a_1,-a_2,-a_3)+\zeta(-a_1-a_3,-a_2)+\zeta(-a_1,-a_2-a_3)\end{aligned} \tag{53}$$

to hold.

Proposition 5.4. *The stuffle relation (53) is violated for some values* $(a_1,a_2,0)\in\mathbb{Z}^3_{\geqslant 0}$ *with* $a_1+a_2>2$.

Proof. Were the stuffle relation (53) to hold for any $a_1,a_2,a_3\in\mathbb{Z}_{\geq 0}$, combining Eqs. (49), (50), (46) and (47), we would have

$$\left(\pi_+\left(\partial_1^{a_1+a_3}\partial_2^{a_2}\left(\frac{\varphi(\varepsilon_1)-\varphi\left(\frac{\varepsilon_1-2\varepsilon_2}{3}\right)}{\varepsilon_1+\varepsilon_2}\right)+\partial_1^{a_1}\partial_2^{a_2+a_3}\left(\frac{\varphi(\varepsilon_1)-\varphi\left(\frac{2\varepsilon_1-\varepsilon_2}{3}\right)}{\varepsilon_1+\varepsilon_2}\right)\right)\right)_{|\vec{\varepsilon}=\vec{0}}$$

$$=\left(\pi_+\left(\partial_1^{a_1+a_3}\partial_2^{a_2}\left(\frac{\varphi(\varepsilon_1)-\varphi\left(\frac{\varepsilon_1-\varepsilon_2}{2}\right)}{\varepsilon_1+\varepsilon_2}\right)+\partial_1^{-s_1}\partial_2^{-s_2-s_3}\left(\frac{\varphi(\varepsilon_1)-\varphi\left(\frac{\varepsilon_1-\varepsilon_2}{2}\right)}{\varepsilon_1+\varepsilon_2}\right)\right)\right)_{|\vec{\varepsilon}=\vec{0}}$$

for any $a_1,a_2,a_3\in\mathbb{Z}_{\geq 0}$, in particular, for $a_3=0$, in which case

$$\left(\pi_+\left(\partial_1^{a_1}\partial_2^{a_2}\left(\frac{\varphi(\varepsilon_1)-\varphi\left(\frac{\varepsilon_1-2\varepsilon_2}{3}\right)}{\varepsilon_1+\varepsilon_2}+\frac{\varphi(\varepsilon_1)-\varphi\left(\frac{2\varepsilon_1-\varepsilon_2}{3}\right)}{\varepsilon_1+\varepsilon_2}\right)\right)\right)_{|\vec{\varepsilon}=\vec{0}}$$

$$=\left(\pi_+\left(\partial_1^{a_1}\partial_2^{a_2}\left(\frac{\varphi(\varepsilon_1)-\varphi\left(\frac{\varepsilon_1-\varepsilon_2}{2}\right)}{\varepsilon_1+\varepsilon_2}+\frac{\varphi(\varepsilon_1)-\varphi\left(\frac{\varepsilon_1-\varepsilon_2}{2}\right)}{\varepsilon_1+\varepsilon_2}\right)\right)\right)_{|\vec{\varepsilon}=\vec{0}}$$

for any $a_1,a_2\in\mathbb{Z}_{\geq 0}$. Hence the equality of the following holomorphic functions holds

$$\frac{\varphi(\varepsilon_1)-\varphi\left(\frac{\varepsilon_1-2\varepsilon_2}{3}\right)}{\varepsilon_1+\varepsilon_2}+\frac{\varphi(\varepsilon_1)-\varphi\left(\frac{2\varepsilon_1-\varepsilon_2}{3}\right)}{\varepsilon_1+\varepsilon_2}=\frac{\varphi(\varepsilon_1)-\varphi\left(\frac{\varepsilon_1-\varepsilon_2}{2}\right)}{\varepsilon_1+\varepsilon_2}+\frac{\varphi(\varepsilon_1)-\varphi\left(\frac{\varepsilon_1-\varepsilon_2}{2}\right)}{\varepsilon_1+\varepsilon_2}$$

which would imply that

$$\varphi\left(\frac{\varepsilon_1-2\varepsilon_2}{3}\right)+\varphi\left(\frac{2\varepsilon_1-\varepsilon_2}{3}\right)=2\varphi\left(\frac{\varepsilon_1-\varepsilon_2}{2}\right).$$

But this does not hold for the function φ as in Eq. (45). Note however, that the Taylor expansions at zero agree on either side up to order 1. □

5.4. *Discussions and outlook*

There are by now various renormalization methods to evaluate multiple zeta values at non-positive integers, which all use an algebraic Birkhoff Hopf factorization, namely

(a) the present geometric approach which uses a heat-kernel type regularization on the summands (here polynomials) and a coalgebra on the domains (here cones),
(b) the analytic approach adopted in [MP] which uses a zeta type regularization as well as a coalgebra on the summands given by tensor products of pseudodifferential symbols,
(c) the number theoretic approach adopted in [GZ] to the renormalization of multiple zeta values at non-integer arguments, which uses a zeta type regularization as well as a coalgebra on the summands given by functions $(x_1, \cdots, x_k) \mapsto x_1^{-s_1} \cdots x_1^{-s_k}$.

There are also other methods which do not use algebraic Birkhoff Hopf factorization such as the approach adopted in [Sa] based on a formula expressing the polynomial integrand Q in terms of the integrated polynomial $P(a) = \int_{[0,1]^n} Q(a+t)dt$ or yet a different approach in [FKMT] based on Mellin-Barnes integrals to desingularize the multiple zeta functions. The diversity of the existing approaches calls for the need to relate them conceptually, yet a holy grail for us at this stage. Understanding the relation among these approaches would be a step towards a better understanding of the renormalization group in this context. Whether, from one of those methods, one can by means of a mere change of regularization obtain the renormalized values derived by another method is a first question we hope to address in some future work.

Acknowledgements: This work was supported by the National Natural Science Foundation of China (Grant No. 11071176, 11221101 and 11371178) and the National Science Foundation of US (Grant No. DMS 1001855). The authors thank the hospitalities of the Kavli Institute for Theoretical Physics China and Morningside Center of Mathematics. The second author thanks Sichuan University, the Lanzhou University and Capital Normal University for their kind hospitality. She is especially grateful to the group of students at the Lanzhou University to whom she delivered lectures on parts of this survey, for their questions and comments that helped her improve this presentation.

References

[A] T. Apostol, **Introduction to analytic number theory**, *Undergraduate Texts in Mathematics*, Springer-Verlag (1976)

[B] A. Barvinok, **Integer Points in Polyhedra** *Zurich Lectures in Advanced Mathematics*, EMS Lecture notes, 2008.

[BV] N. Berline and M. Vergne, Euler-Maclaurin formula for polytopes, *Mosc. Math. J.* **7** (2007) 355-386.

[Ca] P. Cartier, A primer on Hopf algebras, *IHES* /M/06/40 (2006)

[CK] A. Connes and D. Kreimer, Hopf algebras, renormalisation and Noncommutative Geometry, *Comm. Math. Phys.* **199** (1988) 203-242.

[F] L. Foissy, Algèbres de Hopf combinatoires `http://loic.foissy.free.fr/pageperso/cours3.pdf`

[FKMT] H. Furusho, Y. Komori, K. Matsumoto, H. Tsumura, Desingularization of complex multiple zeta-functions, fundamentals of p-adic multiple L-functions, and evaluation of their special values, arXiv:1309.3982.

[GP] S. Garoufalidis and J. Pommersheim, Sum-integral interpolators and the Euler-Maclaurin formula for polytopes, *Trans. Amer. Math. Soc.* **364** (2012) 2933-2958.

[G] L. Guo, **An Introduction to Rota-Baxter Algebra**, International Press, 2012.

[GPZ1] L. Guo, S. Paycha and B. Zhang, Renormalisation by Birkhoff factorisation and by generalized evaluators; a study case, in **Noncommutative Geometry, Arithmetic and Related Topics**, (Ed. A. Connes, K. Consani) John Hopkins Univ. Press p. 183-211 (2011).

[GPZ2] L. Guo, S. Paycha and B. Zhang, Conical zeta values and their double subdivision relations, *Adv. in Math.* **252** (2014) 343-381.

[GPZ3] L. Guo, S. Paycha and B. Zhang, Renormalisation and the Euler-Maclaurin formula on cones, arXiv:1306.3420.

[GPZ4] L. Guo, S. Paycha and B. Zhang, Decompositions and residues for meromorphic functions with linear poles in the light of Euler-Maclaurin formulae on cones, arXiv:1501.00426.

[GPZ5] L. Guo, S. Paycha and B. Zhang, Renormalization on Cones, in preparation

[GZ] L. Guo and B. Zhang, Renormalisation of multiple zeta values *J. Algebra* **319** (2008) 3770-3809.

[Ha] G. Hardy, **Divergent series**, Oxford University Press, 1967.

[Ho] M. E. Hoffman, Multiple harmonic series, *Pacific J. Math.* **152** (1992), 275-290.

[Je] P. Jeanquartier, Transformation de Mellin et développements asymptotiques, *L'Enseignement Mathématique* **25** (1979), 285-308.

[M1] D. Manchon, Hopf algebras, from basics to applications to renormalisation, *Comptes-rendus des Rencontres mathématiques de Glanon* 2001 (2003) and *Hopf algebras in renormalisation*, Handbook of algebra, Vol. 5 (M. Hazewinkel ed.) (2008).

[MP] D. Manchon and S. Paycha, Nested sums of symbols and renormalised multiple zeta values, *Int. Math. Res. Not.* **2010** issue 24, 4628-4697 (2010).

[P] S. Paycha, **Regularised integrals, sums and traces**, *Amer. Math. Soc. Lecture Notes* **59** 2012.

[Sa] B. Sadaoui, Multiple zeta values at the non-positive integers, to appear in *Comptes Rendus de l'Académie des Sciences.*

[Sw] M. E. Sweedler, **Hopf algebras**, Benjamin, New York (1969).

[Za] D. Zagier, Values of zeta functions and their applications, First European Congress of Mathematics, Vol. II (Paris, 1992), *Progr. Math.* **120**, 497-512, Birkhuser, Basel, 1994.

THE BV-ALGEBRA STRUCTURE OF THE HOCHSCHILD COHOMOLOGY OF THE GROUP RING OF CYCLIC GROUPS OF PRIME ORDER

Andrés Angel and Diego Duarte

Mathematics Department,
Universidad de los Andes,
Carrera 1 No. 18A-12,
Bogotá, Colombia

We construct a different Batalin-Vilkovisky algebra structure on the Hochschild cohomology of the algebra of truncated polynomials with coefficients in a field of p elements with p prime. We accomplish this by transfering the Frobenius form of the group ring of cyclic groups. We also explicitly calculate the Batalin-Vilkovisky algebra structure of the Hochschild cohomology of the group rings of cyclic groups of prime order. The algebra structure, even the Gesterhaber structure has been calculated before, but the BV structure that we calculate is a new one.

Keywords: Batalin-Vilkovisky algebras; Hochschild cohomology; cyclic groups of prime order.

1. Introduction

The aim of these notes is to present the calculations of the Hochschild cohomology of truncated polynomials and group rings of cyclic groups. The Hochschild cohomology of truncated polynomials $HH^*(\mathbb{K}[x]/(x^n);\mathbb{K}[x]/(x^n))$ over commutative rings has been studied extensively as well as the Hochschild cohomology of group rings $HH^*(\mathbb{K}[G];\mathbb{K}[G])$. In this article we will concentrate on calculations over $\mathbb{F}_p$, the field with p elements for p a prime.

It is possible to endow with many algebraic structures the Hochschild cohomology of an associative algebra A. It is a graded algebra given by the cup product. In [9], Gerstenhaber proves that the cup product in Hochschild cohomology is graded commutative and even more that there exists a Lie bracket that endows $HH^*(A;A)$ with a structure of Lie algebra, these two structures satisfy some compatibility conditions that are now known to define a *Gerstenhaber algebra*.

When the algebra A satisfies some sort of Poincaré duality the Hochschild cohomology has a richer algebraic structure. Tradler in [19] proves that if A is a symmetric algebra up to homotopy then $HH^*(A; A)$ is a *Batalin-Vilkovisky* algebra.

Menichi in [13] presents another proof for Tradler's result for symmetric differential graded algebras (dga A with a degree d quasi-isomorphism of A-bimodules $\Theta : A \to A^\vee$). The idea is to use an isomorphism from the algebra to its dual to transfer the Connes boundary operator on Hochschild homology to Hochschild cohomology.

We are interested in the algebra structure and the BV-structure on the Hochs-child cohomology of truncated polynomial rings because of the relation with string topology.

Let M be a closed, connected, oriented manifold of dimension n. In [3], Chas and Sullivan defined a Batalin-Vilkovisky algebra structure on the re-graded homology of the loop space of M, $H_{*+n}(LM)$. These algebraic operations were defined at the level of chains using the intersection theory of the manifold and the concatenation and rotation of loops.

In [6], Cohen and Jones reinterpreted these operations at the level of stable homotopy theory and constructed an isomorphism of graded algebras between the homology of the loop space and the Hochschild cohomology of the singular cochains of the manifold

$$H_{*+n}(LM; \mathbb{K})) \cong HH^*(C^*(M; \mathbb{K}); C^*(M; \mathbb{K})),$$

for M a closed, simply connected manifold.

Given that both sides admit the structure of BV-algebras it was expected that these BV-structures were isomorphic.

For spheres and complex projective spaces the algebra structure on the homology of the loop space was calculated in [5] by Cohen, Jones and Yan. In [22], C. Westerland calculates the BV-structure of $HH^*(C^*(\mathbb{S}^n, \mathbb{F}_2); C^*(\mathbb{S}^n, \mathbb{F}_2))$ and $HH^*(C^*(\mathbb{CP}^n; \mathbb{F}_2); C^*(\mathbb{CP}^n; \mathbb{F}_2))$.

The BV-structures of $HH^*(C^*(\mathbb{S}^n, \mathbb{F}_2); C^*(\mathbb{S}^n, \mathbb{F}_2))$ and $H_{*+2}(LS^2; \mathbb{F}_2)$ are calculated by Menichi in [15], where it has been shown that the two BV-algebra structures are not isomorphic, even though the underlying Gesternhaber structures are isomorphic.

The calculations of Westerland and Menichi of $HH^*(C^*(M), C^*(M))$ for the cochains of spheres and complex projective spaces are based on the fact that these manifolds are formal and the cohomology rings are graded truncated polynomial rings.

According to Félix, Menichi and Thomas [7], when M is a formal manifold over a field $\mathbb{K}$, there is an isomorphism of Gerstenhaber algebras

$$HH^*(C^*(M,\mathbb{K});C^*(M,\mathbb{K})) \cong HH^*(H^*(M;\mathbb{K});H^*(M;\mathbb{K})),$$

and, since we know that the cohomology of these spaces are truncated polynomial rings

$$H^*(\mathbb{CP}^n;\mathbb{F}_2) \cong \mathbb{F}_2[x]/(x^{n+1}) \quad \text{with} \quad |x| = 2,$$

$$H^*(S^n;\mathbb{F}_2) \cong \mathbb{F}_2[x]/(x^2) \quad \text{with} \quad |x| = n,$$

we can use a periodic resolution for calculations (see [11], [18]). In [20], Yang uses this resolution to calculate the BV-structures of the Hochschild cohomology of truncated polynomials over $\mathbb{F}_p$. These BV-structures come from the Frobenius form induced by Poincare duality $\varepsilon\left(\sum_{i=0}^n \alpha_i x^i\right) = \alpha_n$ and $\varepsilon(a+bx) = b$ respectively.

The Gesternhaber structure only depends on the algebra structure but the BV-structure depends on the duality isomorphism, by using a different Frobenius form coming from the group ring we get a different BV-structure with the same underlying Gesternhaber algebra.

For the cyclic group of order p, $\mathbb{Z}/p\mathbb{Z}$, we have that the group ring $\mathbb{F}_p\left[\mathbb{Z}/p\mathbb{Z}\right] = \mathbb{F}_p[\sigma]/(\sigma^p - 1)$ is naturally isomorphic, as algebra, to a truncated polynomial ring $\mathbb{F}_p[x]/(x^p)$. By transferring the Frobenius form of the group ring to the truncated polynomial ring we get the following Frobenius form

$$\widetilde{\varepsilon}\left(\sum_{i=0}^{p-1} \alpha_i x^i\right) = \sum_{i=0}^{p-1}(-1)^i \alpha_i.$$

The new results in these notes are:

Theorem 1.1. *Let $A = \mathbb{F}_p[x]/(x^p)$ with p an odd prime and $|x| = 0$. Then $HH^*(A;A)$ has a BV-structure induced by the canonical Frobenius form of the group ring defined as follows*

$$HH^*(A;A) = \mathbb{F}_p[x,v,t]/(x^p,v^2),$$

$$\Delta(t^l x^k) = 0$$

$$\Delta(t^k v x^{2l}) = 2lt^k x^{2l-1} + \sum_{i=2l}^{p-1}(-1)^{i+1}t^k x^i$$

$$\Delta(t^k v x^{2l+1}) = (2l+1)t^k x^{2l} + \sum_{i=2l+1}^{p-1}(-1)^i t^k x^i,$$

where $x \in HH^0(A;A)$, $v \in HH^1(A;A)$ *and* $t \in HH^2(A;A)$ *with* $|x| = 0$, $|v| = 1$ *and* $|t| = 2$.

Theorem 1.2. *Let* $A = \mathbb{F}_2[x]/(x^2)$ *with* $|x| = 0$. *Then* $HH^*(A;A)$ *has a BV-structure induced by the canonical Frobenius form of the group ring defined as follows*

$$HH^*(A;A) = \mathbb{F}_2[x,v,t]/(x^2, v^2 - t) \cong \Lambda(x) \otimes \mathbb{F}_2[v],$$

$$\Delta(v^k x^l) = k(1+x)v^{k-1},$$

where $x \in HH^0(A;A)$, $v \in HH^1(A;A)$ *with* $|x| = 0$ *and* $|v| = 1$.

2. Hochschild (Co)homology

Definition 2.1. Let A be a graded commutative R-algebra with R a commutative ring, suppose that $A = R\langle e\rangle \oplus \bar{A}$. Let

$$T(s\bar{A}) = \bigoplus_{k\geqslant 0}(s\bar{A})^{\otimes k}$$

be the universal tensor algebra of $s\bar{A}$, where we denote by sA the suspension of A, *i.e.*, $(sA)_i = A_{i+1}$, so for $x \in A$ we have $|sx| = |x| - 1$. The *two-sided normalized bar resolution* of A is the differential graded (A,A)-bimodule $\bar{B}(A) = A \otimes T(s\bar{A}) \otimes A$ defined as follows

- For $k \in \mathbb{N}$ let

$$\bar{B}_k(A) := A \otimes (s\bar{A})^{\otimes k} \otimes A$$

  for an element $a_0\,[a_1|\cdots|a_k]\,a_{k+1} \in \bar{B}_k(A)$ of Hochschild degree k and total degree $\sum_{i=0}^{k+1}|a_i|$. We define the *internal degree* as $|a_0| + \sum_{i=1}^{k}|sa_i| + |a_{k+1}| = \sum_{i=0}^{k+1}|a_i| - k$ for $k \geqslant 1$ and we have

$$\textit{Internal degree} = \textit{Total degree} - \textit{Hochschild degree}$$

  We also define $a\,[]\,a' = a \otimes 1 \otimes a' \in A \otimes R \otimes A$ for $k = 0$.
- The differential $d : \bar{B}_*(A) \to \bar{B}_{*-1}(A)$ is defined by

$$\begin{aligned} d_k : \bar{B}_k(A) &\longrightarrow \bar{B}_{k-1}(A) \\ a_0\,[a_1|\cdots|a_k]\,a_{k+1} &\longmapsto (-1)^{|a_0|}a_0a_1\,[a_2|\cdots|a_k]\,a_{k+1} \\ &\quad + \sum_{i=2}^{k}(-1)^{\epsilon_i}a_0\,[a_1|\cdots|a_{i-1}a_i|\cdots|a_k]\,a_{k+1} \\ &\quad - (-1)^{\epsilon_k}a_0\,[a_1|\cdots|a_{k-1}]\,a_ka_{k+1} \end{aligned}$$

where $\epsilon_i = |a_0| + \sum_{j<i} |sa_j|$. This differential lowers the Hochschild degree by one and preserves the total degree since A is a graded algebra.

Definition 2.2. Denote by A^{op} the opposite algebra of A and by A^e the enveloping algebra $A \otimes A^{op}$. Recall that any (A, A)-bimodule can be considered as a left (or right) A^e-module. Let M be a graded (A, A)-bimodule. The *Hochschild chain complex* is the (graded) tensor product $C_*(A; M) = M \otimes_{A^e} \bar{B}(A)$ with

$$C_n(A; M) = \bigoplus_{p+q=n} M_p \otimes_{A^e} \bar{B}_q(A)$$

and differential the (graded) tensor $D = id_M \otimes d$. The *Hochschild homology* with coefficients in M is the homology of the Hochschild chain complex

$$HH_*(A; M) := H(C_*(A; M), D).$$

The *Hochschild cochain complex* is the complex

$$C^*(A; M) = \underline{Hom}_{A^e}(\bar{B}_*(A), M) = \underline{Hom}(T(s\bar{A}), M)$$

and differential D defined by $Df = -(-1)^{|f|} f \circ d$. $\underline{Hom}$ denotes the internal Hom in the category of graded modules, i.e for U, V graded modules, the maps from U to V of homogeneous degree k are:

$$\underline{Hom}^k(U, V) = \prod_m Hom(U^m, V^{m+k}).$$

So,

$$C^n(A; M) = \bigoplus_k \prod_m Hom(\left(s\bar{A}^{\otimes n}\right)^m, M^{m+k})$$

The *Hochschild cohomology of A with coefficients in M* is the homology of the Hochschild cochain complex

$$HH^*(A; M) := H(C^*(A; M), D).$$

Remark 1. For an n-cochain $\varphi \in \prod_m Hom(\left(s\bar{A}^{\otimes n}\right)^m, M^{m+k})$, we say that the internal degree is k, $|\varphi| = k$ and we have $|\varphi(a_0 [a_1| \cdots |a_n] a_{n+1})| = |\varphi|$ + internal degree of $a_0 [a_1| \cdots |a_n] a_{n+1}$, which is $|\varphi| + \sum_{i=0}^{n+1} |a_i| - n$. The differential D raises the Hochschild degree by one and since

$$|D\varphi(a_0 [a_1| \cdots |a_{n+1}] a_{n+2})| = |\varphi(d((a_0 [a_1| \cdots |a_{n+1}] a_{n+2})|$$

this is $|\varphi|$ + internal degree of $d((a_0 [a_1| \cdots |a_{n+1}] a_{n+2})$ which is $|\varphi|$ + total degree $d((a_0 [a_1| \cdots |a_{n+1}] a_{n+2}) - n$. Since d preserves the total degree, we have $|\varphi|$ + total degree $a_0\,[a_1|\cdots|a_{n+1}]\,a_{n+2}-n$, so

$$|D\varphi(a_0\,[a_1|\cdots|a_{n+1}]\,a_{n+2})| = |\varphi| + \sum_{i=0}^{n+2}|a_i| - n$$

which shows that $|D\varphi| = |\varphi|+1$. We define the total degree of φ by $|\varphi|-n$; therefore,

$$Total\ degree = Internal\ degree - Hochschild\ degree$$

and D preserves the total degree.

Remark 2. Hochschild (co)homology can also be defined as derived functors

$$\begin{aligned} HH_*(A;M) &:= Tor_*^{A^e}(M;A) \\ HH^*(A;M) &:= Ext_{A^e}^*(A;M). \end{aligned}$$

If we take $M = A$ the Hochschild cohomology $HH^*(A;A)$ has a graded algebra structure induced from the cup product

$$(\varphi \smile \psi)(a_0\,[a_1|\ldots|a_k]\,a_{k+1}) = \sum_{j=0}^{k}(-1)^{|\psi|\varepsilon_j}\varphi(a_0\,[a_1|\ldots|a_j]\,1)\psi(1\,[a_{j+1}|\ldots|a_k]\,a_{k+1})$$

It also preserves the internal degree, i.e $|\varphi \smile \psi| = |\varphi|+|\psi|$. For convenience we will write $\varphi\psi$ instead of $\varphi \smile \psi$.

In [9], Gerstenhaber proves that the cup product on Hochschild cohomology is graded commutative and that there exists a Lie bracket that endows $HH^*(A;A)$ with a structure of Lie algebra. The Gerstenhaber bracket on $HH^*(A;A)$ is defined as follows

$$\{\varphi,\psi\} = \varphi\circ\psi - (-1)^{(|\varphi|-1)(|\psi|-1)}\psi\circ\varphi$$

where $\circ$ is defined by

$$\begin{aligned} (\varphi\circ\psi)(a_0\,[a_1|\ldots|a_k]\,a_{k+1}) &= \sum_{j=0}^{|\varphi|-1}(-1)^{j(|\psi|-1)}\times \\ &\varphi\left(a_0\left[a_1|\ldots|a_j|\psi(a_{j+1}\left[a_{j+2}|\ldots|a_{j+|\psi|-1}\right]a_{j+|\psi|})|\ldots|a_k\right]a_{k+1}\right). \end{aligned}$$

The cup product and the Lie bracket satisfy the following compatibility conditions.

Definition 2.3. A *Gerstenhaber algebra* is a graded commutative algebra A with a linear map $\{-,-\}: A_i \otimes A_j \to A_{i+j-1}$ of degree -1 such that

(a) The bracket $\{-,-\}$ endows A with a structure of graded Lie algebra of degree 1, *i.e.*, for all a, b and $c \in A$

$$\{a,b\} = -(-1)^{(|a|+1)(|b|+1)}\{b,a\}$$
$$\{a,\{b,c\}\} = \{\{a,b\},c\} + (-1)^{(|a|+1)(|b|+1)}\{b,\{a,c\}\}.$$

(b) The product and the Lie bracket satisfy the Poisson identity, *i.e.*, for all a, b and $c \in A$

$$\{a, bc\} = \{a,b\}c + (-1)^{(|a|+1)|b|}b\{a,c\}.$$

If there is a differential of degree -1 of a Gerstenhaber algebra such that the Gerstenhaber bracket is the obstruction of the operator to be a graded derivation, then the Gerstenhaber algebra is called a Batalin-Vilkovisky algebra.

Definition 2.4. A *Batalin-Vilkovisky* algebra is a Gerstenhaber algebra A with a linear map of degree -1, $\Delta : A_i \to A_{i-1}$ such that $\Delta \circ \Delta = 0$ and

$$\{a,b\} = (-1)^{|a|}(\Delta(ab) - \Delta(a)b - (-1)^{|a|}a\Delta(b))$$

for all a and $b \in A$.

The way to construct BV-structures on Hochschild cohomology is to dualize the Connes operator on the Hochschild chains. Recall that the Hochschild homology is the homology of the complex $A \otimes_{A^e} \bar{B}(A) \cong A \otimes T(s\bar{A})$. The *Connes boundary operator*

$$B : A \otimes T(s\bar{A}) \to A \otimes T(s\bar{A})$$

is defined on the n-chains, $B : A \otimes (s\bar{A})^{\otimes n} \to A \otimes (s\bar{A})^{\otimes n+1}$, by

$$B(a_0 \otimes (a_1, \ldots, a_n)) = \\ = \sum_{i=0}^{n} (-1)^{\sum_{k=0}^{i-1}|sa_k|\sum_{k=i}^{n}|sa_k|} 1 \otimes (a_i, \ldots, a_n, a_0, a_1, \ldots, a_{i-1}).$$

The dual of this operator, $B^\vee : \underline{Hom}(A \otimes T(s\bar{A}), R) \to \underline{Hom}(A \otimes T(s\bar{A}), R)$, defines by adjunction an operator on $\underline{Hom}(T(s\bar{A}), A^\vee) \cong \underline{Hom}(A \otimes T(s\bar{A}), R)$.

When the algebra A is a Poincaré duality algebra (for example a Frobenius algebra) the non-degenerate bilinear form of A induces a chain complex isomorphism

$$\underline{Hom}(T(s\bar{A}), A) \cong \underline{Hom}(T(s\bar{A}), A^\vee)$$

which defines a Δ operator on the Hochschild cochains which gives a BV-structure on $HH^*(A;A)$.

Definition 2.5. Let $\mathbb{K}$ be a field. A *Frobenius algebra* is a $\mathbb{K}$-algebra A of finite dimension with a linear function $\varepsilon : A \to \mathbb{K}$ called a *Frobenius form* (or *augmentation*), such that its kernel does not have non trivial ideals.

In a Frobenius algebra A with Frobenius form ε, a non-degenerate associative bilinear form, called *pairing*, is defined by

$$\begin{aligned}\langle , \rangle : A \otimes A &\longrightarrow \mathbb{K} \\ a \otimes b &\longmapsto \varepsilon(ab)\end{aligned}$$

which induces an isomorphism between A and its dual $A^\vee$.

Definition 2.6. We say that a Frobenius algebra is symmetric if for every $a, b \in A$

$$\langle a, b \rangle = \langle b, a \rangle$$

equivalently if for every $a, b \in A$ we have $\varepsilon(ab) = \varepsilon(ba)$.

Example 1. If G is a finite group then the group ring $\mathbb{K}[G]$, with $\mathbb{K}$ a field, is a symmetric Frobenius algebra with Frobenius form given by

$$\varepsilon : \mathbb{K}[G] \to \mathbb{K}, \quad \varepsilon\left(\sum_{g \in G} \alpha_g g\right) = \alpha_e.$$

Definition 2.7. An augmented graded commutative algebra A is called a *Poincaré duality algebra* of dimension N if there exists an R-module homomorphism $\varepsilon : A^N \to R$ such that the induced bilinear forms $\langle , \rangle : A^k \otimes A^{N-k} \to R$, $a \otimes b \mapsto \varepsilon(ab)$ are non-degenerate. If $\{a_1, \ldots, a_N\}$ is a homogeneous basis of A, then the unique basis $\{a_1^\vee, \ldots, a_N^\vee\}$ of A characterized by $\langle a_i, a_j^\vee \rangle = \delta_{ij}$ is called the *Poincaré dual basis*.

Example 2. Let M be a closed, connected, oriented manifold of finite dimension n. The cohomology ring of M with coefficients in a field $\mathbb{K}$ is a Poincaré duality algebra of dimension n. The augmentation is defined by

$$\begin{aligned}\varepsilon : H^*(M;\mathbb{K}) &\to \mathbb{K} \\ \alpha &\mapsto \alpha([M])\end{aligned}$$

where $[M]$ is the homological fundamental class of M. This augmentation induces a pairing $\langle , \rangle : H^*(M;\mathbb{K}) \otimes H^*(M;\mathbb{K}) \to \mathbb{K}$ defined by

$$\langle \alpha, \beta \rangle = \varepsilon(\alpha \smile \beta) = (\alpha \smile \beta)([M]) = \alpha([M] \frown \beta).$$

Recall that Poincaré duality theorem gives us the following isomorphism

$$\Phi : H^{n-k}(M;\mathbb{K}) \xrightarrow{h} Hom_{\mathbb{K}}(H_{n-k}(M;\mathbb{K}),\mathbb{K}) \xrightarrow{D^*} Hom_{\mathbb{K}}(H^k(M;\mathbb{K}),\mathbb{K})$$

where h is the map induced by the evaluation of cochains on chains, and D^* is the dual of Poincaré duality. Then

$$\Phi(\alpha)(\beta) = \alpha([M] \frown \beta)$$

and the pairings are non-degenerate.

In the case when A is a Poincaré duality algebra (Frobenius algebra), the BV-operator, Δ, is defined as follows

Proposition 2.1.
The operator $\Delta : Hom(T(s\bar{A})^{\otimes n+1}, A) \to Hom(T(s\bar{A})^{\otimes n}, A)$ *is given by*

$$\Delta(f)(a_1,\ldots,a_n) = \sum_{j=1}^{N} (-1)^{|f|+|a^j|\sum_{k=i}^{n}|sa_k|} \times$$
$$\sum_{i=0}^{n} (-1)^{(|sa_0|+\sum_{k=0}^{i-1}|sa_k|)\sum_{k=i}^{n}|sa_k|} \langle 1, f(a_i,\ldots,a_n,a^j,a_1,\ldots,a_{i-1})\rangle {a^j}^{\vee}$$

where $\{a^1,\ldots,a^N\}$ *is a homogeneous basis of* A.

The Δ operator also lowers the Hochschild degree and also the internal degree by 1, this is because $\langle , \rangle : A^k \otimes A^{N-k} \to R$ and $|1| = 0$ therefore the terms in the sum are zero except when $|f(a_i,\ldots,a_n,a^j,a_1,\ldots,a_{i-1})| = N$ but this is equivalent to

$$|f| + \sum_{i=1}^{n} |a_i| - n + |a^j| - 1 = N$$

since $|{a^j}^{\vee}| = N - |a^j|$, we have

$$|\Delta(f)(a_1,\ldots,a_n)| = N - |a^j| = |f| + \sum_{i=1}^{n} |a_i| - n - 1$$

which proves that $|\Delta(f)| = |f| - 1$.

In [14] Menichi proves that Δ induces a BV-structure on $HH^*(A;A)$, which furthermore induces the Gerstenhaber structure of $HH^*(A;A)$.

3. Truncated Polynomials over $\mathbb{F}_p$

From example 2, the cohomology of spheres and complex projective spaces with coefficients in $\mathbb{F}_p$ are Poincaré duality algebras. Moreover, these algebras are truncated polynomial rings

$$H^*(\mathbb{S}^n;\mathbb{F}_p)=\mathbb{F}_p[x]/(x^2)\cong\Lambda(x)\quad\text{with}\quad|x|=n$$
$$\varepsilon:H^*(\mathbb{S}^n;\mathbb{F}_p)\to\mathbb{F}_p,\quad\varepsilon(a+bx)=b$$

and

$$H^*(\mathbb{C}P^n;\mathbb{F}_p)=\mathbb{F}_p[x]/(x^{n+1})\quad\text{with}\quad|x|=2$$
$$\varepsilon:H^*(\mathbb{C}P^n;\mathbb{F}_p)\to\mathbb{F}_p,\quad\varepsilon\left(\sum_{i=0}^{n}\alpha_i x^i\right)=\alpha_n.$$

Sometimes is convenient to grade negatively the cohomology groups so we will consider the cases $|x|=\pm n$.
From now on, we assume that A is $\mathbb{F}_p[x]/(x^p)$ with $|x|=n\in\mathbb{Z}$. Let

$$P(y,z)=y^{p-1}+y^{p-2}z+\cdots+z^{p-1}\in A\otimes A=R[y,z]/(y^p,z^p)$$

with $|y|=|z|=|x|$. Note that $P(y,z)(y-z)=y^p-z^p$. Since the Hochschild (co)homology of an algebra can be calculated using projective A^e-resolutions and the bar construction is not convenient to make explicit calculations, we are going to use the following 2-periodical resolution [18], [11]. For $|x|=0$, see [12] exercise E.4.1.8.

Proposition 3.1. *The following is a A^e-projective resolution of A*

$$P_*(A):\ \cdots\to A\otimes A\xrightarrow{P(y,z)}A\otimes A\xrightarrow{y-z}A\otimes A\xrightarrow{\mu}A\to 0$$

with

$$d_{2k}(a\otimes b)=y^{p-1}a\otimes b+y^{p-2}a\otimes zb+\cdots+a\otimes z^{p-1}b$$

and

$$d_{2k+1}(a\otimes b)=ya\otimes b-a\otimes zb.$$

For $|x|=n$ with $n\in\mathbb{Z}$ the graded version is the following, see [20],[21].

Proposition 3.2. *The following is a A^e-projective resolution of A*

$$P_*(A):\ \cdots\to\Sigma^{np}(A\otimes A)\xrightarrow{P(y,z)}\Sigma^{n}(A\otimes A)\xrightarrow{y-z}A\otimes A\xrightarrow{\mu}A\to 0$$

with $P_{2k}(A)=\Sigma^{knp}(A\otimes A)$, $d_{2k}(a\otimes b)=y^{p-1}a\otimes b+y^{p-2}a\otimes zb+\cdots+a\otimes z^{p-1}b$ and $P_{2k+1}(A)=\Sigma^{nkp+n}(A\otimes A)$, $d_{2k+1}(a\otimes b)=ya\otimes b-a\otimes zb$.

Here we denote by ΣA the de-suspension of A (the inverse functor of s), *i.e.*, $(\Sigma A)_i = A_{i-1}$ so for $x \in A$ we have $|\Sigma x| = |x|+1$. With this regrading, the differential is of degree zero, since $d_{2k+1} : \Sigma^{nkp+n}(A \otimes A) \to \Sigma^{knp}(A \otimes A)$ and $d_{2k+1}(a \otimes b) = ya \otimes b - a \otimes zb$ with $|y| = |z| = n$. and similarly with d_{2k}.

Taking $Hom_{A^e}(\_\_, A)$ of $P_*(A)$, we get the following 2-periodical cochain complex

$$P^*(A) : 0 \to A \xrightarrow{0} A \xrightarrow{px^{p-1}} A \xrightarrow{0} A \xrightarrow{px^{p-1}} \cdots \tag{1}$$

To calculate the algebra structure of $HH^*(A; A)$, we use a chain map between $P_*(A)$ and the bar construction $\bar{B}(A)$

$$\varphi_* : P_*(A) \to \bar{B}(A)$$

for $|x| = 0$, $\varphi_{2k} : A \otimes A \to A \otimes (s\bar{A})^{\otimes 2k} \otimes A$ is defined by

$$\varphi_{2k}(1 \otimes 1) = \sum 1 \left[x^{p-a_1-1}|x|x^{p-a_2-1}|x| \cdots |x^{p-a_k-1}|x\right] x^{\sum_{k=1}^{k} a_k}$$

and $\varphi_{2k+1} : A \otimes A \to A \otimes (s\bar{A})^{\otimes 2k+1} \otimes A$ is defined by

$$\varphi_{2k+1}(1 \otimes 1) = \sum 1 \left[x|x^{p-a_1-1}|x|x^{p-a_2-1}|x| \cdots |x^{p-a_k-1}|x\right] x^{\sum_{k=1}^{k} a_k},$$

where the sum is taken over $0 \leqslant a_1, \ldots, a_k < p-1$ and $\sum_{k=1}^{k} a_k \leqslant p-1$. With the same formulas for $|x| = n$ with $n \in \mathbb{Z}$, we define

$$\varphi_* : P_*(A) \to \bar{B}(A)$$

where

$$\varphi_{2k} : \Sigma^{nkp}(A \otimes A) \to A \otimes (s\bar{A})^{\otimes 2k} \otimes A$$

and

$$\varphi_{2k+1} : \Sigma^{nkp+n}(A \otimes A) \to A \otimes (s\bar{A})^{\otimes 2k+1} \otimes A.$$

By a direct computation we have,

Lemma 3.1 (Lemma 3.3 [20]).
$\varphi^* = Hom_{A^e}(\varphi, A) : C^*(A; A) \to P^*(A)$ is a cochain map.

Using φ we can give the algebra structure of $HH^*(A; A)$.

Proposition 3.3 (Yang). *Let $A = \mathbb{F}_p[x]/(x^p)$ with p an odd prime and $|x| = n$ with $n \in \mathbb{Z}$. Then $HH^*(A;A)$ as an algebra is*

$$HH^*(A;A) = \mathbb{F}_p[x,v,t]/(x^p, v^2)$$

where $x \in HH^0(A;A)$, $v \in HH^1(A;A)$ and $t \in HH^2(A;A)$ with $|x| = n$, $|v| = 1-n$ and $|t| = -np+2$.

Proof. The assertion follows from the 2-periodical resolution and the cochain map φ^* between $C^*(A;A)$ and $P^*(A)$. Recall that $P^*(A)$ is (1)

$$0 \to A \xrightarrow{0} A \xrightarrow{0} A \xrightarrow{0} A \xrightarrow{0} A \xrightarrow{0} \cdots$$

where $A \cong Hom_{A^e}(A^e, A)$. Consider the following elements

$$\begin{aligned} x \in P^0(A), &\quad x(1) = x, \\ v \in P^1(A), &\quad v(1) = 1, \\ t \in P^2(A), &\quad t(1) = 1. \end{aligned}$$

Via φ^*, these elements correspond to Hochschild cochains,

$$\begin{aligned} \bar{x} \in Hom(\mathbb{F}_p, A) = A, &\quad \bar{x}(1) = x, \\ \bar{v} \in Hom(s\bar{A}, A) = A, &\quad \bar{v}(x^i) = ix^{i-1}, \\ \bar{t} \in Hom(s\bar{A}^{\otimes 2}, A) = A, &\quad \bar{t}(x^i, x^j) = x^{i+j-p}. \end{aligned}$$

Note that the internal degree of $1 \in \mathbb{F}_p$ is 0 and the degree of $x \in A$ is n, so $|x| = n$. The internal degree of $x^i \in s\bar{A}$ is $ni-1$ and the degree of $ix^{i-1} \in A$ is $n(i-1)$, so $|v| = n(i-1) - (ni-1) = 1-n$, and the internal degree of $x^i \otimes x^j \in s\bar{A}^{\otimes 2}$ is $ni + nj - 2$ and the degree of $x^{i+j-p} \in A$ is $n(i+j-p)$ which gives $|t| = n(i+j-p) - (ni+nj-2) = -np+2$.

See [20] for the proof that these elements generate and give the desired isomorphism. □

In particular for $|x| = 0$, we have

Proposition 3.4. *Let $A = \mathbb{F}_p[x]/(x^p)$ with p an odd prime and $|x| = 0$. Then $HH^*(A;A)$ as an algebra is*

$$HH^*(A;A) = \mathbb{F}_p[x,v,t]/(x^p, v^2)$$

where $x \in HH^0(A;A)$, $v \in HH^1(A;A)$ and $t \in HH^2(A;A)$ with $|x| = 0$, $|v| = 1$ and $|t| = 2$.

Proof. The proof is the same, note only that now the internal degree of $1 \in \mathbb{F}_p$ is 0 and the degree of $x \in A$ is 0, so $|x| = 0$. The internal degree of $x^i \in s\bar{A}$ is -1 and the degree of $ix^{i-1} \in A$ is 0, so $|v| = 1$, and the internal degree of $x^i \otimes x^j \in s\bar{A}^{\otimes 2}$ is -2 and the degree of $x^{i+j-p} \in A$ is 0 which gives $|t| = 2$. □

For $p = 2$ similarly we have,

Proposition 3.5. *Let $A = \mathbb{F}_2[x]/(x^2)$ with $|x| = n \in \mathbb{Z}$. Then $HH^*(A; A)$ as an algebra is*

$$HH^*(A;A) = \mathbb{F}_2[x,v]/(x^2) \cong \Lambda(x) \otimes \mathbb{F}_2[v]$$

where $x \in HH^0(A;A)$, $v \in HH^1(A;A)$, $|x| = n$ and $|v| = 1 - n$.

4. Group Ring of Prime Cyclic Groups over $\mathbb{F}_p$

It is well-known that group cohomology and Hochschild cohomology are related by Eckmann-Shapiro lemma [2]. From this lemma it can be proved directly that

$$HH^*(\mathbb{K}[G]; M) \cong H^*(G; {}_\chi M)$$

for G a finite group, $\mathbb{K}$ a commutative ring and M a G-module. Here ${}_\chi M$ denotes the left G-module with conjugation action $g \cdot x = gxg^{-1}$. In particular, for $M = \mathbb{K}[G]$ with G abelian Cibils and Solotar in [4] prove that this isomorphism is an isomorphism of algebras

$$HH^*(\mathbb{K}[G]; \mathbb{K}[G]) \cong H^*(G; {}_\chi\mathbb{K}[G]) \cong \mathbb{K}[G] \otimes H^*(G; \mathbb{K}).$$

Also, for groups not necessary abelian, they define the product formula using conjugation classes and establish the conjecture that the isomorphism

$$HH^*(\mathbb{K}[G]; \mathbb{K}[G]) \cong H^*(G; {}_\chi\mathbb{K}[G])$$

is an isomorphism of algebras for any finite group. Siegel and Witherspoon in [17] prove this conjecture using the product formula defined by Cibils and Solotar. In [16] Sánchez-Flores transfers the Gerstenhaber structure of the Hochschild cohomology to the group cohomology of cyclic groups and proves that is highly non-trivial over fields of positive characteristic.

For the cyclic group of order p, $\mathbb{Z}/p\mathbb{Z}$, we have $\mathbb{F}_p[\mathbb{Z}/p\mathbb{Z}] = \mathbb{F}_p[\sigma]/(\sigma^p - 1)$ and

$$H^*(\mathbb{Z}/p\mathbb{Z}; \mathbb{F}_p) \cong \mathbb{F}_p[v,t]/(v^2) \cong \Lambda(v) \otimes \mathbb{F}_p[t] \quad \text{with} \quad |v| = 1,\ |t| = 2$$

Over $\mathbb{F}_p$ there is a natural isomorphism of algebras between the truncated polynomial ring and the group ring of a cyclic group of prime order p

$$\begin{aligned} \psi : \mathbb{F}_p[x]/(x^p) &\longrightarrow \mathbb{F}_p[\sigma]/(\sigma^p - 1) \\ x &\longmapsto \sigma - 1 \end{aligned}$$

Putting together all these, $\mathbb{F}_p\left[\mathbb{Z}/p\mathbb{Z}\right] \otimes H^*(\mathbb{Z}/p\mathbb{Z};\mathbb{F}_p) \cong \mathbb{F}_p[x]/(x^p) \otimes \mathbb{F}_p[v,t]/(v^2)$, we have another proof of

Proposition 4.1. *Let $A = \mathbb{F}_p[x]/(x^p)$ with p an odd prime and $|x| = 0$. Then $HH^*(A;A)$ as an algebra is*

$$HH^*(A;A) = \mathbb{F}_p[x,v,t]/(x^p,v^2) \cong \mathbb{F}_p[x]/(x^p) \otimes \mathbb{F}_p[v,t]/(v^2)$$

where $x \in HH^0(A;A)$, $v \in HH^1(A;A)$ and $t \in HH^2(A;A)$ with $|x| = 0$, $|v| = 1$ and $|t| = 2$.

Notice that the BV-operator, Δ, strongly depends on the Frobenius structure of the algebra. So, it is natural to study what happens with Δ if we change the Frobenius form of the algebra.

According to Yang, when we consider the Frobenius structure given by the Frobenius form that takes the coefficient α_{p-1}

$$\varepsilon : \mathbb{F}_p[x]/(x^p) \longrightarrow \mathbb{F}_p \qquad \varepsilon\left(\sum_{i=0}^{p-1} \alpha_i x^i\right) = \alpha_{p-1}$$

we have,

Theorem 4.2. *[20] Let $A = \mathbb{F}_p[x]/(x^p)$ with p an odd prime and $|x| = n$ with $n \in \mathbb{Z}$. Then as BV-algebra*

$$\begin{aligned} HH^*(\mathbb{F}_p[x]/(x^p);\mathbb{F}_p[x]/(x^p)) &= \mathbb{F}_p[x,v,t]/(x^p,v^2) \\ \Delta(t^k x^l) &= 0 \\ \Delta(t^k v x^l) &= l t^k x^{l-1} \end{aligned} \tag{2}$$

where $x \in HH^0(A;A)$, $v \in HH^1(A;A)$ and $t \in HH^2(A;A)$ with with $|x| = n$, $|v| = 1 - n$ and $|t| = -np + 2$.

Similarly for $p = 2$,

Theorem 4.3. *Let $A = \mathbb{F}_2[x]/(x^2)$ with $|x| = n$. Then as a BV-algebra,*

$$\begin{aligned} HH^*(A;A) &= \mathbb{F}_2[x,v,t]/(x^2, v^2 - t) \cong \Lambda(x) \otimes \mathbb{F}_2[v] \\ \Delta(v^k) &= 0 \end{aligned} \tag{3}$$

$$\Delta(xv^k) = kv^{k-1}$$

where $x \in HH^0(A;A)$, $v \in HH^1(A;A)$ *with* $|x| = n$ *and* $|v| = 1 - n$.

By using the formality of spheres, we have the following result

Theorem 4.4 (Westerland). *As BV-algebras,*

$$HH^*(C^*(\mathbb{S}^n, \mathbb{F}_2); C^*(\mathbb{S}^n, \mathbb{F}_2)) \cong \mathbb{F}_2[x,v]/(x^2)$$
$$\Delta(x^k v^l) = klx^{k-1}v^{l-1}$$

where the topological dimensions of x *and* v *are* $-n$ *and* $n-1$, *respectively.*

We want to define a new Frobenius form over the truncated polynomial ring by pulling-back the natural Frobenius form of the group ring

$$\varepsilon_0 : \mathbb{F}_p[\sigma]/(\sigma^p - 1) \longrightarrow \mathbb{F}_p \qquad \varepsilon_0\left(\sum_{i=0}^{p-1} \alpha_i \sigma^i\right) = \alpha_0.$$

By doing this, we get a new Frobenius form on the truncated polynomial ring,

$$\widetilde{\varepsilon} : \mathbb{F}_p[x]/(x^p) \longrightarrow \mathbb{F}_p \qquad \widetilde{\varepsilon}\left(\sum_{i=0}^{p-1} \alpha_i x^i\right) = \sum_{i=0}^{p-1} (-1)^i \alpha_i. \tag{4}$$

To calculate the new BV-structure over the Hochschild cohomology of $\mathbb{F}_p[x]/(x^p)$, we need to find the dual basis with respect to the new pairing $\langle a,b\rangle = \widetilde{\varepsilon}(ab)$. By taking the set $\{1, x, \ldots, x^{p-1}\}$ as a basis for $\mathbb{F}_p[x]/(x^p)$, the matrices that represent the pairing and its inverse are

$$[\langle,\rangle] = \begin{pmatrix} 1 & -1 & \cdots & -1 & 1 \\ -1 & 1 & \cdots & 1 & \\ \vdots & \vdots & \iddots & & \\ -1 & 1 & & 0 & \\ 1 & & & & \end{pmatrix} \qquad [\langle,\rangle]^{-1} = \begin{pmatrix} & & & & 1 \\ & 0 & & 1 & 1 \\ & & \iddots & 1 & \\ & 1 & \iddots & & \\ 1 & 1 & & 0 & \end{pmatrix}.$$

Therefore, the dual basis is the set $\{x^{p-1}, x^{p-1} + x^{p-2}, x^{p-2} + x^{p-3}, \ldots, x + 1\}$.

Theorem 4.5. *Let* $A = \mathbb{F}_p[x]/(x^p)$ *with* p *an odd prime and* $|x| = 0$. *Then* $HH^*(A;A)$ *has a BV-structure induced by the canonical Frobenius form of the group ring defined as follows*

$$HH^*(A;A) = \mathbb{F}_p[x,v,t]/(x^p, v^2)$$

$$\Delta(t^l x^k) = 0$$

$$\Delta(t^k v x^{2l}) = 2l t^k x^{2l-1} + \sum_{i=2l}^{p-1} (-1)^{i+1} t^k x^i$$

$$\Delta(t^k v x^{2l+1}) = (2l+1) t^k x^{2l} + \sum_{i=2l+1}^{p-1} (-1)^i t^k x^i$$

where $x \in HH^0(A;A)$, $v \in HH^1(A;A)$ *and* $t \in HH^2(A;A)$ *with* $|x| = 0$, $|v| = 1$ *and* $|t| = 2$.

Proof. According to Yang [20] (with a sign correction on the last equation) $HH^*(A;A)$, as a Gerstenhaber algebra, is

$$\begin{aligned} HH^*(A;A) &= \mathbb{F}_p[x,v,t]/(x^p, v^2) \\ \{t^{k_1}x^{l_1}, t^{k_2}x^{l_2}\} &= 0 \\ \{t^{k_1}x^{l_1}, t^{k_2}vx^{l_2}\} &= l_1 t^{k_1+k_2} x^{l_1+l_2-1} \\ \{t^{k_1}vx^{l_1}, t^{k_2}vx^{l_2}\} &= (l_1 - l_2) t^{k_1+k_2} v x^{l_1+l_2-1}. \end{aligned} \tag{5}$$

Since in a BV-algebra, we have the following equations

$$\Delta(ab) = \Delta(a)b + (-1)^{|a|} a\Delta(b) + (-1)^{|a|}\{a,b\} \tag{6}$$

and

$$\begin{aligned} \Delta(abc) = {} & \Delta(ab)c + (-1)^{|a|} a\Delta(bc) + (-1)^{(|a|-1)|b|} b\Delta(ac) \\ & - \Delta(a)bc - (-1)^{|a|} a\Delta(b)c - (-1)^{|a|+|b|} ab\Delta(c). \end{aligned} \tag{7}$$

We only need to calculate $\Delta(x)$, $\Delta(v)$, $\Delta(t)$ $\Delta(tx)$ and $\Delta(vx)$. Taking $\{1, x, \ldots, x^{p-1}\}$ as a basis for A and $\{x^{p-1}, x^{p-1}+x^{p-2}, x^{p-2}+x^{p-3}, \ldots, x+1\}$ as the dual basis induced by the new Frobenius form (4), we have

$$\boxed{\Delta(x) = 0} \text{ by degree.}$$

$$\begin{aligned} \Delta(v)(1) &= \sum_{i=0}^{p-1} -\langle 1, v(x^i)\rangle (x^{p-i} + x^{p-i-1}) \\ &= \sum_{i=0}^{p-1} -\langle 1, ix^{i-1}\rangle (x^{p-i} + x^{p-i-1}) \\ &= \sum_{i=0}^{p-1} -(-1)^{i-1} i (x^{p-i} + x^{p-i-1}) = \sum_{i=1}^{p-1} (-1)^i x^{p-i} - 1 \end{aligned}$$

$$\boxed{\Delta(v) = \sum_{i=0}^{p-1}(-1)^{i+1}x^i}$$

$$\begin{aligned}\Delta(vx)(1) &= \sum_{i=0}^{p-1} -\langle 1, vx(x^i)\rangle(x^{p-i} + x^{p-i-1})\\ &= \sum_{i=0}^{p-1} -\langle 1, ix^i\rangle(x^{p-i} + x^{p-i-1})\\ &= \sum_{i=0}^{p-1} -(-1)^i i(x^{p-i} + x^{p-i-1})\end{aligned}$$

$$\boxed{\Delta(vx) = \sum_{i=0}^{p-1}(-1)^i x^i}$$

$$\begin{aligned}\Delta(t)(x^k) = \sum_{i=0}^{p-1}&\langle 1, t(x^i, x^k)\rangle(x^{p-i} + x^{p-i-1})\\ &- \sum_{i=0}^{p-1}\langle 1, t(x^k, x^i)\rangle(x^{p-i} + x^{p-i-1})\end{aligned}$$

$$\boxed{\Delta(t) = \ 0}$$

$$\begin{aligned}\Delta(tx)(x^k) = \sum_{i=0}^{p-1}&\langle 1, tx(x^i, x^k)\rangle(x^{p-i} + x^{p-i-1})\\ &- \sum_{i=0}^{p-1}\langle 1, tx(x^k, x^i)\rangle(x^{p-i} + x^{p-i-1})\end{aligned}$$

$$\boxed{\Delta(tx) = \ 0.}$$

Note that if we calculate $\Delta(t^2)$ and $\Delta(tv)$ then Δ will be totally determined using equation (7) and induction on powers of x and t. From equation (6) and the bracket, we obtain that

$$\Delta(t^2) = \Delta(t)t + t\Delta(t) + \{t, t\}$$

$$\boxed{\Delta(t^2) = 0}$$

$$\Delta(tv) = \Delta(t)v + t\Delta(v) + \{t, v\}$$

$$\boxed{\Delta(tv) = \sum_{i=0}^{p-1}(-1)^{i+1}tx^i.}$$

Therefore, the new BV-algebra structure on the Hochschild cohomology of A is

$$\Delta(t^l x^k) = 0$$

$$\Delta(t^k v x^{2l}) = 2lt^k x^{2l-1} + \sum_{i=2l}^{p-1}(-1)^{i+1}t^k x^i$$

$$\Delta(t^k v x^{2l+1}) = (2l+1)t^k x^{2l} + \sum_{i=2l+1}^{p-1}(-1)^i t^k x^i.$$

Using equation (6), a straightforward calculation shows that this new BV structure gives the Gerstenhaber structure calculated by Yang. □

Similarly, for $p = 2$,

Theorem 4.6. *Let $A = \mathbb{F}_2[x]/(x^2)$ with $|x| = 0$. Then as a BV-algebra,*

$$HH^*(A; A) = \mathbb{F}_2[x, v, t]/(x^2, v^2 - t) \cong \Lambda(x) \otimes \mathbb{F}_2[v]$$

$$\Delta(v^k x^l) = k(1 + x)v^{k-1}$$

where $x \in HH^0(A; A)$, $v \in HH^1(A; A)$ with $|x| = 0$ and $|v| = 1$.

Proof. In the special case when $p = 2$, as a Gerstenhaber algebra the Hochschild cohomology of A is

$$HH^*(A; A) = \mathbb{F}_2[x, v, t]/(x^2, v^2 - t) \cong \Lambda(x) \otimes \mathbb{F}_2[v]$$
$$\{v^k, v^l\} = 0 \quad \{xv^k, v^l\} = lv^{k+l-1} \quad \{xv^k, xv^l\} = (k - l)xv^{k+l-1}$$

Following the same idea as in the case p an odd prime, we only need to calculate $\Delta(x)$, $\Delta(v)$, $\Delta(vx)$ and $\Delta(v^2)$. Taking $\{1, x\}$ as a basis for A and $\{x, x+1\}$ as the dual basis induced by the new Frobenius form (4)

$$\begin{aligned}
\Delta(x) &= 0 \quad \text{by degree}\\
\Delta(v)(1) &= \langle 1, v(1)\rangle x + \langle 1, v(x)\rangle(x+1) = x+1\\
\Delta(vx)(1) &= \langle 1, vx(1)\rangle x + \langle 1, vx(x)\rangle(x+1) = x+1\\
\Delta(v^2)(x^i) &= \langle 1, v^2(1, x^i)\rangle x + \langle 1, v^2(x, x^i)\rangle(x+1)\\
&\quad + \langle 1, v^2(x^i, 1)\rangle x + \langle 1, v^2(x^i, x)\rangle(x+1) = 0.
\end{aligned}$$

Now, using equation (7) and induction, the new BV-algebra structure of the Hochschild cohomology of A is

$$\Delta(v^{2k}) = 0 \quad \text{and} \quad \Delta(v^{2k+1}) = (1+x)v^{2k}$$
$$\Delta(xv^{2k}) = 0 \quad \text{and} \quad \Delta(xv^{2k+1}) = (1+x)v^{2k}.$$

Using the formula that relates the Gerstenhaber structure and the BV-structure (6), it can be checked that this new BV-structure preserves the Gerstenhaber structure

$$\begin{aligned}
\{v^{2k}, v^{2l}\} &= 0 \\
\{v^{2k}, v^{2l+1}\} &= 0 \\
\{xv^{2k}, v^{2l}\} &= 0 = 2lv^{2k+2l-1} \\
\{xv^{2k+1}, v^{2l}\} &= 0 = 2lv^{2k+2l} \\
\{xv^{2k}, v^{2l+1}\} &= v^{2k+2l} = (2l+1)v^{2k+2l} \\
\{xv^{2k+1}, v^{2l+1}\} &= v^{2k+2l+1} = (2l+1)v^{2k+2l+1} \\
\{xv^{2k}, xv^{2l}\} &= 0 = 2(k-l)xv^{2k+2l-1} \\
\{xv^{2k+1}, xv^{2l+1}\} &= 0 = 2(k-l)xv^{2k+2l+1} \\
\{xv^{2k}, xv^{2l+1}\} &= xv^{2k+2l} = (2(k-l)-1)xv^{2k+2l} \\
\{xv^{2k+1}, xv^{2l}\} &= xv^{2k+2l} = (2(k-l)+1)xv^{2k+2l}.
\end{aligned}$$

□

References

[1] Brown, K.S. *Cohomology of groups*, Springer, 1982.

[2] Cartan, H.; Eilenberg, S. *Homological algebra*, Princeton Mathematical Series, **19**, Princeton University Press, 1956.

[3] Chas, M.; Sullivan, D. *String Topology.* Preprint:math.GT/9911159, 1999.

[4] Cibils, C.; Solotar, A. *Hochschild cohomology algebra of abelian groups*, Arch.Math. (Basel) **68**, 17-21, 1997.

[5] Cohen, R.L.; Jones, J.D.S.; Yan J. *The loop homology algebra of spheres and projective spaces*, arXiv preprint arXiv:math.AT/0210353 v1, 2002.

[6] Cohen, R.L; Jones, J.D.S. *A homotopy theoretic realization of string topology*, Math. Annalen, **324**, 773-798, 2002.

[7] Félix, Y.; Menichi, L.; Thomas, J.C. *Gerstenhaber duality in Hochschild cohomology*, J. Pure Appl. Algebra **199**, no. 1-3, 43-59, 2005.

[8] Félix, Y.; Thomas, J.C. *Rational BV-algebra in string topology*, arXiv preprint arXiv:0705.4194, 2007.

[9] Gerstenhaber, M. *The cohomology structure of an associative ring*, Ann. of Math. (2), **78**, 267-288, 1963.

[10] Hochschild, G. *On the cohomology groups of an associative algebra*, Annals of Mathematics. Second Series **46**, 58-67, 1945.

[11] Holm, T. *The Hochschild cohomology ring of a modular group algebra: the commutative case*, Comm. Algebra **24**, no. 6, 1957-1969, 1996.

[12] Loday, J.L. *Cyclic Homology*, Springer-Verlag , Berlin, Heidelberg, New York, 1971.

[13] Menichi, L. *Batalin-Vilkovisky algebra structures on Hochschild cohomology*, arXiv preprint arXiv:0711.1946, 2007.

[14] Menichi, L. *Batalin-Vilkovisky algebras and cyclic cohomology of Hopf algebras*, K-Theory **32**, no. 3, 231-251, 2004.

[15] Menichi, L. *String topology for spheres*, preprint: math.AT/0609304, to appear in Comment. Math. Helv.

[16] Sanchez-Flores, S. *The Lie structure on the Hochschild cohomology of a modular group algebra*, Journal of Pure and Applied Algebra (in press).

[17] Siegel, S.F.; Witherspoon, S.J. *The Hochschild cohomology ring of a group algebra*, Proceedings of the London Mathematical Society, **79**(1), 131-157, 1999.

[18] The Buenos Aires Cyclic Homology Group *Cyclic Homology of Algebras with One Generator*, K-Theory **5**, 51-69, 1991.

[19] Tradler, T. *The BV algebra on Hochschild cohomology induced by infinity inner products*, preprint: math.QA/0210150v1, 2002.

[20] Yang, T. *A Batalin-Vilkovisky Algebra Structure on the Hochschild Cohomology of Truncated Polynomials*, Topology and its Applications **160**, 1633-1651, 2013.

[21] Westerland, C. *Dyer-Lashof operations in the string topology of spheres and projective spaces*, Math.Z. **250**, 711-727, 2005.

[22] Westerland, C. *String homology of spheres and projective spaces*, Algebr. Geom. Topol. **7**, 309-325, 2007.